Transboundary Water Cooperation

China and its neighbours face a series of water security issues, in which international law plays a vital role. Paramount to both policymakers and researchers in the field of water law, the current status of transboundary water cooperation schemes and how these operate in China is of global significance.

Grounded in international experience, this comprehensive volume provides readers with an up-to-date overview of current international transboundary water resource sharing policies and practices, including detailed case studies at both domestic and international levels. The authors discuss existing international laws, treaties and principles that may stimulate transboundary water cooperation and dialogue, and then analyse a number of international experiences with treaties in North America, Eastern Europe and Central Asia. They take stock of China's water resource issues, legal practices and options; examine case studies of China's southern shared rivers; and explore some innovative approaches to cooperative management of shared waters within China.

The articles in this book were originally published in the journal *Water International.*

Patricia Wouters is a Professor in the School of Law at Xiamen University, China, and Director of the China International Water Law Programme (CIWL).

Huiping Chen is a Professor in the School of Law at Xiamen University, China.

James E. Nickum is an institutional economist affiliated to the International Water Resources Association (IWRA), the Centre for Water and Development at SOAS, UK, and the University of Hong Kong.

Routledge Special Issues on Water Policy and Governance

https://www.routledge.com/series/WATER

Edited by
Cecilia Tortajada (IJWRD), Third World Centre for Water Management, Mexico
James Nickum (WI), International Water Resources Association, France

Most of the world's water problems, and their solutions, are directly related to policies and governance, both specific to water and in general. Two of the world's leading journals in this area, the *International Journal of Water Resources Development* and *Water International* (the official journal of the International Water Resources Association), contribute to this special issues series, aimed at disseminating new knowledge on the policy and governance of water resources to a very broad and diverse readership all over the world. The series should be of direct interest to all policymakers, professionals and lay readers concerned with obtaining the latest perspectives on addressing the world's many water issues.

Water Management and Climate Change
Dealing with Uncertainties
Edited by Cecilia Tortajada, Asit K. Biswas and Avinash Tyagi

Integrated Water Resources Management
From Concept to Implementation
Edited by Cecilia Tortajada

Water Infrastructure
Edited by Cecilia Tortajada and Asit K. Biswas

Frontiers of Land and Water Governance in Urban Regions
Edited by Thomas Hartmann and Tejo Spit

The Water-Energy-Food Nexus in the Middle East and North Africa
Edited by Martin Keulertz and Eckart Woertz

Sustainability in the Water-Energy-Food Nexus
Edited by Anik Bhaduri, Claudia Ringler, Ines Dombrowsky, Rabi Mohtar and Waltina Scheumann

Transboundary Water Cooperation
Principles, Practice and Prospects for China and its Neighbours
Edited by Patricia Wouters, Huiping Chen and James E. Nickum

Water Reuse Policies for Potable Use
Edited by Cecilia Tortajada and Choon Nam Ong

Hydrosocial Territories and Water Equity
Theory, Governance and Sites of Struggle
Edited by Rutgerd Boelens, Ben Crow, Jaime Hoogesteger, Flora E. Lu, Erik Swyngedouw and Jeroen Vos

Energy for Water
Regional case studies
Edited by Christopher Napoli

The Water Legacies of Conventional Mining
Edited by James E. Nickum, David B. Brooks, Anthony Turton and Surina Esterhuyse

Transboundary Water Cooperation

Principles, Practice and Prospects for China and Its Neighbours

Edited by
Patricia Wouters, Huiping Chen and James E. Nickum

Routledge
Taylor & Francis Group
LONDON AND NEW YORK

IWRA

China International Water Law
中国国际水法

First published 2018
by Routledge

2 Park Square, Milton Park, Abingdon, Oxfordshire OX14 4RN
52 Vanderbilt Avenue, New York, NY 10017

Routledge is an imprint of the Taylor & Francis Group, an informa business

First issued in paperback 2019

British Library Cataloguing in Publication Data
A catalogue record for this book is available from the British Library

ISBN 13: 978-1-138-28883-6 (hbk)
ISBN 13: 978-0-367-88678-3 (pbk)

Typeset in TimesNewRomanPS
by diacriTech, Chennai

Publisher's Note
The publisher accepts responsibility for any inconsistencies that may have arisen during the conversion of this book from journal articles to book chapters, namely the possible inclusion of journal terminology.

Contents

PART 2: International experiences

Citation Information

The chapters in this book were originally published in various special issues of *Water International*. When citing this material, please use the original page numbering for each article, as follows:

Chapter 2

One step at a time: international law and the duty to cooperate in the management of shared water resources
Christina Leb
Water International, volume 40, issue 1 (2015) pp. 21–32

Chapter 3

The human right to water and foreign investment: friends or foes?
Huiping Chen
Water International, volume 40, issue 2 (2015) pp. 297–311

Chapter 4

Ten building blocks for sustainable water governance: an integrated method to assess the governance of water
Marleen van Rijswick, Jurian Edelenbos, Petra Hellegers, Matthijs Kok and Stefan Kuks
Water International, volume 39, issue 5 (2014) pp. 725–742

Chapter 5

Transboundary water law and vulnerable people: legal interpretations of the 'equitable use' principle
Karina Vink
Water International, volume 39, issue 5 (2014) pp. 743–754

Chapter 6

The governance of wetland ecosystems and the promotion of transboundary water cooperation – opportunities presented by the Ramsar Convention
Jing Lee
Water International, volume 40, issue 1 (2015) pp. 33–47

Chapter 7
The UNECE Water Convention and the development of transboundary cooperation in the Chu-Talas, Kura, Drin and Dniester River basins
Bo Libert
Water International, volume 40, issue 1 (2015) pp. 168–182

Chapter 8
Do good fences make good neighbours? Canada-United States transboundary water governance, the Boundary Waters Treaty, and twenty-first-century challenges
Emma S. Norman and Karen Bakker
Water International, volume 40, issue 1 (2015) pp. 199–213

Chapter 9
Transboundary water management: lessons learnt from North America
Velma I. Grover and Gail Krantzberg
Water International, volume 40, issue 1 (2015) pp. 183–198

Chapter 10
An analysis of Turkey's water diplomacy and its evolving position vis-à-vis international water law
Aysegül Kibaroglu
Water International, volume 40, issue 1 (2015) pp. 153–167

Chapter 11
Identifying China's transboundary water risks and vulnerabilities – a multidisciplinary analysis using hydrological data and legal/institutional settings
Yan Feng, Daming He and Wenling Wang
Water International, volume 40, issue 2 (2015) pp. 328–341

Chapter 12
Contemporary legal analysis of China's transboundary water regimes: international law in practice
Yu Su
Water International, volume 39, issue 5 (2014) pp. 705–724

Chapter 13
China's practice on the non-navigational uses of transboundary waters: transforming diplomacy through rules of international law
Yanmei He
Water International, volume 40, issue 2 (2015) pp. 312–327

Chapter 14
Benefit-sharing and upstream/downstream cooperation for ecological protection of transboundary waters: opportunities for China as an upstream state
Owen McIntyre
Water International, volume 40, issue 1 (2015) pp. 48–70

Chapter 15
Approaches to investment in Chinese transboundary waters
Xiuli Han
Water International, volume 40, issue 1 (2015) pp. 71–86

Chapter 16
Mapping Asia's trans-boundary waters, with a focus on China
Vivian Louis Forbes
Water International, volume 40, issue 1 (2015) pp. 100–112

Chapter 17
China's southbound transboundary river basins: a case of asymmetry
Mirja Kattelus, Matti Kummu, Marko Keskinen, Aura Salmivaara and Olli Varis
Water International, volume 40, issue 1 (2015) pp. 113–138

Chapter 18
Water security in Himalayan Asia: first stirrings of regional cooperation?
Bjørn-Oliver Magsig
Water International, volume 40, issue 2 (2015) pp. 342–353

Chapter 19
Transboundary water cooperation on the Yarlung Zangbo/Brahmaputra – a legal analysis of riparian state practice
Yang Liu
Water International, volume 40, issue 2 (2015) pp. 354–374

Chapter 20
Benefit sharing in the Mekong River basin
Seungho Lee
Water International, volume 40, issue 1 (2015) pp. 139–152

Chapter 21
A new perspective on water governance in China: Captain of the River
Liping Dai
Water International, volume 40, issue 1 (2015) pp. 87–99

Chapter 22
Exploring China's approach to implementing 'eco-compensation' schemes: the Lake Tai watershed as case study considered through a legal lens
Liping Dai
Water International, volume 39, issue 5 (2014) pp. 755–773

Notes on Contributors

Karen Bakker is a Professor and Canada Research Chair in Political Ecology at the Department of Geography/Program on Water Governance, University of British Columbia, Canada.

Huiping Chen is a Professor in the School of Law at Xiamen University, China.

Liping Dai is a postdoctoral researcher at the Utrecht Centre for Water, Oceans and Sustainability Law, Faculty of Law, Economics and Governance, Utrecht University, the Netherlands.

David Devlaeminck is a PhD candidate at China International Water Law Programme (CIWL), School of Law, Xiamen University, China.

Jurian Edelenbos is Professor and academic director at the Department of Public Administration, Erasmus University, the Netherlands.

Yan Feng is based at the Asian International Rivers Centre at Yunnan University, China.

Vivian Louis Forbes is an Adjunct Research Professor at the National Institute for South China Sea Studies, China.

Velma I. Grover is based at the Centre for Engineering and Public Policy in the School of Engineering Practice, McMaster University, Canada.

Xiuli Han is based at the China International Water Law, School of Law, Xiamen University, China.

Daming He is based at the Asian International Rivers Centre at Yunnan University, China.

Yanmei He is based at the Economic Law School, Shanghai University of Political Science and Law, China.

Petra Hellegers is based at the Water Resources Management Group, Wageningen University, the Netherlands.

Mirja Kattelus is a doctoral student at the Water and Development Research Group, Aalto University, Finland.

Marko Keskinen is a University Lecturer at the Water and Development Research Group, Aalto University, Finland.

Aysegül Kibaroglu is a Professor at, and Chair of, the Department of Political Science and International Relations, Faculty of Economics, Administrative and Social Sciences, MEF University, Turkey.

Matthijs Kok is part-time professor of Flood Risk at the Faculty of Civil Engineering and Geosciences, Delft University of Technology, the Netherlands.

Gail Krantzberg is Professor of Civil Engineering at the Centre for Engineering and Public Policy in the School of Engineering Practice, McMaster University, Canada.

Stefan Kuks is Professor of Innovation and Water Policy Implementation at the School of Management and Governance, Institute for Innovation and Governance Studies, Twente Water Centre, University of Twente, the Netherlands.

Matti Kummu is Assistant Professor at the Water and Development Research Group, Aalto University, Finland.

Christina Leb is Senior Water Resources Specialist at the World Bank, USA.

Jing Lee is a PhD candidate at the Institute for Environment and Development (LESTARI), Universiti Kebangsaan Malaysia (UKM), Malaysia.

Seungho Lee is Professor at the Graduate School of International Studies, Korea University, Republic of Korea.

Bo Libert is Regional Adviser at the United Nations Economic Commission for Europe, Switzerland.

Yang Liu is a PhD student at Ghent University, Belgium.

Bjørn-Oliver Magsig is research fellow at the Department of Environmental and Planning Law, Helmholtz Centre for Environmental Research, Germany.

Owen McIntyre is Senior Lecturer at the Faculty of Law, University College Cork, National University of Ireland.

James E. Nickum is an institutional economist affiliated to the IWRA, the Centre for Water and Development at SOAS, UK, and the University of Hong Kong, China.

Emma S. Norman is Department Chair at the Department of Science/Native Environmental Science Program, Northwest Indian College, USA.

Marleen van Rijswick is Professor of European and Dutch Water Law at the Utrecht Centre for Water, Oceans and Sustainability Law, Department of Law, Utrecht University, the Netherlands.

Aura Salmivaara is based at the Water and Development Research Group, Aalto University, Finland.

Yu Su is based in the China International Water Law Team, School of Law, Xiamen University, China.

Olli Varis is Professor at the Water and Development Research Group, Aalto University, Finland.

Karina Vink is a postdoctoral researcher at the National Institute for Material Science, Japan.

Wenling Wang is based at the Asian International Rivers Centre at Yunnan University, Kunming, China.

Patricia Wouters is a Professor in the School of Law at Xiamen University, China, and Director of the China International Water Law Programme (CIWL).

Acknowledgements

The editors would like to sincerely thank the Xiamen University School of Law for hosting the first international water law symposium in China. Thanks are especially due to the former Dean, Professor Chongli Xu; Vice-Dean Professor Yansheng Zhu; Professor Huaqun Zeng, former Director of the International Economic Law Institute of Xiamen University and now Dean of the Faculty of Law, Macau University of Science and Technology; and all of our colleagues at Xiamen University School of Law for their strong support of China International Water Law (CIWL) and its ongoing research.

We the editors are also enormously grateful to the CIWL research team and to all of the participants at the symposium, whose contributions have made this book possible. Special thanks to Professor Owen McIntyre, Faculty of Law, University College Cork, who helped with preparing this book version and who continues to assist our CIWL research scholars, and to Dr. Sergei Vinogradov, Centre for Energy, Petroleum and Mineral Law and Policy, University of Dundee, for substantive contributions, editorial support and assistance.

We are grateful to Professor Marleen van Rijswick, Director of the Utrecht University Centre for Water, Oceans and Sustainability Law, and all of the Utrecht law team for co-convening and contributing directly to the success of the international water law symposium. Their enthusiastic support for our CIWL researchers and generous assistance with this collection are much appreciated.

Thanks to the Water International editorial team for their helpful guidance throughout the process, and to Carolyn Haynes, Emily Ross and others at Taylor & Francis who have brought our manuscripts to printed fruition. Deep thanks go to the many peer reviewers who generously gave their time and effort to help us to improve the quality of the manuscripts.

A final note of gratitude to our CIWL research team for their assistance in many important ways: Dr. Yu Su, Assistant Professor, Xiamen University, School of Law, and our PhD candidates, Nabaat Tasnima Mahbub, Flavia Loures and especially David Devlaeminck. Last but not least, a big thanks to Dr. Liping Dai, Associate Professor, Hubei University of Economics and Postdoctoral Researcher at Utrecht University, School of Law, who collected and processed these manuscripts along the way. It has been a pleasure to work with each and all of you.

Introduction

Patricia Wouters, Huiping Chen, James E. Nickum and David Devlaeminck

Connected and divided by its international rivers, lakes and aquifers, Asia is home to some of the world's greatest transboundary river systems: the Yarlung Tsangpo/Brahmaputra, Mekong, Ganges and the Indus, to name a few – vast resources serving more than 600 million people across Bangladesh, China, India, Pakistan and Southeast Asia. Asia is home also to the world's second largest economy – China – which shares more than 40 major transboundary waters with its (mostly) downstream riparian neighbours (Wouters & Chen, 2013). Remarkably, the significant array of water resources shared across Asia lack adequate governance arrangements. In particular, legal frameworks are limited in number and substantive content. This situation undermines meaningful upstream–downstream cooperation to resolve competing water and power demands on transboundary rivers.

This growing thirst for freshwater and hydropower should compel even sovereign riparian nations to find ways to cooperate, both within and beyond national borders. Yet Asia's diverse political, economic and cultural foundations and traditions, layered on to universal politics of space, pose significant challenges to achieving effective transboundary water cooperation. As the source of most of Asia's transboundary rivers, China plays a key role in addressing these challenges.

China's transboundary waters: facing the upstream dilemma

China's upstream location on most of the international water resources that it shares with its numerous neighbours affects its approach to transboundary water regimes (Nickum, 2008; He et al., 2014). An upstream state has a hydro-geographic advantage that can lead to its invocation of territorial sovereignty or even national security to justify unilateral appropriation of waters for domestic demands, whether economic, social or environmental, while at the same time stoking distrust on the part of downstream riparians. This distrust can spill over into other domains of international and economic relations with downstream riparians – in effect, nearly all countries bordering China. These 'linkage effects' can lead to an 'upstream dilemma': How can an upstream riparian nation meets domestic imperatives (especially economic, social and environmental concerns) while taking into account the imperatives of downstream riparians needs?

China's 'upstream dilemma' poses difficult challenges, especially as the nation moves rapidly towards greater engagement with its neighbours on a number of fronts, while developing its transboundary rivers for national hydropower and irrigation needs. China's relative insularity in these areas for most of the past 70 years or more puts it on a learning curve both in drawing on precedent and establishing new *modi vivendi* as a 'good neighbour' in its expanding foreign policy.

For some 65 years, the lodestone of China's foreign policy has been the 'Five Principles of Peaceful Co-existence': (1) mutual respect for each other's territorial integrity and sovereignty, (2) mutual non-aggression, (3) mutual non-interference with the internal affairs of the other party, (4) equality and mutual benefit and (5) peaceful coexistence. First codified in a 1954 agreement between China and India, the principles have been incorporated into the Preamble of China's Constitution as the basis for China's foreign policy.[1] More recently, these principles have been supplemented by a call for greater 'win–win cooperation'[2] as China has made significant commitments to engagement outside its borders, including the ambitious 'One Belt and One Road' initiative to enhance connectivity between China and ultimately Europe through the development of land and sea corridors running through many of its neighbours.

Global standards: the United nations watercourses convention and the nations economic commission for europe transboundary waters convention

As a shared transboundary fugitive resource, water poses a particular challenge to the Five Principles of Peaceful Coexistence. How can one reconcile respect for national sovereignty with equality and mutual benefit when an upstream use can adversely affect the quantity, quality or timing of flows downstream? Here, useful reference could be made to two global frameworks developed under the United Nations (UN) – the United Nations Watercourses Convention (UNWC) (Loures & Rieu-Clarke, 2013) and the United Nations Economic Commission for Europe Transboundary Waters Convention (UNECE TWC) (Tanzi et al., 2015). Although these instruments entered into force without endorsement by states in non-arid Asia (or the Americas),[3] they are global agreements, elaborating normative rules, principles and processes that can be tailor-made to address many of the issues arising in the Asian transboundary water context, including the 'upstream dilemma'.

Although China (together with Burundi and Turkey) voted against the 1997 UN Resolution that adopted the UNWC on the grounds that it impinged on national sovereignty, a closer examination reveals that it has endorsed the fundamental principles of the convention – the substantive rule of 'equitable and reasonable use', the duty to cooperate and the peaceful settlement of transboundary water disputes (Wouters, 2014). These notions resonate with the Five Principles of Peaceful Coexistence and have found expression in China's water-related treaty practice (Wouters & Chen, 2013; Su, 2014).

The 'duty to cooperate' is the operational bridge between the substantive and procedural rules in the UNWC. It is elaborated in Article 8 as follows:

> **General obligation to cooperate**
> Watercourse States shall cooperate on the basis of sovereign equality, territorial integrity, mutual benefit and good faith in order to attain optimal utilization and adequate protection of an international watercourse.
>
> In determining the manner of such cooperation, watercourse states may consider the establishment of joint mechanisms or commissions, as deemed necessary by them, to facilitate cooperation on relevant measures and procedures in the light of experience gained through cooperation in existing joint mechanisms and commissions in various regions.

This provision corresponds closely to the Five Principles of Peaceful Coexistence, and to China's procedural practice in the field. On the latter, China has established several joint commissions on transboundary waters with its northern and western riparian neighbours, and less formal institutional arrangements on its southern rivers, where technical cooperation is advanced primarily through data-sharing agreements. In November 2015, a framework for cooperation was established with the launching of the Lancang-Mekong Cooperation Mechanism (Ministry of Foreign Affairs [MFA], 2016a). This arrangement includes all riparians and, although not specifically focussed on water, seeks cooperation based on consensus, mutual benefit and equality (Biba, 2016; MFA, 2016a). In March 2016, the riparians signed the Sanya Declaration, recognizing the interconnected nature of the riparian states and devoting themselves to future cooperation (MFA, 2016b). These incremental steps enhance the opportunities for improved cooperation on transboundary waters.

Dispute settlement for China, generally, focuses on diplomatic exchange, negotiation and consultation, aligned with the Five Principles of Peaceful Coexistence and China's state practice. While some commentators challenge this approach as reflecting China's general unwillingness to engage, upon closer scrutiny, this practice aligns with historic approaches to law and dispute resolution in China (Pan, 2011; Wang, 1990). Hence the Chinese government's position vis-à-vis the UNWC is consistent with a long-standing approach to international law generally (Holland, 2012; Saul, 2013).

The legal research by two of the present authors (Vinogradov & Wouters, 2013; Wouters & Chen, 2013) suggests that China's state practice to transboundary waters is characterized by three remarkable trends: (1) its consistency with China's general approach to international law; (2) the relative conciseness and limited coverage of treaty practice in this field, both in substantive normativity and with respect to geographic reach and (3) a recent move towards increased, albeit, incremental bilateral hydro-diplomacy with riparian neighbours.

Within this context it is clear that China is now well placed to consider more closely the guidance offered in global instruments, especially the UNWC and UNECE TWC, which offer practical approaches to win–win solutions regarding the uses of shared international water resources. The 1992 Convention on the Protection and Use of Transboundary Watercourses and International Lakes (in force from 1996) concluded under the auspices of the UN Economic Commission for Europe initially for the Pan-European region (UNECE TWC). It was amended from July 2014 to open its membership, thus extending its global reach, a move aimed at enhancing water cooperation among the countries sharing the same watercourse (rivers, lakes or groundwater) regardless of their geographical location. The UNECE TWC, which has just celebrated 20 years of successful implementation, currently has 41 parties (the former Yugoslav Republic of Macedonia joining as 41st party on 28 July 2015) and 26 signatories, including the European Union, providing extensive varied state practice from which to learn.

The influence of these two UN global water treaties cannot be overestimated. Mr Michael Møller, (then) Under-Secretary-General of the UN stated recently

> As 60 per cent of the world's freshwaters are transboundary, solid international legal frameworks for water cooperation are critical to prevent conflicts and ensure sustainable management of shared resources. The United Nations Watercourses Convention and the UNECE Water Convention will together provide firm ground for cooperation in transboundary basins around the globe. These two Conventions complement each other and should be implemented in a coherent manner, as the United Nations Secretary-General already stressed in the past.
>
> (UNECE, 2014)

Although few Asian countries have signed up to either of these UN multilateral conventions,[4] each of these instruments offer important guidance on how transboundary water cooperation might be implemented across Asia. Each is based upon the international rule of law – 'the duty to cooperate', the fundamental cornerstone of international relations.

Each sets forth and elaborates significant legal rules, procedures and mechanisms, which evolved as a result of state practice around the world and proved their effectiveness in managing transboundary water resources in different regions and basins.[5] This rich body of experience is readily available for consideration in the Asian context, ripe for tailor-made adjustment and application.

An overview of this book

The contributions to this book are divided into five parts. Part 1 'The Building Blocks of Cooperation', explores international law principles on the issue of international riparian cooperation. Part 2 'International Experiences', looks at practices outside China. The focus then turns specifically to China, with 'China's International Waters and the Potential for Cooperation' (Part 3) and, even more specifically, to 'China's Southbound Basins' (Part 4). Part 5 'Innovative Approaches to Cooperation of China's Shared Domestic Waters,' looks at two domestic innovations to promote cooperation among the stakeholders of Lake Tai. The collection finishes with a conclusion drawing together some of the findings from this study.

Part 1: The building blocks of cooperation

The works presented here provide new insights into China's 'upstream dilemma' through legal and governance analysis from China, other parts of Asia and from comparative case studies across the globe. Divided into five parts, this book first explores the building blocks of cooperation. Chapter 2, by Christina Leb, sets the backdrop for the first section by examining transboundary water cooperation through the lens of international law. She adopts a working definition of cooperation, and suggests that cooperative behaviour to achieve mutual benefits in international relations rests on two key building blocks: reciprocity and good faith. 'The expectation of reciprocity leads states to enter into legal relationships; at the same time they accept restraints to their sovereignty based on the good faith expectation, or trust, that other states will follow suit'. These relationships are consolidated and confirmed in treaties. She demonstrates how international law is a vehicle for incremental cooperation in this field, noting the considerable contribution of the UN. Through a selection of international case studies – ranging from the Brahmaputra, Rhine, Senegal River and Danube to the La Plata – she illustrates the evolutionary nature of transboundary water cooperation, highlighting the role of international law in this process. She notes that, '[t]he conclusion of an international treaty is just one step in the iterative and most often incremental process of international cooperation'. Interdependence between states, generally, and especially with regard to shared freshwater resources, leads states 'to make first concessions in good faith so that the concession is reciprocated and the full potential of mutual benefits becomes achievable'. Thus, loosening the chains of sovereignty through agreed cooperative arrangements between riparian nations provides new opportunities for joint gains – a 'win–win' approach.

In Chapter 3, by Huiping Chen, several fields of international law – international economic law, international water law and international human rights – are examined to evaluate how these combine to support national governments in meeting their duties with respect to the human right to water. The focus on human rights is an understudied field, attracting recent interest (Meshel, 2015; UPEACE Lecture, 2015). Highlighting both the possible benefits and potential harms that foreign investment in water resources might have on a national government's duty to implement the human right to water (in accordance with UNGA, 2010; UN Human Right to Water, 2010),

Chen identifies three international legal obligations connected with the human right to water: the duties to respect, protect and fulfil.

> [These] impose two types of operational obligations on national governments – the obligation to seek ways to realize the right to water, and the obligation to prevent third parties (including foreign investors) from violating or interfering with the right to water, both including the availability (quantity), quality and accessibility of water.

Following a brief survey of relevant state practice, doctrine and jurisprudence in this field, Chen reviews legal measures that a host state might use and concludes

> International practice appears to be changing to include human rights issues in IIAs [international investment agreements], which would support the more complete realization of the human right to water. This could balance the need for investment and investors interests with those universal duties of host states.

Chapter 4, jointly authored by Jurian Edelenbos, Petra Hellegers, Matthijs Kok, Stefan Kuks and Marleen van Rijswick, uses an interdisciplinary method to devise a framework for assessing collaborative and integrative approaches to deal with water management (water shortage, water quality and flood risks), applying different disciplinary approaches – water system analysis, economics, law and public administration. As a European example, it reflects possible state of the art for building a transboundary water management regime. Although regional experiences cannot be automatically transposed, with the clear need for improved transboundary water governance across Asia, this chapter offers some insights on an integrating approach based on considerable practice in the field.

Chapter 5 attempts to link several areas of international law, seeking support for vulnerable people through the principle of 'equitable and reasonable use'. In 'Transboundary water law and vulnerable people – legal interpretations of the principle of "equitable and reasonable use"', Karina Vink defines vulnerable people as those 'who have one or more characteristics that make them more susceptible than others in a community and who therefore require additional measures in order for them to have the same water capabilities as others'. This chapter offers a first study of its kind in this area. The author attempts to knit together the various threads of legal principles across several fields of international law, including transboundary water law and human rights.

Chapter 6, by Jing Lee, provides a legal analysis that goes deeper into the issue of ecosystem protection, an important emerging issue in China. The chapter considers the potential role of the Ramsar Convention (to which China is party) to promote transboundary water cooperation and wetland conservation. Lee's analysis is linked closely with the 'duty to cooperate' provision in the UNWC (Article 8[1] see above), noting also the corresponding provision in the UNECE TWC (Article 2[6]), 'Riparian Parties shall cooperate on the basis of equality and reciprocity, in particular through bilateral and multilateral agreements in order to develop harmonised policies, programmes and strategies covering relevant catchment areas or part thereof.' Identifying the important role that water-related global environmental governance can play in fostering common understandings and as a catalyst for enhancing transboundary water cooperation, Lee then focuses on the substantive and procedural obligations under Ramsar, with a view to exploring how these might facilitate enhanced transboundary water cooperation. How are China's transboundary Ramsar sites shared with Russia, the Kyrgyz Republic, India, Pakistan and others being managed under this global instrument and what impact does this have on transboundary water cooperation? While issues remain, Lee concludes that the institutional framework established under Ramsar enables and promotes cooperation and peace-making through the collaborative management of transboundary wetland ecosystems.

Part 2: International experiences

Although the duties to promote water security often fall onto states, Part 1: The Building Blocks of Cooperation clearly illustrates that transboundary water cooperation is absolutely fundamental. This part analyses various international experiences, beginning by asking the question: Can the UN help? Bo Libert, a UN practitioner, makes an affirmative case in looking at how the UNECE TWC might contribute to promoting transboundary cooperation in Chapter 7. Beginning with an overview of the UNECE Convention and its programme of work, this chapter then presents a succinct overview of the current transboundary water cooperation situations in each of the four basins. The Chu and Talas transboundary water cooperation (Kazakhstan and Kyrgyzstan) is a story about building bilateral cooperation through joint projects – water infrastructure with water allocation also on the table. Now the challenge is to broaden the remit of the joint commission to deal with environmental issues. The Kura-Aras, shared by Armenia, Azerbaijan, Georgia, Iran and Turkey, is moving forward through bilateral agreements, in an incremental manner, with draft texts still being negotiated. The approach on the Drin system, composed of several major rivers and lakes, includes Albania, Greece, Macedonia, Kosovo and Montenegro, is based on the 'Drin Dialogue', which aspires to basin-wide cooperation. An important first step toward this objective – the Drin Memorandum of Understanding (MoU) – provides the institutional and operational framework for progressing towards a more formal cooperation framework. The situation on the Dniester, shared by the Ukraine, Moldova (including Transnistria) and Poland, demonstrates yet another regional approach to transboundary water cooperation. The significant outcome is a rather comprehensive bilateral agreement (Ukraine/Moldova) that has yet to enter into force, but which has fostered three projects that have enhanced cooperation. Libert's analysis of these very diverse experiences yields the following general conclusions: (1) the legal and institutional framework of the UNECE Water Convention is a good platform for the promotion of transboundary water cooperation in individual basins; (2) concluding basin agreements is important but one should not aim to agree on every detail of the future cooperation, involving other international actors in the respective basins is essential; (3) involving non-governmental organizations (NGOs) and other stakeholders should be an objective from an early stage and (4) it is not expensive to support transboundary water cooperation, but cofunding from riparians is essential. He suggests that UN regional commissions can play an important role in facilitating transboundary water cooperation in their respective regions.

This part then turns to how Canada–United States transboundary water cooperation has evolved through 'rescaling'. In Chapter 8, Emma Norman and Karen Bakker distil lessons learned from the century-long relationship. They identify the notion of 'rescaling' as a possible mitigating approach to the competing sovereign rights of the two riparian nation states. Few studies have attempted to analyse comprehensively the rescaling of water governance. The research deploys both quantitative and qualitative data to analyse the effectiveness of the changes in governance in the Canada–United States case study. The key findings are as follows: symmetry in governance frameworks matters, a need for in-built regime flexibility to deal with changing circumstances (such as environmental and social contexts) and concerns over sovereignty might limit basin-wide management approaches. The authors assert that, in rethinking water governance, there is a need to support communities and activities that have been historically absent in formal bi-national treaties. Local stakeholders provide opportunities for forging new bridges for transboundary water cooperation through joining forces to tackle common problems. Given China's diverse ethnic populations, particularly in border regions, this approach might offer some new insights in the transboundary water context (Nickum, 2008).

Some lessons for China and its neighbours may be drawn, judiciously, from over a century of cooperation between Canada and the United States over their transboundary waters, as demonstrated in the final two chapters in this part. Chapter 9, by Velma Grover and Gail Krantzberg, shows an evolving, learning process, requiring tincture of time. The evolution of the Canada–USA International Joint Commission (IJC) and the 1909 Boundary Waters Treaty (which established the IJC) has included a transformation from an allocation-based approach to one that now takes on board changing environmental concerns. The chapter also examines the Great Lakes Water Quality Agreement of 1972 (GLWQA) and explains how the ecosystem approach was applied and remedial action plans were developed and implemented to deal with pollution issues in the Great Lakes (shared by Canada and the United States). The evolutionary nature of both a historic water treaty and the more than century-old river basin commission (the IJC) is revealed in a number of scenarios. Examining how water quality issues have been handled under the Boundary Waters Treaty, the GLWQA and under the auspices of the IJC, this chapter highlights the practical approaches to transboundary water cooperation that have evolved in this bilateral relationship. The authors conclude by listing some seven lessons from the IJC practice that might be replicated. Whether or not these might have traction in the Chinese context will need further study.

China is far from the only country with an 'upstream dilemma' in its relationships with neighbours. In Chapter 10, Aysegul Kibaroglu explores the legal and institutional issues related to Turkey's water policy. 'Turkey, like China was seen as a "hegemon" in its riparian settings'. It is interesting to note that Turkey's MFA is responsible for 'formulating as well as executing Turkey's transboundary water policy'. Would this be a way to consolidate China's approach – to identify a lead governmental agency as the overarching home for transboundary waters issues? The chapter provides a useful overview of Turkey's approach to the UNWC, which, remarkably, has strong parallels to China's views. Thus, any perceived constraint on national sovereignty is rejected and provisions, such as Article 8 of the UNWC, which preserves 'sovereign equality', 'territorial integrity', 'mutual benefit' and 'good faith', find support from Turkey (and China). While the EU Water Framework Directive (WFD) has resulted in Turkey taking steps to harmonize its national water policy with EU approaches, there remains considerable work to be done with respect to its transboundary water policy. Kibaroglu urges the 'international community' to establish

> an effective dialogue with Turkey in order to systematically reconcile its water policy objectives with the global norms that exist in this field, such as the UNWC, and also with the need for a predictable framework for conflict management in transboundary basins.

Part 3: China's international waters and the potential for cooperation

As mentioned earlier, China faces an urgent and complex series of water-related issues. This part aims to explore these related issues while identifying potential solutions. In Chapter 11, Yan Feng, Daming He and Wenling Wang use scientific and policy data to shed further light on these dynamic issues. They argue that China's transboundary water cooperation is hindered by a number of factors, in particular underdeveloped data sharing and fragmented legal regimes (at the national and international levels). The authors suggest that these gaps result in transboundary water risks and vulnerabilities that need to be redressed.

With a particular concern for ecological issues, the chapter calls for increased cooperation on scientific research and joint undertakings, more transparency, and a more coherent approach by China's national transboundary water institutions.

Taking China's water situation into consideration, Chapter 12, by Yu Su, uses a legal analytical framework to examine the elements fundamental to understanding a transboundary water cooperation regime, including scope, substantive rules, procedural rules, institutional mechanisms and dispute settlement. Although China is upstream on almost all of its watercourses, is often considered a hydro-hegemon and voted against the UNWC in the General Assembly, this chapter tells a different story. Through his analysis the author argues that

> China, in its existing water treaty practice has embraced the fundamental principles of IWL, albeit with some shortcomings. And it would appear that things are moving in the right direction, given recent developments both with respect to China's interstate relations and in the field of IWL more generally [. . .]

Such shortcomings of China's treaty practice might be addressed using as a guide the two main watercourse conventions in this field: the 1997 UN Watercourses Convention and the 1992 UNECE Transboundary Watercourses Convention, and 'cross fertilization' from these international regimes could provide 'important normative connectivity', which China's 'bilateral and sometimes inconsistent water treaty practice might benefit from'. The author concludes that

> The analysis of recent developments in both IWL and China's IWL practice supports the view that China is taking part in the communicative process of the making of IWL. While China can become more engaged with IWL by drawing insights from the global conventions and participating in joint mechanisms, it can also become engaged by contributing its own understanding of the legal norms to the international community.

Chapter 13, by Yanmei He, analyses Chinese diplomatic practice in this field. The author finds that China has adopted a 'responsive' diplomatic approach, with the country 'responding' to other riparian demands. She argues in favour of 'preventive' diplomacy, where China anticipates and addresses the transboundary water-related issues arising for its riparian neighbours. This would be more in keeping with China's current foreign policy strategy of seeking 'win–win' solutions and its express objective of being a 'good neighbour'. China's transboundary waters offer a compelling platform for implementing its foreign policy goals and 'preventive' diplomacy would be one way forward. In order to implement a policy of 'preventive diplomacy'

> China may take appropriate policy measures to control pollution and protect the ecosystems of international watercourses, while also developing new rules to balance the rights and obligations of upstream states. Moreover, China ought to cooperate with its neighbours through various means and on multiple levels, including prior notification of planned measures, exchange of information, water allocation, water rights trading, the conclusion of bilateral or multilateral agreements, the establishment of bilateral or multilateral management institutions, and so forth.

Chapter 14, by Owen McIntyre, tackles the hard problem of the upstream/downstream challenges to protecting ecosystems. Using a range of legal arguments, the author asserts at the outset that the view that the UNWC favours downstream states is based on a misunderstanding of the rules of general international water law. The cardinal rule of international water law, the principle of equitable and reasonable utilization, is inherently flexible and quite capable of taking account of a very wide range of needs and interests of riparian states, including potential and future uses and the need to protect the entire watercourse ecosystem. McIntyre posits that 'the continuing evolution of the closely related concepts of "environmental flows", "ecosystem services" and "eco-compensation" provides the conceptual and methodological frameworks to help facilitate such cooperation'. Following elaboration of these notions through a legal discourse, McIntyre concludes by suggesting that 'China should engage fully in regional cooperative initiatives where the feasibility of broad benefit-sharing arrangements could be explored with its southern neighbours'.

Chapter 15, by Xiuli Han, reviews current Chinese practice in the field of international development, identifying three methods of investment in transboundary waters: national investment in national waters, cooperative investment in a riparian state's water resources and cooperative investment in contiguous border waters. Han provides an overview of China's transboundary water investments, including on specific hydropower projects, in the context of China's 12th Five-Year Plan. Following an exploration of the possible approaches to dispute settlement in this field, Han notes the use of arbitration as the primary option for dispute resolution in China's bilateral investment treaties, highlighting this as a possible avenue for the peaceful settlement of transboundary water investment-related disputes. Han closes by urging China to 'strive for investor protection in its transboundary water treaties and IIAs' (international investment agreements) (Chen, 2013). This study, with its focus on bilateral treaty practice, identifies a key research theme that requires more investigation. China's rich bilateral engagement around the world could provide insights for transboundary water cooperation.

Transboundary waters are, for the most part, natural systems. The boundaries they cross are not, and are often contested. Hence map-making, the fundamental activity underlying the determination of boundaries, can be a political act. As Vivian Forbes states in Chapter 16, 'Borders reflect a nation's conception of its own territorial and sovereign limits [. . .].' Tracing the history of China's mapping, Forbes elaborates on some of the challenges to map-making and provides an overview of how this is regulated in China. He concludes by emphasizing the importance of accurate mapping as a means to address some of China's transboundary water issues.

Part 4: China's southbound basins

Even if all parties can agree on borders and turf, there remains the problem of upstream advantage, reflected in greater vulnerability of downstream riparians, resulting in often highly asymmetric bargaining positions. In Chapter 17, Mirja Kattlelus, Matti Kummu, Marko Keskinen, Aura Salmivaara and Olli Varis show how these vulnerabilities work out in six major transboundary river basins that drain southern China: Red River, Mekong, Salween, Irrawaddy, Ganges-Brahmaputra-Meghna and Indus. The study examines the relevant societal and environmental issues through a multidimensional river basin vulnerability (RBV) analysis. Two critical questions are addressed in the chapter: 'how transboundary cooperation could be facilitated in the study basins, and how to ensure that China – that often is the least vulnerable but at the same time a major player in water infrastructure development – could find an incentive to participate in it'. The conclusion finds that China has 'participated fairly weakly in transboundary water cooperation' and calls for 'increased responsibility' on the part of all riparian countries on the six river basins studied. The authors suggest that there are a 'plethora of opportunities for transboundary cooperation' in this context. This chapter received the *Water International* Best Paper Award for 2015.

Taking a regional perspective, Bjørn-Oliver Magsig, in Chapter 18, considers riparian state practice and finds evidence for incremental transboundary water cooperation across the region. Examining the central question of how regional approaches to transboundary water cooperation might be better developed, Magsig first surveys the status of regional water cooperation in Himalayan Asia. He then explores how a 'novel understanding' of common water security and its legal foundations might contribute to strengthening regional transboundary water cooperation. Considering the roles of regional institutions, and relevant soft and hard laws that apply to the region, Magsig observes that the Association of Southeast Asian Nations' (ASEAN) role in regional transboundary water cooperation is increasing, especially with the 'slowly emerging interest of both hydro-hegemons, China and India [. . .]'. Nonetheless, he concludes that

> Himalayan Asia, will not eagerly implement a perception of cooperative sovereignty which comes from 'outside'. It is much more likely that states within the region will develop their own concepts – based on their regional identity and specific political and cultural environment.

Following this analysis of the Himalayan region, Chapter 19, by Yang Liu, uses an international legal lens to find new ways of promoting enhanced transboundary cooperation on this watercourse. The research examines each basin state's approach to national sovereignty, suggesting that this influences practice on the Yarlung Zangbo/Brahmaputra. The treaty regimes reveal diverse arrangements across the basin, reflecting different sovereign approaches. Two specific legal paradigms are identified as possible avenues for increased transboundary water cooperation on the basin. Taking inspiration from the relatively successful series of hydropower agreements between Bhutan and India, the author ponders whether the replication and scaling up of this model, using an integrated hydropower approach, might be one way forward. Another possible model for cooperation could be through the multilateral environmental agreements to which Bangladesh, Bhutan, China and India are already party. Specifically, the author reviews the Convention on Biological Diversity (CBD) (1992) as a vehicle for enhancing transboundary water cooperation on the Yarlung Zangbo/Brahmaputra.

In Chapter 20, Seungho Lee is cautiously optimistic about the prospects on one of those basins, the Lancang/Mekong. He considers the current state of play on this transboundary watercourse, which has been the source of some concern for China and its downstream neighbours. Lee analyses

> the extent to which the cooperative relationship between China and the downstream countries in the Mekong River Basin (Myanmar, Lao PDR, Thailand, Vietnam and Cambodia) has evolved due to a variety of benefits shared by the riparian countries from the 1990s to the present day.

Lee deploys a 'benefit-sharing mechanism' as the analytical framework to evaluate this changing situation. Defining benefit-sharing as 'any action designed to change the allocation of costs and benefits associated with cooperation', the author introduces and explores also the four types of benefit-sharing (to/from/for/beyond the river) set out by Sadoff and Grey. China's upstream dam-building, leading to tensions with the Mekong basin states (all downstream), is characterized as 'China's Benefit Monopoly', a source of distrust across the basin in the 1990s. Lee finds that this scenario is changing, and has evolved as a result of China's increased commitment to cooperation, evidenced is a series of actions on the Mekong.

> China [. . .] is no longer hegemonic. [. . .] Cooperation has been accelerated between China and the downstream countries in terms of ..gradual socio-economic tie[s], development aid flow, and increasing energy security aspects [. . .] achieved through a wider cooperation framework related to the GMS (Greater Mekong Subregion) program.

Lee concludes by suggesting that China's growing willingness to cooperate on the Mekong could be the start of a new era across the region, with shared benefits for all, including supporting the sustainable development of the Mekong River Basin.

Part 5: Innovative approaches to cooperation of China's shared domestic waters

Chapter 21, by Liping Dai, focuses on China's domestic governance approaches, using the notion of stewardship as a management regime aimed at protecting the river. The 'Captain of the River' concept was pioneered in Wuxi City to tackle blue algae bloom in Lake Tai. By 2014, more than 700 rivers in Jiangsu province have their own captain, responsible for dealing with water pollution under a government objective responsibility system (GORS). Local 'captains' are assigned to meet pollution targets in line with China's national policy. Dai explores the strengths and weaknesses of the Captain of the River concept and concludes that it is an efficient and adaptable tool to promote environmental protection, at least in the short term, and during the transition period of legal reform now being undertaken in China.

Chapter 22, also by Liping Dai, focuses on the sub-national level in China, examining how transboundary water resources are managed through an 'eco-compensation' scheme. Dai distinguishes the Chinese approach from the international counterpart, payment for ecosystem services (PES), and provides a case study in the Chang Jiang (Yangtze) region. Identifying four typologies of eco-compensation schemes, Dai analyses how each works in practice, and finds that not every scheme qualifies as a true 'eco-compensation' mechanism. With China taking increased measures to address serious water issues across the nation – pollution, scarcity, floods and droughts – how the nation moves forward with 'eco-compensation schemes' could mean the difference between sustainable and unsustainable water resources management.

All the perspectives, insights and wisdom gathered here from many quarters, notably including China, show that transboundary water cooperation, at any level, is not easy. It takes time and a great deal of effort and good will. Nonetheless, the payoff of cooperation is high, and failure to attain it can portend disaster, even for those who hold the high ground. Fortunately, there are frameworks available, and signs of a willingness by China to seek win–win solutions. We return to these themes in Chapter 23, by Patricia Wouters, Huiping Chen, James Nickum and David Devlaeminck.

Notes

1 Constitution of the People's Republic of China, 4 December 1982, Preamble.

2 'Carry forward the Five Principles of Peaceful Coexistence to build a better world through win–win cooperation'. People's Daily Online, 10 July 10 2014. Retrieved from http://english.peopledaily.com.cn/n/2014/0710/c90883-8753393.html/.

3 At present, some 39 states have endorsed the convention including 36 parties and 3 signatory states yet to ratify it. The current parties and signatories are Benin, Burkina Faso, Chad, Côte d'Ivoire, Denmark, Finland, France, Germany, Greece, Guinea-Bissau, Hungary, Iraq, Ireland, Italy, Jordan, Lebanon, Libya, Luxembourg, Montenegro, Morocco, Namibia, Netherlands, Niger, Nigeria, Norway, Portugal, Qatar, South Africa, Spain, State of Palestine, Sweden, Syria, Tunisia, UK, Uzbekistan and Vietnam. Signatories include Paraguay, Venezuela and Yemen. Only two of these are from Asia outside the Middle East, and Vietnam (a downstream riparian) is the only one from East, South or Southeast Asia.

4 Only a handful of Asian riparian nations have signed up to one or both of these legal framework instruments: Azerbaijan, Kazakhstan, Russia, Turkmenistan, Uzbekistan and Vietnam.

5 In addition to the two water-use-related UN water conventions, the UN Resolution on the Human Right to Water and Sanitation recognizes the duty to cooperate; requires reporting on these obligations; and 'calls upon states and international organizations to provide financial resources, capacity-building and technology transfer, through international assistance and cooperation, in particular to developing countries, in order to scale up efforts to provide safe, clean, accessible and affordable drinking water and sanitation for all' (UNGA, 2010). In a transboundary water context, these water-related human rights introduce national duties that should be considered as relevant factors in the context of increased or new uses of international waters.

References

Biba, S. (2016). China drives water cooperation with Mekong countries (Chinadialogue, 2016 February 1). Retrieved from www.chinadialogue.net/article/show/single/en/8577-China-drives-water-cooperation-with-Mekong-countries

Chen, A. (2013). *The voice from China. An CHEN on International Economic Law*. Berlin and Heidelberg: Springer.

He, D., Ruidong W., Feng, Y., Li, Y., Ding, C., Wang, W., & Yu, D. W. (2014). China's transboundary waters: New paradigms for water and ecological security through applied ecology. *Journal of Applied Ecology*, 51 (5), 1159–1168. doi: 10.1111/1365-2664.12298

Holland, C. (July 2012). Chinese attitudes to International Law: China, the Security Council, sovereignty, and intervention. *NYU Journal of International Law & Politics* (Online Forum).

Loures, F. R., & Rieu-Clarke, A. (Eds.). (2013). *The UN Watercourses Convention in Force: Strengthening International Law for Transboundary Water Management*. Abingdon: Routledge.

Meshel, T. (2015). Human rights in investor–state arbitration: The human right to water and beyond. *The Journal of International Dispute Settlement*, 6(2), 277–307. doi: 10.1093/jnlids/idv007. Retrieved from https://ssrn.com/abstract=2554086

Ministry of Foreign Affairs (MFA), Peoples Republic of China (2016a). Five features of Lancang-Mekong river cooperation. 17 March, 2016. Retrieved from www.fmprc.gov.cn/mfa_eng/zxxx_662805/t1349239.shtml

Ministry of Foreign Affairs (MFA), People's Republic of China (2016b). Sanya declaration of the first Lancang-Mekong cooperation (LMC) Leaders. 23 March, 2016. Retrieved from www.fmprc.gov.cn/mfa_eng/zxxx_662805/t1350039.shtml

Nickum, J. E. (2008). The upstream superpower: China's international rivers. In O. Varis, A. K. Biswas, & C. Tortajada (Eds.), *Management of Transboundary Rivers and Lakes* (pp. 227–244). Berlin: Springer.

Pan, J. (2011). Chinese philosophy and international law. *Asian Journal of International Law*, 1(2), 233–248. doi: 10.1017/S2044251310000354

Saul, B. (2013). China, natural resources, sovereignty and international law. *Asian Studies Review*, 37(2), 196–214. doi: 10.1080/10357823.2013.769497.

Su, Y. (2014). Contemporary legal analysis of China's transboundary water regimes: International law in practice. *Water International*, 39(5), 705–724. doi: 10.1080/02508060.2014.950856.

Tanzi, A., McIntyre, O., Kolliopoulos, A., Rieu-Clarke, A., & Kinna, R. (Eds.). (2015). *The UNECE Convention on the Protection and Use of Transboundary Watercourses and International Lakes: Its Contribution to International Water Cooperation*. Leiden and Boston: Brill-Nijhoff.

UNECE. (2014). UNECE welcomes the entry into force of the United Nations Watercourses Convention. Retrieved from www.unece.org/index.php?id=35610

UNGA. The human right to water and sanitation (A/RES/64/292). 3 August, 2010. Retrieved from www.un.org/ga/search/view_doc.asp?symbol=A%2FRES%2F64%2F292+&Submit=Search&Lang=E

UPEACE/Peace palace library lecture: Judge Kenneth Keith on international water disputes (30 January 2015). Retrieved from www.peacepalacelibrary.nl/2015/01/upeacepeace-palace-library-lecture-icj-judge-kenneth- keith-on-international-water-disputes/?utm_source=Peace+Palace+Library+List&utm_campaign=355dfb794e-Peace+Palace+Library+Newsletter+January+2015&utm_medium=email&utm_term=0_b06a03629a355dfb794e-308611069

Vinogradov, S., & Wouters, P. (2013). *Sino-Russian Transboundary Waters: A Legal Perspective on Cooperation*. Nacka, Sweden: Institute for Security and Development Policy. Retrieved from www.isdp.eu/publications/index.php?option=com_jombib&task=showbib&id=6276

Wang, T. (1990). *International Law in China: Historical and Contemporary Perspectives*. 221 Recueil des Cours. 195. Academie de Droit International de la Haye. Leiden: Brill.

Wouters, P. (2014). The yin and yang of international water law: China's transboundary water practice and the changing contours of sovereignty. *Review of European Comparative & International Environmental Law*, 23(1), 67–75.

Wouters, P., & Chen, H. (2013). China's 'soft-path' to transboundary water cooperation examined in the light of two global UN water conventions – Exploring the 'Chinese way'. *Journal of Water Law*, 22, 232.

One step at a time: international law and the duty to cooperate in the management of shared water resources

Christina Leb

Analyzing the dynamics of international cooperation, this article describes the role of international law in the process. It illustrates that international law is a tool to achieve mutually beneficial relations and can be a driver for cooperation. Describing how these dynamics apply in the context of cooperation on transboundary rivers, the article traces the emergence of the general duty to cooperate in international water law as a legal response to the interdependence and coordination requirements in the management and development of shared water resources.

Introduction

Shared water bodies – rivers, lakes and aquifers – create linkages and interdependencies between states. Across the globe, 276 rivers connect the territories of two or more countries (UN Water, 2013a). About 40% of the world's population in 148 countries depend on the freshwater resources from these river basins (UN Water, 2013a). Freshwater is becoming an increasingly scarce resource; additionally, population growth and rapid urbanization are putting increasing pressure on global freshwater resources. It is expected that by 2025, 1.8 billion people will live in countries or regions under conditions of absolute water scarcity (UN Water, 2013b), partly because water use increases at more than twice the rate of population growth. At the river-basin level, the use of the transboundary water system by one state usually impacts on other states sharing the same system, through changes in water quality, alteration of river flow or change of water availability over time. Uses can affect another state's physical territory, for example through flow velocity changes that impact river bed stability and the geomorphology of the river, or reduction of flood risk and damage through flow regulation. They can affect existing uses or foreclose future water development opportunities in other parts of the basin, for instance by changing the hydropower potential downstream through upstream development. These impacts may cause harm, or they may be beneficial.

In 2013, the International Year of Water Cooperation, several joint declarations from governments, international and regional organizations and civil society were issued that highlight international agreement on the need for cooperative development and management of shared water resources. The Declaration of the High Level International Conference on Water Cooperation, held in Dushanbe in August 2013, recognizes cooperation as essential to ensure "broad-based sustainable economic growth, overall human development, poverty and hunger eradication, public health, food security, energy security, environmental protection, disaster prevention, urban and rural development, resilience to climate change" and stability (INBO, 2013). Similar conclusions were drawn at the World Water Week (Water Cooperation – Building Partnerships) held in Stockholm one month later (Stockholm International Water Institute, 2013). Furthermore, the Sustainable Development Goals, which will define the post-2015 development agenda, recognize the need for cooperation to ensure continued and adequate availability of water for human consumption (United Nations General Assembly [UNGA], 2014). Of these, Goal 6, which concerns ensuring the availability and sustainable management of water and sanitation for all, includes as Target 6.5 that by 2030 integrated water resources management be implemented at all levels, including through transboundary cooperation as appropriate. This article analyzes the dynamics of when states decide that cooperation is appropriate and the tools they use to govern their cooperative relationships.

International law as an expression of cooperation between states

Cooperation is defined as the process by which states take coordination to a level where they work together to achieve a common purpose that produces mutual benefits that would not be available to them with unilateral action alone (Zartman, 2008). Thus, when states expect to derive benefits that would otherwise be unachievable, they will seek engagement and collaboration with those states that are in a position to assist them in order to realize additional gains. To shape their process of engagement, states use a variety of tools, including: dialogue at the technical, diplomatic or highest government levels; the exchange of notes, data and information; collaboration between national entities; the establishment of joint institutions; and international treaties.

Building-blocks of cooperation

People and, later, states that did not own all the resources they needed to fulfil their basic needs or enhance their economic conditions – food supplies, medicinal herbs, wood, coal, metals and ores – engaged in war or trade to gain access to those resources. Over the course of history, a complex web of trade relations and rules of peaceful engagement emerged that in addition to resource access created many other benefits through the exchange of knowledge and goods: scientific and technological advances; higher life expectancy; and increasing resource efficiency to support more and more people. The intention here is not to claim that there were only winners, but to illustrate that cooperative behaviour over time generated multiple benefits that might not have been available otherwise. Cooperative behaviour to achieve mutual benefits rests on two key building-blocks: reciprocity and good faith. To elicit cooperation a first concession must be made; it takes place in good faith that the concession will be reciprocated. If the concession is not reciprocated, good faith is frustrated, and cooperation and the benefits that can be derived therefrom may never occur.

Over time, the interest in sustained mutually beneficial relationships leads to consistent behaviour of reciprocal concessions and self-restraint to not abuse a relationship for one's own benefit, risking the frustration of the good faith and cooperation of the other participating parties. Such continued concordant actions establish general rules of behaviour. In the international legal system, rules that emerge from widespread state behaviour that is consistent over time and accompanied by the belief that it is required by law are considered customary law (International Court of Justice [ICJ], 1969, 1986). Thus, law emerges as the result of state cooperation and the interests of states in continued mutually beneficial relationships (Brunnée, 1989). This does not imply that all customary law emerges from relationships between states that seek mutual benefits. Customary rules have equally formed based on consistent state behaviour that sought to delimit a state's sphere of influence and control. The customary rules on the nonuse of force and nonintervention in the internal affairs of another state came into being in this manner (ICJ, 1986). In a generous understanding of the concept of benefits, one could argue that these equally represent mutual benefits. The key issue here is that the building-blocks of these customary rules and of cooperation are the same: namely, reciprocity and good faith.

The expectation of reciprocity leads states to enter into legal relationships; at the same time, they accept restraints to their sovereignty based on the good-faith expectation, or trust, that other states will follow suit. In the majority of instances states meet these expectations of good faith and reciprocity in their mutual relationships. Throughout history, this has led to increasing economic, political and legal ties between them. To formalize these relationships states use international treaties as a tool to give these relations more stability and predictability.

From coexistence to the duty to cooperate in international law

Over time, state relations developed from defining rules on their respective spatial, temporal and personal limits and obligations of noninterference to developing rules on mutual assistance and joint action. Alongside these rules of coexistence, a growing body of norms of the so-called law of cooperation emerged (Sahovič, 1972).

With an increase in economic ties and interdependence between states and their subjects, the law of coexistence proved no longer sufficient to regulate the increasingly multifaceted relations between states. These developments required coordination and joint action. Positive obligations of assistance and action increasingly entered into the body of rules applied by states to their relations (Verdross & Simma, 1984). Increasing commercial relations required cooperation to maintain and access transport and shipping routes. After World War II, concerted international efforts were made to help war-torn economies in their recovery and facilitate their renewed participation in global markets and financial systems with the establishment of two treaty-based organizations (the Bretton Woods institutions), the International Bank for Reconstruction and Development and the International Monetary fund, which later also supported the economic development of countries coming out of colonization. Similarly, the United Nations was established to avoid the destructive consequences of major wars in the future. International cooperation received a new stimulus when the negative environmental effects of human economic activity, including their transboundary impact, became apparent. The recognition that the sustainable management of the natural environment and related issues required cooperation between states and at the universal level led to a significant number of treaties on international cooperation and joint action, most notably those adopted in the context of the 1992 UN Conference on Environment and Development in Rio (United Nations Conference on Environment and Development [UNCED], 1992), including the Convention on Biological Diversity (CBD, 1992) and the UN Framework Convention on Climate Change (UNFCCC, 1992). These conventions contain rules on the settlement of disputes and establish financial mechanisms to assist those countries which do not have the necessary means to comply with the obligations set out in the conventions.

In legal terms, these developments continuously broadened the reach of the law of cooperation and the emergence of a general principle of cooperation in international law.

The set of principles and rules included in the Charter of the United Nations (UNC, 1945) is commonly considered one of the key treaty sources from which this general principle can be derived. The charter reflects agreement of its now 193 member states on the legally binding rules governing their conduct. In addition to its mandate to maintain international peace and security, the UN has as its objective the achievement of "international cooperation in solving international problems of an economic, social, cultural, or humanitarian character, and in promoting and encouraging respect for human rights and for fundamental freedoms for all without distinction as to race, sex, language, or religion" (Article 1(3)). The articles of the charter, in particular Articles 2, 55 and 56, outline specific cooperation duties, including the pledge of states to act jointly and in cooperation with the UN to achieve social and economic development and higher standards of living.

Within the institutional framework of the UN, government representatives have reconfirmed their commitment to international cooperation several times, including through General Assembly resolutions, such as the Declaration on Principles of International Law Concerning Friendly Relations and Cooperation among States (UNGA, 1970) and the Declaration on the Establishment of a New International Economic Order (UNGA, 1974). These two resolutions, which were adopted at the height of the Cold War and towards the end of the long process of decolonization, endorsed the principle of international cooperation among equals for economic development.

Law as a tool for incremental cooperation

The UN system more generally provides examples that illustrate the role of international law in the iterative process of cooperation. The UN was established, based on an international treaty, with a mandate to maintain international peace and security and promote international cooperation (Charter, Article 1). The Charter establishing the UN is in itself an outcome of state cooperation: governments convened for negotiations with the purpose of reaching an agreement that would be mutually beneficial, i.e. to avert the harmful impacts of war and to progressively improve the living conditions of their populations.

The structure of the UN Security Council demonstrates that international treaties are at the same time outcomes of and reflections of the political realities at the time of their conclusion. The victorious countries, the Soviet Union (replaced in 1991 by Russia), China, the United Kingdom, the United States and France were granted permanent seats and hold veto power. At the same time, the example of the Security Council reflects that international treaties can be subject to modification to better reflect changes in underlying circumstance. Recognizing the increase in UN membership, the number of nonpermanent Security Council seats was increased from 6 to 10, raising the total membership to 15, through an amendment to the Charter which became effective in 1965 (UNGA, 1963). And finally, the continued discussions around permanent membership and Security Council reform also highlight the general 'stickiness' of international treaties, which provides stability and at the same time can prevent treaties from being adequately adjusted to changing circumstances.

Another aspect that is common in the process of using law for international cooperation is the agreement by states to relinquish some of their sovereignty by establishing or joining a treaty-based institutional framework. The intent of such agreements is most often to curtail one's sovereignty as the price of achieving the intended mutual benefits of cooperation more effectively. In the case of the UN, this voluntary self-restriction of sovereignty includes one of its most drastic forms, giving significant decision-making power to the Security Council in Chapter 8 of the Charter. Transferring decision-making rights to a treaty-based mechanism remains a relatively rare way of restricting sovereignty for the benefit of cooperation; more common are 'softer' forms such as limiting the mandate of joint mechanisms to recommending policy directions.

An additional beneficial role treaty-based institutions or joint mechanisms play in the cooperation process is the creation of space to incrementally increase the scope and breadth of issue areas of cooperation. States have used the UN system to steadily increase the issue areas of cooperation and to deepen cooperation through establishing new rights and obligations. In the field of human rights law, for example, the Commission on Human Rights, a subsidiary body of the UN Economic and Social Council (Chapter 10 in the Charter), first drafted the Universal Declaration on Human Rights. In an expression of the international political commitment to these rights, the declaration was adopted by the General Assembly in 1948 in a non-legally binding resolution (UNGA, 1948) and followed up with two treaties, the 1966 International Covenant on Economic, Social and Cultural Rights and the International Covenant on Civil and Political Rights, which outline more detailed rights and obligations of states and individuals. These rights were further defined and the covenants followed up by a series of other human rights treaties, including those defining in more detail the rights of women and children. To strengthen the institutional framework, the Office of the High Commissioner for Human Rights was established in 1993 as a UN agency, and the Commission on Human Rights was replaced in 2006 by the UN Human Rights Council, a subsidiary body of the General Assembly.

Finally, the UNFCCC framework provides an example of how international law can also be used to support or strengthen the role of relatively 'weaker' states in the process of international cooperation. It establishes a principle of common but differentiated responsibilities, and it mandates more developed countries to take the lead in combating climate change and its harmful effects, including through providing technical and financial assistance to developing countries to meet the costs of adaptation (Articles 3 and 4).

The above section does not intend to argue that international law is the panacea for the world's challenges, only to highlight the many ways in which legal tools can be used to promote cooperation. While the willingness of national governments will continue to determine to a large extent the degree of cooperation across borders, international law plays an important role by providing rules that govern the conduct of individual countries (Wouters, 2013a) and give stability and predictability to the relationships among them.

Law and cooperation on international waters

Similar processes of incremental cooperation (Wouters, 2014) that have deepened and widened the scope of cooperation, and continue to do so, take place in many of the world's transboundary river basins.

Use of law in cooperation on transboundary waters

States do not always refer to international treaties, but may conduct cooperative engagements through regular meetings or administrative agreements concluded between ministries or provinces. In the Brahmaputra Basin, for example, India initiated a joint working group on water and power with Bhutan and Bangladesh, the so-called BIB Initiative (CGI, 2014). Furthermore, based on a memorandum of understanding (MoU) concluded between the respective Ministries of Water in May 2013, China agreed to provide hydrological data to India during the flood season (Foundation for Non-violent Alternatives [FNVA], 2014). One year later the two countries agreed in the implementation plan for this MoU to extend the period during which data will be provided. Such non-treaty-based engagements are typical in particular during the early stages of cooperative processes in transboundary basins. In cases where cooperation continues over multiple years or decades, it is common for countries to increasingly formalize their relationships through international treaties to achieve more stability and predictability in their mutual engagements.

In the Rhine River basin, scientific cooperation to develop a pollution-monitoring strategy eventually evolved into a comprehensive legal and institutional framework for river basin management among five riparian countries, including a joint institution, the International Commission for the Protection of the Rhine. This international scientific commission was set up in 1950 by exchange of diplomatic notes among the five riparian countries to investigate the extent and origin of pollution (Frijters & Leentvaar, 2003). With the expressed intention to strengthen cooperation, the participating countries later decided to provide the commission a treaty basis and a clearer mandate on pollution monitoring (Preamble 1963 Agreement on the International Commission for the Protection of the Rhine against Pollution). This 1963 treaty also included a clause that provided for extending the competence of the commission to "all other matters which the signatory Governments refer to it by consensus". In 1999, Germany, France, Luxemburg, Switzerland and the Netherlands, together with the EU, broadened the scope of their cooperation, and with it the mandate of their joint institution, to ecosystem protection, species diversity conservation, flood prevention and flow management, among other concerns. Today, the commission also ensures coordination with the remaining four riparian countries that are not members of the commission, as well as with the Central Commission for Navigation on the Rhine.

A similar process of incrementally broadening cooperation occurred in the La Plata Basin. In this case, the widening scope of cooperation occurred also with respect to geographic scope. The process grew from an initially mostly bilateral engagement to study hydropower development on contiguous stretches of key tributaries in the 1940s and 1950s. Technical commissions were set up between Uruguay and Argentina and between Brazil and Paraguay to study the feasibility of exploitation of the power potential at Salto Grande on the Uruguay River and at Itaipu and Yacyretá on the Paraná River. The three dams were eventually completed in 1979, 1982 and 1994, respectively, and are managed by jointly owned mechanisms. In 1967, at a meeting of the foreign ministers of the basin countries, a Coordinating Intergovernmental Committee was created to study integrated development of the basin (Biswas, 2013). Based on the work of the committee, the five basin countries signed the La Plata Basin Treaty in 1969. The treaty is a framework treaty in which the countries agree to identify areas of common interest and carry out studies, programmes and works on a wide range of issues, including comprehensive basin and water resources management, ecosystem conservation, interconnection of transport systems (including by river), integration of power production, and regional cooperation on education and health. The treaty furthermore leaves open the opportunity to adopt additional bilateral or multilateral agreements for specific projects or deepened cooperation on specific issue areas to achieve the general objectives of "harmonious development and physical integration" of the basin (Articles 1 and 6). The committee, the Conference of Foreign Affairs Ministers and the General Secretariat continue to provide an institutional framework for regular dialog and coordination; this led, for example, to the adoption of a comprehensive Framework Programme for the La Plata Basin in 2001

(Pochat, 2011). To finance the cost of joint studies and development, a joint banking mechanism, the Financial Fund for the Development of the La Plata Basin (FONPLATA), was established in 1976.

Similar processes of incremental cross-border cooperation, accompanied by the establishment of institutions and the conclusion of various, sometimes subsequent, legal agreements to establish rules of engagement and actionable rights and obligations among riparian countries, have occurred in a number of other basins, including the Danube, Nile, Mekong and Senegal River basins (Leb, 2013a). In the case of the Senegal River basin, decade-long cooperation reached the unique level at which basin countries agreed to co-own infrastructure assets that are located not on contiguous river stretches but exclusively on the territory of individual riparian countries.

Emergence of the duty to cooperate in international water law

As described in an earlier article published in this journal (Leb, 2013b), the engagement of states in cooperative management, and the development of transboundary river basins and their use of international treaties to formalize their relationships, have led to the emergence of the general duty to cooperate as a principle of international water law. Similar to the development in international law, there was a trend away from rules determining coexistence towards establishing principles and norms of cooperation (Boisson de Chazournes, 2005). Most early water-related treaties focused on the drawing of political boundaries along rivers or across lakes, allocating exclusive fishing rights, and regulating transit rights and river navigation for ships flying the flag of one state across the territory of another state.

A significant step towards rules on cooperation and joint action occurred, first, mostly with the increased trend of harnessing the hydropower potential of transboundary rivers, and later, when impacts of economic activity on riverine ecosystems negatively impacted human water uses. Because rivers have been used as landmarks to draw international boundaries, a number of important hydropower sites are located on contiguous river stretches, requiring cross-border infrastructure development for the exploitation of their potential. The treaty regime established in the La Plata River basin described above, as well as agreements on other joint projects, such as the 1963 agreement relating to the Iron Gates Water Power and Navigation System and the 1977 treaty concerning the Gabčíkovo-Nagymaros System of Locks, both on the Danube River, are well-known examples.

Heavy industrialization along rivers and their usage as natural sewage drains eventually led to a new set of water treaties focusing on integrated basin management and furthering rules of cooperation. The cumulative effects of water pollution and river regulation works for navigation and power development had significantly impacted availability of safe water supplies for drinking and other domestic uses, reduced fish stock (or led to localized extinction) and increased flood risks along transboundary rivers. Comprehensive basin management treaties, such as the 1994 Danube Protection Convention and the 1999 Convention on the Protection of the Rhine, with their strong focus on water quality restoration and ecosystem protection, were a response to the transboundary impacts of economic activity along rivers. Based on the latter convention, the ministers in charge of water resources adopted Rhine 2020 (Program on the Sustainable Development of the Rhine – Conference of Rhine Ministers, 2001), which had significant success in returning salmon to the river and reducing flood risk through the restoration of natural inundation areas and flood plains. This new kind of basin agreement, introducing cooperation duties with respect to ecosystem protection and restoration, happened around the same time as political commitments for cooperation on the global environment started to be formalized in multilateral conventions such as the CBD and the UNFCCC in the early 1990s and onwards.

Codification as a general principle

Based on the large number of water treaties – almost 450 agreements were signed between 1820 and 2007 (UN Water, 2013c) – international expert commissions drafted several instruments codifying the rules of international water law that were considered as universally accepted. Interestingly, the appreciation of cooperation as a recognized duty in the management of resources that link state territories and establish hydrologic interdependence between all countries sharing the waters occurred relatively late in these various codification efforts. The Institute of International Law (IIL) pioneered such efforts with its 1911 Declaration on International Regulation Regarding Uses of International Watercourses for Purposes Other Than Navigation (Wehberg, 1957). In the tradition of the then still prevalent law of coexistence, the regulation focused more on obligations to refrain from interference than on cooperation. Reflecting the evolution towards an international law of cooperation, the 1961 IIL Articles on Utilization of Non-maritime International Waters (Except for Navigation) included specific cooperation obligations on notification, consultation and negotiation among states planning water uses that might seriously affect the possibility of utilization of the same waters by other countries. In 1966, the International Law Association (ILA) contributed to the codification efforts with its landmark Helsinki Rules (International Law Association [ILA], 1967). It is a seldom-noted fact that the Helsinki Rules, while firmly establishing the principle of equitable and reasonable utilization as one of the general principles of international water law, are fairly guarded on cooperation obligations; only specific obligations limited to the issue areas of pollution, navigation for humanitarian purposes, dispute prevention and timber floating are included. Explicit recognition or confirmation of a general principle of cooperation is absent.

The confirmation of general acceptance of a general duty to cooperate on transboundary water resources management is attributed to codifications of universal rules on pollution prevention. Both the IIL and ILA included a general obligation to cooperate in their respective documents on this topic. The 1979 Athens Resolution of the IIL additionally spelled out specific measures for the implementation of this general obligation, including notification, consultation, regular data exchange, coordination of research and monitoring, and technical and financial assistance to developing countries. And striking a similar note, in its commentary to its 1982 Montreal Rules on Water Pollution in an International Drainage Basin, the ILA reasoned that cooperation could now be considered a fundamental principle of international water law (ILA, 1982). Yet, the recognition and legal nature of the general duty to cooperate was again a subject of debate during the long drafting process of the 1994 Draft Articles that became the basis for the 1997 UN Convention on the Law of the Non-navigational uses of International Watercourses (UNWC). While some members of the International Law Commission (ILC), which undertook the drafting, doubted that the cooperation presented anything more than a principle of international law from which obligations could be derived but that was not a duty in and of itself, others argued that it did indeed represent a legal duty, in particular with respect to the management of shared water resources (International Law Commission [ILC], 1987). In recognition of this second argument, the general duty to cooperate has been integrated into Part 2 (General Principles) of the UNWC. This part also includes other general principles of international water law that are in effect obligations, such as the due-diligence obligation not to cause significant harm (Article 7) and the obligation of regular exchange of data and information (Article 9).

The codification of the duty to cooperate among the general principles of international water law acknowledges the fundamental importance of the rule within this body of norms. Cooperation is not only essential for transboundary water resources management, it is also indispensable for bringing other principles and rules of international water law to fruition. For instance, it is not possible to effectively implement the principle of equitable and reasonable utilization without cooperation. Information exchange is required among states with regard to their respective uses, use interests and dependency on the shared water resources, to assess equitable and reasonable use at the basin level.

The UNWC interweaves cooperation duties with substantive provisions, and its articles on the principle of equitable and reasonable utilization are an illustrative example. The second paragraph of Article 5, codifying this principle, spells out the right and obligation to participate in the use, development and protection of an international watercourse in an equitable and reasonable manner. And Article 6, which provides a non-exhaustive list of criteria to consider in assessing the equity and reasonableness of uses, requires watercourse states to enter into consultations whenever the application of the principle of equitable and reasonable use so requires. Specific obligations to cooperate through consultation, negotiation or exchange of information are also included in the rules on other issue areas for watercourse management and development regulated in the UNWC, including with respect to: eliminating or mitigating significant harm where it occurred (Article 7); implementation of planned measures (Part 3); protection of ecosystems (Article 20); prevention of pollution (Article 21); protection of the marine environment (Article 23); watercourse management (Article 24); flow regulation (Article 25); and in case of emergencies (Part 5). In this sense, the UNWC provides a comprehensive and operational framework for the duty to cooperate (Wouters, 2013b).

Developed by the ILC, which brings together the world's most renowned experts of international law from 34 countries, and adopted by the UN General Assembly with 103 country votes in favour, the UNWC represents today the most authoritative codification of international water law. Its entry into force on 17 August 2014 provides additional legitimacy to the customary principles and rules of international water law that are codified in the UNWC, including the general duty to cooperate and its expression through various specific cooperation obligations.

Conclusion

The conclusion of an international treaty is just one step in the iterative and most often incremental process of international cooperation. It is not the first step, and it is not an indispensable step, for cooperation between states. Collaboration among states also exists in the absence of international treaties. However, this form of cooperation tends to be less stable than treaty-based cooperation, because it is dependent on the willingness of parties that are not obliged by any formalized rules to continue the cooperation and thus provides little guarantee of continuation when individual actors change. States use international treaties as tools to provide the process of cooperation with predictability and stability, because a treaty binds a state, not just the individual government that agreed to it. At the same time, international treaties reflect the political realities at the time of their conclusion. They reflect the relative power of the parties at the time of conclusion and provide tools on the basis of which the relatively weaker or less endowed countries receive assistance. They reflect political, economic and scientific trends and their impacts. Over time, this reflection of trends has led to the emergence of the law of cooperation alongside the law of coexistence, and to the continuous expansion of the law of cooperation. The universally felt negative impacts of human economic activity have led to increased treaty-based cooperation in general. And in the context of the management of transboundary waters, they have led to the recognition of cooperation as a general principle in international water law and continue the trend of environmentalization of international law through increased focus on integrated ecosystem management (Boisson de Chazournes, 2013).

States establish rules of cooperation and agree to limit the scope of their sovereignty because cooperation creates new opportunities. The existence of commonly accepted rules of conduct and the establishment of joint mechanisms can be mutually beneficial. Experience in transboundary basins shows that, more often than not, cooperation between riparian states leads to greater benefits for all basin states. This is one of the reasons why once riparian countries start to engage in cooperation the process tends to be widened in scope over time, both in terms of issue areas covered by cooperation as well as geographically. It is the interdependence between states in the international order more generally, and more particularly with respect to shared water resources, that leads them to make first concessions in good faith so that the concession is reciprocated and the full potential of mutual benefits becomes achievable.

Acknowledgements

This article is based on a presentation given at the CIWL symposium Promoting Transboundary Water Cooperation in Asia, at the University of Xiamen Law School, May 2014. I thank the organizers of the symposium for the opportunity to publish in this volume and the anonymous reviewer for the thoughtful comments.

Disclaimer

The findings, interpretations and conclusions expressed in this article are entirely those of the author, and do not represent the views of the World Bank, its executive directors, or the countries they represent.

References

Agreement on the International Commission for the Protection of the Rhine against Pollution. Adopted on April 29, 1963. 994 *United Nations Treaty Series (UNTS)* 3.

Biswas, A. (2013). *Managing transboundary waters of Latin America.* New York, NY: Taylor Francis.

Boisson de Chazournes, L. (2005). Eaux internationales et droit international : Vers l'idée de gestion commune. In L. Boisson de Chazournes & S. Salman (Eds.), *Les ressources en eau et le droit international – Water Resources and International Law* (pp. 10–30). Leiden/Boston: Martinus Nijhoff Publishers.

Boisson de Chazournes, L. (2013). *Fresh water in international law.* Oxford: Oxford University Press.

Brunnée, J. (1989). "Common interest" – Echoes from an empty shell. *Zeitschrift für ausländisches öffentliches Recht und Völkerrecht, 49*, 791–808.

Charter of the United Nations. (1945). Adopted in San Francisco, 26 June 1945. *United Nations Treaty* Series. 1. XVI. Retrieved from http://www.un.org/en/documents/charter/

Conference of Rhine Ministers. (2001). *Rhine 2020: Program on the sustainable development of the Rhine*. Koblenz: ICPR.
Consulate General of India, Perth (CGI). (2014). Foreign secretaries speech on 'Infrastructure Development in the North Eastern region – leveraging opportunities'. Retrieved from http://www.cgiperth.org/scrollingnewsdetail.htm?26/
Convention on Biological Diversity. Adopted on June 5, 1992. 1970 UNTS 79. Retrieved from http://www.cbd.int/doc/legal/cbd-en.pdf
Convention on Cooperation for the Protection and Sustainable Use of the Danube River (Danube Protection Convention). Adopted on June 29, 1994. Retrieved from http://www.icpdr.org/main/icpdr/danube-river-protection-convention
Convention on the Law of Non-Navigational Uses of International Watercourses. Adopted on May 21, 1997. UN Doc. A/RES/51/299.
Convention on the Protection of the Rhine. Adopted on April, 12, 1999. Retrieved from http://www.iksr.org/fileadmin/user_upload/Dokumente_en/convention_on_tthe_protection_of__the_rhine.pdf
Final Act relating to the Establishment and Operation of the Iron Gates Water Power and Navigation System. Adopted on November 30, 1963. 513 UNTS 12.
Frijters, I. D., & Leentvaar, J. (2003). *Rhine Case Study*. UNESCO-IHP. Retrieved from http://www.ce.utexas.edu/prof/mckinney/ce397/Topics/Rhine/The_Rhine_case_study.pdf
Fondo Financiero para al Desarrollo de la Cuenca del Plata (FONPLATA). Retrieved from http://www.fonplata.org/
Foundation for Non-violent Alternatives. (2014). *Implementation Plan between the Central Water Commission, Ministry of Water Resources, River Development and Ganga Rejuvenation, the Republic of India and the Bureau of Hydrology and Water Resources, Tibet Autonomous Region, the People's Republic of China upon Provision of Hydrological Information of the Yarlung Zangbu/Brahmaputra River in Flood Season by China to India*. Retrieved from http://fnvaworld.org/download/resource-center/2014-July-Implementation-Plan-between-India-and-PRC-for-YTB.pdf
Institute of International Law. (1961). *Annuaire de l'Institut de Droit International – Session de Salzbourg, September 1961*. Basle: Verlag für Recht und Gesellschaft AG.
Institute of International Law. (1979). *Annuaire de l'Institut de droit international, Session d'Athènes 1979*. Paris: S. Karger.
International Court of Justice. (1969). *North Sea continental shelf case (Federal Republic of Germany/Denmark, Federal Republic of Germany/Netherlands), judgment 20 February 1969*. ICJ Reports. New York, NY: United Nations.
International Court of Justice. (1986). *Military and paramilitary activities in and again Nicaragua (Nicaragua v. United States of America), judgment 27 June 1986*. ICJ Reports. New York, NY: United Nations.
International Covenant on Civil and Political Rights (ICCPR). Adopted on December 16, 1966. 999 UNTS 171. Retrieved from http://www.ohchr.org/EN/ProfessionalInterest/Pages/CESCR.aspx
International Covenant on Economic, Social and Cultural Rights (ICESCR). Adopted on December 16, 1966. 993 UNTS 3. Retrieved from http://www.ohchr.org/en/professionalinterest/pages/ccpr.aspx
International Law Association. (1967). Report of the Fifty-second Conference, Helsinki, August 14**th** to August 20**th**, 1966. London: ILA.
International Law Association. (1982). *Report of the Sixtieth Conference, Montreal*. London: ILA.
International Law Commission. (1987). Summary records of the meetings of the thirty-ninth session 4 May-17 July 1987. In ILC. *Yearbook of the International Law Commission - 1987 vol. I*. UN Doc. A/CN.4/SER.A/1987.
International Network of Basin Organizations (INBO) News. (2013). *Declaration of the High Level International Conference on Water Cooperation*. Retrieved from http://www.inbo-news.org/IMG/pdf/Dushanbe_Declaration_all_comments_-3.pdf
Leb, C. (2013a). *Cooperation in the law of transboundary water resources*. New York, NY: Cambridge University Press.
Leb, C. (2013b). The UN Watercourses Convention: the *éminence grise* behind cooperation on transboundary water resources. *Water International*, *38*(2), 146–155. doi:10.1080/02508060.2013.781474

Pochat, V. (2011). International agreements, institutions and Projects in La Plata River Basin. *International Journal of Water Resources Development, 27*(3), 497–510. doi:10.1080/07900627.2011.597830

Sahovič, M. (1972). Codification of the law of friendly relations between states. *Recueil de Cours de l'Académie de Droit International (RCADI), 137*, 243–310.

Stockholm International Water Institute. (2013). *2013 Overarching conclusions – World Water Week 2013: Water cooperation – building partnerships*. Retrieved from http://www.siwi.org/wp-content/uploads/2013/11/2013-Overarching-Conclusions-web.pdf

Treaty concerning the Construction and Operation of the Gabčíkovo-Nagymaros System of Locks. Adopted on, September 16, 1977. 1109 UNTS 236

Treaty of the River Plate Basin. Adopted on April 23, 1969. 875 *UNTS* 11

UN Water. (2013a). *World Water Day 2013 international year of water cooperation*. Retrieved from http://www.unwater.org/water-cooperation-2013/water-cooperation/water-cooperation/en/.

UN Water. (2013b). *Factsheet: Water scarcity.* Retrieved from http://www.unwater.org/fileadmin/user_upload/watercooperation2013/doc/Factsheets/water_scarcity.pdf.

UN Water. (2013c). *Factsheet: Transboundary waters*. Retrieved from http://www.unwater.org/fileadmin/user_upload/watercooperation2013/doc/Factsheets/transboundary_waters.pdf.

United Nations Conference on Environment and Development. (1992). *Rio declaration on environment and development*. UN Doc. A/CONF.151/26 (Vol. I).

United Nations Framework Convention on Climate Change. Adopted 9 May 1992. 1771 UNTS 165. Retrieved from http://unfccc.int/files/essential_background/background_publications_htmlpdf/application/pdf/conveng.pdf.

United Nations General Assembly. (1948). *International bill of human rights. GA Res. 217 (III)*. Retrieved from http://daccess-dds-ny.un.org/doc/RESOLUTION/GEN/NR0/043/88/IMG/NR004388.pdf?OpenElement.

United Nations General Assembly. (1963). *Question of equitable representation on the Security Council and the Economic and Social Council. GA Res. 1991 (XVIII)*. Retrieved from http://www.un.org/en/ga/search/view_doc.asp?symbol=A/RES/1991(XVIII)&Lang=E.

United Nations General Assembly. (1970). *Declaration on principles of international law concerning friendly relations and cooperation among states. GA Res. 2625 (XXV)*. UN Doc. A/80821970.

United Nations General Assembly. (1974). *Declaration on the establishment of a new international economic order*. GA Res. 3201 (S-VI). UN Doc. A/RES/S–6/3201.

United Nations General Assembly. (2014). *Report of the Open Working Group of the General Assembly on Sustainable Development Goals*. UN Doc. A/68/970. Retrieved from http://sustainabledevelopment.un.org/focussdgs.html

Verdross, A., & Simma, B. (1984). *Universelles Völkerrecht* (3rd ed). Berlin: Duncker & Humblot.

Wehberg, H. (1957). *Tableau Général des Résolutions (1873–1956)*. Basle: Verlag für Recht und Gesellschaft AG.

Wouters, P. (2013a). *International law – facilitating transboundary water cooperation*. Global Water Partnership (GWP) Background Paper No. 17.

Wouters, P. (2013b). Dynamic cooperation in international law and the shadow of state sovereignty in the context of transboundary waters (Part 2). *Environmental Liability, 4*, 131–144.

Wouters, P. (2014). The yin and yang of international water law: China's transboundary water practice and the changing contours of state sovereignty. *Review of European, Comparative & International Environmental Law, 23*(1), 67–75. doi:10.1111/reel.12069

Zartman, I. W. (2008). Concept: Cooperation. *PIN Points, 30*, 5–7.

The human right to water and foreign investment: friends or foes?

Huiping Chen

States have a duty to respect, protect and fulfil the human right to water. Water-related foreign investment, governed by international investment agreements (IIAs) that do not yet incorporate human rights provisions, can have both positive and negative impacts on human rights. It is argued here that the right to water could be used as an argument to justify measures taken against water-related investments in disputes between host states and foreign investors. While arbitral tribunals have not yet accepted this argument, it might be accepted in the near future as IIAs incorporate human rights provisions.

Introduction

The right to water is a human right deriving from international human rights treaties and policy statements, closely linked to the right to life. Each state has an international obligation to safeguard its people's right to water. While most countries endeavour to comply with this duty, sometimes fully meeting it is beyond their capability. In this regard, foreign investment in water-related services might be seen to contribute to the development of host states, including the realization of the human right to water.

In practice, foreign investment in a country can contribute to the realization of the human right to water through the provision of water services to promote human well-being, including economic development. However, foreign investment can also sometimes cause problems in the field of water. Host states, according to their obligation on the right to water, must take measures to regulate foreign investment to address, prevent or redress such problems. In this situation, foreign investors' rights under international investment agreements (IIAs) may be affected.[1] Such effects may lead to a dispute and its submission to an international arbitration tribunal for resolution.

In such cases, the author argues that the host state may rely on the human right to water as a strong argument to justify its action against foreign investment; and when the state's international obligations on the right to water conflict with its international obligations under international investment treaties, the former obligations should prevail. Unfortunately, the international tribunals for the water-related cases have not yet accepted the right to water argument in deciding the merits of cases adjudicated to date. The resulting chilling effect and consequent 'freezing of law' in this field could indirectly cause a secondary harm to the human right to water.

This article first examines the emergence of the human right to water, highlighting the justification for its special attention in international law. Next, it examines a state's obligation to the human right to water and the role that foreign investment plays in this regard. In the third part, the paper considers the measures that have been taken by host states to regulate foreign investment so as to guarantee the right to water, and the justifications put forward for such measures. Unfortunately, the right to water argument was not accepted by arbitral tribunals in the past cases. The article concludes that foreign investment plays a supplementary role in the realization of the human right to water in host states, but negative impacts have been caused and should be minimized. Accordingly, a host state has some degree of regulatory power to take measures to prevent those negative impacts where such measures could be justified, and tribunals' failure to take the human right to water argument could cause secondary harm to the implementation of the human right to water.

Emergence of the human right to water

Freshwater is essential to sustain life and for ensuring human dignity. Water is also indispensable to the social and economic development of a state. As noted by a US Court, water 'is a *sine qua non* of the happiness, health, welfare, and agricultural and industrial progress of the state' (*State v. Martin*).

Because water lies at the core of human welfare, environmental sustainability and economic development, it has been argued that the human right to water emanates from other fundamental human rights – first and foremost, the inherent right to life of every human being (McCaffrey, 1992), as enshrined in the International Covenant on Civil and Political Rights (ICCPR).

The human right to water is also closely related to the right to an adequate standard of living and the continuous improvement of living conditions, codified in Article 11 of the International Covenant on Economic, Social and Cultural Rights (ICESCR). In its turn, Article 12 codifies the right to the highest attainable standard of physical and mental health, indicating the steps parties must take to comply with this provision. Those steps involve such matters as the reduction of infant mortality and children's healthy development; environmental hygiene; and disease prevention and control.

Around the world, many states have recognized the importance of water for human security in one way or another. Many countries have codified the right to water in their constitutions or passed comprehensive domestic legal frameworks for the utilization and protection of water. South Africa's 1996 Constitution provides for everyone's right to sufficient water (section 27(1)(b)). In South America, Uruguay approved a constitutional reform in a 2004 referendum defining water as a public good and a human right.

At the legislative level, the United States passed its federal Clean Water Act in 1972, while China promulgated its Water Law in 2002. According to Article 21 of China's Water Law, the development and utilization of water resources should first meet the daily water needs of the population in urban and rural areas. Arguably, this provision incorporates and refines the scope of the human right to water into China's domestic legal framework.

At the international level, the existence of a human right to water was the subject of a heated and lengthy debate in the international community until it achieved widespread recognition in recent years (Leb, 2012). As early as 1977, a reference to the right to water appeared in the 1977 Mar del Plata Action Plan. In concert with this, the 1992 Dublin Statement on Water and Sustainable Development calls attention to 'the basic right of all human beings to have access to clean water and sanitation at an affordable price'. Finally, in 2000, the world's nations endorsed the Millennium Development Goals (MDGs), which include a target on halving, by 2015, the proportion of people without sustainable access to safe drinking water and basic sanitation.

With the multiplication of international human rights treaties, at both the global and regional levels, the right to water and, more generally, issues of water access for vital human needs also began to gain prominence in codification endeavours. Several such treaties, including those on discrimination against women, the rights of the child, and persons with disabilities, implicitly or explicitly reflect the right to water (Leb, 2012).

A growing number of international policy documents recognize expressly that states bear international human rights obligations in relation to access to water and sanitation provision for their nationals. In 2002, the UN Committee on Economic, Social and Cultural Rights (CESCR) issued its General Comment No. 15, titled 'The Right to Water'. Lacking binding force, the General Comment aims primarily to assist parties to the ICESCR in implementing their obligations under the Covenant's aforementioned Articles 11–12. Dealing with the right to water in detail, the document addresses its normative content, obligations of state parties and other actors, instances of violations, and its implementation at the national level.

The General Comment opens with a solemn statement:

> Water is a limited natural resource and a public good fundamental for life and health. The human right to water is indispensable for leading a life in human dignity. It is a prerequisite for the realization of other human rights.

In paragraph 2, the document goes on to indicate the normative content of the right to water:

> The human right to water entitles everyone to sufficient, safe, acceptable, physically accessible and affordable water for personal and domestic uses. An adequate amount of safe water is necessary to prevent death from dehydration, to reduce the risk of water-related disease and to provide for consumption, cooking, personal and domestic hygienic requirements.

In July 2010, the UN General Assembly (UNGA) Resolution on the Human Right to Water and Sanitation reinforced the notion of a self-standing, justiciable human right in this area (Leb, 2012). It expressly 'recognizes the right to safe and clean drinking water and sanitation as a human right that is essential for the full enjoyment of life and all human rights' (UNGA Resolution, 2010).

Only two months later, the Human Rights Council (HRC) passed a resolution titled 'Human Rights and Access to Safe Drinking Water and Sanitation'. This resolution further clarified the legal foundations of the human right to water.

The above documents have strengthened the status of the human right to water in international law and serve as evidence of the international community's commitment to securing its universal enjoyment.

With the emergence of the human right to water, however, numerous questions arise as regard its precise normative content and implementation. What does the right to water specifically entail? According to General Comment No. 15, the right encompasses water availability (quantity), quality and accessibility. Availability means that the water supplied to each person must be sufficient and continuous for personal and domestic uses (Committee on Economic, Social and Cultural Rights (CESCR), 2002, para. 12).

Quality means that the water required for each personal or domestic use must be safe and thus free from microorganisms, chemical substances and radiological hazards that constitute a threat to human health. Furthermore, water should be of an acceptable colour, odour and taste for each personal or domestic use (CESCR, 2002, para. 12).

Accessibility means that water and water facilities and services have to be accessible to everyone without discrimination, within the jurisdiction of the state. Accessibility has four overlapping dimensions: physical accessibility, economic accessibility, non-discrimination and information accessibility (CESCR, 2002, para. 12).

The state's obligation on the human right to water and the role of foreign investment

The human right to water enshrined in both national and international law imposes upon states parties clear obligations with respect to this rule. According to international human rights law, a state's obligation on human rights consists of duties to 'respect, protect and fulfil'. The General Comment, although not legally binding, but considered to contain authoritative interpretations of the ICESCR (Leb, 2012), explains these three obligations.

Under the obligation to 'respect', states should take steps to prevent its citizens or companies (including foreign companies) based within its jurisdiction (i.e. in its territory) from violating the right of access to water (CESCR, 2002, para. 33). Under the obligation 'to protect', states are obliged to prevent third parties from interfering in any way with enjoyment of the right to water (para. 23). The obligation to 'fulfil' requires states to consider the requirements of equity and affordability, which might involve legislating for appropriate pricing policies or the introduction of a system of income supplements for socially disadvantaged groups (paras 25–29).

These three aspects of international legal obligations regarding the right to water actually impose two types of operational obligations on national governments: the obligation to seek ways to realize the right to water, and the obligation to prevent third parties (including foreign investors) from violating or interfering with the right to water, both including the availability (quantity), quality and accessibility of water.

Including foreign investment in the context of a nation state's obligation on the human right to water, the first type of duty might lead to a state to invite foreign investment to play a supplementary role in meeting this obligation; as regards the second type of obligation, a host state may need to take measures to regulate foreign investment where it might adversely affect realization of the human right to water. The paper elaborates these two points as follows.

The state's primary obligation to fulfil the human right to water – foreign investment as enabling platform

Responsibility for the realization of the right to water is imposed upon the state in the first instance. The HRC Resolution in paragraph 6, '*reaffirms* that States have the primary responsibility to ensure the full realization of all human rights [...]'. Leb (2012, pp. 642–43) elaborates as follows:

> In General Comment No. 15, the CESCR confirms the existence of human rights obligations with respect to water. [...] States have an immediate obligation to provide a minimum, essential level of safe water and to take meaningful steps towards the realization of the right without discrimination of any kind. In addition, they have a continued duty to achieve the highest attainable standard of realization of the right based on the resources available to them.

In general, states finance and operate water infrastructure and water services as public goods and public utilities. Until the late 1970s, the public sector in most countries was judged to be in the best position to provide water supply and sanitation services (Fauconnier, 1999). By 2004, public bodies still supplied over 90% of the world's water and financed around 90% of the developing world's investment in water and sewage systems (Petrova, 2006).

Unfortunately, not all states can take on this responsibility by means of their own public sectors. Many public sectors provide inadequate water services due to financial restraints and, especially in the case of some developing countries, the lack of advanced technology and capacity. Public sector water providers are sometimes run inefficiently.

In 2010, approximately 884 million people across the world lacked access to safe drinking water (UNGA Resolution, 2010). The Millennium Declaration commits the international community to reducing by half the proportion of people without sustainable access to safe drinking water by 2015. One estimate suggests that meeting this commitment on safe drinking water alone will require an extra annual investment of at least US $10 billion (Commission on Human Rights (CHR), 2003). As part of this effort, many actors are increasingly considering private sector participation as a source of investment in the water and sanitation sector.

The HRC (2010) suggests that realizing the right to water will need public and private investment, as well as aid funding. UNGA Resolution (2010) calls upon states and international organizations to provide financial resources, capacity-building and technology transfer through international assistance and cooperation, in particular to developing countries, in order to scale-up efforts to provide safe, clean, accessible, and affordable drinking water and sanitation for all.

In response to this call, private investment, including foreign investment, is invited and encouraged. Since the 1990s, foreign investment has increased in the water sector, especially in developing countries. Sufficient water supply guarantees realization of the right to water and foreign investments are encouraged in this field. According to China's Catalogue of Industries for Guiding Foreign Investment (2011 updated), foreign investment is encouraged for the utilization of seawater including desalination, the establishment and operation of water supply companies, and the establishment and operation of water treatment facilities.

There are two major ways that foreign investment may make some contribution to the realization of the human right to water for local populations. The first way is the building of water treatment facilities or water infrastructure, including the building of reservoirs, the digging of wells and the building of water-related infrastructure so as to supply water for local people. For instance, Chinese investors drilled 598 wells, provided and installed a water pump, water tank and generator room, and laid water pipes in Nigeria during 2004–07, according to an exchange of notes between China and Nigeria in 2004 (Zhang, 2014).

The second way that foreign investment might help to support a state to realize its human right to water duty is through participation in various models of public–private water supply schemes. This approach is based on the premise that public–private partnerships (PPPs) enhance a national governments capacity to meet its right to water needs. Thus, some developing countries were required to privatize water supply systems as a condition for development assistance (in the 1990s). Latin American countries such as Argentina, Bolivia and Chile, as well as some Asian countries, such as the Philippines and Indonesia, have executed intensive and wide-ranging privatization programmes involving the opening of their water supply and sanitation services to foreign investment during that period (Solanes & Jouravlev, 2007). The privatization programmes make some contribution to the human right to water. By 2007, roughly 10% of global water consumers received their water from private enterprises (Williams, 2007). However, the privatization programmes also caused serious problems and resulted in several investor-state investment disputes, which will be further examined below.

In sum, although states bear the primary obligation with respect to the human right to water, it is sometimes beyond their capacity (especially for developing countries). Foreign investment's participation in the water supply and water services in the host states helps to promote the right to water, although the role it plays is limited and supplementary to the primary duty of the host state.

The state's obligation to take measures to regulate foreign investment that prevents or interferes with the enjoyment of the right to water

Although states may invite foreign investment to share its obligation with respect to the human right to water, this does not detract from the overarching duty on the national government – as the HRC Resolution in paragraph 6 '*reaffirms* [. . .], the delegation of the delivery of safe drinking water and/or sanitation services to a third party does not exempt the State from its human rights obligations'. In complying with such a duty, per paragraph 9:

> States should ensure that non-State service providers: (*a*) Fulfil their human rights responsibilities throughout their work processes, including by engaging proactively with the State and stakeholders to detect potential human rights abuses and find solutions to address them; (*b*) Contribute to the provision of a regular supply of safe, acceptable, accessible and affordable drinking water and sanitation services of good quality and sufficient quantity.

As discussed above, the privatization programmes implemented around the world were not entirely successful and gave rise to many challenges, including disputes. In some instances they were unable to provide 'sufficient, safe, acceptable, physically accessible and affordable' drinking water as contracted and required by the human right to water duty of the national government. For instance, some water provided by Azurix Corp. in Argentina was cloudy and hazy in appearance, with a musty and earthy flavour, due to the failure to repair the filtering system. Water tariffs also dramatically increased after the privatization challenging the reasonable pricing requirements (i.e. in the Aguas del Tunari case) (Schultz, 2000).

Many privatization programmes were conducted by means of concession contracts. Some of these contracts forced water users to shut down private wells, or confiscated local wells and community-built 'water houses', so as to force them to buy their drinking water from the privatized concessionaires (i.e. Indonesia) (Solanes & Jouravlev, 2007). This actually harms the availability and accessibility of water for local people. The supply of naturally occurring water resources was actually controlled by foreign investors, which caused tension between foreign investors and local stakeholders (Porten, 2007).

Facing the water crisis and human right abuses arising from the privatization programmes, as well as for some other reasons such as the financial crises in Argentina, host states are under obligation to address and redress the problems that arise. Generally this meant that the host governments terminated the concession contract or lease contract and repatriated control of the water supply (i.e. Argentina, Bolivia and Tanzania).

Foreign investment in non-water-related services sectors could also prevent or interfere with the enjoyment of the right to water, especially regarding the quantity and quality of water. Foreign investment in agricultural, industrial, energy and other activities where water is an important input or essential element may cause overuse of fresh water in host states. Foreign investment in bottled water industries may also over-extract underground water resources and thus threaten local use of water, raising concerns about the degradation of springs, rivers, lakes and the depletion of aquifers (i.e. the US Coca-Cola Company in India) (Mylamma, 2007), because water is a finite and vulnerable resource (Dublin Statement). Foreign investment in chemical manufacturing and waste landfill activities can cause pollution or contamination of water resources (the Metalclad case). The exploitation of oil or minerals, or the use of gasoline additives, can also cause contamination of drinking water (the Chevron case and the Methanex case).

As discussed above, when foreign investment causes negative impacts on water, host states are under international obligations to take measures to deal with the issues arising thereof, so as to protect people's right to water.

National law, in accordance with rules of international law, imposes obligations on states to tackle these issues. For example, according to Article 33 of China's Water Law, all levels of government should take measures to prevent the depletion or pollution of water sources so as to guarantee the safety of drinking water. Article 36 of China's Water Law provides that local governments should take measures to strictly control the exploitation of groundwater if it is over-extracted.

In addition, water is a highly localized resource. Access to it depends on local geographical features, such as proximity to large bodies of water, underground aquifers or water springs, and interruptions to service or contamination is often limited to local communities (Allen, 2010). The contamination or over-extraction of ground water, the failure to provide sufficient quality water, and a dramatic increase in water tariffs will often cause social unrest in the local community, including street protests and riots, just as in the cases of Coca-Cola companies in India, Aguas del Tunari in Bolivia and Metalclad in Mexico.

To address and redress the negative impacts on water by foreign investment in non-water-related services industries, and to prevent and minimize further damage to water resources, together with some other reasons, host governments had to cancel the company's licence or permit it to operate bottle water enterprises or waste landfill facilities, or others (i.e. the Coca-Cola case in India and the Metalclad case).

There is another scenario in which the right to water could be affected by foreign investment. The 1998 *Sun Belt Inc. v Her Majesty the Queen (Canada)* case concerned investment and trade in fresh water in Canada. In the 1990s, Canada realized that water would become the most essential environmental source in the next century, and was concerned that the shipping abroad of huge quantities of water might cause environmental problems. As a result, it banned massive exports of fresh water under a 1995 legislative measure (TED Case Studies). Sun Belt claimed that the ban amounted to expropriation of its investment, and a violation of the requirement regarding fair and equitable treatment under the North America Free Trade Agreement (NAFTA), and sought compensation. The dispute was eventually settled between the parties, and the conditions of the settlement and any possible compensation remain unknown (TED Case Studies). What is known is that Canada later cancelled the ban, meaning that it felt itself compelled to conduct free trade in fresh water. The possible negative impacts of the case, as far as the right to water is concerned, include possible adverse effects on the right to water of Canadian future generations.

To sum up, foreign investment in either water services sectors or non-water-related services sector could interfere with the full realization of the human right to water. Host states have a duty to regulate foreign investment in ways that address the negative impacts vis-à-vis the human right to water.

Water-related investment disputes and the human right to water – justifying national government measures

Water-related investment disputes

Foreign investors and their investment are protected by the IIAs concluded between their home countries and the host states; provisions in these agreements generally include: protection from expropriation, or unfair or unequitable treatment, among others. Most IIAs also provide that foreign investors have the right to submit investment disputes with host states to international arbitration for resolution (mostly before the International Centre for the Settlement of Investment Disputes (ICSID)). By the end of 2012, some 514 investor–state investment disputes were submitted to international arbitral tribunals (UNCTAD, 2014).

The measures taken by host states discussed above (termination of concession contracts or least contracts, and cancelling of licence or permit) can minimize or prevent adverse impacts upon the realization of the human right to water (generally assessed at a local level). However, in terms of foreign investment, these measures may affect foreign investment and foreign investors. Foreign investors draw on IIAs to challenge host governments by arguing that their measures constitute indirect expropriation, or, as unfair and inequitable practice, attempting to identify breach of the IIAs. This paper considers the water-related disputes to examine what role the human right to water and its duty upon host governments plays in these investment cases. Most of the measures taken by host states referred to here (except for the Coca-Cola case, where India did not have an investment treaty with the United States) were challenged by foreign investors under IIAs before international arbitral tribunals. To date, the tribunals have made a relatively small number of decisions on water-related cases, including 10-plus cases concerning the provision of water in the privatization programmes, and several cases related to the negative impact on water by foreign investment in waste landfill industries, mining or others (see the Appendix). These disputes involved primarily developing countries such as Argentina, Bolivia and Tanzania.

The water-related investment disputes pose two broad questions with respect to the human-right-to-water legal domain (explicitly, in the privatization of water services or implicitly, in the investment in waste treatment sector or export of fresh water): firstly, whether host states measures against foreign investors may be justified by the argument of the human right to water; and secondly, whether the host state's human right to water obligation (based on universal human rights treaties) may be found to prevail over the international obligation of protection owed to foreign investors (based on investment treaties), in the event of conflict between the two host-state legal obligations.

In reality, the disputes between foreign investors and host states are quite complicated, with layers of legal issues. For example, in some instances, other factors, separate from water-related issues, play significant roles in the dispute – such as the financial crisis in Argentina in 1999; other defects in the concession contracts; and the national policy change of the host government. The present paper focuses only on the water-related aspects of the investment disputes covered here, but with the caveat that other issues should also be considered given the nature of investment disputes.

The human-right-to-water argument as justification for host-state measures

In water-related investment disputes, the human right to water should be taken into consideration as a persuasive argument for a number of reasons. Firstly, the human right to water has emerged as a universally accepted international obligation, binding on all states, and thus central to any water-related dispute that touches this norm. Secondly, the human right to water is a prerequisite for the realization of other human rights (universal connectivity of the body of human rights), which makes water itself a public good and the of supplying water is in the public interest, thus a supra-norm. Thirdly, human rights obligations should prevail over international investment obligations where there is a conflict between them, given the significance of the universal principles embodied in human rights norms.

As established above, the right to water has emerged as a universal human right. Nonetheless, the debate as to the existence of an independent human right to water remains open, with numerous states still uneasy about accepting it, as seen above. In this regard, Leb (2012, p. 643) explains:

> While there is no universal agreement on the existence of a self-standing human right to water, the developments with regard to the right to water have been significant: state obligations towards individuals with regard to water went from not being considered at all in earlier human rights treaties to a clearly stated acknowledgement by the international community that the right to water is indeed fundamental to the fulfilment of all other rights and can be derived from existing human rights.

The human right to water can be understood as setting a basic requirement on national governments with respect to water service providers. In essence, as they take on the right to deliver water, they should also take on the obligation to meet basic human rights in relation to the services provided (Solanes & Jouravlev, 2007). In the water-related investment disputes, one of the reasons that host states take measures against foreign investment is to perform its obligation with respect to the human right to water. States need to meet the legitimate expectations of foreign investors but also the legitimate expectations of the public (Solanes & Jouravlev, 2007). As a result, the right to water may be used as an argument to justify a host government's measures impacting against foreign investment.

Before the General Comment on the Right to Water in 2002, respondent states did not realize that the right to water might serve as a persuasive argument in support of their measures. In the Metalclad case, the Aguas del Aconquija and Vivendi case, the Aguas del Tunari and the Sun Belt case, which were heard prior to 2002, no argument on the human right to water was made. They relied instead on arguments based on environmental concerns since pollution or contamination of water as well as the shortage of water is covered broadly under the environment, which received considerable global support during this period (McCaffrey, 1992). If the right to water were taken into consideration, things might have progressed a bit differently. In the Sun Belt case in 1998, Canada realized the importance of fresh water for its future generations, but did not have a strong legal basis upon which to justify its ban. Possibly inspired by the concept of human right to water explicitly expressed in the General Comment in 2002, Maude Barlow, National

Chairperson of the Council of Canadians, stated in September 2002 that 'The commodification of water is ethically, environmentally and socially wrong. It ensures that decisions regarding the allocation of water centre on commercial considerations, leaving aside fundamental environmental, social and human rights' considerations' (Barlow & Clarke, 2002). A legal approach based upon the human right to water obligation might have led to a different outcome in the Sun Belt case.

The significance of the emergence of the right to water in recent years is that host states may explicitly rely on it to justify their measures in water-related investment disputes. Most of the cases heard after 2002 included arguments based upon the right to water. Argentina was involved in many cases concerning the privatization of water services, including the above-mentioned Aguas del Aconquija case, where Argentina did not rely on the right to water to defend its actions. However, in the 2001 Azurix case, Argentina raised the right to water for the first time, and then in later cases it did so in a more sophisticated manner. In the Suez case and the Suez and Vivendi case, Argentina relied heavily on the right to water to justify its measures affecting foreign investors.

Argentina relied on this position by arguing that it adopted the measures in order to safeguard the human right to water of the inhabitants of the country (Suez and Vivendi Decision, para. 22). It also directly cited General Comment No. 15 on the 'Right to Water' in support of its 'overriding responsibility to ensure the availability of water to all members of society'. In view of such compelling motives, Argentina maintained that its actions were a legitimate and proportionate response rather than an act of indirect expropriation (Peterson).

Non-governmental organizations (NGOs) supported the right to water argument by contending that human rights law required that Argentina adopt measures to ensure access to water by the population, including physical and economic access, and that its actions in confronting the crisis fully conformed to human rights law. Since human rights law provides a rationale for the crisis measures taken, they argued that the tribunal should consider that rationale in interpreting and applying the provisions of the bilateral investments treaties (BITs) in question (Suez and Vivendi Decision, para. 256).

Some foreign investors also acknowledged the existence and importance of the right to water, stating that 'every man, woman and child has the right to a reliable system of clean water and good sanitation' (BiWater Award, para. 379).

While the investment arbitration tribunals recognized, in general, that people enjoy a human right to water (Suez Order, para. 18), they did not accept it as an effective argument in deciding the cases.

The human right to water argument may imply another argument which is often relied on by respondent host states. That is, water is a public good and supplying water is in the public interest. Actually, water is very important, and national governments always pay considerable attention to it. For example, after *Escherichia coli* bacterium contamination of the municipal drinking water supply system in Walkerton, Canada, caused thousands of local residents to become ill and led to a number of deaths, the Walkerton Inquiry was established to investigate and make policy recommendations (Report of the Walkerton Inquiry). This immediate investigation demonstrates the high level of attention paid by the government to matters of public interest.

This approach can be found elsewhere. Due to the importance of water to the life and health of the population, water cannot be treated as an ordinary commodity (Suez and Vivendi Decision, para. 252). The Uruguayan constitution defines water as a public good. General Comment No. 15 also states that 'Water is a limited natural resource and a public good fundamental for life and health' (CESCR, 2002, para. 1). In Wisconsin, raw water resources are a purely public good and belong to the whole of society. Because water services address 'basic needs', just as electricity, telecommunications and public transport services do, the provision of water is regarded as, or closely related to, the public interest. Water (and water services) is not different – somehow it is considered more fundamental to the public, as in the BiWater case.

Because water is not an ordinary commodity, the NAFTA specifically excludes it from its rules. The 1993 Joint Statement by the Governments of Canada, Mexico and the United States stated that

> The NAFTA creates no rights to the natural water resources of any Party to the Agreement [...] nothing in the NAFTA would oblige any NAFTA Party to either exploit its water for commercial use or to begin exporting its water in any form. Water in its natural state in lakes, rivers, reservoirs, aquifers, water basins and the like is not a good or product, it is not traded, and therefore is not and never has been subject to the terms of any trade agreement.

Unfortunately, Canada cancelled its ban on water exports and felt compelled to allow such exports after the settlement of the Sun Belt case.

The public interest argument is referred to by the High Court of Kerala in the Coca-Cola case in India, which ruled that excessive extraction of a public resource (i.e. water) by a private actor is against the public interest (Weiss, 2009). It is also recognized by some international tribunals which have held that 'The factor that gives this case particular public interest is that the investment dispute centers around the water distribution and sewage systems. [...] Those systems provide basic public services to hundreds of thousands of people [...]' (Suez Order and Suez and Vivendi Order). However, the tribunals only accepted this argument in accepting NGOs' application for participation in the cases, but not in its decision in the merits.

The human right to water argument may also be used to resolve the conflicts between the state's human rights obligations and its international investment obligations in some situations.

As discussed above, host states face the dilemma of respecting its population's human right to water, and at the same time meeting its foreign investment commitments. Where a conflict arises there is legal foundation to argue that human rights obligations may (and should) prevail over the investment treaty obligations.

As Argentina argued, 'a conflict between a BIT and human rights treaties must be resolved in favor of human rights because the consumers' public interest must prevail over the private interest of a service provider. (Azurix Award, para. 254). Argentina suggested in other cases that its human rights obligations to assure its population the right to water trumped its obligations under BITs, and also that the existence of the human right to water implicitly gives Argentina the authority to take actions contrary to its BIT obligations (Suez Decision, para. 240). It stressed that the BIT

> must be construed in a manner which does not affect the fulfillment of other international obligations between the states signatory of such BITs. According to Argentina, such an approach would ensure that BIT obligations would be read in light of other rules of international law linking Argentina, the United Kingdom, France and Spain, including 'any treaty on human rights contemplating the human right to water'. (Peterson)

Unfortunately, in these water-related cases, tribunals did not accept these human-right-to-water-based arguments and decided against the host states, thus possibly causing indirect negative impacts on the right to water. With the resulting chilling effect and 'freezing of law' on the part of host states in dealing with water-related law or foreign investment, such negative impacts can have a ripple effect. For example, probably having learned lessons from the failures of the privatized water services in Argentina, Bolivia and Tanzania, and the experience of these states in losing their cases before arbitral tribunals, during the negotiations of the Dominican–Central America Free Trade Agreement, many Nicaraguans showed deep concern that it would limit citizens' access to water. Costa Rica explicitly listed water services for exclusion from foreign investment. That means foreign investors are not permitted to invest in water services. Some NGOs also strongly advocate host states to refuse foreign investment in water service (Withanage, n.d.).

The major reason that these tribunals did not take water into special consideration lies in the fact that most IIAs did not contain any provisions on the human right to water, or human rights. Typically the tribunals relied primarily on the IIAs' provisions in deciding these cases. This situation could be changing, as in recent years United Nations Conference on Trade and Development (UNCTAD) (in its Investment Policy Framework for Sustainable Development) and some other regional organizations (such as Southern African Development Community in its SADC Model Bilateral Investment Treaty Template) are promoting the incorporation of human rights provision into IIAs. In this context, since the right to water has emerged as a human right, the tribunals under these new types of IIAs will be required to take into account these legal provisions in deciding cases.

The human right to water argument may lead to more impartial and acceptable decisions in the water-related cases. In the Metalclad case, the tribunal did not consider water as a significant aspect and made broad interpretations of expropriation provisions, which was criticized by the UN High Commissioner for Human Rights that

> the broad interpretations could have direct consequences for regulations intended to promote and protect human rights. While this case focused on environmental protection, government action in relation to toxic wastes has flow-on effects in relation to the enjoyment of human rights such as the right to water. The decisions raise questions about the assumptions of responsibility – moral or legal – for actions that could negatively affect human rights. (CHR, 2003)

If the right to water argument were made and accepted by tribunal, the outcome may have been different.

Conclusions

All states are under a duty to meet their obligations with respect to the universal human right to water. For many countries, developing countries in particular, it may be difficult to meet these duties. In such situations, foreign investment might assist national governments in implementing the human right to water. Examples include support for the supply of water services, infrastructure, capacity development support and so forth. While water-related foreign investment can have positive impacts on the human right to water, it might also have negative impacts, which compromise achieving this duty. In the final analysis it is for national governments to take measures in compliance with their international legal obligations. It is not an easy path. International investment needs to be regulated, but problems can nonetheless arise. Foreign investors, relying on the IIAs (BITs) have a right to take host governments to arbitration where they consider the treaty has been breached due to host-state measures. This paper has demonstrated how the human right to water might be used to justify the legality of host-state actions in certain cases. International practice appears to be changing to include human rights issues in IIAs, which would support the more complete realization of the human right to water. This could balance the need for investment and investors' interests with those universal duties of host states.

Acknowledgements

The author would like to thank sincerely Professor Patricia Wouters, an expert on international water law appointed under China's Thousand Talent Program and my dear counterpart at Xiamen University School of Law, for her inspiring and continuing encouragement and careful editing of this paper. Gratitude is given also to Dr Owen McIntyre and the anonymous peer reviewers for their constructive and valuable comments made on an earlier draft of this article. Thanks also to Flavia Rocha Loures (PhD candidate, Xiamen School of Law) for her kind assistance in research and language editing of one section of this paper.

Note

1. IIAs include bilateral investment treaties, trilateral investment agreements, the investment chapter in free trade agreements (i.e. the North-American Free Trade Agreement) and others (i.e. the Energy Charter Treaty).

References

Aguas del Tunari. Aguas del Tunari S.A. v Bolivia. ICSID Case No. ARB/02/3.

Allen, D. (2010)."This Business Will Never Hold Water:" International Investment Arbitration on Public–Private Water Service Provision – A Comment on Biwater Gauff (Tanzania) Limited v. United Republic of Tanzania. Retrieved from http://ssrn.com/abstract=1540256 [Accessed 10 September 2014].

Azurix Award. Azurix Corp. v. Argentina, ICSID Case No. ARB/01/12, Award.

Barlow, M., & Clarke, T. (2002, September, 2). Who owns water? Retrieved from http://www.ratical.org/co-globalize/whoOwnsWater.html [Accessed 16 October 2014].

BiWater Award. BiWater Gauff Limited v United Republic of Tanzania, ICSID Case No. ARB/05/22, Award.

Committee on Economic, Social and Cultural Rights. (2002). 29th Session, *General Comment No. 15: The Right to Water* (Arts. 11 and 12 of the Covenant), UN Doc E/C.12/2002/11 (20 Jan 2003).

Commission on Human Rights. (2003). *Report of the High Commissioner for Human Rights on Human Rights, Trade and Investment*, United Nations Economic and Social Council, Commission on Human Rights, E/CN.4/Sub.2/2003/9, 2 July 2003.

Fauconnier, I. (1999). The privatization of residential water supply and sanitation services: Social equity issues in the California and international contexts. *Berkeley Planning Journal*, *13*, 1999.

HRC (Human Rights Council). (2010). *Report of the independent expert on the issue of human rights obligations related to access to safe drinking water and sanitation*, Catarina de Albuquerque, A/HRC/15/31, 15th Session, June 29, 2010.

Leb, C. (2012). The right to water in a transboundary context: Emergence of seminal trends. *Water International*, *37*(6), 640–653. doi:10.1080/02508060.2012.710950

McCaffrey, S. (1992). A human right to water: Domestic and international implications. *Georgetown International Environmental L aw Review*, *5*(1), 1–24.

Metalclad. Metalclad v. Mexico, ICSID Case No. ARB(AF)/97/1.

Methanex. The Methanex Corporation v. United States of America, UNCITRAL, 1999.

Mylamma. (2007).Kerala's Anti-Coke Crusader. *Plachimada*, *May*, 2007. Retrieved from http://www.irenees.net/bdf_fiche-acteurs-332_en.html [Accessed 8 October 2014].

Peterson, L. E. Human Rights and Bilateral Investment Treaties: Mapping the role of human rights law within investor-state arbitration, Rights and Democracy, International Centre for Human Rights and Democratic Development.

Petrova, V. (2006). At the frontiers of the rush for blue gold: Water privatization and the human right to water. *Brooklyn Journal of International Law, 31*, 2006.
Porten, S. (2007). Aguas del Tunari in bolivia: The water war, September 2007. Retrieved from http://www.caseplace.org/d.asp?d=1456 [Accessed 20 August 2014].
Report of the Walkerton Inquiry: A Strategy for Safe Drinking Water, Vol. *2*. Toronto: Queen's Printer for Ontario. (2002).
Schultz, J. (2000). Bolivia's water war victory. *Earth Island Journal, Autumn* 2000. Retrieved from http://thirdworldtraveler.com/South_America/Bolivia_WaterWarVictory.html [Accessed 20 August 2014].
Solanes, M., & Jouravlev, A. (2007, November 2007). Revisiting privatization, foreign investment, international arbitration, and water, United Nations. *CEPAL*. SERIE Recursos naturales infra-estructura. Retrieved 8 January 2015 from http://www.iatp.org/files/451_2_101503.pdf.
State v. Martin. 168 Ohio St. 37, 151 N.E.2d 7 (1958).
Suez and Vivendi Decision. Suez, Sociedad General de Aguas de Barcelona S.A. and Vivendi Universal S.A v. Argentine Republic, ICSID Case No. ARB/03/19, the *Decision on Liability*.
Suez and Vivendi Order. Suez, Sociedad General de Aguas de Barcelona S.A. and Vivendi Universal S.A v. Argentine Republic (ICSID Case No. ARB/03/19), *Order in response to a Petition for Transparency and Participation as Amicus Curiae* (19 May 2005).
Suez Decision. Suez, Sociedad General de Aguas de Barcelona S.A. and Interagua Servicios Integrales de Agua S.A. v. Argentine Republic (ICSID Case No. ARB/03/17, *Decision on Liability*.
Suez Order. Suez, Sociedad General de Aguas de Barcelona S.A. and Interagua Servicios Integrales de Agua S.A. v. Argentine Republic (ICSID Case No. ARB/03/17), *Order in Response to a Petition for Participation as Amicus Curiae.*
TED Case Studies, US–Canada water case. Retrieved from http://www1.american.edu/TED/water.htm [Accessed 12 September 2014].
UNCTAD. (2014). *Investor–State Dispute Settlement: A sequel* – UNCTAD Series on Issues in International Investment Agreements II, UNCTAD/DIAE/IA/2013/2, 23 July 2014.
UNGA Resolution. (2010). UN General Assembly, The human right to water and sanitation, Resolution adopted by the General Assembly on 28 July 2010, A/RES/64/292, 3 August, 2010.
Weiss, E. (2009). The Evolution of International Water Law, *Collected courses of the hague academy of International Law 331*. The Netherlands: Martinus Nijhoff Publishers.
Withanage, H. (n.d.) Water; an economic good or a Public good. Retrieved from www.elaw.org/system/files/sri+lanka.**water**+a+**public**+**good**.doc. [Accessed 20 January 2015].
Zhang, Y. (2014 September 10). Wells, doctors and railways – Stories by Chinese People in the African Desert. Retrieved from http://www.doc88.com/p-2169066600031.html

Appendix

Cases concerning the provision of water (in chronological order)

Compañía de Aguas del Aconquija S.A. and Vivendi Universal S.A. v Argentina (ICSID (Case No. ARB/97/3)

Azurix Corp. v. Argentina (ICSID CASE No. ARB/01/12), Aguas del Tunari S.A. v Bolivia(ICSID Case No. ARB/02/3)

Suez, Sociedad General de Aguas de Barcelona S.A. and Interagua Servicios Integrales de Agua S. A. v. Argentine Republic (ICSID Case No. ARB/03/17)

Aguas Cordobesas S.A., Suez, and Sociedad General de Aguas de Barcelona S.A. v. Argentine Republic (ICSID Case No. ARB/03/18)

Suez, Sociedad General de Aguas de Barcelona S.A. and Vivendi Universal S.A v. Argentine Republic (ICSID Case No. ARB/03/19)

Anglian Water Group v. Argentine Republic (AWG Group Ltd. v. The Argentine Republic, UNCITRAL)

Azurix Corp. v. Argentine Republic (ICSID Case No. ARB/03/30)

SAUR International v. Argentine Republic (ICSID Case No. ARB/04/4)

Biwater Gauff Limited v United Republic of Tanzania (ICSID Case No. ARB/05/22)

Impregilo S.p.A. v. Argentine Republic (ICSID Case No. ARB/07/17)

Urbaser S.A. and Consorcio de Aguas Bilbao Bizkaia, Bilbao Biskaia Ur Partzuergoa v. Argentine Republic (ICSID Case No. ARB/07/26)

Gelsenwasser AG v. People's Democratic Republic of Algeria (ICSID Case No. ARB/12/32)

Cases concerning negative impact on water (in chronological order)

Metalclad v. Mexico (ICSID CASE No. ARB(AF)/97/1)

Branimir Mensik v. Slovak Republic (ICSID Case No. ARB/06/9)

Chevron Corporation and Texaco Petroleum Corporation v. The Republic of Ecuador (UNCITRAL, PCA Case No. 2009–23)

Ten building blocks for sustainable water governance: an integrated method to assess the governance of water

Marleen van Rijswick, Jurian Edelenbos, Petra Hellegers, Matthijs Kok and Stefan Kuks

A three-step interdisciplinary method to assess approaches to water shortage, water quality and flood risks is presented. This method, based on water system analysis, economics, law and public administration, seeks to create common understanding based on newly developed concepts and definitions. First, generating content knowledge about the water system and about values, principles and policy discourses. Second, providing an organizational process with sufficient stakeholder involvement, insight into the trade-off between social objectives, and attribution of responsibilities in addition to regulations and agreements. Finally, implementing the agreed service level through adequate infrastructure, enforcement and conflict resolution.

Introduction

Well-managed water resources can be a significant driver for growth and can generate huge benefits for human health, the environment and the economy. On the other hand, badly governed water resources can significantly hinder growth, reduce opportunities for further development, put ecosystems at risk, cause societal disruption, create political instability and impose economic costs (IPCC-WGII, 2007).

Due to the complex nature of water systems (multilevel, multi-scale and multi-actor), a water governance approach is needed in which different values, interests and uses of water are interconnected so that water policy and measurements are developed and implemented with the support of different stakeholder groups (Edelenbos, Bressers, & Scholten, 2013; Pahl-Wostl 2006; Tortajada, 2010; UNESCO, 2006). This governance perspective is also sometimes referred to as polycentric approaches. Polycentric governance systems are defined as systems in which 'political authority is dispersed to separately constituted bodies with overlapping jurisdictions that do not stand in hierarchical relationship to each other' (Skelcher, 2005, p. 89).

The importance of water governance is growing as changes in population, diet, land use and economic activities exacerbate competition between water users to access the resource they need (Edelenbos et al., 2013; OECD, 2011). Moreover, climate change generates uncertainty about water availability and safety in the future. Finally, water resources touch upon issues of land use, such as economic and urban development, ecology, nature development and agriculture (Tortajada, 2010). Water governance assessment methods explicitly deal with this multifaceted and integral aspect of water resources in countries and regions (OECD, 2011, 2014).

Integration in water management and water governance is widely acknowledged. The concept of integrated water resources management is therefore warmly embraced by many disciplines and in theory as well as practice. The definition of the Global Water Partnership (2003) is most commonly used: 'Integrated water resources management is a process which promotes the coordinated development and management of water, land and related resources, in order to maximize the resultant economic and social welfare in an equitable manner without compromising the sustainability of vital ecosystems.' However, it has become clear that this concept is broad, all-encompassing and impressive, but also has little practical resonance now and in future water management practices. It leaves too many questions unanswered and is therefore fashionable but unimplementable (Biswas, 2005, p. 323). As Biswas (2005) states, there is a need to develop an objective, impartial and undogmatic assessment of the applicability of integrated water resources management.

Although it is presumed that it is not possible to use a concept practically – and a related assessment method – that covers developed and less developed countries, countries with different physical, hydrological and climatologic circumstances, countries that have different cultures and values and different institutional, legal and environmental regimes (Biswas, 2005, p. 335), a start will be attempted here. The proposed assessment method is built in a multidisciplinary way, although it is realized that not all relevant disciplines are yet included explicitly. However, the method is based on many scientific background papers that have been written with other disciplines as well. The method can be broadened with and used by other disciplines such as psychology, history, political science, sociology, geography and so on, but at this moment we chose to stick to those fields of which we have expertise and knowledge, trying not to make the same mistakes as have been made previously, developing a multidisciplinary method without having the appropriate knowledge available.

Because water is an integral issue, it touches upon different fields and disciplines, such as water system analysis, economics, water law and public administration, which all need balanced attention and scrutiny in the assessment approach. At this moment assessments are often fragmented and mono-disciplinary in nature and do not deal with this integral aspect of water resources. However, an increasing amount of integral assessment approaches appear, but these approaches often lack scientific substantiation and grounding (OECD, 2011, 2014). The information and knowledge base on which they rely can be very weak and fragmented. Integral and interdisciplinary assessment methods are scarce, partly for the reason that such integral and interdisciplinary assessments are particularly complex to develop and implement. It is argued that more knowledge and experience are needed in developing and implementing such integral assessment methods.

The main objective of this study is therefore to develop an interdisciplinary method and approach grounded in the scientific literature and which can approach water issues in a holistic and integral way. The developed method is of a diagnostic nature. It will create the opportunity to identify strengths and especially weaknesses in water governance capacity that need further attention in enhancing water governance in order to deal with water issues in an efficient, effective and legitimate way. Various disciplinary methods (Arts & Leroy, 2006; Perry, 2003) have been reviewed and expert knowledge of various disciplines, respectively water system analysis, economics, law and public administration, has been used to arrive at shared values and assessment criteria.

Beyond the conventional discussions about water governance, this paper sheds light on the interdisciplinary nature of water governance in particular (Brouwer, 2013; De Vries, Zonneveld, & Buijze, 2014). The integrated method to assess the governance of water management is presented in the second section. The 10 building blocks to assess weaknesses in water management and governance are discussed in detail in the third section. The fourth section concludes and shows that expert knowledge from various disciplines is required for water governance assessment. Finally, an appropriate role of various disciplines in this debate is suggested.

Integrated assessment method

The strength of water management and governance can be judged on the basis of the assessment method shown in Figure 1. It has been developed to assess the main gaps in the (1) knowledge base, (2) weaknesses in the organization process, and (3) problems that may arise when implementing the agreed service level. The proposed interdisciplinary method consists of 10 building blocks and is of a diagnostic nature.

Sound water management requires knowledge about the water system in time and space and about values, principles and policy discourses. This knowledge is required for the organizational process to come to an agreed service level. The organizational process requires sufficient stakeholder involvement, insight into the trade-off between social objectives, attribution of responsibilities, authorization and the associated means as well as regulations and agreements. Finally, the agreed service level has to be implemented, which requires engineering of infrastructure, enforcement and conflict resolution.

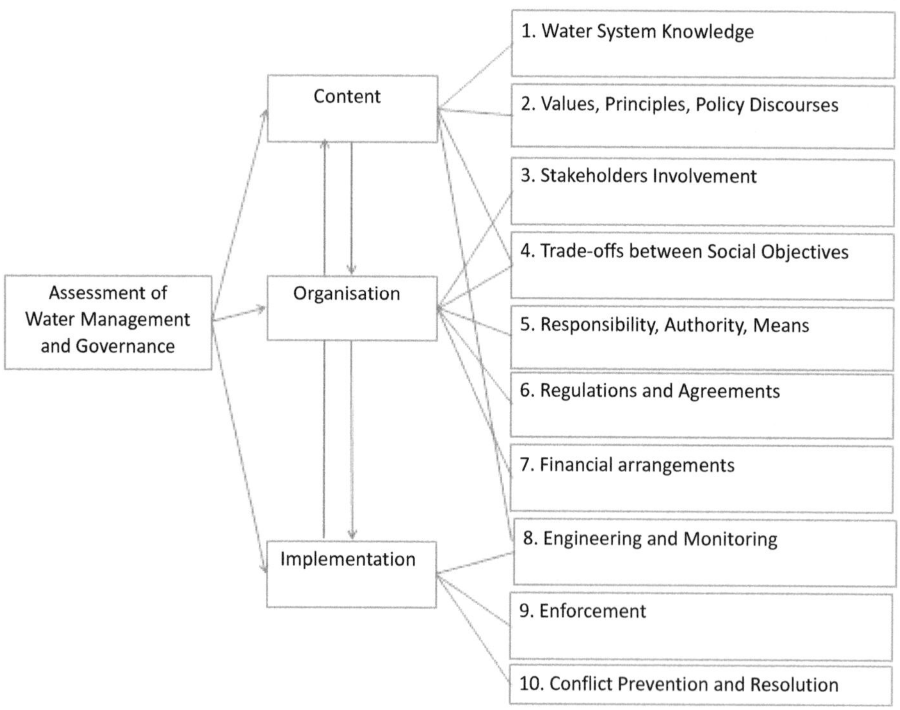

Figure 1. Multiple dimensions of water management and governance.

The assumption is that water governance is sound when all three main dimensions and corresponding 10 building blocks are dealt with. The sequence and interdependence among these dimensions and elements have important implications. Sound water management and governance will only eventuate when there is congruence between these dimensions and corresponding elements. For example, implementation without attribution of responsibilities is, for instance, doomed to fail, and enforcement only can take place as regulations and mutual agreements are in place. This illustrates the integral and interdisciplinary aspect of the assessment method. Finally, Figure 1 illustrates the cyclic aspect of the method. Once the agreed service levels have been implemented, the loop might go back to the content as it can have consequences for the state of the water system, or when monitoring shows the need for revision of one of the building blocks. The building blocks are the result of a two-year collaboration project on interdisciplinary approach of water governance in which the authors participated. First, the researchers developed building blocks from their own specific discipline and research, and subsequently the buildings blocks were merged and interrelated in various interdisciplinary workshop meetings, of which Figure 1 was the result. New wordings for shared concepts were developed to avoid a too strong focus on one of the individual disciplines, which often leads to misunderstandings when using the method.

Building blocks to assess the governance of water

Each of the 10 building blocks is described below. A criterion to judge the strength of each building block is defined. For example, whether there is sufficient knowledge with regard to the water system or sufficient institutions in place depends on the kind of problem. So the judgment is case specific.

Water system knowledge

The water system is defined as the combination of natural physical resources (such as rivers, rainfall, seas, lakes) and man-made infrastructure (such as canals, pumping stations, reservoirs, flood defences). The system supports societal functions, such as domestic and industrial water use, irrigation, shipping, hydropower, safety, etc., and includes the ecosystems related to water. Knowledge of this system refers to knowledge of the natural processes, but also knowledge of the properties of the infrastructure. In order to know the water system, it is essential to measure the important characteristics. For example, in order to assess the availability of water, the precipitation at several locations in the river basin has to be measured. And to assess the risk of flooding, water levels and changes in the topology in rivers and seas have to be known.

The necessary knowledge of the water system depends on its societal functions. These functions often change in time, and hence the requirements of the water system will also change. Moreover, changes in the natural environment will also influence the performance of the water system. It is important to know the impact of these changes, and to support decisions by showing the impact (e.g. with respect to costs, economic benefits and risk) of these decisions. Because these future impacts cannot be measured, it is important to have mathematical–physical models in order to assess the impact of relevant measures. Water systems show a high natural variability. Uncertainties include not only this natural variability, but also the epistemic uncertainties. For example, an uncertainty analysis of emergency areas along the River Rhine in the Netherlands shows that this measure is not effective when all (knowledge) uncertainties are taken into account, where this measure is effective in the world without uncertainties (Kok, Stijnen, & Silva, 2003).

Water resources involve the demand, supply and quality of fresh water. Green growth requires that water allocation is framed by the amount of water that is available within sustainable limits at a specific time. This requires good knowledge of water availability and water uses, and their interactions. The linkage between upstream consumption and downstream availability calls for a basin approach. Water accounting and hydrological modelling are suitable analytical tools in this respect. Remote sensing is a consistent method to provide impartial data on actual evapotranspiration, i.e. water consumption and biomass production and hence differences in water productivity. Besides knowledge about the water system in time and space, insight is also required into the impact of investments in water resources development on water availability. This knowledge is required to frame the organization process. In many places, water is currently over-allocated and heavily polluted, i.e. more entitlements have been issued for a system than can be sustained.

Flood risk involves the assessment of risk: the probability of failure and the consequences of failure (such as casualties and economic losses). Recent examples are the catastrophic flood of New Orleans in 2005, the tsunami in Japan in 2010, the extensive river floods in Thailand in 2011, Sandy hurricane in 2012 and the floods in the Balkan region in spring 2014. Flood risks can be mitigated in many ways and knowledge of these risks is essential in order to implement measures that do mitigate these risks. Measures include land-use planning, building requirements, flood defences, evacuation planning and taming rivers along waterfronts.

> *Assessment criteria*: Is there sufficient knowledge of the existing water system in order to deliver the required service level of societal functions? If not, what are the gaps; is sufficient knowledge available to assess the impact on the water system because of changes in environment and societal functions?

Values, principles and policy discourses

Water issues touch upon different values, principles and narratives, and water management has a normative dimension that is often neglected (Driessen & Van Rijswick, 2011).

Values

Values may differ per country or society, and depend strongly on historical, cultural, normative and political views. Notwithstanding the existing of different values, it contributes to good water management and governance which are based on shared values because this enables the finding of legitimate solutions that are easier to implement (Ambrus, Gilissen, & Van Kempen, 2014; Bressers & Kuks, 2004). Trust in the integrity of partners involved and the aim to find solutions that are accepted by all is a crucial factor in good water governance. Common interests may play a positive role in finding shared values. Well-known values are the recognition of human rights, equity, human dignity, justice, trust and solidarity or self-determination. Values can also be more specified for the water domain like flood protection for all against acceptable societal costs, or the availability of sufficient and clean (drinking) water, an equitable, sustainable and fair use of resources or the value that no significant harm should be done to others (Warner & Van Buuren, 2009). Values can be further elaborated in principles.

Principles

Principles are also of a more general character and should not be confused with rules and regulations. Several groups of principles can be distinguished, such as (1) institutional principles, among others decentralization, subsidiarity, river basin management, or integration; (2) principles of good governance, proportionality and public participation; (3) specific environmental principles, e.g. the precautionary principle, the polluter-pays principle, the prevention principle, the principle that pollution should preferably be tackled at the source; and (4) technical principles, e.g. from global design to detailed design (see the eighth building block). Values and principles play a guiding role in enhancing creativity by developing new alternatives in policy development and decision-making (Keeney, 1992).

Narratives

The content of a water problem is shaped in different policy discourses. An example of two different and opposing policy discourses in the Netherlands is protection against extreme water events (dams, dikes) and living with water (providing more room for the river, for example). Policy discourses can be considered different ensembles of actors with specific story lines, frames, values and principles that emphasize certain aspects of water problems and policy measures.

> Discourses frame certain problems; that is to say, they distinguish some aspects of a situation rather than others. [...] As such, discourse provides the tools with which problems are constructed. Discourses at the same time from the context in which phenomena are understood and thus predetermines the definition of the problem.
>
> (Hajer, 1993, pp. 45–46)

By constructive dialogue a process of social learning is established (Mostert et al., 2007; Pahl-Wostl, 2006). In dialogue mutual understanding is realized about the stakeholder's interests, values and viewpoints (Meijerink, 2008). Some explicitly speak of *adaptive* water governance in order to stress the need for using adaptive management approaches to deal with complex water issues (Huitema et al., 2009). These approaches can be characterized as flexible, learning and reflexive and are needed to take into account the true complexity of water systems (Brunner & Steelman, 2005; Folke, Pritchard, Berkes, Colding, & Svedin, 2007).

To understand the content of a water issue one needs to understand not only the facts (as under the first building block), but also the values, cultural, social and political background, and images that actors have developed in the course of time dealing with water issues. Water issues are oftentimes approached from highly institutionalized angles and discourse that predetermines certain courses of action, and excluding certain viewpoints, contents and courses of action at the same time. This is also called framing. Framing is a three-fold process of selection, focusing and embedding (Dewulf, Mancero, & Cardenas, 2011). People frame issues by bringing certain aspects of a complex problem domain into the picture (a process of selection), by putting certain aspects on the foreground and others on the background (a process of focusing) and by using certain aspects as the overarching elements within which the rest fits (a process of embedding). In practice different discourse coalitions (Bulkeley, 2000; Dewulf et al., 2011; Hajer, 1993) emerge that often fight with each other in order to becoming dominant (opposed to recessive) policy discourse coalitions in determining what the exact problem is (and sometimes ignoring the actual state of the water system) and which type of solutions seem appropriate in handling water issues.

> *Assessment criteria*: Is there sufficient knowledge of shared or conflicting values, viewpoints and principles (represented by different policy discourse coalitions) for water issues and their consequences for facing water management issues?

Stakeholder involvement

Water management and governance can be considered complex processes in which various actors with different and oftentimes opposing values, viewpoints and interests discuss, deliberate and negotiate problem analysis and solution-finding to water issues. Water governance points at situations where decision-making and implementing take place in networks of public, private and semi-private actors and where governments increasingly use horizontal forms of steering to achieve results within these actor networks (Edelenbos & Teisman, 2011). Governments are often dependent on many stakeholders with different resources (knowledge, money, etc.). In this context, governments sometimes deliberately, sometimes forced by circumstances, give more room to stakeholders to influence decision-making and implementation.

The deliberate attempt by governments to include stakeholders can be found in many countries and is described under many labels like public participation, interactive decision-making, consumer panels, public–private partnerships, stakeholder charters etc. The assumption is that the involvement of stakeholders in general enhances the content of policy proposals (because more knowledge becomes available) and creates more support legitimacy for water policy measures. Stakeholder involvement would improve the quality of decision-making by opening up the decision-making process and making better use of the information and knowledge that is available in society. Moreover, it would improve public understanding of the management issues at stake, make decision-making more transparent, and might stimulate the different government bodies involved to coordinate their actions more in order to provide serious follow-up to the inputs received. Water policy measures can become less controversial and implementation of decisions can be much smoother.

Inspired by Dahl's 'preconditions for a polyarchy', Berry et al. (1993, p. 55) formulated two dimensions of participation that are important for a system of *strong participation*. These are *width* and *depth* of participation, which together determine the strength of stakeholder involvement. The width of participation is the degree to which each member of a community is offered the chance to participate in each phase of the (water) policy process. This is also referred to as inclusiveness (Young, 2000). However, the inclusion of stakeholders in not without problems, as often usual suspects get involved in integrated water management processes as they got the skills, education and knowledge to become dominant in these kind of processes. As a result, the representativeness or broad inclusion of stakeholders is often not realized in water management projects (Sabatier et al., 2005; Van Buuren, Klijn, & Edelenbos, 2012).

The *depth* of participation is determined by the degree to which stakeholders have the opportunity to determine the final outcome of the governance process. In the analysis of width and depth of participation, it is important to distinguish the process on the one hand and the final outcomes of that process on the other. The realized depth of participation is often a problem as it stresses the aspect of power and authority in water management; often responsible government agencies want to stay in charge (Edelenbos & Teisman, 2011; Sabatier et al., 2005).

In the analysis of width of participation, we consider how governments have opened up the process for representation and participation by various stakeholder groups, like NGOs, citizens, private companies, farmers, etc. In short, what opportunities for stakeholder have been made available? An evaluation of the width of participation during the governance process is focused on the articulation of interests. The analysis of the depth of participation in the outcome is focused on the degree and type of influence stakeholders have had in shaping opinions and the realization of outcomes. In what way did the stakeholder have meaningful participation and did they have a say (voice) in the analysis of the (water) problem and the search for finding solutions for that problem?

> *Assessment criteria*: Are all relevant stakeholders involved? Are their interests, concerns and values sufficiently balanced considered in the problem analysis, solution search process and decision-making?

Trade-offs between social objectives: service-level agreements

Allocation

The economics of water management is about the allocation of scarce resources, which can be water quantity, water quality as well as safety against flooding. The water and safety allocation process is a political bargaining process, where pros and cons of several sharing options between legitimate claimants at a specific time are weighted, and winners and losers make their case heard (Backes & Van Rijswick, 2013; Van Rijswick, 2008). It is often not clear what objectives, such as food security and food self-sufficiency, guide allocation decisions and how the principles of equitable access, economic efficiency, sustainability and customary norms and values can be reconciled. To support this process, insight is required into various social (potentially conflicting) objectives and into the trade-offs between the impacts on various objectives of alternative measures. What are, for instance, the foregone benefits or opportunity costs of allocating water to ensure food self-sufficiency instead of allocating water in the most productive manner? Hydro-economic models can provide insight into the trade-offs of various between social objectives.

Reallocation

Many countries recognize the need for reforming their water allocation. The implementation of new water-allocation mechanisms will have various implications. For instance, changes in water allocation can have an impact on the competitiveness of some industries; their costs can be unevenly distributed across social groups. Such considerations can make the political economy of the reform of water policies complex. In practice, the transition of existing service levels towards new agreed service levels will take place when the gains will outweigh the transaction costs of the reform and the difficulties related to challenging the beneficiaries of status quo.

Allocation mechanisms

The service-level agreements have to be translated into rules, regulations and procedures, such that the water service delivered to each user is specified under different hydrological circumstances, as will be elaborated in the sixth building block. The service-level agreements can be achieved by way of water-allocation mechanisms (e.g. rationing, pricing or markets), as well as by water quality and safety standards (e.g. norms) combined with the instruments needed to achieve these standards (Jolink, 2010; Van Rijswick, 2008, 2011; Van Rijswick & Havekes, 2012). The suitability of certain rules depends on the time and geographical scales over which they have to operate. For instance, formal water entitlements provide holders with a secured right to water for many years; they are generally set at national, state or basin level; temporary water reallocations generally apply at a yearly basis only, and at a very local scale.

Assessment criteria: Are agreed service-level decisions based on trade-offs of costs, benefits and distributional effects of various alternatives?

Responsibility, authority and means

Property rights

The identification of responsibilities and authorities with respect to water starts with the determination of property rights. Property rights can be understood as social relations that define the titleholder with regard to water resources in relation to all others. Such rights only have an effect when there is some authority system that agrees to defend a rights holder's interest (Bromley, 1991). Four traditional types of ownership are identified in the literature (e.g. Ostrom, 1990): *private property* (the exclusive property title is in the hands of private individuals or corporations); *common property* (to be seen as the groups' private property, access by others can be excluded); *state (public) property* (ownership and use control is in the hands of the state, individuals and groups may be able to make use of resources, but only at the forbearance of the state, public agencies have the right to determine use or access rules, individuals have a duty to observe these rules); and *no property* (*res nullius*, the access to resources is not based on formally regulated property rights).

At the same time, property rights always include restrictions on the use of property by owners themselves, and such restrictions become more extensive in modernized and densely populated societies (Young, 1982). In general, a strongly increased complexity of bundles of property rights can be noticed as well as a strongly evolving *public domain* in terms of a communalization of water ownership and use rights. Bottom-up-organized common property arrangements exist for drainage and irrigation, often already for centuries. In the first half of the 20th century a growing concern for public health and sanitation results in the development of infrastructure for water supply and sewerage (Kissling-Näf & Kuks, 2004).

Allocating authority and responsibilities

To restrict property rights, the public domain needs authority at various administrative levels (central/decentral), needs to assign responsibilities to public and non-public actors, and creates *means* to empower authority (Van Rijswick & Havekes, 2012; Van Rijswick & Tappeiner, 2014). The authority should be able to guide stakeholder processes, to guide the process that should lead to powerful collective choices, to guide other relevant public authorities, to guide redistributions of property rights, to control progress and to take the right measures for effective operations.

Means: participative capacity

The participative capacity of the public domain refers to the input structures of the policy process on which it depends if all water uses have an equal opportunity to become expressed and recognized. Decentralization and strong local communities are seen as a favourable condition for participation, which is especially the case in countries that have adopted subsidiarity, a multilevel governance structure. *The integrative capacity* of the public domain refers to intra-policy coordination (i.e. the internal integration of the water policy field), to inter-policy coordination (i.e. the cross-sectoral integration of water policy with related policy fields), and to external coordination between water policy institutions and non-governmental actors, including consultations with target groups (Jänicke & Weidner, 1997).

Assessment criteria: Are authorities, responsibilities and means well-organized to deal with water issues at the appropriate administrative scale(s) in a participative and integrative way?

Regulations and agreements

Regulations and agreements can be considered pivot points between the first five building blocks of the assessment methodology and the last four blocks and more specifically between content and implementation. There are several ways to translate the service-level agreements that have been made as trade-offs of the social objectives in rules, regulations, agreements and procedures (De Burca & Scott, 2006).

Appropriateness

The appropriateness of rules and agreements will depend on actual circumstances, depending on the cultural, political, institutional and economic circumstances (developing/developed countries, rural/urban areas, religion, political/philosophical traditions); the legal traditions (common/civil law/traditional/indigenous law systems); the governmental organization (centralized/decentralized/river basin management approach); the parties that are involved (agreements between states, regulations of governments within a country to protect and develop public works, to protect ecosystems, to regulate private activities that have an impact on water systems, agreements between market parties, public–private partnerships, involvement of NGOs and citizens); the leading values and principles upon which the bargaining is based (see the second building block); the relevant and local water system characteristics; the actual water problem that has to be solved; and last but not least, the intention of the parties. The main assessment criterion of regulations and agreements is *legitimacy.*

Legitimacy

However, legitimacy has several aspects and is a concept that is used in several disciplines (Van Buuren, Driessen, Teisman, & Van Rijswick, 2013). It is related to the following aspects:

- Based on shared or agreed values and principles including those who refer to vulnerable values and groups in society to enhance effectiveness and to avoid conflicts (Addink, Anthony, Buyse, & Flinterman, 2010; Misiedjan & Gupta, 2014).
- In conformity with the rule of law (Hayek, 1973–1979; Brunnée and Toope, 2010).
- Offering (legal) certainty with regard to rights, duties and accountability to provide a base for action (Fuller, 1964; OECD, 2014; Van Holten & Van Rijswick, 2014; Van Rijswick & Salet, 2012).[1]

- Formulated in a way that they are enforceable and effective (able to achieve the intended goals) (Backes & Van Rijswick, 2013; Bardach & Kagan, 1982; Buijze, Salet, & Van Rijswick, 2014; Coglianese & Kagan, 2007);
- Decision-making at the most appropriate level and based on transparent rules, sufficient and relevant information and taking all interests that are at stake into account (also vulnerable and minority interests) (Howarth, 2009; Curtin & Meijer, 2006).
- Offering the right mix of public and private instruments for the objective at stake (Bardach & Kagan, 1982; Coglianese & Kagan, 2007; Diver, 1983; Mees et al., 2014).
- Taking distributional effects into account to avoid damage to the water system, other interests and policy fields, and in this way avoid conflicts (Lindhout & Van den Broek, 2014; Van Doorn-Hoekveld, 2014).

Legal certainty and adaptiveness

Regulations and agreements can be based on a procedural, a substantive or a combined approach, depending on the need for flexibility and adaptability on the one hand and (legal) certainty and an enforceable protection level on the other. Flexibility and adaptiveness can be achieved by open (less detailed) norms (that can be further substantively developed in mutually communication and interaction), the use of principles and duties of care, while certainty on duties, rights and accountability can better be achieved by substantive norms and standards and a clear allocation of responsibilities (Westerman, 2007a, 2007b). Regulations and agreements should be based on a right balance between adaptivity and flexibility on the one hand and legal certainty at the other (Green, Garmestani, Van Rijswick, & Keessen, 2013; Keessen & Van Rijswick, 2012).

Assessment criteria: Are regulations and agreements legitimate and adaptive, and if not, what are the main problems with regard to the above mentioned legitimacy aspects?

Financing water management

Empowerment with financial means is needed and a crucial element for good water governance. Without discussing the ways water management is financed, a sustainable and equitable financing of water management may hamper reaching the service-level agreements (SLAs) that have been agreed upon or may lead to instability because of equity concerns (Lindhout, 2012; Lindhout & Van den Broek, 2014). There are several ways to finance water management or principles upon which to base the financing system. One can think of cost recovery through a solidarity principle, which means that the costs of water policy are recovered from the national budget or budgets of decentralized authorities. One can also think of cost recovery through a profit principle, which means that those who have an interest in water services and profit from it also pay for it. Last but not least, is the finance system according to international agreed principles as, for example, the polluter pays principle?

Assessment criteria: Is the financial arrangement sustainable and equitable?

Engineering and monitoring

The design and management of the existing infrastructure may not be suitable to fulfil the societal functions. For instance, the capacity of irrigation canals may be insufficient, flood defences may not be strong enough, water rights can only be traded between interconnected basins, and volumetric pricing only operates where water use is metered.

Improving water infrastructure

SLAs are used to determine whether the existing infrastructure needs to be improved, and which improvements are needed. These SLAs are determined by the responsible authorities and agencies after consultation of stakeholders, and are among others based upon the trade-offs between (competing) societal objectives.

Often, the SLAs can be implemented in many different ways, using different technologies, different materials and resulting in different (ecological and spatial) consequences. Focusing on only one alternative does not represent the interests of the people involved. In engineering textbooks, a 'from global to detailed' principle is often used: first a global design, then a detailed design and next the implementation. Such principles are important, since it makes efficient use of available resources and does enable involvement of the relevant authorities and stakeholders.

Economic analysis has a role to play in assessing the returns on investment in constructing infrastructure To what extent does the present value of future benefits exceed the present value of current fixed and future variable costs of investments (and what discount rate is used)? It also has a role to play in assessing the cost-effectiveness of alternative infrastructural measures. An important issue here is to include the maintenance actions in the design, and to make the trade-off between less investment costs or less maintenance costs, because less investment costs often results in high maintenance costs.

Monitoring

Monitoring of the water system is needed in order to assess whether the system meets the agreed SLAs (Beijen, Van Rijswick, & Tegner Anker, 2014). Monitoring can be done in many ways, and can be available online, but that is not necessary. However, monitoring is not a goal in itself, but the data have to be used in order see whether the (water) system meets the requirements.

> *Assessment criteria*: Are SLAs sufficiently available (implicit or explicit) in order to redesign the existing infrastructure? Are the design and consequences of different alternatives sufficient available? Is there sufficient monitoring of the system and are the data analysed?

Enforcement

Enforcement as an often forgotten critical issue in the policy process

Water management and water governance assessments often focus mainly on the beginning of the policy process. There is ample attention for public participation, formulating goals, rules and standards, and the process of decision-making. However, good water management and governance should pay attention to the whole policy process from goal-setting to the actual achievement of goals. Therefore more attention is needed for the implementation of regulations and agreements and the possibilities to enforce the agreements that have been made. A lack of enforcement will hamper the effectiveness of water management and governance and may in the end lead to conflicts and decreasing legitimacy.

Enforceability of rules and agreements

Enforceability of agreements starts right at the beginning of the water management process and will be influenced by all prior blocks of the assessment methodology. Regulations and agreements that cannot be enforced will suffer from a lack of credibility and, in the end, legitimacy. This is only different in case the intention of the parties involved is not focused on achieving the objectives upon which they agreed.

There are several ways to improve the enforceability of rules and agreements. It is argued that rules and agreements that are based on shared values and principles will be easier to enforce because parties have the strong conviction that they should behave in conformity with the rules (Buijze et al., 2014; Van Rijswick & Salet, 2012). Nevertheless, there will always be the risk that in an individual case it will be more profitable not to comply with rules and agreements. Furthermore, it is not always possible to make regulations and rules that fully fit with the values of all parties, which is reflected in the need for bargaining.

In general, clear substantive norms and standards as well as clear process norms and standards regarding allocation of responsibilities and resources are useful when it comes to enforceability and in the end the effectiveness of water management (Bardach & Kagan, 1982; Coglianese & Kagan, 2007; Van Rijswick & Salet, 2012). Especially in cases, parties want clarity on their rights and duties because they want to protect their interests and know who is accountable for achieving the goals set. Enforceability will be an important issue. The same applies in cases where vulnerable values (water and ecosystems and the rights of vulnerable groups) are at stake. In case vulnerable values are at stake, their role should be recognized, their interests should be sufficiently protected and there should be ways for enforcement. More procedural and open norms are often used when the interests at stake are hard to agree upon, or in case agreements are made with regard to long-term and more general formulated objectives or in cases of large uncertainties with regard to system knowledge or social, economic, technical or physical developments.

Public and private enforcement

Depending on the regulations and agreements (public, private or a mix), enforcement can be undertaken by public or private parties. In general, private agreements and regulations will be enforced by private parties, but public regulations can be enforced by both public and private parties. Important with regard to enforcement are the available remedies to achieve the objectives.

> *Assessment criteria*: Are regulations and agreements enforceable by public and/or private parties, and are there appropriate remedies available?

Conflict prevention and resolution

Conflict prevention

Shared waters can be either a source of conflict or they may offer opportunities for cooperation, prosperity and stability. It is therefore important to identify the potential economic, social and political benefits of cooperation. The concept of 'water valuation for water dispute resolution' shows the advantages of 'benefit-sharing' rather than 'water-sharing'. This approach involves *thinking about water in terms of its value*, rather than just in terms of its quantities, quality and ownership. By focusing on the value of water, all benefits (private as well as socio-political) to be obtained from water can be explicitly optimized by reallocating water from low- to high-value usage. Such powerful information can be used by the parties to negotiate about the best water allocation and about benefit-sharing. This alternative form of dispute resolution will lead to cooperation and confidence-building, which is beneficial for all. If disputants think about water allocation in a cooperative manner, then the risk of violent conflict about water can be considerably reduced. Furthermore, from the above-described building blocks it follows that conflicts can be prevented by clear norms, standards, responsibilities and enforceable rules, standards, instruments and agreements.

Conflict resolution

In case conflicts do occur, parties need an independent mediator, arbiter or court to decide on the conflict and who is able to force parties to act in conformity with the final ruling. Conflicts can be prevented or solved in a proper way if stakeholders have formulated mutually accepted rules and procedures that prescribe how to handle or follow procedures in the case conflict of interests arise in water governance and management. Moreover, conflict resolution and mediation focused on positive-sum solutions are often stressed to find ways out of conflict as people are stimulated to connect and combine different concerns and interest (Wolf, 1997).

Assessment criteria: Are there sufficient conflict prevention and resolution mechanisms in place?

Conclusions

Developing integrated and sustainable water governance is a complex challenge that needs the input of all interested stakeholders. Moreover, the knowledge and experiences of several disciplines is needed. A one-dimensional approach, which only focuses on, for example, technical, or legal, or economic or public administration aspects to achieve good water management will not bring the integrated protection of the world's water resources, equitable access to clean water and sanitation, protection against flooding, the protection of vulnerable ecosystems and the necessary economic sustainable development. In our method we combined knowledge from several disciplines and combined this knowledge in an integrated way. The method makes clear that most building blocks are based on the knowledge of several disciplines, which all play their particular role in developing sustainable water management and which all should be taken into account. It is this integrated approach within each building block and between the several building blocks that can contribute to the development of sustainable water management. Of course, each river basin, region, country or water problem asks for tailor-made solutions. There is no 'one size fits all solution', but discussing the criteria set out in the several building blocks, in which insights and knowledge from technical, legal, economic and public administration perspectives are combined, helps to obtain a clear overview of the problems, challenges, pitfalls and success factors, thus enabling the development of sustainable water management.

We developed building blocks that were elaborated from simultaneously different disciplines, water systems analysis, economics, law and public administration. Table 1

Table 1. Relationship between the building blocks and disciplines.

	Water systems analysis	Economics	Law	Public administration
1. Water system knowledge	×	×		
2. Values, principles, policy discourses	×	×	×	×
3. Stakeholders involvement		×	×	×
4. Tradeoffs between social objectives	×	×		
5. Responsibility, authority and means			×	×
6. Regulation and agreements			×	×
7. Financial arrangements	×	×	×	×
8. Engineering and monitoring	×	×		
9. Enforcement			×	×
10. Conflict prevention and resolution		×	×	×

shows the relationship between the 10 building blocks discussed and to what extent different disciplines contributed to its developed and formulation of assessment criteria.

We believe that the added value of this interdisciplinary work is that building blocks have been identified that transcend and align different disciplines in developing meaningful interdisciplinary and integral assessment of water management praxis. From this, criteria are developed that are both underpinned in theoretical literature and operationalized to be used in many concrete water project and programme assessments in the last years, ranging from several projects in Europe, Asia, Africa, South America and Australia and from flood risk issues to water quality and water scarcity issues.[2] These criteria and corresponding building blocks offer the opportunity to monitor constantly the content, organization and implementation of water management projects and programmes and make appropriate adjustments to water governance in order to realize sustainable water management.

However, we want to stress that the method needs fine-tuning to specific water management situations and contexts. Every water challenge is a unique combination of history, climate, social and economic pressures, resource endowment, trading opportunities, and so on. To this extent, water management and governance must be uniquely designed and implemented. It is important that a water governance assessment tool will recognize the unique and situational aspects of water governance processes in different projects, processes and programmes in various countries, river basins and regions, and at the same time give clear meaning and provide critical assessment of the content, organization and implementation of those water governance processes. This paper develops a framework to achieve these aims.

Notes

1. Fuller developed the following criteria for proper law-making being generality, promulgation, non-retro-activity, clarity, non-contradiction, not asking the impossible, durability, and congruence between rules and official action.
2. See http://www.watergovernancecentre.nl and http://ucwosl.rebo.uu.nl/.

References

Addink, H., Anthony, G., Buyse, A., & Flinterman, C. (2010). *Human rights & good Governance.* SIM special. Utrecht: Utrecht University.

Ambrus, M., Gilissen, H. K., & Van Kempen, J. J. H. (2014). Public values in water law: A case of substantive fragmentation?. *Utrecht Law Review 2014, 10*(2), 8–30.

Arts, A., & Leroy, P. (2006). *Institutional dynamics in environmental change.* Houten: Springer.

Backes, C., & Van Rijswick, H. F. M. W. (2013). Effective environmental protection: Towards a better understanding of environmental quality standards in environmental legislation. In L. Gipperth and C. Zetterberg, *Miljörättsliga perspektiv och tankevändor, Vänbok till Jan Darpö & Gabriel Michanek* (pp. 19–50). Uppsala: Iustus Förlag AB. ISBN 9789176788424

Bardach, E., & Kagan, R. A. (1982). *Going by the book: The problem of regulatory unreasonableness.* Philadelphia, PA: Temple University Press.

Beijen, B. A., Van Rijswick, H. F. M. W., & Tegner Anker, H. (2014). A comparison of monitoring obligations in European environmental directives. *Utrecht Law Review 2014, 10*(2), 126–135.

Berry, J. M., Portney, K. E., & Thomson, K. (1993). *The rebirth of urban democracy.* Washington, DC: Brookings Institution.

Biswas, A. K. (2005). Integrated water resources management: A reassessment. In A. K. Biswas,O. Varis, & C. Tortajada, *Integrated water resources management in South and South-East Asia.* Water Resources Management Series. New York: Oxford University Press.

Bressers, H., & Kuks, S. (2004). *Integrated governance and water basin management: Conditions for regime change towards sustainability.* Dordrecht-Boston-London: Kluwer Academic Publishers.

Bromley, D. W. (1991). *Environment and economy. Property rights and public policy.* Oxford UK/ Cambridge USA: Blackwell.

Brouwer, R. (2013). *On the interdisciplinary nature of the study of law.* Paper presented at the Methodology Seminar, Faculty of Law, Utrecht 2013.

Brunner, R. D., & Steelman, T. A. (2005). *Towards adaptive governance.* In Brunner, R. D., Steelman, T. A. Coe-Juell, L. Cromley, C. M. Edwards, C. M. & Tucker, D. W., *Adaptive governance: Integrating science, policy and decision-making* (pp. 1–21). New York: Columbia University Press.

Brunnée, J., & Toope, S.J. (2010). *Legitimacy and legality in international law, an interactional account.* Cambridge: Cambridge University Press.

Buijze, A. W. G. J., Salet, W. G. M., & Van Rijswick, H. F. M. W. (in press). *How central interventions enable contextualized practices of sustainable development.*

Bulkeley, H. (2000). Discourse coalitions and the Australian climate change policy network. *Environment and Planning C: Government and Policy, 18*(6), 727–748. doi:10.1068/c9905j

Coglianese, C., & Kagan, R. A. (2007). *Regulation and regulatory processes.* Aldershot, UK: Ashgate.

Curtin, D. M., & Meijer, A. (2006). *Does transparency strenghten legitimacy?. Information Policy, 11*, 109–122.

De Burca, G., & Scott, J. (2006). *New governance in the EU and USA.* Oxford: Hart Publishing.

De Vries, J., Zonneveld, W., & Buijze, A. (in press) *Governance and law: It takes two to tango.*

Dewulf, A., Mancero, M., & Cardenas, G.. 2011. Fragmentation and connection of frames in collaborative water governance: A case study of river catchment management in Southern Ecuador, in special issue Water Governance, edited by J.

Diver, C. S. (1983). The optimal precision of administrative rules. *The Yale Law Journal, 93*, 65–109. doi:10.2307/796245

Driessen, P. P. J., & Van Rijswick, H. F. M. W. (2011). Normative aspects of climate adaptation policy. *Climate Law, 2*(4), 1–23.

Edelenbos, J., & Teisman, G. R. (2011). Symposium on water governance. Prologue: Water governance as a government's actions between the reality of fragmentation and the need for integration. *International Review of Administrative Sciences, 77*(1), 5–30. doi:10.1177/0020852310390090

Edelenbos, J. Bressers, N. E. W. & Scholten, P. H. T. (Eds.). (2013). *Water governance as connective capacity.* Farnham, Surrey: Ashgate.

Folke, C., Pritchard, L., Berkes, F., Colding, J., & Svedin, U. (2007). The problem of fit between ecosystems and institutions: Ten years later. *Ecology & Society, 12*(4), 201–220.

Fuller, L. L. (1964). *The morality of law.* New Haven, CT: Yale University Press.

Global Water Partnership. (2003). *Integrated water resources management toolbox*, Version 2, Stockholm, GWP Secretariat, p. 2.
Green, O., Garmestani, A., Van Rijswick, H., & Keessen, A. (2013). EU water governance: Striking the right balance between regulatory flexibility and enforcement? *Ecology and Society, 18*(2), 10. doi:10.5751/ES-05357-180210
Hajer, M. (1993). Discourse Coalitions and the institutionalisation of practice: The case of acid rain in Great Britain. In Fischer, Frank/Forester, John (Hrsg), *The argumentative turn in policy analysis and planning* (pp. 43–67). Durham/London: Duke University Press..
Hayek, F. A. (1973–1979). *Law, legislation and liberty* (Vol. 3). London: Routledge and Kegan Paul.
Howarth, W. (2009). Aspirations and realities under the water framework directive: Proceduralisation, participation and practicalities. *Journal of Environmental Law, 21*(3), 391–417. doi:10.1093/jel/eqp019
Huitema, D., Mostert, E., Egas, W., Moellenkamp, S., Pahl-Wostl, C., & Yalcin, R. (2009). Adaptive water governance: Assessing the institutional prescriptions of adaptive (co-)management from a governance perspective and defining a research agenda. *Ecology and Society, 14*(1), 26. [online] Retrieved from http://www.ecologyandsociety.org/vol14/iss1/art26/
IPCC-WGII. (2007). *Climate change 2007: Impacts, adaptation and vulnerability. Contribution of the Working Group II to the Fourth Assessment report of the Intergovernmental Panel on Climate Change Summary for Policy Making*. New York: Cambridge University Press.
Jänicke M. & Weidner H. (Eds.). (1997). *National environmental policies: A comparative study of capacity-building*. Berlin-Heidelberg: Springer Verlag.
Jolink, A. (2010). *Legal implications of introducing economic instruments in the field of European and Dutch water management*. Utrecht: Utrecht University.
Keeney, R. L. (1992). *Value-focussed thinking, a path to creative decision-making*. Cambridge/London: Harvard University Press.
Kissling-Näf, I. and Kuks, S. (Eds.). (2004). *The evolution of national water regimes in Europe. transitions in water rights and water policies towards sustainability*. Dordrecht-Boston-London: Springer (Kluwer Academic Publishers).
Keessen, A. M., & Van Rijswick, H. F. M. W. (2012). Adaptation to climate change in European water law and policy. *Utrecht Law Review, November 2012*, 38–50.
Kok, M., Stijnen, J. W., & Silva, W. 2003. Uncertainty analysis of river flood management in the Netherlands, In T. Bedford & P.H.A.J.M. van Gelder, (Eds.), Safety and Reliability, *Proceedings of ESREL 2003–European safety and reliability conference* 2003, 15–18 June 2003, Maastricht, The Netherlands, pages 1179–1187, Lisse: Swets & Zeitlinger.
Lindhout, P. E. (2012). A wider notion of the scope of water services in EU water law, boosting payment for water-related ecosystem services to ensure sustainable water management? *Utrecht Law Review, 8*, 86–101. 2012–3.
Lindhout, P. E., & Van den Broek, B. (2014). The polluter pays principle: Guidelines for cost recovery and burden sharing in the case law of the European court of justice. *Utrecht Law Review 2014, 10*(2), 46–59.
Mees, H. L. P., Dijk, J., Van Soest, D., Driessen, P. P. J., Van Rijswick, H. F. M. W., & Runhaar, H. (2014). A method for the deliberate and deliberative selection of policy instrument mixes for climate change adaptation. *Ecology and Society, 19*(2), 58. [online] Retrieved from: http://www.ecologyandsociety.org/vol19/iss2/art58/ doi:10.5751/ES-06639-190258
Meijerink, S. (2008). Explaining continuity and change in international policies: Issue linkage, venue change, and learning on policies for the river Scheldt estuary 1967–2005. *Environment and Planning A, 40*(4), 848–866. doi:10.1068/a3911
Misiedjan, D., & Gupta, J. (2014). Indigenous communities: Analyzing their right to water under different international legal regimes. *Utrecht Law Review 2014, 10*(2), 77–90.
Mostert, E., Pahl-Wostl, C., Rees, Y., Searle, B., Tàbara, D., & Tippett, J. (2007). Social learning in European river basin management: Barriers and supportive mechanisms from 10 river basins. *Ecology and Society, 12*(1), 19. [online] Retrieved from: http://www.ecologyandsociety.org/vol12/iss1/art19/
OECD (2011). *Water governance in OECD countries. A multi-level approach*. Paris: OECD publishing.
OECD (2014). *Water governance in the Netherlands: Fit for the future?* Paris: OECD.

Ostrom, E. (1990). *Governing the commons. The evolution of institutions for collective action.* Cambridge, New York: Cambridge University Press.

Pahl-Wostl, C. (2006). Transitions towards adaptive management of water facing climate and global change. *Water Resources Management, 21*(1), 49–62. doi:10.1007/s11269-006-9040-4

Perry, P. (2003). Non-state actors and water resources development – An economic perspective. *Non-State Actors and International Law, 3*(1), 99–110. doi:10.1163/156771203322428395

Sabatier, P. A., Focht, W., Lubell, M., Trachtenberg, Z., Vedlitz, A., & Matlock, M. (2005). *Swimming upstream: Collaborative approaches to watershed management, American and comparative environmental policy series.* Cambridge, MA: The MIT Press.

Skelcher, C. (2005). Jurisdictional integrity, polycentrism, and the design of democratic governance. *Governance, 18*(1), 89–110. doi:10.1111/j.1468-0491.2004.00267.x

Tortajada, C. (2010). Water governance: A research agenda. *International Journal of Water Resources Development, 26*(2), 309–316. doi:10.1080/07900621003683322

UNESCO. (2006). *Water, a shared responsibility.* Barcelona: UNESCO/Bergham Books.

Van Buuren, A., Driessen, P., Teisman, G., & Van Rijswick, M. (2013). Toward legitimate governance strategies for climate adaptation in the Netherlands: Combining insights from a legal, planning, and network perspective. *Regional Environmental Change,* doi:10.1007/s10113-013-0448-0

Van Buuren, A., Klijn, E. H., & Edelenbos, J. (2012). Democratic legitimacy of new forms of water management in the Netherlands. *International Journal of Water Resources Development, 28*(4), 629–645. doi:10.1080/07900627.2011.627756

Van Doorn-Hoekveld, W. J. (2014). Compensation in flood risk management with a focus on shifts in compensation regimes regarding prevention, mitigation and disaster management. *Utrecht Law Review 2014, 10*(2), 216–238.

Van Holten, S., & Van Rijswick, H. F. M. W. (2014). The consequences of a governance approach in European Environmental directives for flexibility, effectiveness and legitimacy. In M. Peeters, & R. Uylenburg (Eds.), *EU environmental legislation: Legal perspectives on regulatory strategies* (pp. 13–47). Cheltenham: Edward Elgar Publishing.

Van Rijswick, H. F. M. W. (2008). *Moving water and the law, on the distribution of water rights and water duties within river basins in European and Dutch water law, inaugural address Utrecht University.* Groningen: Europa Law Publishing.

Van Rijswick, H. F. M. W. (2011). De verdeling van schaarse waterrechten [the allocation of scarce water rights). In F. J. van Ommeren, W. den Ouden & C. J. Wolswinkel (Eds.), *Schaarse publieke rechten* (pp. 133–158). Den Haag: Boom Juridische Uitgevers.

Van Rijswick, H. F. M. W., & Havekes, H. J. M. (2012). *European and Dutch water law.* Groningen: Europa Law Publishing.

Van Rijswick, H. F. M. W., & Salet, W. (2012). Enabling the contextualization of legal rules in responsive strategies to climate change. *Ecology and Society, 17*(2), 18. Retrieved from: http://dx.doi.org/10.5751/ES-04895-170218

Van Rijswick, H. F. M. W., & Tappeiner, I. M. (2014). *Institutional legal framework for resilient, sustainable regional water management in times of climate change.* Cheltenham: Edward Elgar Publishing.

Warner, J., & Van Buuren, W. A. (2009). Multi-stakeholder Learning and fighting on the River Scheldt. *International Negotiation, 14*(2), 419–440. doi:10.1163/157180609X432888

Westerman, P. C. (2007a). Governing by goals: Governance as a legal style. In: *Legisprudence: International Journal for the Study of Legislation* (pp. 51–72).

Westerman, P. C. (2007b). The Emergency of New Types of Norms. In L. J. Wintgens (Ed.), *Legislation in context; essays in legisprudence* (pp. 117–133). Aldershot: Ashgate.

Wolf, A. T. (1997). International water conflict resolution: Lessons from comparative analysis. *International Journal of Water Resources Development, 13*(3), 333–366. doi:10.1080/07900629749728

Young, I. M. (2000). *Inclusion and democracy.* Oxford: Oxford University Press.

Young, O. R. (1982). *Resource regimes. Natural resources and social institutions.* Berkeley, Los Angeles, London: University of California Press.

Transboundary water law and vulnerable people: legal interpretations of the 'equitable use' principle

Karina Vink

Vulnerable people require additional measures to ensure their water capabilities, as they have certain characteristics making them more vulnerable than others. As pointed out by recent studies, transboundary water access laws and policies do not sufficiently address the needs of vulnerable people. The prevailing legal arrangements often only address extrinsic vulnerability and forgo focusing on intrinsic vulnerability, which creates the need for different transboundary water legislation. This paper shows how international treaties can address the right to certain water capabilities by considering not merely the current but also future global populations into the creation of their transboundary agreements.

Introduction: global water issues

Now that predictions of water scarcity are such that developed countries cannot escape its effects, transboundary laws and policies are adapting and becoming more attentive to rights regarding the sharing of water. The question remains: how are the water access rights of vulnerable people taken into account in transboundary water laws and policies? The objective of this study is to examine existing transboundary water laws and agreements from the viewpoint of vulnerable people, looking at both extrinsic and intrinsic vulnerability, and water capabilities. In previous research on flood disaster laws and policies (Vink, Takeuchi, & Kibler, 2014) it was found that even in developed countries, such as the Netherlands, the United States and Japan, insufficient attention is paid to vulnerable people in flood-risk management laws, ranging from national laws to local policies, and there is much room for improvement of laws and policies. Therefore, one is likely to find similar room for improvement for measures for vulnerable people in transboundary water laws and agreements.

Extrinsic and intrinsic vulnerability, and water capabilities

Water capabilities can be defined as those abilities one can perform by means of water. They could be applied to a multitude of scales, including but not limited to: humans, households, ecosystems, animals, cities, nations or even the entire planet. As for humans, what is defined as water capabilities can vary across cultures. For *basic* human life, water is needed as drinking water, for the preparation of food and for hygienic purposes (including, for example, menstrual hygiene). For a *meaningful* human life, more capabilities may be identified and exercised as necessary parts of daily activities. While the previously mentioned capabilities may enable a human being to exist in 'decent health', this may not equate to the standards of 'decent life'. The preamble of the World Health Organization (WHO) states 'health is a state of complete physical, mental and social well-being and not merely the absence of disease or infirmity' (WHO, 1948), indicating that health is more than merely surviving. For instance, water is used for several religious purposes, including ritual washing, holy water and the immersion of statues in water. Some examples of faiths including ritual washing are Islam (before prayers, five times a day), Christianity (baptism), Hinduism (purification in the river Ganges), and Shinto (cleansing). Many more examples of religious or cultural traditions involving water exist worldwide, and while these practices may not be directly necessary for survival, people performing these practices may regard them as part of their required water capabilities.

With regards to the human right to water, vulnerable people may be defined as 'people who have one or more characteristics that make them more susceptible than others in a community and who therefore require additional measures in order for them to have the same water capabilities as others' (based on Vink et al., 2014). People suffering from vulnerability with regard to water capabilities can have one or more characteristics (based on Vink et al., 2014):

- Fewer material and/or financial resources, e.g. people who cannot afford water of safe quality, people without access to water-saving technologies, people who do not have sufficient containers, homeless people.
- Less physically or mentally capable, e.g. people who are shunned from or unable to access proper hygienic facilities due to illness or disability, people who crawl and are thereby exposed to dirt and increased sweating, people who have fever or diarrhoea.
- Less knowledge or experience, e.g. people who are unaware of water quality and hygiene, people who are unaware of water-saving procedures. Education plays a key role in the sustainable and hygienically sound use of water.
- Restricted by commitments, e.g. people who have dependants (family, friends, animals), people with place attachment, people with certain duties. For this characteristic it could be argued that if people do not have any of the other characteristics of vulnerability, people restricted by commitments are not truly vulnerable themselves. However, it is understood that while these people may have all the necessary means to provide for their own water capabilities, they choose to forego some part of these in favour of other commitments, thereby not being fully capable themselves.

Extrinsic vulnerability with regards to water capabilities can be described as unequal water capabilities brought into existence due to external factors. Examples include inequalities brought into existence by upstream countries withholding water of sufficient quantity or quality to downstream countries, unequal power structures between richer farming organizations and poorer individual farmers, or other social and cultural biases towards certain genders, classes, castes, ethnicities, etc.

Intrinsic vulnerability with regard to water capabilities refers to the fact that we as people do not require the same amount of water over the course of our lives, depending on our characteristics. Providing everyone with equal access, or access to an equal quantity of water, may not lead to equal capabilities or may not do justice to their individual characteristics. Vulnerable people may be subject to circumstances and characteristics that affect the minimum amount of water they require to maintain a decent living standard. Examples of groups of potentially vulnerable people with intrinsic vulnerability include children, pregnant and nursing people, people living with certain disabilities or illnesses (temporary/chronically) who require medical care and extra washing, and people working under conditions of extreme environments (unclean environments, extreme heat or physically demanding functions). Due to this inherent difference in people's characteristics, an equitable water law would ideally address people's water capabilities rather than prescribing a fixed amount of water per human being.

Both extrinsic and intrinsic vulnerability can lead to the four characteristics of vulnerability with regard to water capabilities.

Water as a human right

In order for rights to be meaningful, the possibility to defend rights is essential, whether they are imbedded in international laws as *jus cogens* (compelling law) or not (Rogers & Hall, 2003). Water conventions should provide some legitimate forum for resolving the diverging water needs and establishing a legally binding instrument (Gupta et al., 2013). This section discusses how several legal documents discuss water as a human right, focusing on the human right to water for vulnerable people in transboundary water law.

First of all, the Universal Declaration of Human Rights (UN, 1948), created in 1948, contains several articles relevant to water as a human right. It declares all human beings as equal in dignity and rights (Article 1), without distinctions such as race, sex and religion (Article 2), and everyone having the right to life, security of person (Article 3), a standard of living adequate for the health and well-being of themselves and of their family, and the right to security in the event of circumstances beyond their control (Article 25). This can be understood as including the right to water to all people, regardless of vulnerability. The Declaration itself is not a legally binding document; rather it expresses standards for the world to follow.

The Habitat Agenda (United Nations Habitat, 1996) refers to vulnerable people in section IV B, 4, para. 93, as people who are marginalized and excluded 'from the socio-economic mainstream and decision-making processes and the lack of access on an equal basis to resources and opportunities'. To improve this, the Agenda suggests improving decision-making processes within national- and international-enabling environments. The Agenda further states that the main cause of vulnerability is due to 'circumstances, rather than inherent characteristics'. This suggests the Agenda recognizes extrinsic vulnerability in the form of unequal water capabilities brought into existence due to external factors as the main form of vulnerability. In addition, in para. 97 the Agenda calls for the removal of physical constraints so that people with disabilities or older adults might continue to live independently, which implies the Agenda does consider some aspects of intrinsic characteristics of vulnerability. The groups of vulnerable people identified by the Agenda in para. 38 are people living in poverty, the homeless, women, older adults, indigenous people, refugees, displaced persons, people with disabilities, migrants, children, and those belonging to vulnerable and disadvantaged groups. This summation also points to both extrinsic and intrinsic vulnerable people.

Under the auspices of the UN, two legally binding instruments were negotiated and adopted based on the Declaration of Human Rights, namely the International Covenant on

Civil and Political Rights, and the International Covenant on Economic, Social and Cultural Rights (ICESR) (UNOCHCR, 1966), which entered into force in 1976. This latter covenant has been ratified or acceded to by 162 states to date, and contains broadly similar articles referring to equality (Articles 2, 3), and natural wealth and resources (Articles 1, 25). While water is undisputedly included as a part of natural resources, the existence of water rights is not clearly articulated in this covenant. Implementation of the covenant is self-monitored by the governments and reported to the Committee on Economic, Social and Cultural Rights, which subsequently recommends concluding observations on the reported results.

The Dublin Statement on Water and Sustainable Development dedicates one of its four principles to women as playing a pivotal role in managing water (World Meteorological Organization, 1992), which it recommends ought to be reflected in water management policies by including attention to women's needs. Further recommendations include priority to the quarter of the world's population lacking clean water supply and sanitation in order to reduce poverty and disease. The Statement suggests resolving water conflicts by managing river basins between countries, with international agreements as basis.

The 1992 United Nations Economic Commission for Europe (UNECE) Convention on the Protection and Use of Transboundary Watercourses and International Lakes (Water Convention) focuses on environmental protection rather than guaranteeing individual rights (McIntyre, in press a). Nevertheless, there are some sections relevant to vulnerable people, especially when considering the UNECE Guide to Implementing the Water Convention (2013). Article 2,1 of the Water Convention states that all 'Parties shall take all appropriate measures to prevent, control and reduce any transboundary impact'. Transboundary impact is defined in Article 1 as 'any significant adverse effect on the environment resulting from a change in the conditions of transboundary waters caused by a human activity' and includes effects on 'human health and safety, flora, fauna, soil, air, water, climate, landscape and historical monuments', as well as cultural heritage or socio-economic conditions. The UNECE Guide (para. 82) specifies that water quality objectives and criteria should be evaluated on a case-by-case basis. McIntyre (in press b) concludes that this case-by-case basis evaluation leaves no room for generally applicable objectives for equitable and reasonable use. Despite this, one general rule is determined, however ambiguous. Para. 109 of the UNECE Guide states that not every transboundary impact is negotiable, in particular those impacts that irreversibly harm the environment so that present or future vital human needs are endangered. Currently, the consensus about 'vital human needs' is described in para. 108 as comprising 'sufficient water to sustain human life, include both drinking water and water required for the production of food in order to prevent starvation'. The word 'including' suggests it is not limited to these two water capabilities and might encompass additional essential capabilities. The water quality evaluation criteria most likely to change in the future are social or economic factors (McIntyre, in press b). This signifies that a clear understanding of a state's own population, now and in the future, as well as those of the other states in the river basin, is required as a basis for determining the unnegotiable needs. Finally, Article 16,2 of the Water Convention states that all information regarding water quality objectives should be available to the public. The UNECE Guide (para. 334) elaborates that this information should be available without discrimination as to citizenship, nationality or domicile. This would give concerned citizens regardless of country of origin the necessary information to start a case when anyone's vital human needs are not met.

The 1997 UN Watercourses Convention (UN, 2005) specifically requires all countries part of a watercourse to take into consideration the social and economic needs of all involved states: (Article 6,1b), 'the population dependent on the watercourse in each watercourse State' and 'the effects of the use or uses of the watercourse in one watercourse State on other watercourse States' (Article 6,1(b), (c), and (d). It also requires countries to prevent harm to other watercourse states (Article 7). While this international water convention, like the Economic Commission for Europe Transboundary Water Convention, contains many articles referring to the governing rule of equitable and reasonable use of water courses, as explained by Wouters (2014), it is only recently entering into force and opened to accession by non-UNECE states. In its current form this convention does not unambiguously define equitable or reasonable use of water courses, and thereby foregoes an important implementation step. However, there are several statements as to how equitable and reasonable use might be defined, which may be interpreted to prioritize the human right to water. Article 6 mentions certain factors should be taken into account when determining what equitable and reasonable use might be. Article 6,3 states that all these factors should be considered, but also that weights should be applied to these factors, which are 'to be determined by its importance in comparison with that of other relevant factors'. Article 10,1 then states that no use might have inherent priority over other uses, while Article 10,2 prioritizes those uses that fulfil vital human needs. While the convention itself remains unclear on how to apply weights to the various identified uses, the statement of 'vital human needs' can be interpreted as including the human right to water on an individual basis. The UN Watercourses Convention User's Guide (para. 5.1.2) states that equity should not be taken to mean the water course should be divided into equal shares. The guide also refers to the relevant factors matrix, which has a more detailed description of the factors to be evaluated for determining whether uses are equitable and reasonable (para. 6.1.2). This matrix considers six categories of data, including: population details, recreational and cultural water uses, social benefits and case-specific details. This promises to encompass a greater scope of water capabilities than drinking water and sanitation. Furthermore, Article 32 on non-discrimination may be interpreted to apply not merely to the citizens of one watercourse state, but also to the other watercourse states in the basin, thus allowing for people in different states judicial procedures, compensation or relief when an activity causes them harm. McCaffrey (2001) is convinced the convention will be successful regardless of implementation, as it has obtained the 'status as the most authoritative statement of general principles and rules', which may be applied to future agreements. However, as pointed out by the UN Watercourses Convention User's Guide (Rieu-Clarke, Moynihan, & Magsig, 2012), if the concerned states are not parties in the convention and the states view administrative law as territorial, it may not be possible to protect foreign citizens on an individual basis. The ambiguity surrounding equitable and reasonable use, what exactly constitutes as no harm, and how disputes should be settled while allowing for sovereignty of states, is what caused Turkey to vote against the convention as one of three countries to do so (UN, 1997).

General Comment No. 15 on the Right to Water (CESCR, 2003) states unequivocally that 'the human right to water is indispensable for leading a life in human dignity. It is a prerequisite for the realization of other human rights'. Many groups of vulnerable people are mentioned in the General Comment (e.g. women, children, minority groups, indigenous peoples, refugees, asylum-seekers, internally displaced persons, migrant workers, prisoners, detainees (Article 16)), along with measures states should take on a national level to ensure the right to water for these vulnerable groups (e.g. no discrimination on ground of people with different race, colour, sex, age, language, religion, political or other opinion, national or social origin, property, birth, physical or mental disability, health status (including HIV/AIDS), sexual orientation and civil, political, social or other status

(Article 13)). Para. 12 (a) links the amount of water to which one is entitled to the WHO guidelines, but adds that 'some individuals and groups may also require additional water due to health, climate, and work conditions', which could be interpreted to include several forms of intrinsic vulnerability. On a transboundary level, para. 30 explains the role of international cooperation, assistance and joint action to fulfil people's rights. Para. 33 states that countries should not inhibit the right to water of individuals and communities in other countries. In addition, countries should guarantee the right to water in their international agreements (para. 35). Para. 55 discusses that violations of the right to water should be able to be addressed at both national and international levels, which is a promising step for accountability.

In July 2010 the UN General Assembly recognized the human right to water and sanitation for the full enjoyments of life and all human rights by means of Resolution 64/292 (UN, 2010). At this point, the resolution was still non-binding, however in September 2010 this resolution was adopted as a human right by the Human Rights Council (United Nations Human Rights, 2010), stating that:

> the human right to safe drinking water and sanitation is derived from the right to an adequate standard of living and inextricably related to the right to the highest attainable standard of physical and mental health, as well as the right to life and human dignity.

Article 8 (c) furthermore calls attention to vulnerable and marginalized groups, and encourages non-discrimination and gender equality. As a human right, access to safe drinking water and sanitation have become legally binding, according to the UN. Yet, as described by Leb (2012), while there is a certain amount of international commitment to the human right to water, it is in fact not a legally binding agreement, and about 25% of the present state representatives did not vote on the resolution. Tremblay (2013) explains that the right to water has not attained the status of customary international law because the states are not consistently enforcing and implementing this right.

The Millennium Development Goals (MDGs), to be reached by 2015, did not include water access, let alone water capabilities, as one of its eight main development goals. This despite the importance of water access in affecting the eight goals, such as eradicating extreme poverty and hunger, combating diseases, reducing child mortality and improving maternal health. Rather, water access was a sub-target of goal number seven, ensuring environmental sustainability (Target 7.C Halve by 2015, the proportion of the population without sustainable access to safe drinking water and basic sanitation; UN, 2013a). While many of the goals have been reached, undeterred by a growing global population, most progress has been achieved in respect of issues that were easy to tackle. For instance, goal number one of eradicating extreme poverty and hunger had as targets to halve the proportion of people whose income is less than US$1.25 per day, and to halve the proportion of people who suffer from hunger. The choice to only halve the proportion of people suffering has led to the attempt to reach this 50% in the easiest way possible, thus often excluding the most vulnerable people. Furthermore, while the UN is eager to point out how progress in this area has exceeded the target, Goff & Crow (2014) point out that the way water access and sanitation are measured grossly misconstrue the actual situation in practice, and are likely to continue to do so in the proposed upcoming post-2015 Development Agenda (UN, 2013b). This Agenda contains new goals from 2015 onwards and represents a transformative shift so as to leave no one behind, focusing on ending hunger and ensuring human rights for all people regardless of ethnicity, gender, geography, disability, race or other status. It has 12 goals, of which item number six is universal access to water and sanitation, which are important water capabilities. Its sub-goals are to:

- 6a. Provide universal access to safe drinking water at home, and in schools, health centres, and refugee camps.
- 6b. End open defecation and ensure universal access to sanitation at school and work, and increase access to sanitation at home by *x*%.
- 6c. Bring freshwater withdrawals in line with supply and increase water efficiency in agriculture by *x*%, industry by *y*% and urban areas by *z*%.
- 6d. Recycle or treat all municipal and industrial wastewater prior to discharge.

If achieved, these goals might provide wider access to wholesome water, allowing for the fulfilment of water capabilities. However, the problems with measuring progress of the development goals include poor water quality, and water source control by large-scale vendors or organized crime remains unaddressed by these goals. As with the previous MDGs, leaving the exact percentages up to the countries in question will again promote the resolution of easier problems first, leaving the most vulnerable people without water access, despite the nobly formulated transformative shift.

Transboundary water rights and equitable access of populations to water have recently been discussed by Goff & Crow (2014), Wouters (2014), and Zwarteveen & Boelens (2014), among others. These studies mainly cover transboundary water rights from the viewpoint of extrinsic vulnerability, i.e. an inequality brought into existence by external factors. This is also clearly recognized by General Comment No. 15 on the Right to Water (CESCR, 2003); Resolution 64/292 (UN, 2010); and the Human Rights Fact Sheet No. 35, The Right to Water, by UNOCHCR (2010). These types of extrinsic vulnerability constitute one of the ways people can be vulnerable with regard to water capabilities. Without transboundary water agreements it is possible that states in downstream areas, or states with less of an active voice, have restricted or no access to water of quantities or of a quality suited to guaranteeing decent human life to its citizens. Certain groups of people might be systematically excluded from full water access due to cultural norms or lack of legal protection, such as refugees, migrants and illegal settlers. This might be exacerbated in transboundary water conflicts, especially when a more politically powerful country controls upstream areas of river basins. Aspects of intrinsic vulnerability have not yet been given such attention.

How to define equitable and reasonable use and vital human needs?

The time-consuming processes for the development, alteration and implementation of transboundary law creates the need to get it right from the start, notably by including the rights of vulnerable people in transboundary water laws. The 1.8 billion people expected to live in absolute water scarcity by 2025 make up a significant number and justify a closer examination of the water access of populations in transboundary areas, which are prone to conflict concerning water. Water is a human right for all people (CESCR, 2003; United Nations Human Rights, 2010). As mentioned, this does not equate to all people having the right to access an equal amount of water.

Exactly how to address equitable and reasonable water use remains subject to debate. Issues of scale and quantities versus capabilities arise. When determining equitable and reasonable use, should we look at a country level, water basin level, ecosystem level, local or household level? Should we differentiate between types of use, e.g. environmental, industrial, agricultural and household use? Should we instead examine the capabilities people have given a certain amount of water access, and take a rights-based approach? From the viewpoint of vulnerable people, do we subsequently need to estimate the number of potentially vulnerable people that might require more water than the average self-reliant person? What we see in the UNECE Water Convention, the UN Watercourses convention, General Comment No. 15 on the Right to Water, and in the development goals is that states have certain duties towards their citizens, and the needs of individual states have to be taken into account in international relations. States have a duty to cooperate in the global community. To represent their citizens accurately, including vulnerable people, it is crucial that states value and respect the voices of their own and of all global citizens, regardless of their chosen discourse.

Taking the viewpoint of a set minimum daily amount of water per person, several issues arise. The standard minimum amount of water required on a daily basis in order to guarantee a decent life has been calculated at 50–100 litres (WHO, 2011). This standard includes water for drinking and food preparation and consumption, personal hygiene, and cleaning the living environment, and also sets out certain quality guidelines as well. However, this amount of 100 litres is an average that does not take individual differences into account. Looking at developed countries that apply water-saving technologies (e.g. for the three most consuming activities of toilet flushing, showering and washing with laundry machines), people use roughly 20% more than the minimum amount of 100 litres. As this is very close to the calculated minimum amount of 100 litres, it could be argued to apply a slightly higher minimum amount to all people, including a buffer for vulnerable people. For instance, in the Netherlands the average household daily use per person is calculated at 120 litres (Vitens, 2010).

Goff & Crow (2014) critique the use of a minimum amount of water per person, and explore the concept of water use for capabilities rather than set quantities per person or household, and warn that, to resolve water inequity, other disciplines must be addressed to resolve inequity on a scale beyond water alone. Merely replacing a set minimum amount of water with a term such as 'sufficient water' may leave room for individuals with characteristics requiring more water, but it may likewise be abused to mean as little as possible for all people affected by the policy. Moreover, in considering capabilities and different disciplines, it becomes ambiguous and culturally dependent which capabilities and disciplines should be considered as most important. Zwarteveen & Boelens (2014) posit that in discussing water and justice, the chosen discourse will determine how equitable and reasonable access can be discussed, and therefore certain unequal power balances are predetermined. The UN Watercourses Convention User's Guide (para. 5.1.2) mentions that distributive justice (equitably dividing both burdens and benefits) can only be reached by procedural fairness (including minority groups in decision-making).

Besides recognizing different water capabilities varying per culture, enabling equitable and reasonable use among all global citizens also necessitates knowledge of the number of vulnerable people, now and in future (at least one future generation, in order to anticipate demographic changes). Ideally, these people should be directly involved in the *formation* of key decision-making processes, and not merely the decision-making process itself, to guarantee their voice is heard. The question imbedded into the creation of transboundary water agreements should be always to check which potentially vulnerable groups (e.g. having a certain age, economic status, ethnicity) cannot get sufficient water, or water of a certain quality, for their desired daily activities, and if there is systematic discrimination and exclusion of certain groups. If this is the case, representatives of these groups should be directly involved in resolving the issue. Walters (2014) points to implementation issues in India, where courts ruled the right to water can be derived from the Constitution, but the government nevertheless remains indifferent to the cause of homeless people, as they lack voting rights. In India, including homeless in the policy formation process could make a significant change for their water capabilities.

Including vulnerable people in policy formation may help in resolving another issue that arises when considering allocating water resources for more human rights than merely the right to water. Tremblay (2010) highlights the right to development and the right to a clean, healthy and sustainable environment, which are both necessary for survival and have transboundary causes and effects. As all human rights are equally important, the human right to water, even if prioritizing human needs, can then still be contested by these other rights. By including vulnerable people themselves in the policy formation process, the difficult burden of prioritizing human rights may be somewhat shifted to the end users.

It should not be overlooked that water quantity and quality available for human water capabilities by themselves are not the only issues to be discussed in transboundary agreements. Other issues involving water management include: water quality of the environment affecting human health, an overload of water quantity leading to floods, and changes in the flow contents of a natural river (organisms such as fish, insects, zooplankton; nutrients; natural structures creating necessary wildlife habitats such as stones and trees). There is a strong correlation between the human water security threat and the biodiversity threat (Vörösmarty et al., 2010). Loss of biodiversity will inevitably negatively impact human life. Dams or navigation channels can affect or alter the natural flow regimes by changes in (Richter, Baumgartner, Powell, & Braun, 1996):

- Magnitude: determines the availability or suitability of an aquatic habitat; influences the position of the water table relative to the rooting zone of riparian vegetation.
- Timing: influences whether life cycle requirements related to flow are met, and the degree of stress or mortality from extreme events.
- Frequency: influences population dynamics by cuing life cycle events (reproduction, migration, mortality).
- Duration: determines whether certain life cycle phases may be completed, and the degree to which stressful conditions may accumulate.
- Rate of change: cues preparedness for hydrologic change (e.g. refuge), the stranding of mobile organisms in floodplains, and the adaptation of rooting zones to lowering water tables.

Changes in these five aspects may lead to negative effects on rivers and aquatic life, including: a loss of ecosystem integrity and services, biodiversity, fisheries, groundwater recharge, and water quality; sedimentation, erosion and channel changes; invasion by non-native species, disappearance of sensitive species; and more frequent/intense flooding. As such, changes in the natural flow content of a river can also affect the standards of life of downstream communities, both environmental and human. The livelihoods and health of people dependent on certain environmental conditions should be taken into account in transboundary agreements, as stated in the UNECE Water Convention.

Recommendations by Kent (2012, p. 12) regarding disaster risk reduction and rights-based planning might also improve the current legal issues surrounding the human right to water. Kent writes that:

> Rights-based planning for disaster risk reduction would clarify rights, obligations, and accountability mechanisms at every level of governance. Since disasters sometimes mean

> that local jurisdictions cannot cope, there should be clear rules regarding the provision of assistance from outside, alongside self-help from within the jurisdiction. There are some promises that governments might not be willing to make, with regard to their own people, and with regard to assistance to others outside their jurisdiction. Nevertheless, the clear articulation of at least some rights and obligations, together with effective accountability mechanisms, would be likely to improve the performance of disaster risk reduction programs.

Applying this to water law, both national and international, it is clear that governments may choose not to endorse the water rights of certain people within their own country or that of neighbouring countries. Governments may also choose to favour certain types of economic development linked to greater revenues and development across the country overall, rather than certain local farming practices. However, the few water capabilities identified as vital human needs are to be respected not only by a state itself, but also by all states in the river basin. This may require expanding the understanding of 'vital human needs' to be akin to the view of the WHO preamble, including not merely capabilities for pure survival, but those necessary for a decent human life. As stated in the human right to water and sanitation (Resolution 64/292), the right to water is inextricably related to the right to life and human dignity. Which water capabilities are comprised by human dignity are best described by the people themselves.

Conclusions

This paper has touched upon current and future global water issues, and has evaluated available legal documents covering water as a human right regarding water capabilities and vulnerable people. It has explored debates on how equitable and reasonable and vital human needs can be measured and found that the effects on vulnerable people have not yet been fully represented. By giving vulnerable people a voice within the formation of the understanding of the right to water and which water capabilities are unnegotiable in transboundary river basins, we can better ensure the right to water reaches all.

Acknowledgments

The author would especially like to thank Professor Patricia Wouters, Dr Owen McIntyre, Associate Professor Md. Nasif Ahsan, and an anonymous reviewer for their valuable comments and suggestions which improved the quality of the paper.

References

Goff, M., & Crow, B. (2014). What is water equity? The unfortunate consequences of a global focus on 'drinking water'. *Water International, 39*(2), 159–171. doi:10.1080/02508060.2014.886355

Gupta, J., Akhmouch, A., Cosgrove, W., Hurwitz, Z., Maestu, J., & Ünver, O. (2013). Policymakers' reflections on water governance issues. *Ecology and Society, 18*(1), 35. doi:10.5751/ES-05086-180135

Kent, G. (2012). *Rights-based disaster planning*. Retrieved form http://www2.hawaii.edu/~kent/RightsBased.docx

Leb, C. (2012). The right to water in a transboundary context: Emergence of seminal trends. *Water International, 37*(6), 640–653. doi:10.1080/02508060.2012.710950

McCaffrey, S. (2001). The contribution of the UN convention on the law of the non-navigational uses of international watercourses. *International Journal of Global Environmental Issues, 1*(3/4), 250–253. 10.1504/IJGENVI.2001.000980

McIntyre, O. (in press a). The UNECE water convention and the human right to water and sanitation. In A. Tanzi, O. McIntyre, A. Kolliopoulos, A. Rieu-Clarke & R. Kinna (Eds.), *The*

UNECE convention on the protection and use of transboundary watercourses and international lakes: Its contribution to international water cooperation. Leiden: Brill/Martinus Nijhof.

McIntyre, O. (in press b). The UNECE water convention and the principle of equitable and reasonable uitlisation. In A. Tanzi, O. McIntyre, A. Kolliopoulos, A. Rieu-Clarke & R. Kinna (Eds.), *The UNECE convention on the protection and use of transboundary watercourses and international lakes: Its contribution to international water cooperation*. Leiden: Brill/Martinus Nijhof.

Office of the United Nations High Commissioner for Human Rights (UNOCHCR). (1966). *International covenant on economic, social and cultural rights*. Retrieved from http://www.ohchr.org/EN/ProfessionalInterest/Pages/CESCR.aspx

Office of the United Nations High Commissioner for Human Rights (UNOCHCR). (2010). *Human rights fact sheet No. 35 the right to water*. Geneva: UN. ISSN 1014–5567, GE.10-14425. Retrieved from http://www.ohchr.org/Documents/Publications/FactSheet35en.pdf

Richter, B. D., Baumgartner, J. V., Powell, J., & Braun, D. P. (1996). A method for assessing hydrologic alteration within ecosystems. *Conservation Biology, 10*(4), 1163–1174. 10.1046/j.1523-1739.1996.10041163.x

Rieu-Clarke, A., Moynihan, R., & Magsig, B. (2012). *UN watercourses convention user's guide*. Retrieved from http://www.gwp.org/Global/Our%20Approach/Strategic%20Allies/User's%20Guide%20to%20the%20UN%20Watercourses%20Convention%20(2012).pdf

Rogers, P., & Hall, A. (2003). *Effective water governance*. Global Water Partnership Technical Committee, Background Paper no.7.

Tremblay, H. (2010). *A clash of paradigms in the water sector? Tensions and synergies between integrated water resources management and the human rights- based approach to development*. Retrieved from http://dx.doi.org/10.2139/ssrn.1661146

Tremblay, H. (2013). The implications of human needs for human rights-based water allocation: Review of the human right to water. *McGill International Journal of Sustainable Development Law and Policy, 8*(2), 261–271.

United Nations. (1948). *The universal declaration of human rights*. Retrieved from http://www.un.org/en/documents/udhr/index.shtml

United Nations. (1997). *UN General Assembly Resolution A/RES/51/229*. Retrieved from http://www.un.org/ga/search/view_doc.asp?symbol=A/51/PV.99

United Nations. (2005). *Convention on the law of the non-navigational uses of international watercourses 1997*. United Nations. Retrieved from http://legal.un.org/ilc/texts/instruments/english/conventions/8_3_1997.pdf

United Nations. (2010). *64/292. The human right to water and sanitation*. Retrieved from http://www.un.org/es/comun/docs/?symbol=A/RES/64/292&lang=E

United Nations. (2013a). *The millennium development goals report*. New York, NY: UN. ISBN 978-92-1-101284-2. Retrieved from http://www.un.org/millenniumgoals/pdf/report-2013/mdg-report-2013-english.pdf

United Nations. (2013b). *A new global partnership: Eradicate poverty and transform economies through sustainable development*. New York, NY: United Nations Publications. Retrieved from http://www.post2015hlp.org/wp-content/uploads/2013/05/UN-Report.pdf

United Nations Committee on Economic, Social and Cultural Rights. (2003). *General Comment No. 15: The Right to Water (Arts. 11 and 12 of the Covenant)*. Retrieved from http://www.refworld.org/docid/4538838d11.html

United Nations Economic Commission for Europe. (1992). *Convention on the protection and use of transboundary watercourses and international lakes (Water Convention)*. Retrieved from http://www.unece.org/fileadmin/DAM/env/documents/2013/wat/ECE_MP.WAT_41.pdf

United Nations Economic Commission for Europe. (2013). *Guide to implementing the water convention*. Retrieved from http://www.unece.org/fileadmin/DAM/env/water/publications/WAT_Guide_to_implementing_the_WC/ECE_MP.WAT_39_Guide_to_implementing_water_convention_small_size_ENG.pdf

United Nations Habitat. (1996). *The habitat agenda*. Retrieved from http://ww2.unhabitat.org/declarations/ch-4b-4.htm

United Nations Human Rights. (2010). *Human rights council decision 2/104 - human rights and access to water*. Retrieved from http://www.ohchr.org/EN/Issues/WaterAndSanitation/StudyWater/Pages/UsefulLinks.aspx

Vink, K., Takeuchi, K., & Kibler, K. M. (2014). A quantitative estimate of vulnerable populations and evaluation of flood evacuation policy. *Journal of Disaster Research*. Forthcoming.

Vitens. (2010). Hoeveel water verbruiken we per dag? [How much water do we use daily?] *Information in Dutch*. Vitens. Retrieved from https://www.vitens.nl/vragen/Paginas/Wetenswaardigheden-Hoeveel-water-verbruiken-we-per-dag.aspx

Vörösmarty, C. J., McIntyre, P. B., Gessner, M. O., Dudgeon, D., Prusevich, A., Green, P., ... Davies, P. M. (2010). Global threats to human water security and river biodiversity. *Nature, 467*, 555–561. doi:10.1038/nature09440

Walters, V. 2014. Urban homelessness and the right to water and sanitation: Experiences from India's cities. *Water Policy, 16*(4), 755–772. doi:10.2166/wp.2014.164

World Health Organization. (1948). Preamble to the Constitution of the World Health Organization as adopted by the International Health Conference. New York, 19–22 June, 1946.

World Health Organization. (2011). *The human right to water and sanitation: Media brief.* Retrieved from http://www.un.org/waterforlifedecade/pdf/human_right_to_water_and_sanitation_media_brief.pdf

World Meteorological Organization. (1992). *The dublin statement on water and sustainable development.* Retrieved from https://www.wmo.int/pages/prog/hwrp/documents/english/icwedece.html

Wouters, P. (2014). The Yin and Yang of international water law: China's transboundary water practice and the changing contours of state sovereignty. *Reciel, 23*(1), ISSN 2050–0386. doi:10.1111/reel.12069

Zwarteveen, M. Z., & Boelens, R. (2014). Defining, researching and struggling for water justice: Some conceptual building blocks for research and action. *Water International, 39*(2), 143–158. doi:10.1080/02508060.2014.891168

The governance of wetland ecosystems and the promotion of transboundary water cooperation – opportunities presented by the Ramsar Convention

Jing Lee

Based on a consideration of the interrelationships between cooperation, institutions and governance, this article explores the governance of transboundary wetlands, or wetlands within transboundary basins, and the potential role of the Convention on Wetlands of International Importance, especially as Waterfowl Habitat (Ramsar Convention) in the promotion of transboundary water cooperation.

Introduction

Inland waters, such as rivers and wetlands, are the most threatened of all biomes. It is estimated that, since 1990, half of the world's wetlands have been lost, and more than 20% of the world's 10,000 known freshwater species have become extinct, threatened or endangered (Blumenfeld, Lu, Christophersen, & Coates, 2009, p. 8; Kingsford, 2011, p. 217). This has given rise to tensions between water users within and between nations, as they compete for dwindling water resources in a changing climate. A former United Nations Secretary-General has stated, in this sense, that 'fierce national competition over water resources has prompted fears that water issues contain the seeds of violent conflict' (UNDESA, 2014).

Wetlands and water are hydrologically connected in the sense that wetlands regulate the water regime, and yet they are dependent on the waters situated within a wider river basin and coastal and marine systems to which wetlands are connected (Salathe, 2004; Ramsar Convention, Annex C, Resolution IX.1, COP 9, 2005a; Zsuffa et al., 2010, p. 2). The hydrological linkages between water and wetlands are even more important in transboundary basins, where the sustainable management of wetlands depends on the cooperation between riparian states. According to a World Conservation Monitoring Centre (WCMC) Report, among the 955 Ramsar sites included in the study, almost one-third are located within catchment basins shared between two or more countries (Groombridge, 1999, para. 5; Verschuuren, 2008, p. 51).

In such cases, the close ecological linkages between wetlands and water should inform transboundary cooperation and the governance of wetlands, as a basis for dispute prevention, trust-building and joint natural resources management across borders. The Ramsar Convention (1971) and the institutions established there under could play an important role in this respect, including through its science-based technical and policy tools, adopted to guide parties in the management of shared water resources.

The paper starts by outlining the interrelations between governance, institutions and cooperation prior to the analysis of the framework of wetland governance under the Ramsar Convention. Next, the paper illustrates the opportunities presented by such a framework for promoting transboundary water cooperation and wetland conservation. In this context, this paper focuses on the governance of transboundary wetlands, or wetlands within transboundary basins, and the potential role of the Convention on Wetlands of International Importance, especially as Waterfowl Habitat (Ramsar Convention) in this regard.

Governance, institutions and cooperation

The concept of governance pertains to decision-making actors and processes. It refers to the articulation of formal and informal rules by which human behaviour is governed and the means by which society 'determines and acts on goals and priorities', in an institutionalized legal and policy framework (Steiner, Kimball, & Scanlon, 2003, p. 227). The 'power, relationships, responsibility, and accountability' aspects of governance encompass 'the sums of the many ways individuals and institutions, public and private, manage their common affairs', which emphasize the dynamics within the institution (Borrini-Feyerabend, Johnston, & Pansky, 2006, p. 116; Graham, 2003, pp. 2–3; Warning, 2009, p. 4). Underlying the dynamic processes of governance are principles of good governance, such as strategic vision, responsiveness of institutions and processes, legitimacy, transparency and accountability, equity and fairness, and the observation of the rule of law that guide and steer the means (how power is exercised) and the ends (the results of power) of decision-making (adopted and adapted from Davidson, Lockwood, Curtis, Stratford, & Griffith, 2006; Graham, 2003).

The increasingly complex, diverse and uncertain nature of environmental management due to non-linearity or threshold effects of human and natural intervention in ecosystems, coupled with the change in social dynamics, call for a more collaborative, inclusive and adaptive approach to structuring governance at multiple levels. The advancement of a new worldview of the planet as a unitary whole, with the global water cycle interconnecting the territories of all states (Natarajan & Khoday, 2014, p. 577), is congruent with the establishment of supranational governance structures. Such structures are necessary to facilitate coordination and cooperation over the management of natural systems that do not follow political borders, and thus make all nations interdependent (Esty, 2006, p. 1497; Leb, 2013, p. 40).

A prominent feature of governance at the supranational level, especially environmental governance, is the establishment and operation of social institutions charged with: dispute settlement; normative, technical, and standard-setting and development in support of norm implementation; policy coordination; information sharing; and multilateral stakeholder participation. Such institutions have proven effective in enabling access to information and justice, strengthening cooperation, providing technical and financial assistance, facilitating the involvement of actors, managing conflict and, more generally 'alleviating collective-action problems in a world of interdependent actors'. As such, institutions are 'collective identities and cognitions' structured through the codification of formal rules and norms that are capable of furthering the expansive development of rules through regular interactions (Boisson De Chazournes, 2013; Leb, 2013, p. 28; Wendt, 1992, p. 399; Young, 1994, p. 15). Hence, it is firmly established that cooperative arrangements, which include principles and institutions 'encapsulated' in a permanent regime governed by treaty, are necessary to prevent, mitigate, administer and solve problems arising from the use of transboundary natural resources (Caponera, 1985, pp. 563, 572–587).

In this sense, Brierly calls for 'the creation of institutions that will enable the manifold functions of government, with whatever adaptions are necessary [. . .] to be performed internationally' (International Bureau of the Permanent Court of Arbitration, 2003, p. 25), including in relation to the good governance of transboundary natural resources, instead of being largely left to voluntarist devices of international law. In other words, the reciprocity and stability of international relations structured through an institutional setting enhance the process of cooperation (International Bureau of the Permanent Court of Arbitration, 2003; Desai, 2010, p. 13). Cooperation is facilitated through mutual availability of information and closer compliance in a formalized platform, whereby the willingness of states to exercise restraint is assured by the anticipation of similar concessions by other states (Leb, 2013, pp. 29–31). Institutions and cooperation are mutually reinforcing in the sense that an institution provides the platform for interactions among its members; and these interactions develop normative and concordant practices that may crystallize into international law.

In the context of today's global environmental governance, treaty-based international institutions consist of 'a heterogeneous set of actors, which are linked to each other through a variety of cooperative arrangements' (Hey, 2008, pp. 751–752). Such arrangements include conferences or meetings of the parties, secretariats and subsidiary bodies overseeing scientific and technical cooperation or compliance. They have gone beyond the state-centric paradigm of the positivist doctrine of international law by creating formal platforms for states to interact with one another and with non-state actors regularly, in good faith (Desai, 2010, p. 12; Redgwell, 2002, pp. 90–91). These interactions under the umbrella of environmental treaty regimes serve to foster the development of common understandings that strengthen legal and political commitments to a given environmental issue through the organization and the coordination of the 'constellation of interests upon which substantive compromise will eventually be founded' (Brunnée, 2005, p. 104; Gehring, 2008, p. 471).

The prevalent role of multilateral environmental agreements (MEAs) and the mushrooming of environmental treaty regimes in this area reflect a recognition of institutions as appropriate frameworks to facilitate and catalyse cooperation through environmental governance. Hence, the analysis of the governance of wetlands ecosystems, institutionalized under the Ramsar Convention, is called for in exploring the potential of the treaty regime in the promotion of transboundary water cooperation, especially when wetlands and water are inextricably linked.

Governance of wetland ecosystems under the Ramsar Convention

Substantive and procedural obligations

The Ramsar Convention gives 'wetlands' an all-encompassing definition that recognizes the interconnectedness between wetlands and water. Therefore, the convention applies to areas of marsh, fen, peatland or water, whether natural or artificial, permanent or temporary, with water that is static or flowing, fresh, brackish or salt, including areas of marine water the depth of which at low tide does not exceed 6 metres (Article 2, Ramsar Convention). The Ramsar Convention is founded upon the overarching twin pillars of conservation and wise use, which are to be implemented through local and national actions and international cooperation. The concepts of conservation and wise use are located within different domains where conservation entails a more cautious management (i.e., preservation) of Listed wetlands whereas wise use allows the 'sustainable utilisation for the benefit of humankind in a way compatible with the maintenance of the natural properties of the ecosystems' (Farrier & Tucker, 2000, p. 24). The convention recognizes that the loss of wetlands would be irreparable, and combines far-sighted national policies with coordinated international actions to secure healthy wetland ecosystems (Preamble, Ramsar Convention).

In view of the importance for the convention's effectiveness, international cooperation is in itself considered its third pillar. Under the convention, 'international' wetlands include those situated wholly within the boundary of one state but which form part of the migratory route of seasonal waterfowls that are ecologically dependent on wetlands, as well as the wetlands that transcend state boundaries and frontiers (Preamble and Article 1, Ramsar Convention). It is noted that Article 5(1) of the Ramsar Convention imposes the obligation to consult with contracting parties over the implementation of commitments arising from the convention on 'a wetland extending over the territories of more than one Contracting Party *or where a water system is shared by Contracting Parties*' (added emphasis). A careful reading of the provision supported the assertion that a domestic wetland situated within an international river basin could be considered 'international' by virtue of sharing the water system within that particular international river basin.

As per the convention's core implementation mechanism, contracting parties are required to designate wetlands of international significance within their territories to be included in the List of Wetlands of International Importance (Preamble and Article 2, Ramsar Convention; Strategic Framework and Guidelines for the Future Development of the List of Wetlands of International Importance, Annex 2, Resolution XI.8, COP 11, 2012c). Beyond the requirement of designation, the conservation of Listed and non-Listed wetlands takes place through the establishment of nature reserves (Article 4(1), Ramsar Convention; Resolution VIII.14, COP 8, 2002, p. 5, para. 2 and p. 21, para. 69). In this regard, designation to the list is not strictly required, as long as their wise use is promoted through the establishment of such reserves.

The Listed wetlands and nature reserves must be sustainably managed, as per an adequate planning process aimed at ensuring wetland conservation and wise use (Article 3 (1), Ramsar Convention). The continuous process of management planning shall incorporate monitoring of the ecological character of Listed and non-Listed wetlands. Moreover, contracting parties are to inform the Secretariat at the earliest possible time of any changes or likelihood of changes to the ecological character of these wetlands due to technological developments, pollution or other human interference (Article 3(2), Ramsar Convention).

Contracting parties are otherwise encouraged to conduct research, and to exchange data and publications on wetlands and their flora and fauna (Article 4(3), Ramsar Convention). The promotion of the conservation and the adequate wardening of wetlands also encompass efforts to increase waterfowl populations on appropriate wetlands, and to develop capacity in the fields of research, management and wardening (Article 4(4) and (5), Ramsar Convention).

With respect to international wetlands, their management shall be based on regular consultations between contracting parties regarding the convention's implementation, particularly to coordinate and support present and future policies and regulations concerning the conservation of wetlands (Article 5(1), Ramsar Convention).

This mix of substantive and procedural requirements – including designation, monitoring and reporting, research, exchange of information, consultations, and capacity building – facilitated by the institutional mechanisms under the convention, provide a firm foundation on which cooperation on the management of transboundary wetlands and shared water systems could be forged and sustained.

Institutional mechanisms

The institutional mechanisms established under the Ramsar Convention aim to facilitate the implementation of the substantive and procedural obligations examined above, towards achieving the conservation and wise use of wetlands. Such mechanisms include, *inter alia*, decision-making bodies, advisory bodies and national focal points.

Decision-making bodies

The decision-making capacity with regard to the collective-action problem of wetlands encroachment, which is to be addressed by the Ramsar treaty regime, lies with the Conference of the Parties (COP) and the Standing Committee. The latter is an intersessional executive body that executes the functions of the COP in between ordinary COP meetings, based on the decisions adopted by the former. In its turn, the COP has an all-encompassing mandate to review and promote the implementation of the convention (Ramsar Convention, Article 6, Resolution 3.3, COP 3, 1987, para. 1).

An important feature of the Ramsar COP, prevalent in the decision-making bodies of recent environmental treaties, is the active participation of non-state actors ahead of, during and in the follow-up to its meetings, as well as in the working sessions and activities of the COP's subsidiary bodies (Article 7, Ramsar Convention; Rules 6 and 7, Rules of Procedure, Ramsar Convention, COP 11, 2012d). Although the power to make decisions rests exclusively upon states, stakeholders in the Ramsar process have played a crucial role in informing, guiding and influencing the decision-making process.

This pluralization of participation in the supreme body of the treaty, coupled with the expansion in scope and increasing complexity of environmental problems contribute towards a shift in the regulatory nature of international environmental law. The fundamental structures of international law and politics, influenced by participatory, consensus-based COP law-making (Wiersema, 2009, p. 237), are gradually moving towards a 'decline in the importance of the sovereign state and the state system and an accompanying rise of governance by a dynamic global civil society'. This process has contributed to enhancing international environmental cooperation (Raustiala, 1997, pp. 539, 586).

As numerous experts have observed, the participation of non-state actors in the multifarious processes of international environmental law implementation, particularly in the compliance and enforcement mechanisms of environmental treaties, has resulted in the creation of transnational solidarities and the convergence of consensus over certain values. This outcome cumulatively changes the modes by which the 'legal/illegal distinction is perceived and applied by relevant actors in contemporary international society' (Bianchi, 2011, pp. 51, 52; Simma & Paulus, 1998, p. 274).

Advisory bodies

Compliance with the various aforementioned obligations involved in the management of wetlands depends on the availability of a solid scientific foundation as a basis for the convention's effectiveness. The specific references in the convention's text to the exchange of data, information and publications, as well as to capacity building highlight the importance placed on communication, education and awareness-raising where participation by all relevant stakeholders is deemed crucial to achieve the goals of the convention (Articles 4(3) and 6(2), Ramsar Convention).

In 1993, therefore, the Ramsar contracting parties decided to establish the Scientific and Technical Review Panel (STRP) to meet the rising demand for authoritative scientific and technical advice in the convention's implementation. In this context, the STRP's mandate is to provide related guidance to the Standing Committee, the Ramsar Secretariat and the COP (Resolution 5.5, COP 5, 1993c; STRP Modus Operandi 2013–2015, Resolution IX.11, COP 9, 2005b; Resolution XI.18, COP 11, 2012b). The STRP is assisted by a network of National Focal Points (NFPs), appointed by the respective contracting parties (Appendix, Terms of Reference for the STRP National Focal Points, Resolution X.9, COP 10, 2008c, p. 10, para. 2).

In addition, the Tenth COP in Changwon adopted the Programme on Communication, Education, Participation, and Awareness (CEPA) 2009–2015, which supersedes the 1999 and 2003 CEPA Resolutions. The CEPA is monitored by the CEPA Oversight Panel, and provides relevant guidance to the contracting parties, the convention's bodies and stakeholders. This guidance includes the CEPA programmes or action plans on wetland areas of particular interest, which may have international, regional, national or local applicability. The CEPA National Focal Points form part of a global expert network that shares information, promotes the dissemination of resource materials, and supports the development or expansion of programmes for individual, group and community participation in wetland and water resource management (Resolution X.8, COP 10, 2008b, p. 2; Annex, Resolution X.8, COP 10, 2008b, p. 5, para. 3; Annex, Resolution X.29, COP 10, 2008a, p. 4; Resolution VI.19, COP 6, 1996b; Resolution IX.18, COP 9, 2005c, paras 1, 4, 5; Resolution VII.9, COP 7, 1996b, paras 7, 22.c).

The Ramsar Secretariat

The continuing bureau duties shall be performed by the Secretariat, which answers to the Standing Committee and is hosted by the International Union for Conservation of Nature (IUCN) (Article 8, Ramsar Convention). The Ramsar Secretariat is headed by a Secretary General who carries out the day-to-day supervision and coordination of the convention's activities (Resolution XI.1, 2012a, p. 2, paras 12, 14; Articles 6 and 10*bis*(3), Ramsar Convention).

Ramsar Advisory Mission

The Ramsar Advisory Mission is a technical assistance mechanism adopted to support contracting parties in the management and conservation of Listed sites where the ecological character of these sites are threatened (Recommendation 4.7, COP 4, 1990). According to the convention, such sites are to be included in the Montreux Record and the threats against them, properly addressed (Resolution 5.4, COP 5, 1993b; Resolution VI.1, COP 6, 1996a).

National implementing agencies

The Administrative Authority and the National Ramsar/Wetlands Committees (NRC) are national implementing agencies appointed to enable interactions between international and national levels throughout the convention's implementation process.

An Administrative Authority acts as the Ramsar Secretariat's main contact point for communication with the respective contracting party. It is responsible for the national implementation of the convention. The authority is also expected to encourage synergies and cooperation with other relevant conventions, and foster the harmonization of the information management systems there under (Resolution X.29, COP 10, 2008a, p. 1, paras 1–3, p. 2, paras 8–9, p. 4).

The NRCs serve as the permanent national focal points for matters and programmes relating to the convention (Annex, Resolution X.29, COP 10, 2008a, p. 4), and are in charge of coordinating implementation at the national level (Recommendation 5.7, COP 5,1993a). The NRCs play an instrumental role in fostering international cooperation, with both governmental and non-governmental institutions, within and beyond the convention's structure, represented among their members. As such, the NRCs serve as national fora for addressing disputed wetland issues and matters pertaining to the conservation and wise use of wetlands. Their mandate encompasses not only Ramsar sites, but also unlisted wetlands of national importance, including any relevant nature reserves (Annex, Resolution X.29, COP 10, 2008a; Bonells & Zavagli, 2011, p. 1, paras 3–4, 2008, p. 4).

To summarize briefly the flow of processes involved, the Ramsar contracting parties adopt decisions on the implementation of the convention at their periodical meetings and implement such decisions under the supervision of the Standing Committee during the inter-sessional periods. The STRP assists contracting parties on scientific and technical matters, and the CEPA Oversight Panel provides guidance for supporting awareness-raising and capacity-building efforts. The Ramsar Advisory Mission works with contracting parties to safeguard, through a comprehensive Monitoring Procedure, the ecological character of Listed Ramsar sites included in the Montreux Record. The Ramsar Secretariat supports the triennial implementation cycles, assisted by the Administrative Authority that acts as the focal point for communications at the national level. Implementation of the Ramsar Convention, under the oversight of the Administrative Authority, is facilitated by the four NFPs and the NRCs.

Together, these institutional mechanisms facilitate implementation and international cooperation towards the conservation and wise use of wetlands. They do so by serving as platforms where interactions between contracting parties, and between them and multiple stakeholders, progressively lead to a common understanding of wetland matters, as a basis for compromise in the spirit of cooperation.

Ramsar Convention, water and cooperation: a theoretical framework

Water has been a source of tension 'since the Neolithic revolution when humans settled down to cultivate food between 8,000 and 6,000 BCE' (Kramer & Wolf, 2013, p. 1). However, the pervasive pessimistic view that water is always a trigger of conflict, perpetrated by 'sensationalist media pieces and UN Secretary Generals', does not seem to be supported by evidence, which shows more frequent instances of cooperation than of disputes (Mithen, 2013; Zeitoun & Mirumachi, 2008). The shift 'from confrontation to cooperation, from monologue to dialogue and from dissent to consensus' is due to an inclination of states, situated within an increasingly multilateralized international order, to integrate competing demands and the interests of stakeholders, based on political accommodation and a proactive stance to avoid conflict (UN-Water, 2006, p. 388).

In this regard, the Ramsar framework of governance could be a window of opportunity for the integration of competing demands and interests of stakeholders in the alleviation of conflict where the convention represents the consensus of the international community, as reflected in the quasi-universal participation of 168 countries in the meetings of the COP. The vision of the Ramsar Convention is to 'develop and maintain an international network of wetlands which are important for the conservation of global biological diversity and for sustaining human life' (Annex 2, Resolution XI.8, COP 11, 2012c, p. 7).

In order to realize the vision for the Ramsar List of Wetlands of Importance, the contracting parties had committed to work cooperatively with its international organisation partners, local stakeholders and the Ramsar Secretariat towards accomplishing five objectives, the most relevant to cooperation being Objective 3 that aims to foster cooperation in the selection, designation and management of Ramsar Sites (Objective 3, Annex 2, Resolution XI.8, COP 11, 2012c, p. 9).

The objectives agreed upon in COP 11 strengthen the obligation to cooperate and coordinate over the implementation of obligations arising from the Ramsar Convention 'in the case of a wetland extending over the territories of more than one Contracting Party or where a water system is shared by Contracting Parties' (Article 5, Ramsar Convention). Article 5 is to be interpreted broadly, with reference to all obligations arising from the convention (Handbook 20, Ramsar Convention Secretariat, 2010, p. 8, para. 3(a), (b) and (c)). This obligation is paramount in the development and maintenance of an international network of wetlands for the conservation of global biodiversity, in addition to sustaining human life (Annex 2, Resolution XI.8, COP 11, 2012c, p. 7).

Moreover, the commitment of contracting parties to establish national networks of Ramsar Sites, representatives of every natural or near-natural wetland type presented in a bio-geographical region, or critical to the survival of biodiversity, requires cooperation between contracting parties. The explicit requirement to integrate wetland site management within a broad-scale environmental management plan, including river basin and coastal zone management, reflects the interconnectedness between water and wetlands within a broader river basin. That requirement presupposes cooperation amongst contracting parties with wetlands located within transboundary river basins.

Thus, international cooperation is fostered and facilitated through the vehicle of a global network of Ramsar Sites that traverses national and supranational/regional boundaries. Cooperation here is fostered not only among contracting parties and relevant stakeholders, but also with other environmental treaty regimes. Such an expansive view of cooperation supports the vision to conserve global biodiversity, and to contribute towards human health, livelihoods and well-being by safeguarding essential water-related ecological services/benefits (Objectives 4 and 5, Annex 2, Resolution XI.8, COP 11, 2012c, pp. 9–10).

Transboundary water cooperation could be found in the imposition of the duty to cooperate amongst international watercourses states under the UN Convention on the Non-Navigational Uses of International Watercourses (UNWC, 1997) where Article 8(1) states that 'Watercourse States shall cooperate on the basis of sovereign equality, territorial integrity, mutual benefit and good faith in order to attain optimal utilization and adequate protection of an international watercourse.' Similarly, the emphasis on equality and reciprocity in the UNECE Convention on the Protection and Use of International Watercourses and International Lakes (UNECE Convention, 1992) enshrined the spirit of cooperation where Article 2(6) provides that 'Riparian Parties shall cooperate on the basis of equality and reciprocity, in particular through bilateral and multilateral agreements in order to develop harmonized policies, programmes and strategies covering relevant catchment areas or part thereof.' Hence, the explicit reference to the Ramsar Vision to 'develop and maintain an international network of wetlands' should also guide the governance of international water resources, and could contribute to the effective application of the basic norms of international water law within international wetlands. Such norms include equitable and reasonable use and participation, and its related corollaries of harm prevention and the duty to cooperate (Objective 3, Annex 2, Resolution XI.8, COP 11, 2012c; Caponera, 1985, p. 566, 2012, p. 9).

The governance architecture provided under the Ramsar Convention framework enhances the substantive duty to cooperate whereas the principle of equitable and reasonable use and participation under international water law, in its turn, could contribute to advancing the vision of the Ramsar Convention and the promotion there under of international cooperation on the establishment and management of an international network of protected wetlands. The application of the principles of systemic integration and mutual supportiveness in the interpretation of the duty to cooperate as provided under the Ramsar Convention, the UNWC, and also the UNECE Convention, strengthens the normativity of the duty to cooperate over transboundary water resources (Broude, 2008, p. 206; Lee, 2014, p. 9; Pavoni, 2010, pp. 678–679).

Thus, the Ramsar Convention can be a catalyst for cooperation based on environmental management goals and the sharing of benefits associated with healthy ecosystems. Cooperation on such issues can contribute to building trust and alleviating tensions between riparian states, laying down the path for cooperation on more controversial matters. An example of how a shared conservation initiative may facilitate trust-building through cooperation over the management of transboundary wetlands under the Ramsar framework is the termination of a 400-year-old dispute over alluvial islands at the confluence of the Amur and Ussuri rivers and a small island on the Argun River in December 2008 (Griffin, 2012, p. 27).

Such a role for the convention has been recognized in the concrete application of international law in the *San Juan River case*, where tensions between Costa Rica and Nicaragua escalated over the dredging and construction of a canal in 2010 and the commencement of two new artificial *caños* in 2013 by the latter. The judgment of the International Court of Justice (2011) alluded to the potential for the Ramsar Convention to foster cooperation between riparian states on the conservation and wise use of water and wetlands (Article 1, Ramsar Convention).

Crystallizing the emergence of an ecosystem approach in international law (McIntyre, 2014, p. 92), the court ruled that the 'very close environmental connection' between the wetland in the disputed area and the waters of the Lagoon requires cooperation between Costa Rica and Nicaragua, together with the Ramsar Secretariat, to guard the wetland against the risks of irreparable environmental damage. The court underscored that the requirement for such cooperation would be even stronger in relation to those water bodies listed as Wetlands of International Importance (Declaration of Judge Greenwood, ICJ, 2011, p. 51, para. 15). Judge Greenwood was of the opinion that the disputing parties should devise and implement a set of protective measures, taking into account both the Ramsar Convention and the guidelines on cooperation adopted there under, which suggested a bigger role to be played by the Ramsar Convention Secretariat (Declaration of Judge Greenwood, ICJ, 2011, p. 51, para. 15).

Although the court did not specifically spelt out the particular guidelines on cooperation adopted under the Ramsar Convention, it could be deduced that the court referred to the Volume 20 *Handbook on International Cooperation: Guidelines and Other Support for International Cooperation under the Ramsar Convention on Wetlands* (Ramsar Convention Secretariat, 2010). This deduction could be supported by the aim of the Ramsar Strategic Plan 2009–2015, adjusted for the 2013–15 triennium by Resolution XI.3 (2012e) to enhance the conservation and wise use of wetlands using effective international cooperation through the active application of the Guidelines for International Cooperation under the Ramsar Convention (Goal 3, The Ramsar Strategic Plan 2009–2012, Resolution XI.3, COP 11, adjusted 2012, pp. 6, 13, 14).

In the Order made on 22 November 2013 subsequent to Costa Rica's request for the indication of new provisional measures in particularly to order the cessation of work on two new *caños*, the court reiterated that the two new *caños* are situated within the disputed territory in the Humedal Caribe Noreste wetland where Costa Rica bears obligations under the Ramsar Convention. According to the decision, the court ordered that Costa Rica may, with prior notice given to Nicaragua, take any necessary measures to prevent irreparable prejudice to the environment of the disputed area, so as to avoid adverse effects on the San Juan River *following consultation* with the Secretariat of the Ramsar Convention (added emphasis) (ICJ, 2013, p. 13, para. 54 and p. 15). The 2013 Order reiterated the suggestion of Judge Greenwood in his Declaration in 2010 to provide Ramsar Convention with a bigger role in facilitating cooperation between Costa Rica and Nicaragua, in recognition of the inseparability of the disputed area with the lagoon.

Ramsar Convention and the promotion of transboundary water cooperation in China

Along the lines drawn in the above ICJ decision, the Ramsar Convention could play a particularly valuable role in relation to China. A party to that convention, China controls the headwaters of at least 10 major rivers, sharing 18 transboundary basins with neighbouring states. All those basins are situated within a region riven with political tension over issues that threaten water security, such as water scarcity and stress, and armed with an escalating programme of dam construction. The region seems to possess all the key ingredients for the creation of what John Beddington, England's Chief Scientific Advisor, once referred to as a 'perfect storm' of conflict (China Water Risk, 2012; Population Institute, undated, p. 2; Sample, 2009).

Griffin has examined 234 wetlands identified as transboundary sites, within a total of 91 countries, ranking them against 23 indicators of the Global Peace Index and categorizing these transboundary wetland sites into four categories of relative peacefulness (Griffin, 2012, p. 24). In particular, she highlights the role of the Ramsar Convention in providing a diplomatic opportunity for cooperation on the joint management of water resources and wetlands between China and its neighbouring countries with which the former shares transboundary Ramsar sites (Griffin & Ali, 2014, p. 235). The transboundary Ramsar sites shared between Russia and China, for example, are under Category 4, indicating a lack of trust between the countries which may or may not be expressed in violent conflict (Griffin, 2012, pp. 26, 27, 56, 57).

Multiple environmental issues, including desertification and water shortages, which have plagued both countries, especially China, exacerbate that lack of trust. Regarding those sites, therefore, China and Russia should work under the umbrella of the Ramsar

Convention's institutional framework to manage viable transboundary conservation areas jointly, with the aim of reinforcing peaceful relations and environmental sustainability (Griffin, 2012, p. 58). The joint management of viable transboundary conservation areas could be greatly assisted by the CEPA programme that was set up to strengthen knowledge empowerment in addressing issues relating to distrust and lack of capacity. The lack of trust and capacity could also be supplemented by scientific and technical competencies developed under the STRP that are disseminated across the range of actors at the level of implementation through the institutional framework of the Ramsar Convention.

Apart from that, China also shares a transboundary Ramsar site with the Kyrgyz Republic: Chatyr Kul Lake in the Tien Shan Mountains. That site falls under Categories 3 (economic challenges/poverty/need for capacity-building) and 4 (lack of trust), and could thus benefit from enhanced cooperation within the framework of the Ramsar Convention. Even the highly militarized territorial dispute between China, India and Pakistan over the Siachen glacier in Kashmir could benefit from the Ramsar Convention. The convention's objective third-party platform and its reliance on sound scientific and technical inputs would enable those countries to identify mutually agreeable and beneficial solutions for the conservation and wise use of wetlands.

Conclusions

> Cooperation is the bedrock of international law, prescribing limits on absolute state sovereignty in a myriad of ways.
>
> (Wouters, 2013, p. 88)

The fear of a 'perfect storm', brewing over the threat of water insecurity, exemplifies the co-existing potential of water as a seed of conflict or a catalyst for cooperation. A legal platform for interactions to facilitate the development of normative practices that could crystallize into international law, and ensure compliance with the norms created, provides the foundation for effective cooperation. The duty to cooperate has emerged as a global imperative. This duty has been increasingly promoted in the institutionalized setting of MEAs. In the case of the Ramsar Convention, the inseparability of wetlands and water provides a strong basis for transboundary water cooperation there under. The Ramsar Convention incorporates an overarching duty on states to consult with one another regarding transboundary wetlands, including *water systems shared* by contracting parties, in a quest to enhance cooperation over international waters (added emphasis).

The marked alteration of the face of contemporary international society, characterized by the 'breakthrough of "supra-nationalism", the progress made in terms of institutionalisation, not to say integration and "globalisation" of international society', has witnessed the gradual replacement of the traditional international law of co-existence with the international law of cooperation (Declaration of President Bedjaoui, ICJ, 1996, p. 270, para. 13). In the same paragraph, Judge Bedjaoui declared that:

> The resolutely positivist, voluntarist approach of international law [...] has been replaced by an objective conception of international law, a law more readily seeking to reflect a collective juridical conscience and respond to the social necessities of States organised as a community.

The institutionalization of the governance of the common heritage of mankind, which could include wetlands, freshwater biodiversity and ecosystems (UN System Task Team on the Post-2015 UN Development Agenda, 2013, p. 5), reflects the 'collective juridical conscience' of the duty to cooperate over transboundary natural resources. That institutionalization responds to the 'social necessities of States organised as a community' to cooperate in the joint-management of these transboundary natural resources.

In this regard, the institutional framework presented by the Ramsar Convention for the governance of wetlands enables and promotes cooperation, confidence building, and ultimately peace-making between states sharing transboundary basins through the collaborative management of transboundary wetland ecosystems. The willingness of contracting parties sharing a transboundary Ramsar site to initiate formal collaboration is evidenced by the 16 Transboundary Ramsar Sites that have been formally established, pursuant to Article 5 of the Ramsar Convention and Resolution VII.19 (1999) on International Cooperation (Griffin & Ali, 2014, p. 231; Resolution VII.19, COP 7, 1999a).

Similar cooperative action is foreseeable in the promotion of transboundary water cooperation between China and its neighbouring countries. The transboundary Ramsar sites shared between China, Russia, Kyrgyz Republic, India, Pakistan and others, shrouded in varying degree of tension and sustainable development challenges, could be better managed through the cooperative framework institutionalized under the Ramsar treaty-regime.

Hence, the science/ecosystem-based governance architecture available under the Ramsar treaty-regime provides the cooperative agenda, the means, the ends and the appropriate forum where trust could be fostered and cooperation over transboundary waters be promoted.

Acknowledgements

This article is based on a presentation given at the 2014 Transboundary Water Cooperation Symposium and Workshop, 'Promoting Transboundary Water Cooperation in Asia', Xiamen, China, May 2014, jointly sponsored by Xiamen Law School, China and Universiti Kebangsaan Malaysia (UKM) [GGPM-2013-083]. The author wishes to thank the organizers of the symposium and workshop for the opportunity to publish in this volume and also the anonymous reviewers for their thoughtful comments made on an earlier draft of the paper.

References

Bianchi, A. (2011). The fight for inclusion: Non-state actors and international law. In U. Fastenrath, R. Geiger, D. Khan, A. Paulus, S. Von Schorlemer, & C. Vedder (Eds.), *From bilateralism to community interest: Essays in honour of Judge Bruno Simma* (pp. 39–57). Oxford: Oxford University Press.

Blumenfeld, S., Lu, C., Christophersen, T., & Coates, D. (2009). *Water, wetlands and forests. A review of ecological, economic and policy linkages* (CBD Technical Series No. 47). Montreal and Gland: Secretariat of the Convention on Biological Diversity and Secretariat of the Ramsar Convention on Wetlands.

Boisson De Chazournes, L. (2013). *Fresh water in international law.* Oxford: Oxford University Press.

Bonells, M., & Zavagli, M. (2011). *National Ramsar/Wetlands Committees across the six Ramsar regions: Diversity and benefits.* Gland: Ramsar Convention Secretariat.

Borrini-Feyerabend, G., Johnston, J., & Pansky, D. (2006). Governance of protected areas. In M. Lockwood, G. Worboys, & A. Kothari (Eds.), *Managing protected areas: A global guide* (pp. 116–145). London: Earthscan.

Broude, T. (2008). Principles of normative integration and the allocation of international authority: The WTO, the Vienna Convention on the Law of Treaties, and the Rio Declaration. *Loyola University Chicago International Law Review, 6*(1), 173–208.

Brunnée, J. (2005). Reweaving the fabric of international law? Patterns of consent in environmental framework agreements. In R. Wolfrum & V. Röben (Eds.), *Developments of international law in treaty making.* Berlin: Springer.

Caponera, D. A. (1985). Patterns of cooperation in international water law: Principles and institutions. *Natural Resources Journal*, *25*, 563–587.
China Water Risk (2012, February 9). Geopolitical risks: Transboundary rivers. Retrieved September 12, 2014, from http://chinawaterrisk.org/resources/analysis-reviews/geopolitical-risks-transboundary-rivers/
Davidson, J., Lockwood, M., Curtis, A., Stratford, E., & Griffith, R. (2006). *Governance principles for regional natural resource management* (Report No. 1 of the Project "Pathways to Good Practice in Regional NRM Governance"). Tasmania: School of Geography and Environmental Studies, University of Tasmania.
Desai, B. H. (2010). *Multilateral environmental agreements. Legal status of the secretariats*. Cambridge: Cambridge University Press.
Esty, D. C. (2006). Good governance at the supranational scale: Globalizing administrative law. *The Yale Law Journal*, *115*(7), 1490–1562. doi:10.2307/20455663
Farrier, D., & Tucker, L. (2000). Wise use of wetlands under the Ramsar Convention: A challenge for meaningful implementation of international law. *Journal of Environmental Law*, *12*(1), 21–42. doi:10.1093/jel/12.1.21
Gehring, T. (2008). Treaty-making and treaty evolution. In D. Bodansky, J. Brunnée, & E. Hey (Eds.), *The Oxford handbook of international environmental law*. Oxford: Oxford University Press.
Graham, J. (2003). *Governance principles for protected areas in the 21st century prepared for the fifth World Parks Congress, Durban, South Africa*. Ontario: Institute on Governance.
Griffin, P. (2012). *The Ramsar Convention: A new window for environmental diplomacy*. Burlington: Institute for Environmental Diplomacy and Security, University of Vermont.
Griffin, P., & Ali, S. H. (2014). Managing transboundary wetlands: The Ramsar Convention as a means of ecological diplomacy. *Journal of Environmental Studies and Sciences*, *4*, 230–239. doi:10.1007/s13412-014-0173-0
Groombridge, B. (1999). *Shared wetlands and river basins of the world: Preliminary findings of a GIS analysis*. World Conservation Monitoring Centre. Report prepared in support of Ramsar COP 7 Technical Session V (Ramsar COP 7 DOC.20.1).
Hey, E. (2008). International institutions. In D. Bodansky, J. Brunnée, & E. Hey (Eds.), *The Oxford handbook of international environmental law*. Oxford: Oxford University Press.
ICJ (2011). *Certain activities carried out by Nicaragua in the border area (Costa Rica v. Nicaragua)*. (Judgment of 8 March 2011) ICJ Report 6. ("*San Juan River*" case).
ICJ (2013). *Certain activities carried out by Nicaragua in the border area (Costa Rica v Nicaragua), construction of a road in Costa Rica along the San Juan River (Nicaragua v Costa Rica)* (Judgment of 22 November 2013).
International Bureau of the Permanent Court of Arbitration (Ed.). (2003). *Resolution of international water disputes*. The Hague: Kluwer Law International.
International Court of Justice (ICJ) (1996). *Legality of the threat or use of nuclear weapons* (Advisory Opinion of 8 July 1996).
Kingsford, R. (2011). Conservation management of rivers and wetlands under climate change – A synthesis. *Marine and Freshwater Research*, *62*, 217–222. doi:10.1071/MF11029
Kramer, A., & Wolf, A. T. (2013). The key to managing conflict and cooperation over water. *World of Science*, *11*(1), 4–12.
Leb, C. (2013). *Cooperation in the law of transboundary water resources*. Cambridge: Cambridge University Press.
Lee, J. (2014). *Preservation of ecosystems of international watercourses and the integration of relevant rules: An interpretative mechanism to address the fragmentation of international law*. Leiden, Boston: Brill/Nijhoff.
McIntyre, O. (2014). The protection of freshwater ecosystems revisited: Towards a common understanding of the 'ecosystems approach' to the protection of transboundary water resources. *Review of European, Comparative & International Environmental Law*, *23*(1), 88–95. doi:10.1111/reel.12068
Mithen, S. (2013). *Thirst: Water and power in the ancient world*. Phoenix: London.
Natarajan, U., & Khoday, K. (2014). Locating nature: Making and unmaking international law. *Leiden Journal of International Law*, *27*(3), 573–593. doi:10.1017/S0922156514000211

Pavoni, R. (2010). Mutual supportiveness as a principle of interpretation and law-making: A watershed for the 'WTO-and-Competing-Regimes' debate? *The European Journal of International Law, 21*(3), 649–679. doi:10.1093/ejil/chq046

Population Institute (undated). 2030: The 'perfect storm' scenario. Retrieved September 12, 2014, from http://www.populationinstitute.org/external/files/reports/The_Perfect_Storm_Scenario_for_2030.pdf

Ramsar Convention (1971). *Convention on wetlands of international importance especially as waterfowl habitat* (adopted 2 Feb 1971, entered into force 21 Dec 1975) *United Nations Treaty Series*, 996, 245.

Ramsar Convention (1987). *Resolution 3.3: Establishment of a standing committee*. Regina: COP 3.

Ramsar Convention (1990). *Recommendation 4.7: Mechanisms for improved application of the Ramsar Convention*. Montreux: COP 4.

Ramsar Convention (1993a). *Recommendation 5.7: National committees*. Kushiro: COP 5.

Ramsar Convention (1993b). *Resolution 5.4: The record of Ramsar sites where changes in ecological character have occurred, are occurring, or are likely to occur (Montreux Record)*. Kushiro: COP 5.

Ramsar Convention (1993c). *Resolution 5.5 Establishment of a scientific and technical review panel*. Kushiro: COP 5.

Ramsar Convention (1996a). *Resolution VI.1: Working definitions of ecological character, guidelines for describing and maintaining the ecological character of listed sites, and guidelines for operation of the Montreux Record*. Brisbane: COP 6.

Ramsar Convention (1996b). *Resolution VI.19 Education and public awareness*. Brisbane: COP 6.

Ramsar Convention (1999a). *Resolution VII.19 Guidelines for international cooperation under the Ramsar Convention*. San José: COP 7.

Ramsar Convention (1999b). *Resolution VII.9: The Convention's Outreach Programme 1999–2002: Actions to promote communication, education, and public awareness to support implementation of the Convention on Wetlands (Ramsar, Iran, 1971)*. San José: COP 7.

Ramsar Convention (2002). *Resolution VIII.14: New guidelines for management planning for Ramsar sites and other wetlands*. Valencia: COP 8.

Ramsar Convention (2005a). *Resolution IX.1 Annex A: A conceptual framework for the wise use of wetlands and the maintenance of their ecological character*. Kampala: COP 9.

Ramsar Convention (2005b). *Resolution IX.11 Revised modus operandi of the Scientific and Technical Review Panel (STRP)*. Kampala: COP 9.

Ramsar Convention (2005c). *Resolution IX.18 establishment of an oversight panel for the CEPA activities of the convention*. Kampala: COP 9.

Ramsar Convention (2008a). *Resolution X.29: Clarifying the functions of agencies and related bodies implementing the convention at the national level*. Changwon: COP 10.

Ramsar Convention (2008b). *Resolution X.8: The Convention's Programme on Communication, Education, Participation, and Awareness (CEPA) 2009–2015*. Changwon: COP 10.

Ramsar Convention (2008c). *Resolution X.9: Refinements to the modus operandi of the Scientific and Technical Review Panel (STRP)*. Changwon: COP 10.

Ramsar Convention (2012a). *Resolution XI.1: Institutional hosting of the Ramsar Secretariat*. Bucharest: COP 11.

Ramsar Convention (2012b). *Resolution XI.18: Adjustments to the modus operandi of the Scientific and Technical Review Panel (STRP) for the 2013–2015 triennium*. Bucharest: COP 11.

Ramsar Convention (2012c). *Resolution XI.8 Annex 2: Strategic framework and guidelines for the future development of the list of wetlands of international importance of the convention on wetlands (Ramsar, Iran, 1971) – 2012 Revision*. Bucharest: COP 11.

Ramsar Convention (2012d). *Rules of procedure for meetings of the conference of the contracting parties to the Convention on Wetlands of International Importance especially as Waterfowl Habitat (Ramsar, Iran, 1971)*. Bucharest: COP 11.

Ramsar Convention (2012e). *Resolution XI.3 Adjustment to the strategic plan 2009–2015 for the 2013–2015 triennium*. Bucharest: COP 11.

Ramsar Convention Secretariat (2010). *International cooperation: Guidelines and other support for international cooperation under the Ramsar Convention on Wetlands*. Gland: Ramsar Convention Secretariat. Vol. 20 Ramsar Handbook for the Wise Use of Wetlands.

Raustiala, K. (1997). The 'participatory revolution' in international environmental law. *Harvard Environmental Law Review, 21*, 537–586.

Redgwell, C. (2002). Multilateral environmental treaty-making. In V. Gowlland-Debbas (Ed.), *Multilateral treaty-making*. The Hague: Kluwer Law International.

Salathe, T., 2004. Transboundary river basin management. ("Integrated Management of Natural Resources in the Transboundary Dniester River Basin", 16–17 September 2004, Chisinau, Republic of Moldova). Gland: Secretariat of Ramsar Convention.

Sample, I. (2009, March 18). *World faces 'perfect storm' of problems by 2030, chief scientist to warn*. The Guardian.

Simma, B., & Paulus, A. L. (1998). The 'international community': Facing the challenge of globalization. *The European Journal of International Law*, *9*, 266–277. doi:10.1093/ejil/9.2.266

Steiner, A., Kimball, L. A., & Scanlon, J. (2003). Global governance for the environment and the role of Multilateral Environmental Agreements in conservation. *Oryx*, *37*(2), 227–237. doi:10.1017/S0030605303000401

UN System Task Team on the Post-2015 UN Development Agenda (2013). Global governance and governance of the global commons in the global partnership for development beyond 2015. Thematic Think Piece.

UNECE Convention (1992). *UNECE Convention on the Protection and Use of International Watercourses and International Lakes* (done at Helsinki, 17 March 1992). *International Legal Materials*, 31, 1312.

United Nations Department of Economic and Social Affairs (UNDESA) (2014). International decade for action 'Water for Life' 2005–2015. Retrieved October 25, 2014, from http://www.un.org/waterforlifedecade/transboundary_waters.shtml

UN-Water (2006). Water a shared responsibility. UNESCO-WWAP. UN World Water Development Report 2, World Water Assessment Programme.

UNWC (1997). *United Nations Convention on the Law of Non-Navigational Uses of International Watercourses* (adopted 21 May 1997, entered into force 17 August 2014). *International Legal Materials*, 36,700.

Verschuuren, J. (2008). The case of transboundary wetlands under the Ramsar Convention: Keep the lawyers out! *Colorado Journal of International Environmental Law and Policy*, *19*(1), 49–127.

Warning, M. J. (2009). *Transnational public governance: Networks, law and legitimacy.* Basingstoke: Palgrave Macmillan.

Wendt, A. (1992). Anarchy is what states make of it: The social construction of power politics. *International Organization*, *46*(2), 391–425. doi:10.1017/S0020818300027764

Wiersema, A. (2009). The new international law-makers? Conferences of the parties to multilateral environmental agreements. *Michigan Journal of International Law*, *31*, 231–287.

Wouters, P. (2013). 'Dynamic cooperation' in international law and the shadow of state sovereignty in the context of transboundary waters. *Environmental Liability*, *3*, 88–97.

Young, O. R. (1994). *International governance: Protecting the environment in a stateless society.* Ithaca, NY: Cornell University.

Zeitoun, M., & Mirumachi, N. (2008) Transboundary water interaction I: Reconsidering conflict and cooperation. *International Environmental Agreements: Politics, law and Economics*, published online 3 September 2008. doi:10.1007/s10784-008-9083-5

Zsuffa, I., D'Haeyer, T., Douven, W., Schwartz, K., Liersch, S., Debels, P., Cools, J., & Van Der Craats, I. 2010. Enhancing the role of wetlands in integrated water resources management for Twinned River Basins in EU, Africa and South America in support of EU Water Initiatives. WETwin. Newsletter Issue No. 1.

The UNECE Water Convention and the development of transboundary cooperation in the Chu-Talas, Kura, Drin and Dniester River basins

Bo Libert

This article describes the support of the United Nations Economic Commission for Europe and partners for developing transboundary cooperation in four basins (Chu-Talas, Kura, Drin and Dniester Rivers) and the progress achieved. The work has been based on the application of international water law as reflected in the UNECE Water Convention and guided by the body of experience accumulated under the convention. Experiences from this work are described and conclusions drawn.

Background

Support for the development of transboundary water cooperation in individual basins is part of the programme of work of the United Nations Economic Commission for Europe (UNECE) Convention on the Protection and Use of Transboundary Watercourses and International Lakes (Water Convention) (1992). At the request of riparians the UNECE Secretariat has been invited to facilitate transboundary water cooperation in several basins. This article documents the resulting processes and outcomes in four transboundary basins in different parts of the UNECE region: Central Asia, Eastern Europe, the South Caucasus and the Western Balkans. In each case the author has been involved in the work since its initiation. From the perspective of a neutral actor, this article critically reviews the experiences and draws conclusions with regard to important qualitative factors that have influenced the development of transboundary water cooperation.

Each basin is experiencing healthy political relations amongst riparians and a principal willingness to cooperate. This does not mean that all sectors of (for example) an upstream country view the cooperation and consultation mechanisms positively. In many cases there is a widely distributed unwillingness to share information or to consult with stakeholders inside and outside riparian countries.

The description of the respective processes in this article is restricted to the development of formalized basin-wide cooperation. Under the Water Convention work programme, several additional pilot projects are underway in some of these basins, for example on climate change adaptation or analyzing the water–energy–food nexus, which also contributes to the development of transboundary cooperation (UNECE, 2013).

Whilst this article focuses on the UNECE contribution to the development of cooperation in these basins and the link to international legislation, the partners referred to, such as the Organization for Security and Co-operation in Europe (OSCE) in the Kura, Chu-Talas and Dniester cases and Global Water Partnership Mediterranean (GWP-Med) in the case of the Drin, have played key roles in the respective cases.

The UNECE Water Convention and its programme of work

The Water Convention was negotiated in the early 1990s by countries in the UNECE region to strengthen the ecologically sound management and protection of transboundary surface and groundwaters. It is a framework convention that obliges its parties to prevent, control and reduce transboundary impacts, to use transboundary waters in a reasonable and equitable way and to ensure their sustainable management. The convention prescribes that parties bordering the same transboundary waters cooperate by entering into specific agreements and through the establishment of joint bodies. The convention includes provisions on monitoring, research and development, consultations, warning and alarm systems, mutual assistance, and exchange of information, as well as access to information by the public. The origin, development and legal regime of the convention is well described by Wouters and Vinogradov (2003).

Initially negotiated as a regional instrument, and with 40 parties at present, the convention was amended in 2003 to allow accession by all United Nations member states. The amendments entered into force 6 February 2013, turning the convention into a global legal framework (Trombitcaia & Köppel, 2014). It is expected that countries outside the UNECE region will be able to join the convention as of 2015.

In previous programmes of work of the convention, as well as in the latest three-year programme, approved by the Meeting of the Parties in 2012 (UNECE, 2013), one central component is to assist parties and non-parties, at their request, to improve cooperation on their transboundary waters and to ratify and implement the convention and its amendments. This entails legal and technical assistance, using the principles of the convention as well as guidance documents developed under the convention. These activities encompass drawing up, revising and/or adapting agreements on transboundary waters, strengthening national and transboundary water institutions (joint bodies) and enhancing the capacity of countries and joint bodies in integrated river basin management.

The four basins

Chu and Talas Rivers

> The Chu-Talas Basins, which are shared by Kazakhstan and Kyrgyzstan, include the basins of three transboundary rivers: the Chu, the Talas and the Assa. Most of the run-off of the Chu, Talas and Assa forms in Kyrgyzstan and the flow of the three rivers is regulated. (UNECE, 2011a)

More than two million people, in Kazakhstan and Kyrgyzstan, live in the territory of the Chu and Talas river basins, with a large part of the population engaged in agriculture. Most of the agricultural production is dependent on irrigation, with the water resources of the Chu and Talas basins being of critical importance for 800,000 ha of irrigated land. The availability of water resources in the two basins for irrigation, for drinking and sanitation and for the stability of aquatic ecosystems is limited and may decrease further (Figure 1).

In 2002 UNECE and the UN Economic and Social Commission for Asia and the Pacific (UNESCAP) were approached by representatives of the Water Department in Kyrgyzstan and the Water Committee in Kazakhstan. As a response to the challenge of

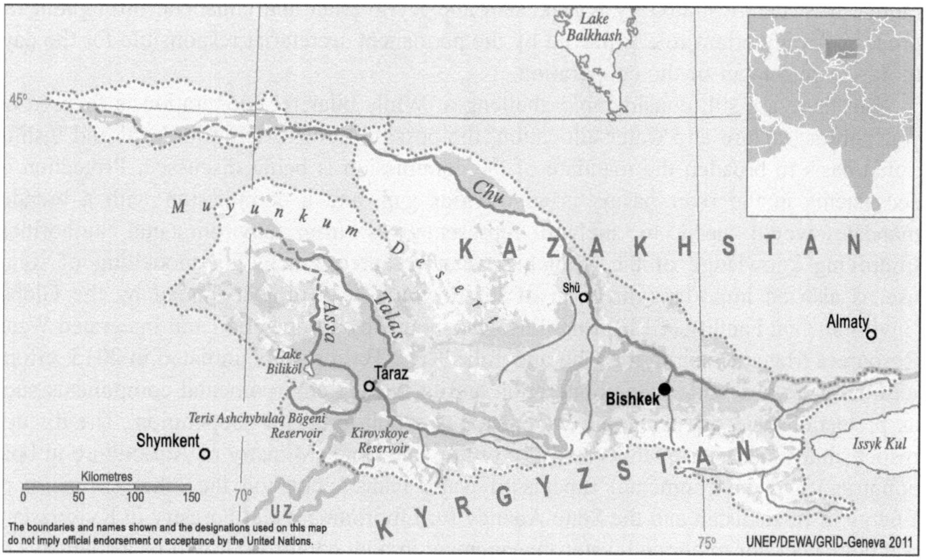

Figure 1. The Chu, Talas and Assa basins (UNECE, 2011a).

managing jointly used water infrastructure and allocation of shared water resources between Kazakhstan and Kyrgyzstan, the countries had signed an intergovernmental agreement in January 2000 on the use of water management infrastructure on the rivers Chu and Talas (Agreement between the Government of the Republic of Kazakhstan and the Government of the Kyrgyz Republic on the Use of Interstate Water Management Installations on the Rivers Chu and Talas) (Grishin, Bernardini, Enderlein, & Libert, 2006). However, there was a need to set up a bilateral commission for the implementation of the agreement, and the countries requested support from the two UN bodies.

The project, Support for the Creation of a Commission on the Chu and Talas Rivers between Kazakhstan and Kyrgyzstan, supported by the two regional UN commissions and the OSCE, aimed to establish the commission and define its procedures and costs for exploitation and maintenance of the water management infrastructure. The project was implemented successfully, and a permanent bilateral commission was established in July 2006 (Libert & Lipponen, 2012). The commission has since been active, and 17 commission meetings had taken place by summer 2014.

An established Donor Consultative Group has been working closely with the commission to ensure that available funding is used as appropriately and efficiently as possible. A number of projects funded by the EU, the Asian Development Bank and Switzerland have contributed to the dynamic development of the Kazakh-Kyrgyz water cooperation. A follow-up project implemented with the support of the OSCE and UNECE resulted in the broadening of the bilateral cooperation, an improved understanding by the two countries on the available water resources, improved access to information, and a draft concept for a transboundary basin council for the involvement of stakeholders (UNECE, 2011b). UNECE, OSCE and the United Nations Development Programme are also implementing a pilot project analyzing possible effects of climate change and the adaptation to climate change that may be needed (UNECE, n.d.).

Several aspects of the Chu-Talas cooperation represent good examples for the whole Central Asian region, but what can be stressed first and foremost is the good will and mutual trust demonstrated by Kazakhstan and Kyrgyzstan and enhanced throughout the process. An important role is played by the permanent secretariat responsible for the day-to-day management of the cooperation.

But there are still considerable challenges. While bilateral cooperation is focused on water infrastructure and water allocation, the possibility of revising its legal and institutional basis to broaden the mandate of the commission is being discussed. Protection of ecosystems in the river basins is an obvious gap, and a commission with a broader mandate would need to include representatives from environmental authorities. Improving knowledge of the available water resources, including modelling of water use, is also an important direction of future work. In a project funded by the Global Environmental Facility (GEF), Enabling Transboundary Cooperation and Integrated Water Resources Management in the Chu and Talas River Basins, to be initiated in 2015, efforts will be made to broaden the cooperation and to include environmental components such as protection of ecosystems and water quality within bilateral cooperation. The divided responsibility of water quantity and use on the one hand (Ministry of Agriculture in both countries) and environmental aspects of water management on the other (Ministry of Energy in Kazakhstan and the State Agency for Environment and Forestry in Kyrgyzstan) is a bottleneck for integrated water management in both countries. It will be a challenge to establish functional cooperation between water and environmental authorities in the two countries in the implementation of the new GEF project.

Kura

> The basin of the river Kura is shared by Armenia, Azerbaijan, Georgia, the Islamic Republic of Iran and Turkey. The 1,515 km long river has its source in Turkey on the north slope of the Allahuekber Mountains Range at the height of 3,068 m a.s.l., and discharges to the Caspian Sea. The basin has a pronounced mountainous and highland character in Turkey, with an elevation between 1,300–3,068 m a.s.l. (UNECE, 2011a)

The Kura-Aras river basin is the most important transboundary water system in the South Caucasus. For much of the twentieth century, bilateral treaties regulated water cooperation between the USSR and Turkey (1927) and between the USSR and Iran (1926) (Grishin et al., 2006). The bilateral agreements were focused on 'boundary waters' (waters that form the border between two countries) and were limited in scope. Nonetheless, these countries agreed to share waters equally and established joint commissions tasked with managing common resources. Presently the Kura is one of the most significant watercourses in the UNECE region *without* a transboundary water cooperation agreement. The development of transboundary water cooperation in the Kura-Aras Basin has been an objective of several international projects and initiatives since the dissolution of the Soviet Union. However, due to the currently hostile relations between Armenia and Azerbaijan and also a reluctance of Turkey to be more actively engaged in transboundary cooperation, it has not been possible to formalize basin-wide water cooperation (Figure 2).

While a multilateral, basin-wide treaty appears unrealistic at present, opportunities exist to move towards a cooperative management framework through bilateral agreements, first and foremost between Azerbaijan and Georgia on a significant part – almost 50% – of the Kura Basin.

Azerbaijan and Georgia rely on the water in the Kura Basin as a key source of drinking water, as well as for sanitation, municipal and industrial water use, hydropower and irrigated agriculture, among other uses. Increased variation in and reduction of water

Figure 2. The Kura-Aras basin (UNECE, 2011a).

flow, deterioration of water quality, ecosystem degradation, flooding and bank erosion are some of the issues that might be alleviated through cooperation between the two riparian states.

After initial consultations with the OSCE and UNECE, the ministers of environment of Azerbaijan and Georgia in 2007 requested support in developing bilateral cooperation on the Kura River. The OSCE and UNECE developed and identified funding for the Environment and Security Initiative (ENVSEC) project, Implementation of the UNECE Water Convention and Development of an Agreement on the Management of Transboundary Watercourses Shared by Georgia and Azerbaijan. The objective of this project was to support Georgia in the ratification and implementation of the UNECE Water Convention and to strengthen transboundary water cooperation between Azerbaijan and Georgia. To achieve the latter objective a process was initiated to draft and negotiate a bilateral agreement between the two countries.

With the facilitation of the OSCE and UNECE, bilateral negotiations were organized in the framework of the project. The latest meeting was held in January 2014, and the resulting draft agreement includes very few remaining unresolved issues. It is expected that the draft will be the basis of final negotiations between the two countries.

The draft agreement under development aims to provide a framework for cooperation between the two countries on water pollution prevention and control, conservation of biodiversity, emergency situations, data and information sharing and public participation. The agreement also envisages the establishment of a joint commission for the protection and rational use of water resources of the Kura River basin.

While the process of negotiations has developed well and a text is close to approval, a few important ministries in Georgia have concerns regarding the conclusion of such an agreement in view of plans to increase irrigation as well as to build new hydropower stations upstream. The draft agreement would not stop such developments, but it would demand sharing of information as well as consultations in advance of development projects.

The OSCE and UNECE are continuing their dialogue with the Georgian authorities to pave the ground for the conclusion of the agreement.

Drin

> The Drin River starts at the confluence of its two headwaters, the transboundary Black Drin and White Drin Rivers at Kukës in Albania. The interconnected hydrological system of the Drin River basin comprises the transboundary sub-basins of the Black Drin, White Drin, and Buna/Bojana (outflow of Skadar/Shkoder Lake in the Adriatic Sea) Rivers, and the sub-basins of Prespa, Ohrid and Skadar/Shkoder Lakes. Albania, Greece, the former Yugoslav Republic of Macedonia, Kosovo (UN administered territory under UN Security Council Resolution 1244) and Montenegro share the Drin Basin. (UNECE, 2011a)

With its three major lakes and several connecting rivers and groundwater flows, the Drin Basin has a complex hydrological structure. Parts of the basin are economically developed, while others are virtually undisturbed. The water resources are important for irrigation and hydropower production, and there are plans for the construction of new dams. Fishing is an important source of income around the three lakes. Tourism is significant, particularly at Lake Ohrid, and there are plans for its development in other parts of the basin. A significant problem is frequent and severe floods, as well as protection of ecosystems. Mining is still important in the Albanian part of the Drin Basin and, along with other industries and sanitation, it contributes to pollution.

Transboundary agreements on water resources and ecosystem management cooperation have been concluded for each of the three international lakes of the Drin system. In the Lake Ohrid basin, the GEF Lake Ohrid Conservation Programme (1998–2004) and the political commitment of the riparians resulted in the signing by Albania and the former Yugoslav Republic of Macedonia of the Agreement for the Protection and Sustainable Development of Lake Ohrid and its Watershed in 2004. The Lake Ohrid Watershed Committee was established in November 2005. The Agreement between Montenegro and Albania for the Protection and Sustainable Development of the Skadar/Shkodra Lake and its Watershed was signed in 2008, and a GEF-supported project on integrated ecosystem management was initiated the same year. The Skadar/Shkodra Lake Commission was established in 2009. Lake Prespa was the first shared lake in the region to be declared a transboundary protected area by the riparians, and an informal multi-stakeholder joint body has facilitated cooperation. Between 2006 and 2011 GEF funded a project on ecosystem management. An Agreement on the Protection and Sustainable Development of the Prespa Park Area was signed in 2010, but the corresponding joint body has not yet been set up. These developments on the three lakes are very positive, but until recently there was no framework for cooperation at the basin level.

The idea of basin-wide cooperation in the Drin Basin was initially raised during the International Roundtable on Integrated Management of Shared Lake Basins in South-Eastern Europe, organized by GWP-Med and partners in Ohrid, Macedonia, in 2006 (more details on the Drin River basin as well as the process described below are given by Scoullos, Faloutsos, & Libert, 2014).

Based on this idea and the expression of interest by stakeholders, a Consultation Meeting on Integrated Management of the Extended Drin River Basin was organized by the Albanian Ministry of Environment, Forestry and Water Administration, UNECE and

GWP-Med in Tirana in November 2008. The meeting gave a mandate to the partners in the Petersberg Phase II/Athens Declaration Process and UNECE to facilitate the establishment of a shared vision for the coordinated management of the basin.

The Drin Dialogue, a multi-stakeholder process for the establishment of a shared vision, was formally launched a year later in Podgorica. The meeting brought together representatives of the riparians, as well as the existing formal joint lake basin management commissions and committees. It was then decided that UNECE and GWP-Med would facilitate the implementation of the Drin Dialogue.

The Drin Core Group was established by the Podgorica meeting as a body to facilitate communication and cooperation among the riparians and coordination of the implementation of the Drin Dialogue.

With the support of GWP-Med and UNECE, the Drin Dialogue ran from 1 May 2010 to 30 November 2011, when the five Drin River riparians signed a Memorandum of Understanding for the Management of the Extended Transboundary Drin River Basin. The decision had been made to develop the MoU as a first step to be signed by the main water-responsible agencies, rather than to aim for an intergovernmental agreement. Because negotiations and ratification of intergovernmental agreements frequently take several years, this alternative approach made it possible to productively use the window of opportunity of political willingness amongst the five riparians to establish a sustainable basis for cooperation in spite of some political complications.

The Drin MoU includes elements commonly found in such agreements, such as the definition of an institutional structure for cooperation, and identifies short-, medium- and long-term actions to address the existing problems. This has made the application of the Drin MoU action-oriented, though it remains a weakness that only one ministry/agency per riparian is formally responsible for the cooperation. One of the long-term actions in the Drin MoU is the negotiation of a comprehensive intergovernmental agreement.

The institutional structure for the implementation of the Drin MoU was established in 2012 and includes:

- the Meeting of the Parties (the first was held in May 2013 in Tirana);
- the Drin Core Group (DCG), which meets twice a year with the mandate to coordinate actions for the implementation of the MoU; it includes representatives of (1) the competent ministries of the riparians, (2) the (pre)existing joint structures for each of the three lakes, (3) UNECE, (4) GWP-Med, and (5) the Mediterranean Information Office for Environment, Culture and Sustainable Development, representing civil society;
- the DCG Secretariat, providing technical and administrative support to the DCG; GWP-Med serves by appointment of the parties through the MoU as the Secretariat; and
- three expert working groups – on EU Water Framework Implementation, Monitoring and Information Exchange, and Biodiversity and Ecosystem Protection.

To facilitate implementation of the Drin MoU, the Drin Action Plan was prepared as an 'evolving document' in accordance with the decisions of the Meeting of the Parties and the DCG. It is an advantage that resources for the full implementation of the Drin Action Plan activities in Albania, Former Yugoslav Republic of Macedonia and Montenegro have been secured from GEF, supporting the project Enabling Transboundary Cooperation and Integrated Water Resources Management in the Extended Drin River Basin, scheduled to start in the spring of 2015.

The main efforts made until 2014 focused on the establishment of a formal cooperation framework. This has been somewhat of a challenge, with as many as five riparians, the specific situation with Kosovo (a UN-administered territory under UN Security Council Resolution 1244), and the complex political relations between Greece and the former Yugoslav Republic of Macedonia. For these reasons the Drin Dialogue was designed to be more formal and structured than the process in the other three basins. The challenge is now to develop efficient substantive cooperation on the basis of the formal structures established. It is expected that the EU approximation in four of the riparians will be an important driver for further processes.

Dniester

> The basin of the 1,362-km long river Dniester is commonly considered shared by Ukraine and the Republic of Moldova, as the share of Poland is very small. The river has its source in the Ukrainian Carpathians, and discharges into the Black Sea. Major transboundary tributaries include the Kuchurhan and the Yahorlyk. The basin is mountainous in the upper part, and lowlands prevail in the lower part. Valuable wetland systems extend along the Dniester Estuary, including some 100 wetland lakes (10–15 of the lakes are major). They play a vital role in maintaining the water balance and supporting the basin's biological diversity. (UNECE, 2011a)

In addition to supplying drinking water for a large part of the basin, including the major city of Odessa in Ukraine, hydropower generation and fisheries are other important sectors the river serves. In expeditions to review the situation in 2011, water quality problems in the lower course of the Dniester and its estuary were noted, as well as alarming declines in fish populations (Moldovan-Ukrainian Joint Field Research of Fish Fauna in the Lower Dniester Basin – 2011, 2012; Moldovan-Ukrainian Joint Hydrochemical Expedition in the Dniester River in 2011, 2012) (Figure 3).

The breakaway region of Transnistria on the eastern side of the Dniester in Moldova gives a specific political context to the transboundary cooperation.

A bilateral agreement between Ukraine and Moldova (both countries are parties to the Water Convention) was signed in 1994 (Agreement between the Government of Ukraine and the Government of Moldova on Joint Boundary Waters Management and Protection) (Grishin et al., 2006), and a meeting of plenipotentiaries was instituted as a cooperative mechanism. The agreement and its institutional mechanism have been functional, but the resulting cooperation has been restricted to a narrow set of objectives mainly related to water flow and release. It has long been in need of revision and modernization taking into account the principles of integrated water resource management (IWRM).

At two joint sessions of the Moldovan and Ukrainian parliamentary committees, held in 1997 and 1999, decisions were made to support the elaboration and signing of a bilateral convention on the Dniester. The Moldovan Environment Ministry submitted in May 2000 a draft Convention on the Use of Water and Biological Resources and Conservation of Biodiversity of the Dniester River to its Ukrainian counterpart (Trombitcaia, 2002).

This was the background of the Dniester Process that was initiated on the request of the two countries to the OSCE, the United Nations Environment Programme and UNECE. Three projects – Dniester I, II and III – have, as part of the Environment and Security Initiative, supported the development of the transboundary cooperation in the Dniester River basin.

Dniester I, Transboundary Cooperation and Sustainable Management of the Dniester River (2004–2006), assisted the Republic of Moldova and Ukraine in the development of a diagnostic study of the transboundary issues in the basin.

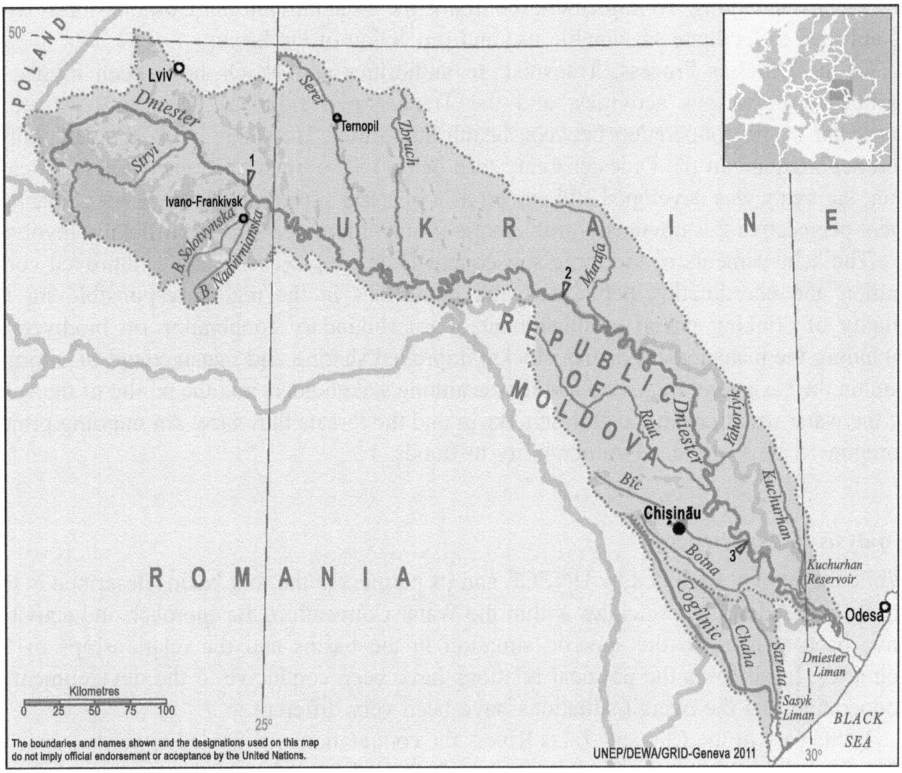

Figure 3. The Dniester basin (UNECE, 2011a).

Dniester II, Action Programme to Improve Transboundary Cooperation and Sustainable Management of the Dniester River (2006–2007), focused on the development of a programme to improve transboundary cooperation.

Dniester III, Transboundary Cooperation and Sustainable Management in the Dniester River Basin, Phase III: Implementation of the Action Programme (2009–2011), assisted (among other things) the two governments in the development and negotiations of the new Dniester treaty.

On 29 November 2012 the minister of environment of the Republic of Moldova, Gheorge Salaru, and the minister of ecology and natural resources of Ukraine, Eduard Stavytskyi, signed the bilateral Treaty on Cooperation on the Conservation and Sustainable Development of the Dniester River Basin. The ceremony took place at the sixth session of the Meeting of the Parties to the UNECE Water Convention, in Rome. As of autumn 2014 the treaty is not yet in force, because it has not been ratified by the Ukrainian Parliament.

The treaty is a pioneering example for the region in the post-Soviet era. It significantly broadens the existing cooperative arrangements to cover the entire river basin and all sectors important for the management and protection of the shared waters. It provides a framework for cooperation on water pollution prevention and control, water flow regulation, conservation of biodiversity and protection of the Black Sea environment. It also addresses the monitoring of data exchange, public participation and cooperation in emergency situations. To implement the treaty, the bilateral Dniester Commission is to be established to facilitate sustainable use and protection of the basin.

In the Dniester Process, Transnistrian authorities and NGOs have been invited to participate in various activities, and the Transnistrian side has been very active, in particular in the cooperation between health authorities. However, while the discussions and negotiations on the Dniester treaty (see below) were open to Transnistrian participation, the treaty was developed and signed as a bilateral agreement. From a water management perspective it is obviously problematic if only a few riparians are officially involved.

The achievements of the three subsequent Dniester projects include improved cooperation and coordination between health authorities in the region responsible for the quality of drinking water; establishment of transboundary cooperation on biodiversity, including the management of fish stocks; improved sharing and management of information on the basin; and increased awareness among stakeholders and the public of the value of the water resources in the Dniester Basin and the threats they face. An ongoing project component aims to reduce vulnerability to floods.

Analysis of experiences

While the support provided by UNECE and its partners in the four basins described in this article is based on the principles within the Water Convention, the approach and activities have been adapted to the specific situation in the basins and the relationships of the countries. In all cases the political relations have been conducive to the development of cooperation, but the overall situations have been very different.

In the case of the Chu and Talas Rivers the cooperation was guided towards a specific objective (sharing of costs for jointly used dams and canals), while in the other cases a broader IWRM approach was applied. Political ups and downs have been frequent. During the Kura Basin discussions there have been periods when individual ministries in Georgia were negative towards the anticipated cooperation. Political crises in Ukraine have had negative effects on progress, for example the procedural difficulty of getting to the stage at which the Parliament can ratify the signed Dniester treaty.

In the Chu-Talas and Dniester Basins, where frameworks for cooperation were already in place, the support has also included practical efforts such as joint monitoring and exchange of information.

On the basis of the experience of the different processes, some general conclusions can be drawn.

1. *The legal and institutional framework of the UNECE Water Convention is a good platform for the promotion of transboundary water cooperation in individual basins.*

The UNECE Water Convention is in many ways a useful framework to provide support for the development of transboundary water cooperation in individual basins. The convention provides the legal basis for efforts to develop specific agreements and establish joint institutions in transboundary basins (Article 9). Its institutional platform has a mandate to provide clear guidelines and to facilitate long-term political and financial support for such efforts. Its programme of work with regular multilateral meetings is an excellent framework for countries – both parties and non-parties – to voice their needs. An active dialogue between the Secretariat and countries with an interest in developing transboundary water cooperation takes place under the auspices of the convention. This makes it possible to establish a realistic dialogue with riparians, to be well informed on political realities and to identify windows of opportunity. The experiences of the Water Convention Secretariat give a good basis for transferring knowledge and capacity between basins. The neutral UN mandate is an important factor in countries' inviting UNECE to play a mediating role.

More generally, it is obvious that neutral partners can play an important role in the initiation and development of transboundary water cooperation. A suggestion from one of the riparians may be seen as an attempt to promote its own agenda; it is easier for a neutral voice to be proactive, to develop specific proposals that are accepted as a basis for discussion.

For UNECE as a facilitating partner, a major strength has been the authority of the Water Convention, which represents international water law. Also, in work with non-parties, principles of international water law are generally not challenged and can be applied.

It is a further advantage if outside neutral support is based on the long-term engagement of an institution rather than a time-limited initiative or project. Development of transboundary water cooperation is a political process and cannot be solely project-based. In the four cases described above, extra-budgetary project funding has been of decisive importance. However, if UNECE and other partners had not been engaged and provided support, also when direct funding has been unavailable, it is likely that progress would have been much more limited. The dialogue and support must be adapted to the political landscape, and experience shows that this landscape changes frequently. The three phased projects under the Dniester Process are a good example, where each phase built on the previous ones and could take account of the short-term political realities.

2. *Concluding basin agreements is important, but one should not aim to agree on every detail of the future cooperation.*

Negotiation of an intergovernmental transboundary water cooperation agreement is usually a long-term process, where the draft text is scrutinized and discussed in detail. While long periods for the negotiations may be good for the understanding and buy-in of the authorities involved, they do not always lead to a better legal result.

The Dniester treaty can be highlighted as an example of a modern and legally well-developed agreement. It significantly broadens the existing cooperation arrangements of 1994 to cover the entire river basin and all sectors important to the management and protection of the shared waters. It provides a framework for cooperation on water pollution prevention and control, water flow regulation, conservation of biodiversity and protection of the Black Sea environment. It also addresses the monitoring of data and information exchange, public participation and cooperation in emergency situations.

It is particularly important that transboundary water cooperation agreements define the procedures, principles and scope of cooperation. The lack of such information in the agreement of 2000 between Kazakhstan and Kyrgyzstan on the Chu and Talas Rivers has made it difficult to establish the corresponding commission. However, agreements do not need to go into further details with regard to cooperation. This is also the approach of the Water Convention, which sees the cooperation agreement and establishment of a joint body as a starting-point for longer-term cooperation. Once a platform for cooperation is established, the direct interaction between countries in the basin will guide the development of cooperation.

There are other approaches to initiating cooperation, such as establishing a process to identify 'shared benefits' to provide the logical basis for cooperation. The Chu-Talas cooperation may be seen as such an example. The specific aim of the establishment of the agreement in 2000 was the shared funding of jointly used water infrastructure. The definition of shared benefits is a frequent approach in processes supported by the World Bank (Sadoff & Grey, 2005). The GEF has a different approach, where the focus is to develop so-called transboundary diagnostic analyses (TDA) as a basis for the development of a strategic action programme (SAP). The SAP then constitutes the document to be signed as an agreement between the riparians (TDA/SAP Manual, 2013). This will be part of the future work in the basins (see below).

In the case of the Drin, the negotiation and conclusion of a memorandum of understanding between individual institutions was chosen rather than a full-fledged intergovernmental agreement. This made it possible to avoid certain political bottlenecks, as described above. The negative aspect is the narrow institutional basis of the cooperation – only the water authorities in each of the riparians are involved. But with the definition of an institutional structure for cooperation and the identification of short-, medium- and long-term actions the MoU has proven to be fully functional.

3. *Involving other international actors in the respective basins is essential.*

In the four basins, efforts were made to establish an inclusive framework with donors and international organizations for the initiated cooperative process. In all four cases positive synergies between different international actors were identified and exploited. The most obvious example is the Chu-Talas Basin cooperation, where a variety of organizations have contributed in different ways and a semi-formal Donor Consultative Group could be established. In the Dniester Process all meetings included a 'coordination session', where all actors were invited to make presentations and exchange information with partners. In the Drin Dialogue, key organizations and donors participate in the coordinating Drin Core Group, and joint funding of activities is frequent. As to the Kura process, a constructive dialogue and cooperation with a GEF project, Reducing Transboundary Degradation in the Kura Ara(k)s River Basin, has emerged.

It is obvious that the alternative, to monopolize a transboundary cooperation process, would lead to inefficient use of available resources, as well as risks of competition. Unfortunately, such cases are not infrequent.

4. *Involving NGOs and other stakeholders should be an objective from an early stage.*

Stakeholders often play a supporting role in the development of transboundary cooperation, and their alternative perspectives can help define its vision. A broad involvement of stakeholders provides better chances to identify and support national champions. The involvement of various stakeholders is also an important objective of IWRM. However, national authorities are sometimes reluctant to apply this principle in transboundary cooperation processes.

NGOs have played significant roles in the Drin and Dniester Basins. In the case of the Dniester, NGOs have provided considerable strength to the process when governments have changed, and when the general political situation in Moldova and in particular in Ukraine has been unstable. The important role of the NGO Eco-TIRAS could be highlighted. Eco-TIRAS was established as a basin umbrella organization bringing together approximately 50 NGOs from all parts of the basin, including Transnistria. With its constituents and leadership it has provided a solid support for the development of cooperation in the Dniester Basin. More generally, the establishment of a basin-wide umbrella NGO is an interesting approach in transboundary basins.

In some Central Asian countries water-related NGOs are not or cannot be so active. The reasons are rooted in the more authoritarian societal structure as well as the high political importance of water resources allocation. This is a challenge in the development of transboundary water cooperation in Central Asia.

5. *It is not expensive to support transboundary water cooperation, but co-funding from riparians is essential.*

In the initial stages of cooperation, external funding of meetings and consultants and organizational support has been an important catalyzing factor in all four basins. This support is not necessarily costly. For example, external funding for the first-phase project in the Chu and Talas Basins to establish the Chu-Talas Commission was limited to €100,000, plus OSCE, UNECE and UNESCAP staff costs of the same order of magnitude.

In later stages more significant external funding and efforts may be needed to broaden and deepen the cooperation. Further work in the four basins has been established with the GEF, which is in the position to provide more significant funding. However, to achieve sustainability, funding at an early stage from the supported countries is important. Under present economic conditions this is a challenge, and it remains a weakness in all four basins that only limited funding from the countries is available.

The involvement and participation of stakeholders as well as co-funding contribute to and are indicators of ownership of the respective processes. As a facilitating partner it is a difficult balance – to not 'take over' but still contribute to leading the process. This risk of a restricted ownership is higher when there are several riparians involved, such as in the Drin Dialogue, and lower when a permanent secretariat has been established, such as in the Chu-Talas Basin.

6. *Weak inter-agency cooperation within countries is frequently a bottleneck.*

IWRM is an important principle in the development of transboundary water cooperation, but it is frequently not applied at the national level – particularly in Central Asia, where water for irrigation is managed separately from environmental and hydropower aspects. Experience in the Chu and Talas Basins shows that it is quite challenging to involve several agencies in one country in transboundary cooperation. An interesting practical approach that will be part of the further development of cooperation in GEF projects is to set up and link inter-ministerial committees to transboundary water cooperation.

Conclusions

The mandate of the Water Convention and the regular interaction with parties and non-parties in its programme of work open up opportunities to identify windows of opportunity and develop cooperation in specific basins. This article has described the developments in four basins demonstrating the facilitation of UNECE and partners of transboundary water cooperation. Applying international water law as reflected in the Water Convention, and guided by the body of experience accumulated under the convention, progress can be demonstrated in these basins.

An important aspect of the processes described is their importance not only in initiating basin-wide cooperation but also in developing capacity on international water law and specifically on the UNECE Water Convention. UNECE works on various practical aspects of transboundary water cooperation in more than 20 countries, and it is obvious that this work contributes to a better understanding of international water law among experts and institutions in the individual countries.

A weakness of UNECE as an actor in facilitating transboundary cooperation in individual basins is its dependence on extra-budgetary funding. While relations with some donors are very good, sometimes new challenges cannot be taken on because of lack of funding. By working closely with partners this aspect is to some extent overcome, and the developing cooperation with GEF may improve this situation.

With the Water Convention becoming a global legal framework, basins outside the UNECE region may benefit from support. Support for bilateral cooperation on hydrology and environment in the Pyanj River between Afghanistan and Tajikistan is an example already in place. However, while the present geographic scope of engagement benefits from a good understanding and knowledge of the individual countries, a global responsibility, involving up to four times as many countries, is a game-changer. If given the mandate and opportunity, other organizations, including UN regional commissions (such as the Economic and Social Commission for Asia and the Pacific) could play a role in facilitating transboundary cooperation in their respective regions.

With two global conventions on transboundary water cooperation (the UNECE Convention as well as the 1997 United Nations Convention on the Law of the Non-navigational Uses of International Watercourses) entering into force in 2013 and 2014, the increasing clarity and strength of international water legislation should give momentum to the promotion of its principles and application globally as well as in individual basins. The approach demonstrated in this article may contribute to this momentum.

Acknowledgements

The author thanks Iulia Trombitcaia and Dimitris Faloutsos for their constructive comments on an earlier draft of this article. Thanks also to the UNECE Water Team, as well as project partners from other organizations that in practice and discussions have contributed to the work done and experiences collected in the four basins.

References

Grishin, N., Bernardini, F., Enderlein, R., & Libert, B. (2006). *Transboundary water cooperation in the newly independent states. Water Series No 4, Transboundary water cooperation, trends in the newly independent states* (pp. 3–33). Geneva: Economic Commission for Europe.

Libert, B., & Lipponen, A. (2012). Challenges and opportunities for transboundary water cooperation in Central Asia: Findings from UNECE's regional assessment and project work. *International Journal of Water Resources Development*, *28*(3), 565–576. doi:10.1080/07900627.2012.684527

Moldovan-Ukrainian Joint Field Research of Fish Fauna in the Lower Dniester Basin – 2011 (Russian). (2012). Retrieved from http://dniester.org/wp-content/uploads/2012/03/report_Dniester_fish_expedition-2011_21March2012_Rus_FINAL.pdf [Accessed 10 September 2014].

Moldovan-Ukrainian Joint Hydrochemical Expedition in the Dniester River in 2011 (Russian). (2012). Retrieved from http://dniester.org/wp-content/uploads/2012/03/report_Dniester_expedition-2011_FINAL_21-March-2012.pdf [Accessed 10 September 2014].

Sadoff, C., & Grey, D. (2005). Cooperation on International Rivers, a continuum for securing and sharing benefits. *Water International*, *30*(4), 420–427. doi:10.1080/02508060508691886

Scoullos, M., Faloutsos, D., & Libert, B. (2014). The Drin coordinated action: Towards an integrated transboundary water resources management. water scarcity. In F. De Châtel, G. Holst-Warhaft, & T. Steenhuis (Eds.), *Security and democracy: A Mediterranean mosaic* (pp. 62–72). Athens: Cornell University and Global Water Partnership Mediterranean.

TDA/SAP Manual. (2013). International Waters Learning Exchange & resource Network. Retrieved from http://iwlearn.net/manuals/tda-sap-methodology/tda-sap-methodology/tda-sap-manual-vol ume-1/view [Accessed 10 September 2014].

Trombitcaia, I. (2002). Transboundary cooperation of Moldova and Ukraine on the Dniester draft convention. *Journal of Environmental Law and Litigation*, *17*(1), 145–160.

Trombitcaia, I., & Köppel, S. (2014). From a regional towards a global instrument – the 2003 amendment to the UNECE Water Convention. In A. Tanzi, O. McIntyre, A. Kolliopoulos, A. Rieu-Clarke, & R. Kina (Eds), *The UNECE convention on the protection and use of transboundary watercourses and international lakes*. Leiden: Brill Academic Publishers (in press).

UNECE. (1992). Convention on the Protection and Use of Transboundary Watercourses and International Lakes. Retrieved from http://www.unece.org/fileadmin/DAM/env/water/pdf/watercon.pdf. [Accessed 12 September 2014].

UNECE. (2011a). *Second assessment of transboundary rivers, lakes and groundwaters*. Geneva: Economic Commission for Europe.

UNECE. (2011b). PROJECT REPORT: Development of cooperation on the Chu and Talas Rivers (Chu – Talas II). Retrieved from http://www.unece.org/fileadmin/DAM/env/water/Chu-Talas/ChuTalas_II_Project_Report_Short_ENG.pdf. [Accessed 15 September 2014].

UNECE. (2013). Programme of work for 2013–2015. ECE/MP.WAT/37/Add.1. Retrieved from http://www.unece.org/fileadmin/DAM/env/water/mop_6_Rome/Official_documents/ECE_MP.WAT_37_Add.1_ENG.pdf. [Accessed 12 September 2014].

UNECE. (n.d.). Chu Talas - Promoting Cooperation to Adapt to Climate Change in the Chu and Talas Transboundary Basin. Retrieved from https://www2.unece.org/ehlm/platform/display/ClimateChange/Chu+Talas. [Accessed 15 September 2014].

Wouters, P., & Vinogradov, S. (2003). Analysing the ECE Water Convention: What Lessons for the Regional Management of Transboundary Water Resources? *Yearbook of International Co-operation on Environment and Development* 2003/04, Fridtiof Nansen Institute, Oslo, Retrieved from http://www.fni.no/ybiced/03_04_wouters-vinogradov.pdf. [Accessed 15 September 2014].

Do good fences make good neighbours? Canada–United States transboundary water governance, the Boundary Waters Treaty, and twenty-first-century challenges

Emma S. Norman and Karen Bakker

This article analyzes the rescaling of transboundary water governance and explores challenges and opportunities for the twenty-first century. The analysis is grounded in the example of the Canada–United States transboundary water governance regime, and asks two questions: What are the lessons learned since Canada and the United States first signed the Boundary Waters Treaty 100 years ago? And what is the potential of rescaling to influence the tension between the 'sovereign rights' of a nation and transboundary water governance protocols based on 'good neighbourliness'?

Introduction

A fundamental tension characterizing transboundary water governance is the interplay between the sovereign rights and responsibilities of the state and incentives to consider one's upstream or downstream neighbour(s). Canada and the United States have more than 100 years of experience in negotiating these issues. The Canada–US transboundary water governance regime – notably, the 1909 Boundary Waters Treaty (BWT) and the subsequent creation of the binational International Joint Commission (IJC) – is frequently praised as a leading global example of constructive governance of transboundary waters. The regime has been of particular interest given the longevity of the BWT and the relatively high degree of cooperation between the two countries (no doubt facilitated by the fact that Canada and United States are both upstream and downstream from each other, depending on the water body under consideration).

This article analyzes key aspects of the Canada–US transboundary water governance regime, with particular attention to the recent trend of the rescaling of transboundary water governance to local (and often informal) actors, which we argue is relevant to global debates about good governance in transboundary water management. The trend of rescaling governance – in which a shift or a widening of actors in the decision-making process occurs – may at first seem surprising. Until recently, scholars of transboundary water governance focused primarily (if not exclusively) on nation-states and formal, legal water governance regimes. Historically, international water law dictates that the relationships between sovereign nations be defined through formal, nation-state-driven governance mechanisms, e.g. treaties and agreements (Furlong, 2010; McCaffrey, 2006). Traditionally, scholars have focused on these nation-state mechanisms, and have often emphasized questions of state security and environmental security (Hirsch & Jensen, 2006).

However, in the last 30 years, the Canada–US regime of transboundary water governance has undergone a shift – or 'rescaling' – downwards to the subnational and upwards to the supranational scale, in a process sometimes termed 'glocalization' (Swyngedouw, 1997, 2004). In more recent years, this rescaling has also taken a postcolonial turn, in which Indigenous–led transboundary mechanisms have been conceived to actively 'decolonize' transboundary water governance mechanisms, both in discourse and practice (Norman, 2015; Norman, Bakker, & Cook, 2012). Accordingly, scholars have begun analyzing the rescaling of environmental governance and policy harmonization (Cash et al., 2006; Lebel, Garden, & Imamura, 2005; Mirumachi, 2013; Sayre, 2005), water governance (Budds & Hinojosa, 2012; Newig & Fritsch, 2009; Norman, Cohen, & Bakker, 2013; Norman, Cook, & Cohen, in press; Perreault, 2005; Sheppard & McMaster, 2004) and, more specifically, the rescaling of *transboundary* water governance (Chen, Rieu-Clarke, & Wouters, 2013; Cohen & Harris, 2014; Cohen & McCarthy, 2014; Feitelson & Fischhendler, 2009; Furlong, 2010; Norman, 2015; Norman & Bakker, 2009; Sneddon & Fox, 2006; Zeitoun, Eid-Sabbagh, Talhami, & Dajani, 2013).

Concurrent with this rescaling, a shift from government to governance has occurred, in which nonstate actors (e.g. environmental non-governmental organizations, citizen groups, regional actors, Indigenous groups) have greater involvement in transboundary water governance (Norman & Bakker, 2009; Norman et al., 2013; United Nations, 2008) through 'soft-path' approaches (Brooks, Brandes, & Gurman, 2009; Gleick, 2002, 2003; Sandford & Phare, 2011). These actors participate in transboundary water governance through a proliferation of regional agreements and programs (provincial–state agreements, citizen coalition groups, intertribal agreements, etc.) and through engagement in supranational mechanisms such as the United Nations Economic Commission for Europe (UNECE). This 'thickening' of the governance regime parallels broader debates in water and environmental management regarding the desirability of delegating decision making to 'lower' scales of governance (state, provincial, municipal) and to nonstate actors (citizen groups, industry groups, etc.) (Bulkeley, 2005). In addition, the thickening has implication for the First Nations and Native American tribes (Indigenous Communities), who have both inherent rights to the land and water, and treaty rights through negotiated federal treaties. However, these inherent and treaty rights were not included in the original transboundary water treaties (such as the Boundary Waters Treaty and the Columbia River Treaty) – which were conceived as bi-national, rather than multi-national.

Given the global relevance of these trends, this paper analyzes the rescaling of transboundary governance in the case of Canada–United States transboundary waters. The analysis builds on earlier analyses of transboundary water cooperation, which quantified the extent and scale of rescaling, to explore changes in governance processes, as well as the challenges, successes and lessons learnt. The analysis is structured around two questions: What lessons can be learned from the trend of rescaling governance over the last 100 years? And what is the potential of rescaling to influence the tension between the 'sovereign rights' of a nation and transboundary water governance protocols based on 'good neighbourliness'?

The Boundary Waters Treaty and the duty to cooperate

When the BWT was signed, on 11 January 1910, by representatives of the US government and the Crown (on Canada's behalf), it reflected a mutual desire of the federal governments to cooperate on transboundary water issues. One hundred years later, much has changed in terms of the make-up of the actors involved in the governance and the expectations for public involvement. The pages that follow explore the widening of participation (or expected participation) in the governance process and link that to wider trends of rescaling transboundary water governance. First, though, it will be useful to provide historical context for the signing of the BWT.

The BWT represented the culmination of years of diplomatic negotiations to ensure friendly relations between countries, as well as increased public pressure to address transboundary water issues. The BWT achieved two crucial goals. It established a core of legal principles to govern the management of internationally shared waters between Canada and the United States, and it established an institutional framework to supervise the implementation of the principles. The treaty is not without its problems, but in the context of its era it represents an important achievement in international cooperation on shared water resources.

A central component of the BWT was the creation of the IJC, a governing body established to mediate conflict between the two federal governments. Given the extraordinary length of the international border, this mediation was considered both inevitable and long-term. As US secretary of state Elihu Root stated prior to signing the treaty (as quoted by Munton, 1981): "I do not anticipate that the time will ever come when this Commission will not be needed. I think that as the two countries along this tremendous boundary become more and more thickly settled the need for it will increase."

Despite the relatively widespread agreement on the need for a formal cooperation arrangement, the settlement of the treaty was delayed as government representatives debated a critical point in the negotiation: the relative strength of the IJC. A key issue was the question of whether to assign the commission "real arbitral power or merely to assign a more limited mandate to recommend courses of action, which the governments could accept or reject as they saw fit" (Carroll, 1981, p. 44). Canada advocated a 'stronger' version (the Gibbons-Clinton draft), which allowed for a commission with authority over the final decision and the creation of a supranational court to adjudicate boundary water disputes. The United States opted for a 'weaker' version (Dreisziger, 1981). Root articulated this preference in a letter to British ambassador James Bryce in 1908:

> The difficulty of the United States in assenting to an agreement that all questions within the broad field described by the Gibbons-Clinton draft shall be referred for final determination to such a commission as is proposed, is in the main that such questions necessarily involve, not merely questions of fact and of law suitable for the determination of a commission or arbitral tribunal, but many questions of policy, of mutual concession and of the give and take which is in so great a number of cases the efficient means of reaching possible settlement of difficult controversies. Such questions of policy, of concession, of discretion make it impossible for the Government of the United States to commit to any commission under our system of government. (quoted in Carroll, 1981, p. 44)

Ultimately, the version with less authoritative power prevailed.

The US position is unsurprising given that during the time of treaty negotiation, the Harmon Doctrine – which maintains that a country is sovereign over the portion of an international watercourse within its borders (McCaffrey, 1996) – was at the forefront of American thinking. US attorney general Judson Harmon, in response to Mexican protests, noted that "the fundamental principle of international law is the absolute sovereignty of every nation, as against all others, within its own Territory" (21 Op. Attorney General,

281–282, 1895). This historical negotiation is important context in the contemporary discussion of 'sovereign reach'. In essence, the tension was around where the power of adjudication lay – in the commission, or in the country itself. The fact that the power balance was seen as an 'either-or' rather than a 'both-and' is an important point to consider for contemporary transboundary water governance mechanisms.

Although the balance between sovereign rights and the need to cooperate with neighbours continues to evolve, Secretary Root's prediction held true: along the Canada–US border, the need remains for federal governments to address issues of shared concern, communicate intentions, and mediate potential as well as actual conflicts. However, some key factors have changed. First, a greater range of types and numbers of actors have become involved in these discussions. At the time of the treaty signing, the governance structure was envisaged as nation-to-nation. Although the six commissioners (three from each country) are asked to act in their professional capacity, rather than representing 'national interests', the structure of the IJC remains distinctly binational. Second, the trend towards rescaling has caused the IJC – which remained largely federally oriented until the past decade – to rethink its approach and role. Below, we deal with each of these issues in turn.

Analyzing the rescaling of transboundary water governance

Few studies have attempted to comprehensively analyze the rescaling of water governance and to use both qualitative and quantitative data to analyze the effectiveness of the changes. To address this gap, a comprehensive survey/historical review was conducted of the governance mechanisms across the Canada–US borderland over the course of 100 years (see Norman and Bakker (2009) for full results). The study found that an increasing number of nonfederal transboundary water governance mechanisms (sub-state, citizen, and regional) have emerged since the 1980s – although the majority (57%) of transboundary governance mechanisms between Canada and the United States remain federally oriented because of the historic legacy of nation-to-nation governance. Table 1 shows the trend of declining new mechanisms designed at the federal level and an increased involvement of local mechanisms, in terms of number of organizations and mechanisms. Even when solely evaluating governance mechanisms (excluding organizations), the trend clearly indicates a rise in local participation over the past two decades.

Another key finding from the analysis is the identification of a trend in the creation of provincial–state agreements, which marked the beginning of subfederal, government-to-government agreements. These agreements appear in the form of memoranda of understanding (MOUs) and memoranda of agreement (MOAs) – nonbinding and collaborative in nature – between provinces and states. Whether the increase in subnational mechanisms is a reaction of nonfederal actors to a downshifting of federal responsibilities, an empowerment of regional actors, or a lessening of authority of federal transboundary mechanisms remains unclear – and can be likened to the question of the chicken and the egg.

The number of provincial–state agreements between Canada and the United States peaked in the mid-1990s, with agreements particularly prevalent along the East and West Coasts. However, despite the increase in number of provincial–state agreements and organizations, the institutional capacity to maintain the binational governance structure is highly variable. For example, the British Columbia–Washington Environmental Cooperation Council, created in 1992, was one of the first binational provincial–state organizations and was initially considered one of the most active groups (Jolly, 1998). The

Table 1. Relationship between scale of governance and type of governance instrument.

	Federal		State–Provincial		Multi-level		Local		Total subnational	
	Number	Percentage	Number	Percentage	Number	Percentage	Number	Percentage	Number	Percentage
Binding	73	76.84	15	42.86	0	0	0	0	15	20.55
Nonbinding	3	3.16	12	34.29	2	6.45	1	0.17	15	20.55
Organization	19	20	8	22.86	29	39.55	5	0.83	42	57.33

Source: Canada–US Transboundary Water Governance Instruments Database (2007); Norman and Bakker (2009).
Note. Binding: a formal mechanism such as treaty, agreement, order, or exchange of notes. Nonbinding: a memorandum of understanding, memorandum of cooperation, or memorandum of agreement. Organization: multi-stakeholder and local (subnational) transboundary groups.

council was developed as part of an MOU signed between the governor of Washington and the premier of British Columbia, who, at the time of signing, had similar political views on the environment. The political climate at the time (in the wake of two disastrous oil spills, one in Greys Harbor, Washington, and the other in Alaska from the infamous Exxon Valdez) paved the way for tremendous support of binational environmental agreements at a 'lower scale' (the provincial–state scale). Thus, this rescaling of transboundary water governance occurred because of a combination of political will, public awareness (piqued by the crisis of the oil spills), and the growing expectation for greater public involvement in environmental governance. However, the political will to support the Environmental Cooperation Council waned, particularly as leadership changed and as the provincial and state governments faced budget cuts and increased demands for border security (which became reprioritized after 9/11). In addition, the lack of consistency of counterparts across the border provided difficult conditions in which to sustain professional relationships. This is particularly detrimental for transboundary water governance because strong personal relationships between counterparts (knowing who to call on the 'other side') has consistently been seen as an antidote to asymmetries in governance mechanisms, where state and provincial regulatory responsibilities often do not match up. Thus, change in leadership, staffing, and reduced funding took a toll on the momentum of the group, and after 15 years of steady biannual meetings, the Environmental Cooperation Council took a several-year hiatus. However, as of January 2014 the council has re-formed under new leadership. Thus, subnational transboundary governance mechanisms can provide flexibility to fill gaps in governance, but their non-treaty nature (MOUs and MOAs) can make the relationship more susceptible to changes in political will and budget priorities.

A counter-example to this, however, is the Atlantic region's Gulf of Maine Council, which has maintained a steady meeting schedule despite changing political climates. The geographic focus of the respective groups may be an important explanatory factor. Where the Atlantic group is focused solely on marine waters, the British Columbia–Washington council includes both marine and freshwater, adding geopolitical and environmental complexity, which adds financial costs. In addition, the border between Washington and British Columbia became heavily scrutinized, particularly after 9/11, in a way that the border in Maine was not (perhaps due to the remote nature of the Maine–New Brunswick border and its distance from urban centres).

A parallel trend has been the increase in application of supranational groups such as the Convention on the Protection and Use of Transboundary Watercourses and International Lakes. An example of rescaling at the supranational level is the 2003 amendments to this convention that allowed states *outside* the UNECE region to become party to the convention. (The UNECE has 56 members, and its region covers more than 47 million km^2. Its member states include the countries of Europe, but also countries in North America (Canada and the United States), Central Asia (Kazakhstan, Kyrgyzstan, Tajikistan, Turkmenistan and Uzbekistan) and Western Asia (Israel) (UNECE, 2013)). The decision to open up the convention beyond the UNECE member countries is reflective of a broader trend: the rescaling of water governance, in which supranational and global actors (as well as local actors) are playing more important roles than in the past (Bulkeley, 2005). According to Norman, Cohen, and Bakker (in press, p. 2), "Rescaling the Convention by including new actors thus holds significant promise both practically – in terms of the creation of a network for information related to transboundary water governance – and conceptually – in terms of envisioning the world's water sources as interconnected, nested, and both politically and hydrologically 'border*less*'."

Rescaling the IJC

With the rescaling of environmental governance over the last 30 years, even federal organizations such as the IJC have begun to shift their governance approaches to reflect greater inclusion of nonstate actors. For example, the creation of the International Watersheds Initiative enabled the IJC to be proactive rather than reactive in its approach, and to include actors that live within specific watersheds in the governance process. That is, rather than wait for a formal request from the federal governments to mediate conflict (through what is called the reference system), the initiative was designed to work proactively at a watershed level, to sustain communication between actors living within the watershed (the need for which is exacerbated at the site of the international border). It is important to note that the BWT and the IJC were not originally set up to look at large-scale watersheds, so this is a distinct shift in focus. Currently, there are four official international watershed boards along the Canada–US border, governing the St. Croix River (spanning New Brunswick and Maine), the Red River (spanning Minnesota, North Dakota and Manitoba), the Rainy River (spanning Minnesota and Ontario) and the Souris River (spanning Saskatchewan, Manitoba, and North Dakota), with other pilot boards under consideration (see Figure 1). Although the initiative is not without its critics (see

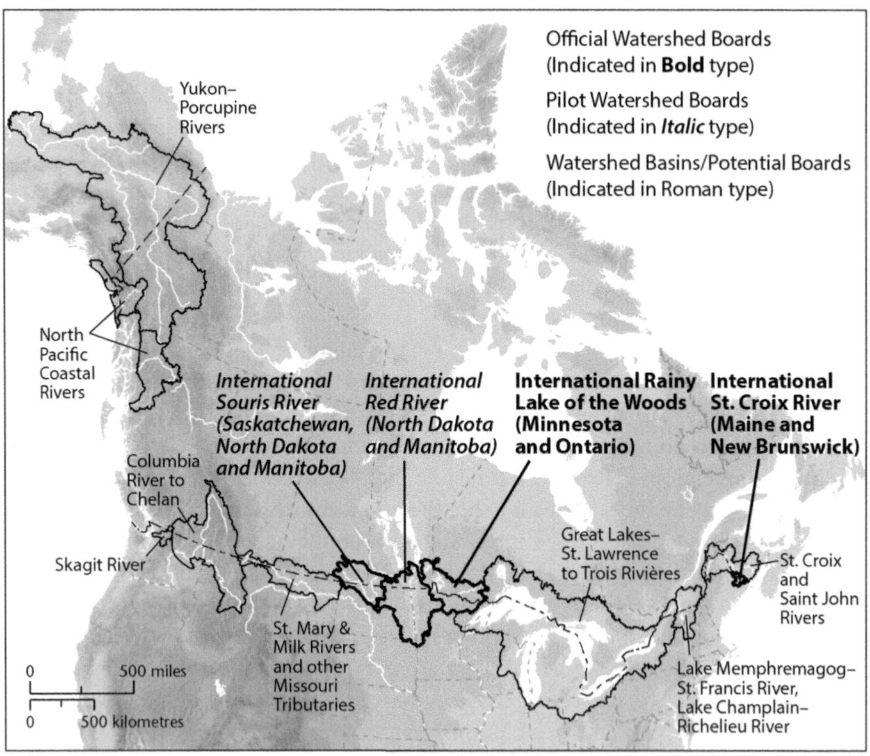

Figure 1. Map of official and proposed international watershed boards.

Source: Adapted from IJC (2014). Reprinted from Norman (2015) with permission from Routledge.

Table 2. Eras of Canada–US transboundary water governance, 1945–2015. Adapted from Norman & Bakker (2009).

Era	Period	Role	Example
Cooperative development	1945–1965	Projects of mutual benefit Federal government encourages hydroelectric development	Columbia River Treaty; St. Lawrence Seaway and Hydropower Development; Niagara Treaty
Comprehensive management	1965–1985	Issue-based Comprehensive river basin planning and more 'environmentally conscious' framework Water expertise built up at federal level	Great Lakes Water Quality Agreement
Sustainable development	1985–2005	Linking of economy and environment Issues more integrative, anticipatory and preventive	Great Lakes Annex
Participatory	2000–present	Increased citizen participation in established frameworks	IJC International Watersheds Initiative
Postcolonial	2000–present	Increased involvement of Indigenous actors in established frameworks Development of Indigenous-led governance in transboundary water issues	Coast Salish Gathering/ Yukon River Inter-tribal Watershed Council

e.g. Cohen & Davidson, 2011), the new governing body represents flexible and innovative institutional thinking that responds to growing expectations for greater involvement of nonstate actors in environmental governance.

A new era of postcolonial water governance?

Surprisingly, analysis of rescaling revealed that the role of federal actors in transboundary water governance peaked in the 1940s, during what Pentland and Hurley (2007) refer to as the Cooperative Development Period (Table 2). During this post-war era, an emphasis on hydroelectric power facilities in both Canada and the United States presented opportunities to create large-scale projects that required binational cooperation. Furthermore, the creation of hydropower facilities in Canada exceeding domestic need (because of 'pre-build' policies) led to creation of export markets in the United States, thereby committing the two countries to decades of cooperation over water- and energy-related issues.

In fact, the trends found in this analysis coincide closely with different eras of management style and priorities, as outlined in Table 2. During the Comprehensive Management Era, in the late 1960s to early 1980s, local governance instruments become more prevalent and the federal role begins to decline. Comprehensive river planning and more 'environmentally friendly' projects typify this period. The creation of the Great Lakes Water Quality Agreement in 1972 (revised in 1978) is a leading example of comprehensive-style governance.

A growing number of Indigenous-led mechanisms have emerged along the Canada–US borderland, which continues to contribute to the growing rescaling and the new era of postcolonial governance discussed in the previous section. In particular, Indigenous, transboundary water governance mechanisms have become established in the Great Lakes basin (Ontario and Minnesota, Wisconsin, Michigan, New York), the Yukon River basin (Yukon Territory and Alaska) and the Salish Sea basin (British Columbia and Washington). In this postcolonial era, not only are Indigenous communities working collaboratively to address environmental issues facing their communities, but they are also working on strengthening traditional lifeways and promoting self-governance. In essence, Indigenous-led initiatives are serving as mechanisms to both decolonize governance and provide means for self-determination. Arguably, the socio-cultural and political impacts of colonial occupation are exacerbated at the borderland, where Indigenous homelands were bifurcated by the demarcation of the international border and the subsequent creation of Native reserves – reservations, in the United States (Norman, 2014). The emergence of a postcolonial era in North American transboundary water governance has international relevance, because it gives question to the context of sovereignty and cooperation.

For example, the recent move to designate membership of First Nations, Metis, and Tribal members in the Rainy Lake board is precedent-setting for the IJC. Although there has been an increasing move to include representation of Indigenous actors on other IJC boards, this practice has not been consistently applied. The explicit inclusion of Indigenous actors in transboundary water governance activities – which had previously been reserved for federal actors – points to the beginning of a new postcolonial era of governance (see Table 2) (Norman, 2014).

International relevance

What are some of the key lessons learned since Canada and the United States first signed the BWT more than 100 years ago? And how might these lessons be internationally relevant? This question is arguably highly relevant to regions, like East Asia, in which rapid development is occurring, transforming both biophysical landscapes and governance frameworks. Below, we explore some key lessons learned, including the importance of symmetry in governance frameworks, the need for built-in flexibility, the importance of 'basin-wide' vision, and the opportunities and challenges posed by the rescaling of water (and indeed environmental) governance.

The first issue – of symmetry in governance frameworks – has been one of the major challenges with which those engaged in administering the BWT have had to grapple. One of the primary limitations of the BWT is the asymmetrical treatment of boundary waters (i.e. waters that form part of the international boundary) versus rivers *crossing* the boundary or waters that are *tributary* to boundary waters. The principle that obtains for boundary waters is one of 'equal and similar rights' on the part of each nation; for tributary or transboundary waters, each nation retains in general a sovereign right to use or divert the waters as it sees fit. In practice, this has meant that integrated water resources management (see Falkemark, 2011) – which is premised on basin-wide water management – cannot be comprehensively implemented within the original scope of the BWT.

A second, related issue is the need for built-in flexibility to deal with changing environmental and social contexts. The hierarchy of water uses established in the BWT reflected the priorities of 1909, which are different from contemporary priorities. For example, the order of precedence listed in the BWT is: domestic and sanitary uses; navigation; power and irrigation. Considerations such as in-stream uses (central to aquatic and wildlife health) are not mentioned. Equally, the emphasis of the BWT on water quantity issues, and the relative neglect of water quality issues, does not reflect contemporary priorities. There is one reference to water quality in the treaty: Article IV provides in part that "waters flowing across the boundary shall not be polluted on either side to the injury of health or property on the other". There is, however, no elaboration on this obligation. Moreover, the high degree of public participation now expected (and in some cases required) for environmental decision making was not envisioned in the BWT. Building in flexibility to deal with changing environmental and social issues is key to the longevity of the instrument.

The third point pertains to the limits potentially imposed by sovereignty concerns on a basin-wide vision for water management. For example, the IJC does not have the power to initiate an investigation of transboundary water issues; its role in investigating and reporting on these matters is limited by the requirement of a formal request from both the Canadian and American governments. (The treaty also provides for the possibility of an arbitral function if both governments consent; in practice, however, this has never been invoked.) This has limited the role of the IJC, which does not have an independent power of investigation, allowing some transboundary disputes to fester, such as that of Devils Lake, in which the US state of North Dakota acted unilaterally to divert water that eventually flows into the Canadian province of Manitoba, raising concerns for water quality and lack of regard for international protocol (Brandson & Hearne, 2013).

The original framework of the IJC thus perpetuates a reactive rather than proactive approach to transboundary water governance, which limits the ongoing cooperation at a subnational/regional level. Now that the 'cooperative development' phase has seemingly ended, the IJC's previously important role as a quasi-judicial body charged with approving (or denying) obstructions or diversions of boundary waters (and, in more limiting circumstances, for works in transboundary or tributary waters) is of diminishing relevance. Pollution and water quality issues have come to the fore; but the sovereignty card trumps basin-wide thinking in some cases. Judicial decisions, such as the Trail Smelter Arbitration in early 1900s, may alter this situation, but creating a treaty that requires recourse to the courts may be suboptimal, given that legal timeframes may not be congruent with the speed required for decision making in ecologically sensitive areas and cases.

This in turn suggests that it might be useful to pursue an explicit strategy of rescaling of water governance, in which a 'thickening' of transboundary water governance (which has happened organically, with little federal oversight or encouragement in the Canada–US case) may be a useful counterweight to the sovereign rights of nation-states – or indeed a useful impetus to fulfil commitments based on legal norms such as good neighbourliness. This is the basis of the common law maxim *sic utere tuo* (which was the legal precedent formed from the Trail Smelter Arbitration in the early 1900s), but the key point is that one should not use one's property in a way that is injurious to another.

This is of course relevant to more general debates over environmental governance (Hirsch, 2001; Jonas & Gibbs, 2003; Kramsch & Mamadouh, 2003; Kramsch 2002; Maddock 2004; Munton, 2003; Paehlke, 2001; Parson, 2001; Verchick & Hulen, 2003; Wismer & Mitchell, 2005), in which there is a growing expectation of public participation and regional participation in environmental issues. The tension for *transboundary* issues, however, remains as a tug of war between sovereign rights and responsibilities and the duty to cooperate.

What international governance regimes and 'soft-path' governance approaches provide, however, is a changing expectation for behaviour at the start. Amory Levins first coined the term 'soft path' in the context of energy use; however, the term has increasingly been applied to describe a movement in water governance, particularly in Canada. Soft-path approaches offer a more integrative and holistic alternative to supply-side water management models, which often focuses on large-scale, centralized, physical infrastructure to meet growing demands. The soft-path approach suggests that water issues should be framed not as 'how to meet growing demand' but as 'how to reduce demand' and 'how to achieve long-term sustainability'. Thus, it adopts a longer-term and more integrated approach, in which addressing water-related issues requires multiple entry points, including policy, education, and economic incentives, in addition to technical solutions.

Thus, changing the culture of how we address water-related issues is as important as changing the culture of water cooperation, particularly in water issues that span international borders. The postcolonial era of governance described above provides an opportunity to rethink transboundary water governance, as integrated water resources management did 25 years ago. But in this case, besides the opportunity to find basin-wide solutions to long-term management of transboundary watersheds, it also actively attempts to decolonize the borderland by creating indigenous-led mechanisms that rely on traditional networks and protocols. In cases throughout the Canada–US borderland this approach is having very real impacts on more traditional federally run mechanisms. In the Pacific Coast region, Coast Salish Gatherings and intertribal Canoe Journeys are raising public awareness of water-related issues – particularly ocean acidification, declining marine habitat and resources such as salmon, and impacts of water pollution. In the Great Lakes, the Grandmother Water Walkers and the Great Lakes Indian Fish and Wildlife Commission are working on issues related to declining water quality and fish contamination due to extraterritorial pollutants such as PCBs and mercury. And in the Yukon Basin, the Yukon River Inter-Tribal Watershed Council is working towards protecting one of North America's last 'wild rivers' and to continue to maintain the priorities of the Indigenous peoples who rely on the river for subsistence. In all three cases, the preservation of these waterways is also explicitly linked to cultural preservation, economic vitality, self-determination and self-governance (Norman, 2015).

Thus, in rethinking environmental governance – and transboundary water governance more specifically – we suggest that there is a need to support communities and activities that have historically been absent from formal binational treaties. Indigenous peoples had little presence in the original treaty-making process in North America related to transboundary waters. Court cases and treaties between tribes and governments have had increased presence in water governance. The North American example provides an opportunity to learn from retrofitting treaties or, as in the Columbia River renegotiation, 'modernizing' the treaty. Thus, for countries and regions that are currently creating or updating their transboundary water governance mechanisms, a key consideration is to look at the cultural politics associated with the act of creation of the border. The design of the mechanism should reflect historical 'upstream' and 'downstream' relationships and provide avenues for regional mechanisms to take leading roles in long-term governance (Cohen, Norman, and Bakker, 2014).

Conclusion and discussion

This paper has analyzed the rescaling of transboundary water governance between Canada and the United States. We asked: What are the lessons learned since the Canada and the United States first signed the Boundary Waters Treaty 100 years ago? And what is the potential of this rescaling to influence the tension between the 'sovereign rights' of a nation and transboundary water governance protocols based on 'good neighbourliness'?

Some of the key lessons gleaned from past experience with transboundary water cooperation across the Canada–US border include: the challenges which arise from asymmetrical governance frameworks; the importance of embedding flexibility into governance frameworks, with recognition that decision-making processes and relationships will evolve in the future; the opportunities (and challenges) posed by rescaling of water governance; and the importance of 'thinking like a watershed'.

In addition, the trend of rescaling governance, in which local and nonstate actors are more present in environmental (particularly water) issues, has the potential to influence how sovereign rights and good neighbourliness play out in transboundary water governance. The Canada–US example suggests that pursuing an explicit strategy of rescaling of water governance, in which a 'thickening' of transboundary water governance may be a useful counterweight to the sovereign rights of nation-states, or indeed a useful impetus to fulfil commitments based on legal norms such as good neighbourliness, is a potentially powerful approach. As explored above, the tension between sovereignty and collaborative cooperation continues to evolve, as federal governments renegotiate (or reconfirm existing) processes for communication, addressing issues of shared concern and mediation of potential as well as actual conflicts.

A key question from the Canada–US example remains: How do the movement from 'government' to 'governance' in transboundary water management and the increased role of nonstate actors impact the ability to cooperate? Our qualitative research found that although regional and subnational mechanisms have increased over time, this trend did not necessarily translate to increased decision-making capacity. Thus, although the trend is promising, we caution against the conflation of creation with capacity. We suggest that the need to support soft-path approaches remains a central component in promoting transboundary water cooperation, particularly in light of rescaling from government to governance.

Another key insight is the possibility that we are entering a new 'decentred' – and, in some cases, potentially postcolonial – era of water governance. Within the IJC, two notable examples of rescaling include the creation of the International Watersheds Initiative and the explicit inclusion of First Nations governments – 'third sovereigns' – in international watershed boards. The former reflects a shift to include actors in the governance process who live and work within the said watershed; the latter reflects a growing movement for the enactment of Indigenous-led governance.

Lastly, the concept of 'soft-path' approaches, in combination with a 'postcolonial' era of transboundary water governance, provides a critical step towards working beyond the inherent tension between sovereign rights and transboundary water governance protocols based on good neighbourliness. The soft path approach is a holistic framework for managing water resources, which differs from traditional ('hard path') approaches in its emphasis on institutional and governance innovation to meet users' water-related needs (rather than fixed water demands) – thereby creating potential for increased water efficiency and improved socio-hydrological relationships. When rooted in a binational approach, in which the locus of power remains with the federal governments, whose responsibilities intrinsically stop at the international border, the ability to work collaboratively, long-term, across the border is limited. However, if the starting point in the collaborative approach deemphasizes the international border rather than reifying it, the starting point is intrinsically more cooperative. Interestingly, the soft-path approaches that have less 'state-run' influence may be in a better position to work in this postcolonial context. For example, even provincial–state agreements wax and wane with political support and climate. The grass-roots, Indigenous-led mechanisms, however, are likely to be less subject to these fluctuations, as the starting point is based on historical, pre-colonial connections and the governance structures take on both environmental and social issues, which widen the approach to address regional issues.

Our survey of 100 years of North American, Canada–United States transboundary governance suggests that the rescaling of governance provides an opportunity to rethink how to best govern transboundary waters. This issue is of international significance, given widespread interest in rescaling (also referred to as 'devolution' or 'delegated water governance').

References

Brandson, N., & Hearne, R. (2013). Devils Lake and Red River Basin. In E. S. Norman, A. Cohen, & K. Bakker (Eds.), *Water without borders: Canada, the United States and shared waters*. Toronto: University of Toronto Press.

Brooks, D. B., Brandes, O. M., & Gurman, S. (Eds.). (2009). *Making the most of the water we have: The soft path approach to water management* (pp. 195–204). London, UK: Earthscan.

Budds, J., & Hinojosa, L. (2012). Restructuring and rescaling water governance in mining contexts: The co-production of waterscapes in Peru. *Water Alternatives*, *5*(1), 119–137.

Bulkeley, H. (2005). Reconfiguring environmental governance: Towards a politics of scales and networks. *Political Geography, 24*(8), 875–902. doi:10.1016/j.polgeo.2005.07.002

Carroll, J. E. (1981). Patterns old and new. In R. Spencer, J. Kirton, & K. R. Nossa (Eds.), *The International Joint Commission seventy years on* (pp. 43–59). Toronto: University of Toronto.

Cash, D. W., Adger, N. W., Berkes, F., Garden, P., Lebel, L., Olsson, P., . . . Young, O. (2006). Scale and cross-scale dynamics: Governance and information in a multilevel world. *Ecology and Socieity, 11*(2), 8. Retrieved from http://www.ecologyandsociety.org/vol11/iss2/art8/

Chen, H., Rieu-Clarke, A., & Wouters, P. (2013). Exploring China's transboundary water treaty practice through the prism of the UN Watercourses Convention. *Water International*, *38*(2), 217–230. doi:10.1080/02508060.2013.782134

Cohen, A., & Davidson, S. (2011). The watershed approach: Challenges, antecedents, and the transition from technical tool to governance unit. *Water Alternatives*, *4*(1), 521–534.

Cohen, A., & Harris, L. M. (2014). Performing scale: Watersheds as 'natural' governance units in the Canadian context. In M. Glass & R. Rose-Redwood (Eds.), *Performativity, politics, and the production of social space.* New York, NY: Routledge.

Cohen, A., & McCarthy, J. (2014). Reviewing rescaling: Strengthening the case for environmental considerations. *Progress in Human Geography*, 7 March, doi:10.1177/0309132514521483

Cohen, A., Norman, E. S., & Bakker, K. (2014). *Renegotiating boundaries: Political ecologies of shared waters*. Paper presented at the Dimensions of Political Ecology Conference, University of Kentucky, Lexington, Kentucky, February 28.

Dreisziger, N. (1981). Dreams and disappointments. In R. Spencer, J. Kirton, & K. R. Nossal (Eds.), *The International Joint Commission seventy years on* (pp. 8–23). Toronto: University of Toronto.

Falkemark M. (2011). Early focus on water strategies for the twenty-first century: IWRA as an interdisciplinary forerunner. *Water International*, *36(7)*, 776–781. doi:10.1080/02508060.2011.630577

Feitelson, E., & Fischhendler, I. (2009). Spaces of water governance: The case of Israel and its neighbors. *Annals of the Association of American Geographers*, *99*, 728–745. doi:10.1080/00045600903066524

Furlong, K. (2010). Neoliberal water management: Trends, limitations, reformulations. *Environment and Society: Advances in Research*, *1*, 46–75.

Gleick, P. H. (2002). Water management: Soft water paths. *Nature*, *418*, 373. 25 July. doi:10.1038/418373a

Gleick, P. H. (2003). Global freshwater resources: Soft-path solutions for the 21st Century. *Science*, *302*, 28 November, 1524–1528. doi:10.1126/science.1089967

Hirsch, P. (2001). Globalisation, regionalisation and local voices: The Asian Development Bank and rescaled politics of environment in the Mekong region. *Singapore Journal of Tropical Geography, 22*(3), 237–251. doi:10.1111/1467-9493.00108

Hirsch, K., & Jensen, M. (2006). *National interests and transboundary water governance in the Mekong.* Australian Mekong Resource Centre. pp. 199. Retrieved from http://www.envirosecurity.org/actionguide/view.php?r=422&m=publications

IJC. (2014). *International Joint Commission's international watershed initiative.* Retrieved from: http://www.ijc.org/en_/IWI

Jolly, C. (1998). The BC-WA Environment and the founding of the ECC. Lecture notes from International Environmental Law Class. Seattle: University of Washington.

Jonas, A. E. G., & Gibbs, D. C. (2003). Changing local modes of economic and environmental governance in England: A tale of two areas. *Social Science Quarterly, 84*(4), 1018–1037. doi:10.1046/j.0038-4941.2003.08404011.x

Kramsch, O., & Mamadouh, V. (2003). Crossing borders of political governance and democracy. *Journal of Borderlands Studies, 18*(1), 39–50. doi:10.1080/08865655.2003.9695600

Kramsch, O T. (2002). Reimagining the scalar topologies of cross-border governance: Eu(ro)regions in the post-colonial present. *Space & Polity, 6*(2), 169–196. doi:10.1080/1356257022000003617

Lebel, L., Garden, P., & Imamura, M. (2005). Politics of scale, position and place in the governance of water resources in the Mekong Region. *Ecology and Society, 10*, 18.

Maddock, T A. (2004). Fragmenting regimes: How water quality regulation is changing political-economic landscapes. *Geoforum, 35*, 217–230. doi:10.1016/j.geoforum.2003.09.004

McCaffrey, S. (1996). An assessment of the work of the International Law Commission. *Natural Resources Journal, 36*, 297–318.

McCaffrey, S. (2006). *Understanding international law.* Elsevier: LexisNexis.

Mirumachi, N. (2013). Transboundary water security: Reviewing the importance of national regulatory and accountability capacities in international transboundary river basins. In B. Lankford, K. Bakker, M. Zeitoun, & D. Conway (Eds.), *Water security: Principles, perspectives and practices* (pp. 166–177). London: Earthscan Series in Water Management.

Munton, D. (1981). Dependence and interdependence in transboundary environmental relations. *International Journal [Canada], 36*, 139–184. doi:10.2307/40201941

Munton, R. (2003). Deliberative democracy and environmental decision-making. In Negotiating environmental change: New perspectives for the social sciences. In F. Berkout, M. Leach, & I. Scoones, (Eds.), *Negotiating environmental change: New perspectives for the social sciences.* (pp.137–152). Cheltenham, UK: Edward Elgar.

Newig, J., & Fritsch, O. (2009). Environmental governance: Participatory, multi-level - and effective. *Environmental Policy and Governance, 19(3)*, 197–214. doi:10.1002/eet.509

Norman, E. S. (2015). *Governing transboundary waters: Canada, the United States and indigenous communities.* London: Routledge, Earthscan Series in Water Management.

Norman, E. S., & Bakker, K. (2009). Transgressing scales: water governance across the Canada–U.S. borderland. *Annals of the Association of American Geographers, 99*(1), 99–117. doi:10.1080/00045600802317218

Norman, E. S., Bakker, K., & Cook, C. (2012). Introduction to the themed section: Water governance and the politics of scale. *Water Alternatives, 5*, 52–61.

Norman, E. S., Cohen, A., & Bakker, K. (2013). *Water without borders? Canada, the United States, and shared waters.* Toronto: University of Toronto Press.

Norman, E. S., Cohen, C., & Bakker, K. (in press). The water convention from a North American perspective. In A. Tanzi, O. McIntyre, A. Kolliopoulos, & A. Rieu-Clarke (Eds.), *The UNECE convention to the protection and use of transboundary watercourses and international lakes: Its contribution to international water cooperation.* Boston: Brill Press.

Norman, E. S., Cook, C., & Cohen, A. (in press). Introduction: Why the politics of scale matter in the governance of water. In E. S. Norman, C. Cook, & A. Cohen (Eds.), *Negotiating water governance: Why the politics of scale matters* (pp. 1–25). London: Ashgate.

Paehlke, R. (2001). Spatial proportionality: Right-sizing environmental decision-making. In E. A. Parson, (Ed.), *Governing the environment: Persistent challenges, uncertain innovations* (pp. 73–125). Toronto: University of Toronto Press.

Parson, E. A. (2001). Environmental trends: A challenge to Canadian governance. In E. A. Parson (Ed.) *The environment: Persistent challenges, uncertain innovations* (pp. 3–30). Toronto: University of Toronto Press.

Pentland, R., and Hurley, A. 2007. Thirsty neighbors: A century of Canada–U.S. transboundary water governance. In K. Bakker (Ed.), *Eau Canada: The future of Canada's water* (163–184). Vancouver: University of British Columbia.

Perreault, T. (2005). State restructuring and the scale politics of rural water governance in Bolivia. *Environment and Planning A*, *37*(2), 263–284. doi:10.1068/a36188

Sandford, R. W., & Phare, M. A. (2011). *Ethical water: Learning to value what matters most.* Toronto: Rocky Mountain Books.

Sayre, N. F. (2005). Ecological and geographical scale: Parallels and potential for integration. *Progress in Human Geography*, *29*, 276–290. doi:10.1191/0309132505ph546oa

Sheppard, E., & McMaster, R. B. (2004). *Scale and geographic inquiry. Nature, society, and method.* Oxford: Oxford University Press.

Sneddon, C., & Fox, C. (2006). Rethinking transboundary waters: A critical hydropolitics of the Mekong basin. *Political Geography*, *25*, 181–202. doi:10.1016/j.polgeo.2005.11.002

Swyngedouw, E. (1997). Neither global nor local: 'Glocalization' and the politics of scale. In K. Cox (Ed.), *Spaces of globalization: Reasserting the power of the local* (pp. 137–166). New York/London: Guilford/Longman.

Swyngedouw, E. (2004). *Social power and the urbanization of water: Flows of power.* New York, NY: Oxford University Press.

United Nations. (2008). United Nations Task Force on Transboundary Waters. *Transboundary waters: Sharing benefits sharing responsibilities*. Geneva, Switzerland.

UNECE. (2013) United Nations Economic Commission for Europe, 'Implementation Committee'. Retrieved from http://www.unece.org/env/eia/implementation/implementation_committee_meetings.html

Verchick, R. R. M., & Hulen, R. M. (2003). Why the global environment needs local government: Lessons from the Johannesburg Summit. *Urban Lawyer*, *35*(3), 471–494.

Wismer, S., & Mitchell, B. (2005). Community-based approaches to resource and environmental management. *Environments: A Journal of Interdisciplinary Studies*, *1* (33), 1–18.

Zeitoun, M., Eid-Sabbagh, K., Talhami, M. & Dajani, M. (2013). Hydro-hegemony in the Upper Jordan waterscape: Control and use of the flows. *Water Alternatives*, *6*, 86–106.

Transboundary water management: lessons learnt from North America

Velma I. Grover and Gail Krantzberg

This article examines the history of the International Joint Commission (IJC) and the 1909 Boundary Waters Treaty (BWT), as well as the evolution of the IJC from considerations of water allocations predominantly to an ecosystem approach via the International Watershed Initiative. It also explores how environmental factors led to the formation of new institutional arrangements such as the Great Lakes Water Quality Agreement (GLWQA) of 1972 and further explains how the ecosystem approach was applied and remedial action plans were developed and implemented to deal with pollution issues in the Great Lakes. The article looks at innovations in cooperation and concludes with lessons that can be learnt from the IJC.

Introduction

Canada and the United States share the longest border in the world, of approximately 8000 km, about 40% of which is shared water. Although most well-known are the Great Lakes, 300 lakes, rivers and streams define the Canadian–US border. Transboundary water resources can easily lead to upstream–downstream or 'tragedy of commons' types of problems. The Boundary Waters Treaty (BWT) between the two countries was signed in 1909 to address these issues. Although the BWT and its related institutions have faced their own challenges, they have survived for over a century alongside the emergence of new treaties, agreements and institutions. In a way, the governance system has evolved over a period of time in response to the ecological, social and economic stresses (Muldoon, 2012). This paper assesses how the International Joint Commission (IJC), the Great Lakes Water Quality Agreement (GLWQA) and the Great Lakes Compact can provide lessons for transboundary water management. In particular, this paper focuses on how the IJC has contributed to improvements in the Great Lakes Basin ecosystem, based on empirical evidence derived from historical records and experiential learning from direct contact with the authors of the IJC, its staff, its advisory boards, task forces and leaders in the Great Lakes environmental and governance regime. However, this paper will also highlight some of the challenges faced by the IJC. For example, its inability to repeat its success in reversing the eutrophication of the 1970s and early 1980s in the face of the concurrent, multiple and overlapping issues of today. This makes it difficult for the IJC to influence effectively Canada and the United States, particularly in the Great Lakes region.

This article also looks at innovations in cooperation and concludes with lessons that can be learnt from IJC.

International Joint Commission and the 1909 Boundary Waters Treaty

The 1909 BWT is one of the earliest innovations in transboundary governance that has greatly influenced international governance and institutions (Muldoon, 2012). In the treaty,

> boundary waters are defined as the waters from main shore to main shore of the lakes and rivers and connecting waterways, or the portions thereof, along which the international boundary between the United States and the Dominion of Canada passes, including all bays, arms, and inlets thereof, but not including tributary waters which in their natural channels would flow into such lakes, rivers, and waterways, or waters flowing from such lakes, rivers, and waterways, or the waters of rivers flowing across the boundary. (preliminary Article of the Treaty, 1909)

The treaty was visionary as it transcended political boundaries to focus attention on ecosystems (Botts & Muldoon, 1996). It was also innovative in the sense that it recognized the environmental, economic and cultural value of the shared Great Lakes basin resources (Crane, 2012).

The 1909 BWT was created for a variety of reasons. First, issue-specific approaches were no longer acceptable due to water management problems related to development of the region, such as the lack of a comprehensive plan for development of hydropower along Niagara and St Mary's river, disputes related to irrigation, country-specific infrastructure development and improvements to support commercial navigation in the Great Lakes. Second, there were no water management rules or regulations to resolve any of these issues and there were no institutional means for developing or coordinating joint plans. Thus there was diplomatic pressure to move towards a comprehensive approach that required both rules and regulations to manage water as well as international agencies to implement such rules (Crane, 2012). Those who negotiated the BWT believed that solutions to boundary problems should be based on deliberations of a permanent binational and equal institution, rather than through bilateral diplomatic negotiations.

The IJC was created under Articles VI and VII of the BWT as a binational body to prevent disputes over transboundary waters. The IJC promises equity without interfering with national sovereignty (Holmes, 1981). The achievement of the common good as a basis for consensus has been the goal of the commission for over a century.

IJC: structure and functions

The commission has three principal functions:

- *Regulatory:* it approves or disapproves applications from governments, companies or individuals for obstructions, uses or diversion of water that can affect the natural level or flow of boundary water.
- *Investigative:* it investigates questions of difference, which are referred to the commission by the two governments, and reports the facts to the two governments with recommendations for action. The governments decide whether or not to act upon the commission's recommendations.

- *Surveillance/coordination:* it monitors compliance with the orders of approval. The IJC can monitor and coordinate actions or programmes that result from the governmental acceptance of recommendations made to them by the commission (Krantzberg, Bratzel, & MacDonald, 2006).

Furthermore, under Article VII of the GLWQA of 1987, the IJC shall *assist* in the implementation of the agreement. The most notable contributions of the IJC to Great Lakes protection and improvement derive from this ability to assist the governments in making progress towards the purpose of the agreement.

The commission has six members, three on behalf of the United States appointed by the president, and three on behalf of Canada appointed by the prime minister on the recommendation of the governor in council.

The commission is considered to be independent of either government. It is not a supranational agency with legal authority. It theoretically operates on the principle of the reciprocal good of the resource. Each commissioner upon the first meeting after being appointed signs a declaration to perform the duties imposed under this treaty impartially (Boundary Waters Treaty, 1909). Further, commissioners declare to seek the best solution to common problems based objectively on results from joint fact-finding studies. For technical information and policy advice in conducting investigations, the IJC depends mainly on boards or task forces with equal membership from each country. Under the GLWQA, membership has been extended to non-governmental experts, including representatives of environmental organizations and industry (Krantzberg et al., 2006).

IJC and the Great Lakes Water Quality Agreement (GLWQA)

Signed by Prime Minister Pierre Trudeau and President Richard Nixon on 15 April 1972, the GLWQA was created with the purpose of enhancing and maintaining the quality of the waters of the Great Lakes basin ecosystem. The parties to the GLWQA are the federal governments of Canada and the United States, which commit to collaborating with other governmental jurisdictions within the Great Lakes basin. The two countries hold the primary responsibility for achieving the objectives of the agreement (Krantzberg et al., 2006).

The majority of the policies and agreements in the Great Lakes region have been reactive. As discussed below, pollution events during the 1960s and the public outcry regarding environmental disasters such as the fires on the Cuyahoga River and the hypoxic condition of Lake Erie led to the signing of the first GLWQA in 1972 (Botts & Muldoon, 2005). Although, the agreement should undergo review every six years, this has not happened. Further amendments to the agreement were designed primarily to respond to toxic substances (1978), sediments, airborne depositions etc. (1987) and climate change (2012).

Nutrients

One of the major achievements of the IJC was the study undertaken in 1960 that led to the signing of the GLWQA in 1972, one of the most significant contributions of the IJC to the Great Lakes revitalization in its history (Krantzberg et al., 2006). This study was the result of a highly influential request in 1964 by Canada and the United States (termed a 'reference') to study pollution in Lake Erie and the lower lakes (LeMarquand & Scott, 1980). The study determined that excessive loading of phosphorus, mainly from anthropogenic sources, was leading to severe eutrophication in Lake Erie and Lake Ontario (Vollenweider, 1968).

The chief objective of GLWQA in 1972 was to reduce the levels of phosphorus loadings to the lakes to curb eutrophication. For the most part, improvements to the degraded lower lakes were the result of enhanced and expanded sewage treatment plans. Programmes to improve the quality of storm water runoff from agricultural applications catalysed conservation tillage. Algal proliferation began to decline and more beaches were open for swimming and recreational use, for longer durations (Krantzberg et al., 2006).

At the same time as when GLWQA was signed, the IJC received another pioneering reference on Pollution from Land Use Activities (PLUARG). This five-year comprehensive study on non-point source pollution in the Great Lakes produced over 100 reports and concluded that the Great Lakes were being polluted from non-point sources by phosphorus, metals, industrial chemicals and pesticides. Atmospheric deposition was also a significant source of pollution to the Great Lakes basin. Intensive agricultural activities were recognized as one of the major pollutant sources (Krantzberg et al., 2006).

The main aim of the 1972 GLWQA was to reverse eutrophication. The chief water quality success was the decline of algae growth and other evidence of slowing of eutrophication that followed reductions of phosphorus loadings (Bertram, Neilson, L'Italien, Glumac, & Williams, 1995). This environmental success stands as an extraordinary example of inter-jurisdictional cooperation fostered by IJC recommendations that were based on science and impartial fact-finding (Krantzberg et al., 2006).

Air quality (Annex 15 of the 1998 GLWQA)

Although the first time the commission investigated air pollution was in 1928 – when the IJC reported the damage caused in Washington state in the United States by a smelter at Trail in British Columbia, Canada – reference for air pollution has come up several times since. The two governments, in a 1966 reference, directed the commission to examine the matter of air pollution in the Detroit–Windsor–Port Huron–Sarnia corridor to determine if air polluted on either side of the boundary caused injury to citizens and property on the other side. The commission reported to governments that there was air pollution in the area, quantified the extent and recommended air quality objectives. Since its establishment in 1966, the International Air Quality Advisory Board has generated 24 Progress Reports to the commission on significant transboundary air quality issues (Krantzberg et al., 2006). In 1975, a reference to the IJC from both governments requested them to monitor air quality in the Detroit–Windsor and Port Huron–Sarnia areas and to report annually on the ambient air quality trends with a focus on sulphur dioxide, suspended particulates and odours. The IJC noted in 1983 that the regulatory framework and control strategies in combination with the decommissioning of some older industrial plants have improved the air quality, especially on the three indicators being reported and informed the governments that the reference is now complete.

In 1988, the IJC was asked by the governments to look into the hazards posed by the airborne emissions to humans and the environment in the Detroit–Windsor region. To look specifically into the problem, the commission formed the International Air Pollution Advisory Board (IAAB) for this region with the mandate to investigate the impact of air pollution on human health. This board identified a list of 125 chemicals for consideration and reported on emissions and air monitoring data for these 125 chemicals. Analysis was completed to estimate the range of exposure of specific pollutants and their toxicity. The board identified 15 chemicals with high carcinogenic potential and also reported that the risk of cancer associated with several of these contaminants in the two regions was significantly higher than the US Clean Air Act benchmark (Krantzberg et al., 2006). Following a review of the recommendations of the board, public comments and written submissions, the commission issued its 1992 Air Quality in the Detroit–Windsor/Port Huron–Sarnia Region Report to the governments (Krantzberg et al., 2006; IJC, 1992).The IJC alerted the governments to matters surrounding transport, deposition and impacts of sulfur dioxide, nitrogen oxides, ozone and particulate matter, as well as some persistent toxic chemicals (Krantzberg et al., 2006). Based on the work carried out by IAAB, the IJC concluded that operation of the International Air Deposition Network (IADN) is an example of a binational programme that generates comparable and compatible data (monitoring) for selected toxic substances on both sides of the border (IJC, 2000; Krantzberg et al., 2006). However, the IJC also recommended that some other pollutants such as mercury be included in the assessment on both sides. In its *Tenth Biennial Report* (IJC, 2000), the IJC recommended that both parties should (1) identify both in- and out-of-basin sources of atmospheric deposition of persistent toxic substances to the Great Lakes, quantify their contribution to the total burden of these substances to the lakes, and use this information to formulate and implement appropriate prevention and control measures; and (2) adopt a source-receptor computer model, improve emissions inventory information, and add dioxin and mercury to the IADN to improve the databases for these two substances (Krantzberg et al., 2006). Both countries have installed instruments to measure and monitor mercury contamination on both sides of the border (Canada–USA, 2002). The IJC was successful in demonstrating the linkages between the impact of emissions of both conventional and hazardous air pollutants and human health. However, the IJC has fallen short when calling for the virtual elimination of mercury emissions from coal-fired utilities (Krantzberg et al., 2006).

Trace chemicals

An important achievement of the IJC was to introduce concepts such as the ecosystem approach and also virtual elimination of persistent substances. Scientific investigations carried under the umbrella of the GLWQA contributed to a broader understanding of the course and toxicity of both organic and inorganic contaminants. Studies have determined how toxic chemicals from both point and non-point source chemicals, including atmospheric pathways, bioaccumulate in the food chain and also lead to an understanding of how toxic chemicals cycle within an ecosystem (Krantzberg et al., 2006; Botts & Muldoon, 1996). Because of this increased knowledge of the pathways of toxic chemicals in the ecosystem, the parties (the governments of Canada and the United States) amended the 1978 GLWQA and codified the use of the ecosystem approach with an objective to eliminate virtually the discharge of all persistent toxic substances (Krantzberg et al., 2006).

During the 1980s the government programmes led to measurable declines in direct discharge of toxic chemicals, which also resulted in the decline of concentrations of such chemicals in both water and fish tissues. However, the initial concern about linkages between fish tumours and abnormal reproduction in wildlife that has been exposed to toxic contaminants changed to a concern about the growth and development in human infants. This led to the identification of hundreds of chemicals in the Great Lakes and a determination of which are most harmful. In 1987 the GLWQA was again amended and a new protocol, which called for more aggressive programmes to address these harmful chemicals, was added. Notably, Annex 2, based on the work of the Water Quality Board (WQB) (see below), and Annex 15, stemming from the work of the International Air Quality Advisory Board, were added, as were other annexes related to research, monitoring and surveillance. By 1991, agreement was reached on a list of 11 critical contaminants designated for virtual elimination (IJC, 1991).

The IJC recommended in its *Fifth Biennial Report* (IJC, 1990) that Lake Superior be designated 'a demonstration area where no point source discharge of any persistent toxic substance will be permitted'. Over the next two years, this recommendation generated enthusiasm and diligence on the part of governments, non-governmental organizations and individuals to develop such a programme. Governments were explicit in their response to the commission and the public. In a 1 October 1991 public release titled *A Binational Program to Restore and Protect the Lake Superior Basin*, governments stated: 'The challenge to designate Lake Superior as a "demonstration area where no point source discharge of any persistent toxic substance will be permitted," is accepted' (Krantzberg et al., 2006).

The binational programme, stimulated by the IJC, contains specific provisions for the reduction and elimination of point-source discharge of persistence toxic substances into Lake Superior, which led to the 1991 revision of the US Clean Air Act with necessary emission standards or control measures to protect Lake Superior. Similarly, Ontario also prepared a new and revised regulation to reduce and eliminate point-source discharges of persistent toxic substances under its Municipal and Industrial Strategy for Abatement (Krantzberg et al., 2006).

The IJC formed a Virtual Elimination Task Force in July 1990 to investigate the requirement of the amended GLWQA to eliminate virtually the input of persistent toxic chemicals into the Great Lakes Basin Ecosystem. The final report of this taskforce in 1993 presented a conceptual framework for a virtual elimination strategy for three chemicals: polychlorinated biphenyls (PCBs), mercury and chlorine as a feedstock. The report also emphasized the need to be guided by a vision to maintain ecosystem integrity, characterized by a clean and healthy Great Lakes basin ecosystem and the absence of injury to living organisms and to society (Krantzberg et al., 2006). Another innovative concept initiated by the IJC is the principle of reverse onus to prevent further releases of persistent toxic substances into the environment. In 1994 the IJC recommended that the end-of-pipe solutions or discharge limits are not acceptable solutions to these toxic chemicals; a vigorous policy is needed to eliminate all sources of persistent toxic substances (with some exceptions such as specialized, unavoidable, controlled and temporary application). The Great Lakes WQB supported this recommendation and concluded that there is no acceptable assimilative capacity for persistent bioaccumulative toxic substances (Krantzberg et al., 2006). In 1994, in the *Seventh Biennial Report* the IJC concluded that there 'is sufficient evidence now to infer a real risk of serious impacts in humans' and recommended that governments should involve industry and other stakeholders to see how these chemicals can be eliminated (IJC, 1994). In 1996 IJC report reiterated the need for a reverse onus approach with emphasis on prevention and precautionary principles to avoid damage caused by toxic substances.

Studies suggesting that Great Lakes fish contain many neurotoxins, including PCBs and methyl mercury, leading to bioaccumulation in organisms at the top of the food chain, including humans, (and impacting development of children) (IJC, 2000) led to the signing of the 1997 Great Lakes binational Toxics Strategy of Canada and the United States (Canada EaE, 1997).

Contaminated sediment

The IJC has been intimately and regularly involved in contaminated sediment management since the GLWQA particularly of 1978 and 1987, which has led to development and implementation of several significant programmes to deal with contaminated sediment. For example, the International Working Group on the Abatement and Control of Pollution from Dredging activities was formed which recommended that a binational Great lakes register of dredging activities be established. In 1978, GLWQA formed the Dredging Subcommittee (reporting to GLWQ board) which was responsible for the establishment of a dredging register and publishing binational reports giving information on sediment contamination, dredging projects and disposal/reuse of dredged spoils. Another Research Advisory Board Expert Committee on Engineering and Technological Aspects of Dredging was established to look at the use of Confined Disposal Facilities for contaminated dredge spoils. This work led to the wider development and use of Confined Disposal Facilities as a more environmentally sound means of dealing with contaminated sediment (SedPAC, 1997). As a result of this work and the ecosystem approach focus called for in the 1978 Agreement, the Science Advisory Board established a Contaminated Sediment Task Force to examine the assessment and potential remediation of contaminated sediment, beyond navigational dredging interest (Krantzberg et al., 2006).In 1986, the terms of reference for the Dredging Subcommittee were expanded to include non-navigational sediment leading to the formation of Sediment Work Group. This group was also involved in the Remedial Action Plans and efforts.

In 1995 the IJC also established the Sediment Priority Action Committee (SedPAC) to examine the Parties progress in managing contaminated sediment. SedPAC was also responsible for the identification of obstacles to solve sediment contamination problem and how the IJC can help the parties in overcoming these obstacles. Since one of the major issues most of the Areas of Concerns (An area of concern (AOC) is defined in the GLWQA as a 'geographic areas that fails to meet the general or specific objectives of the agreement where such failure has caused or is likely to cause the impairment of beneficial use of the area's ability to support aquatic life.' In other words, it is an area on the Great Lakes that is experiencing significant environmental degradation. There are a number of reasons why an area is identified as 'degraded'. There may be excess nutrients in the water, contamination from chemicals or bacteria, or a loss of fish or wildlife habitat.)were facing was related to sediment contamination, SedPAC identified the challenges and grouped them into six categories: limited funding and resources, regulatory complexity, lack of a decision-making framework, limited corporate involvement, insufficient research and technology development, and limited public and local support (Krantzberg et al., 2006). In 1999, SedPAC released a report with the intent to share advances in data interpretation regarding sediment management decision-making with Remedial Action Plan (RAP) practitioners (Krantzberg, 1999).The report concluded that, at present, a great deal of data has been collected on the physical, chemical, and biological elements that modify contaminant bioavailability and ecological effects. It was also noted that there are currently few, if any, simple or proven methods to predict recovery of use impairments based on sediment cleanup (Krantzberg, 1999). The IJC agreed that more research is needed to quantify the relationships between contaminated sediment and known use impairments. The concept of ecological benefit forecasting (i.e., predicting ecological benefits and restoration of beneficial uses) is an important management need, which, if accomplished, would be a substantial step forward (Krantzberg et al., 2006). A greater emphasis needs to be placed on post-project monitoring of the effectiveness of sediment remediation. Based on this work, Krantzberg and Zarull (2001) advised the commission that there continues to be a need for methods to interpret and integrate multiple pieces of information on sediment chemistry, biological information from field monitoring and laboratory sediment bio-assessment in an ecologically meaningful way.

The IJC identified the obstacles to remove sediment contamination and opportunities on how to overcome these obstacles. They did not, however, engage governments or industries in a direct dialogue to stimulate active intervention and ecosystem recovery (Krantzberg, 1999).

Ecosystem modelling

In 1993, the IJC established 'Lake Erie Steering Committee' later also called 'Lake Erie Task Force', to study consequences of various anthropogenic actions on Lake Erie and to determine the adverse effects of numerous stressors on fish communities. The modelling work generated by the IJC created a new understanding of the complex ecosystem dynamics of Lake Erie and contributed to the suite of available models for whole lake decision making and policy formulation (Krantzberg et al., 2006).

Nuclear matters

The 1987 GLWQA sets forth specific objectives for radiological pollution under Annex 1. In 1995, the IJC formed a 'Nuclear Task Force' to review, assess and report on the state of radioactivity in the Great Lakes which released a report 'Inventory of Radionuclides for the Great Lakes' in December 1997. The Task Force report examined in detail the bioaccumulation of elements in freshwater aquatic biota and selected species of terrestrial plants (mosses) and wildlife (caribou). According to the Task Force report, many elements perform their physiological function as part of interconnected metabolic systems, and therefore, one must examine the behaviour of several elements as a suite before assuming that a given bioaccumulation factor is appropriate (IJC, 1995–97). While contributing to the knowledge of the behaviour and effects of radionuclides, in recent years the IJC has not commented on the future of nuclear power generation in the face of peak oil (Krantzberg et al., 2006).

Invasive species

Most of the work of GLWQA focused on chemical contaminants and only in the past decade or two has there been a focus on exotic (alien) species and their import vectors. The IJC has issued alerts against the threat of aquatic invasive species to the Great Lakes–St Lawrence River basin ecosystem and economy. These concerns lead to the 1990 IJC–Great Lakes Fishery Commission joint report on exotic species and the shipping industry.

Research in 1999–2001 by the WQB (2001) supported the conclusion that the Great Lakes invasion by a succession of non-indigenous aquatic species, is displacing important native species, interfering with beneficial uses of the Lakes and costing billions of dollars to control. Sources of alien invasive species to the Great Lakes basin include cargo vessels, aquaculture, escapes from aquaria, ornamental ponds, research and educational facilities, canals and diversion water flows, and the release of live bait (Krantzberg et al., 2006). The *Twelfth Biennial Report* (IJC, 2004) also called on governments to issue a reference on aquatic alien invasive species to the IJC to, among other things, help identify the most effective ways to coordinate binational prevention efforts and harmonize national plans, particularly those dealing with residual ballast water and sediment in ballast tanks and evaluate the effectiveness of current institutional arrangements (Krantzberg et al., 2006).

Management techniques for some invasive species have raised an interesting dilemma for fish management: the often conflicting aims of supporting the economic aspects of sport and commercial fisheries, or the aim of achieving ecosystem health and stability. For example, historically the Great Lakes ecosystems were dominated by the top predator lake trout *Salvelinus namaycush* and a diverse assemblage of coregonine fishes. However, overfishing and invasive species began impacting the native fish community (Stein & Goddard, 2006), some of which became extinct, while the sea lamprey and alewife invaded the upper four Great Lakes (Christie & Goddard, 2003). Once sea lamprey treatment was applied in 1960, it started the process of rehabilitation of lake trout in Lake Superior (Hansen et al., 1995), but fishery managers also supported the development of commercial trawl fisheries to reduce the size of alewife populations. Managers began stocking Pacific salmon to serve as the top predator to control alewife and as a sport fishery. However, with the decline in the alewife population, the salmon population also declined during the 1980s (Hansen & Holey, 2002) and because of some kidney disease mortality escalated and resulted in a serious management concern for the booming sport fishery. 'The recreational fishers became so politically influential that management agencies eliminated commercial fisheries for alewife by 1991 because they competed with salmon for alewife' (O'Gorman & Stewart, 1999). Because of this large economic impact, and the popularity of salmon fishing, management agencies increasingly sought to manage lakes Michigan, Huron and Ontario to preserve the salmon fishery, rather than to control or eliminate the alewife population. Instead, the idea of balance between the Pacific salmon and their alewife prey became the prevailing management philosophy, despite clear evidence that the alewife negatively impact native fishes that previously sustained important fisheries (Madenjian et al., 2008). As a result, it might be argued that it could make sense for managers to direct fewer resources toward active management of Pacific salmon in favour of efforts to restore native species in a changing food web. Conversely, the popular and economically important Pacific salmon sport fishery exerts strong political pressure on managers to maintain management for Pacific salmon, despite a reduced capacity to manage for the very characteristics desired by the anglers: high catch rates and large size. As a result, managers of these lakes must now make the difficult decision about whether to manage for short-term economic success by attempting to maintain a balance between Chinook salmon demand and declining alewife production or to manage the system to encourage rehabilitation of native fishes previously suppressed by alewife.

Remedial Action Plans (Annex 1 of the 2012 Protocol)

In the 1970s and 1980s, the WQB, in its annual assessment of water quality in the Great Lakes, identified specific areas, such as harbours, river mouths and connecting channels, that had serious water quality problems. The WQB, in its 1977 Annual Report, again listed the problem areas, described the nature of the problem, identified dischargers of one or more substances that were likely to be the cause of the problem, and commented on progress towards compliance with jurisdictional enforcement programmes (Water Quality Board, 1977). The report also described remedial programmes in the drainage basin of each problem area and progress towards meeting boundary water quality objectives. In 1983 the WQB determined that classifying areas of concern was difficult due to the lack of specificity of the criteria used to classify the areas and the guidelines to be used for their evaluation. This led to difficulties in data interpretation for the purpose of defining the problems and deducing trends in environmental quality. In order to overcome these difficulties, the board developed a procedure for data assessment and identification of Areas of Concern. The unique experiment in place-based remediation and protection called for in the 1987 Protocol emerged directly from the recommendations of the WQB (Great Lakes Water Quality Board, 1985; Krantzberg et al., 2006).

In 1985, the WQB identified 42 degraded Areas of Concern around the Great Lakes. Areas were characterized by some or all of the following conditions (Krantzberg et al., 2006):

- Restrictions on fish and wildlife consumption.
- Tainting of fish and wildlife flavour.
- Degradation of fish and wildlife populations.
- Fish tumours or other deformities.
- Bird or animal deformities or reproduction problems.
- Degradation of benthos.
- Restrictions on dredging activities.
- Eutrophication or undesirable algae.
- Restrictions on drinking water consumption, or taste and odour problems.
- Beach closures.
- Degradation of aesthetics.
- Added costs to agriculture or industry.
- Degradation of phytoplankton and zooplankton populations.
- Loss of fish and wildlife habitat.

In the language of the GLWQA, these are impaired 'beneficial uses'. With the signing of the 1987 protocol to the GLWQA, Canada and the United States agreed to restore these areas by developing and implementing RAPs. They also agreed to address these impairments for each lake through the development and implementation of Lakewide Management Plans (LaMPs) (Krantzberg et al., 2006). The agreement calls for the federal governments, in cooperation with state and provincial governments, to ensure the public is consulted throughout the development and implementation of the RAPs. Despite organizational and fiscal hurdles, several RAPs are being applied and as a result there are notable advances in remediation and prevention programmes. Essential elements that characterize successful initiatives include true participatory decision-making, a clearly articulated and shared vision, and focused and deliberate leadership (Krantzberg et al., 2006). The first comprehensive review of progress in developing and implementing RAPs was released by the IJC in 2003 (IJC, 2003). The WQB's 1985 recommendation has resulted in a new, participatory regime for Great Lakes recovery that builds the capacity for sustainable communities (Krantzberg et al., 2006).

Latest protocol (2012)

The IJC's *Twelfth Biennial Report* called for the renewal of GLWQA, since it had not been updated in more than 17 years while science and technology had grown substantially. As such, 'we need to keep pace with what we know and review the Agreement with an eye toward a sustainable future' (IJC, 2004; Jetoo and Krantzberg, in press). The governments finally started the review process in 2006 and the revised protocol (after long public consultations and comments) was signed in 2012. The purpose of the 1987 GLWQA is still part of the new protocol, but its scope now includes current issues such as climate change, aquatic alien invasive species, habitat and species, and near-lake shore areas (IJC, 2011) of the Great Lakes Basin. IJC (2011) identified governance as a key issue in its *15th Biennial Report*, noting that 'there is a critical need to modify existing governance to strengthen coordination across jurisdictional lines to address ecological challenges in the nearshore' (Jetoo & Krantzberg, 2014). Another key inclusion is the engagement of First Nations, tribal governments, Metis and municipal governments(Jetoo and Krantzberg, in press).

Legal mechanisms to incorporate the protocol

Both the United States and Canada have existing legal mechanisms that enshrine parts of the protocol in cooperative agreements and law. For example, Canada has relied on the Canadian–Ontario Agreement Respecting the Great Lakes Basin Ecosystem (COA) as a mechanism for cooperation between the Province of Ontario and the federal government for Great Lakes Restoration. According to the Ontario Ministry of Environment website (2013), the Province of Ontario is working on the proposed Great Lakes Protection Act which has the potential to provide tools for setting broad direction for ecological restoration as well as accommodating targeted action in priority degraded areas. Similarly the United States *has recognized the* GLWQA *in the Clean Water Act (Ontario, 2006), the Beaches Act, the Great Lakes Restoration Initiative (GLRI) and several Executive Orders of its presidents. These are visionary precedents that can aid in the implementation of the protocol (*Jetoo and Krantzberg, in press*).*

Public participation has been central to the success of IJC, GLWQA and RAPs programme. As explained by Hall (2009), 'Since the 1970s, the second generation (after the BWT) of environmental agreements between the United States and Canada demonstrate a dramatic growth in the role of citizens in achieving compliance with international environmental law.' The GLWQA relies 'heavily on citizens to ensure compliance and implicitly recognize that the two federal governments may have more in common with each other than with citizens and other stakeholders on both sides of the border when it comes to environmental protection and harm'.

Great Lakes Compact

The original Great Lakes Basin Compact of 1968 was an early attempt to create a framework that would assist with the management of the Great Lakes. The compact created the Great Lakes Commission, but since its recommendations were non-binding it had little effect on water management (Hall, 2006; Palay, 2009). While only a guiding document, the compact provided a model for a congressionally approved compact that might be utilized in future to create a more effective regulatory regime. In the 1980s, concern over water shortages and drought throughout North America and globally raised the possibility of water diversions and the large-scale export of water from the Great Lakes (Bakenova, 2008; Palay, 2009). In response to the possibility of loss of control, the region developed the Great Lakes Charter in 1985. The charter was a cooperative agreement between the eight Great Lakes states with input from Ontario and Quebec. The charter committed each state and province to regulate any new consumption or diversion greater than 2 million gallons per day, to provide prior notice to all signatories when any diversion or consumption would exceed 5 million gallons per day, and to create a database of all consumptive uses greater than 100,000 gallons per day (Hall, 2006).

While the charter represented a comprehensive attempt to manage water diversions and consumption in the basin, the lack of formal mechanisms for rejection or approvals of large diversions and that the charter was only soft law has resulted in a characterization of the agreement as ineffective (Hall, 2006; Palay, 2009).

By 2005, the Great Lakes–St Lawrence River Basin Water Resources Compact and the companion Great Lakes–St Lawrence River Basin Sustainable Water Resources Agreement (2005) were signed. The compact and agreement, which includes the eight Great Lakes states, Ontario and Quebec, created a set of common standards for state water management programmes and out-of-basin diversions, and includes both surface and ground water in an agreement with the full force of federal law (Hall, 2006, 2009; Palay, 2009; Squillace, 2006). Hall (2006) describes the compact as a new and effective form of horizontal federalism that combines the flexibility of state implementation of regulations that will be best suited to regional realities with the power of the federal government to adjudicate between states and create common and enforceable standards of practice. Squillace (2006), though, argues that the compact represents an unnecessarily complex intrusion into state authority and that it will ultimately prove inadequate for managing the complex realities of protecting the Great Lakes. Whether the compact represents a genuine solution to the collective action problem of creating an effective water quantity governance regime without resorting to intervention by the federal government remains to be seen.

Due to their voluntary nature, the Great Lakes Compact (1968) and the Great Lakes Charter (1985) provided almost no incentive for states and provinces to impose water management regimes. The new compact and agreement create a framework that must be implemented by the states and should be implemented by the provinces; it provides a set of standards that must be applied to all new water uses and includes commitments to water efficiency (Hall, 2006). However, management of in-basin consumption is still managed by individual states and provinces, with reporting requirements for most existing uses. The compact is therefore not nearly as successful at meeting the challenge of governing in-basin consumption as out-of-region diversions. This is hardly surprising as all states and provinces have a vested interest in preventing new users accessing the Great Lakes but also have an interest in maintaining the freedom to use those same resources themselves. This divergence of interests will continue to play a central role in future Great Lakes governance.

The evolution of some regional agreements, strategies in the 21st century

The water governance arrangements between Canada and the United States formally began with the international Treaty in 1909 and have evolved over a century. Some of the newer agreements, protocols, compacts and so on that have important implications for the future of governance in the Great Lakes basin are now mentioned.

Great Lakes–St Lawrence River Basin Water Resources Compact (United States) and the Great Lakes–St Lawrence River Basin Sustainable Water Resources Agreement (Canada)

On December 13, 2005, the Great Lakes Governors of Illinois, Indiana, Michigan, Minnesota, New York, Ohio, Pennsylvania and Wisconsin, and the Premiers of Ontario and Québec signed the Great Lakes–St. Lawrence River Basin Sustainable Water Resources Agreement, and the Governors endorsed the companion Great Lakes–St. Lawrence River Basin Water

> Resources Compact. The agreements detail how the States and Provinces will manage and protect the Basin and provide a framework for each State and Province to enact laws for its protection. (13 December 2005; GLSLR Body, 2007)

The agreements call for a ban on new diversions of water from the basin. Limited exceptions could be allowed, such as for public water supply purposes in communities near the basin, but exceptions would be strictly regulated.

Canada Ontario Agreement (COA) Respecting the Great Lakes Basin Ecosystem

> The Canada Ontario Agreement Respecting the Great Lakes Basin Ecosystem (COA) (1994) outlines how the governments of Canada and Ontario will cooperate and coordinate their efforts to restore, protect and conserve the Great Lakes Basin ecosystem. It builds on the actions taken through previous COAs and focuses on priorities for future action. The sixth COA was signed in 2007 and the seventh COA is currently being negotiated. It was set to expire in 2010 but has been extended to 31 March 2011. The COA establishes an action plan and clear roles and responsibilities between federal and provincial ministries, and helps Canada to meet its commitments under the GLWQA. This one-year extension was approved primarily to allow Canada and Ontario to continue their important work to protect and restore the Great Lakes while the governments of Canada and the United States negotiate amendments to the Agreement (Crane, 2012).

The Great Lakes regional collaboration of national significance

The strategy called for federal action: (1) to stop immediately the introduction of aquatic invasive species; (2) to increase habitat conservation and species management, including the restoration of 550,000 acres of wetlands; (3) to protect near-shore waters and coastal areas by cleaning up pollution from sewers; (4) to clean up highly contaminated Areas of Concern; (5) to improve water quality by reducing polluted runoff; (6) to eliminate the discharge of, and reduce exposure to, persistent toxic substances; (7) to increase monitoring and coordination; and (8) to ensure long-term, sustainable growth and economic development (Force USFIT, 2005).

Conclusions

One reason for the success of the IJC is geography. Along the US–Canada boundary, many rivers flow north from the United States into Canada, while others rivers flow south from Canada into the United States. This creates both upstream and downstream interests in both countries. Also, east of the Great Lakes, the international boundary cuts right through the lakes and ultimately along part of the St Lawrence River, making both countries equally concerned about upstream and downstream interests. Since both countries have downstream and upstream interests, there is often an incentive to reach agreement; otherwise, an offending upstream party could face problems in another watershed in which they are the downstream interest. Although the IJC itself might need some restructuring to modernize its governance, there are certain lessons that can be replicated at other places:

- In the Smelter Case in 1928, when sulphur dioxide fumes from a smelter in Trail, British Colombia, was causing damage in the US state of Washington, the IJC recommended remedial measures and provided a formula for payments for damage settlements. This sets a precedent in international law that a country is responsible to pay for environmental damage it caused in another country.

- An important achievement of the IJC was the introduction of concepts such as the ecosystem approach. The 1978 amendment codified the use of the ecosystem approach to virtually eliminate toxic substances.
- The International Air Deposition Network is an excellent example of a binational programme that generates comparable and compatible data monitoring for selected toxic substances on both sides of the border.
- The success of the IJC is based on Scientific Joint Fact Missions by their boards – such as the WQB – illustrating that it is important to have joint scientific missions.
- Principle of Reverse Onus with Emphasis on Prevention and Precautionary Principles for persistent toxic substances.
- Given its limited budget, the IJC can only effectively research an issue if it is referred to them by both parties. Such references come with funding, and in their absence it is hard for the IJC to be proactive. Perhaps one of the lessons learnt is to have independent funding, not linked to a reference, actively to look into emerging issues. The IJC studies and recommends solutions to transboundary issues when asked to do so by the national governments. When it receives a government request, called a reference, it appoints a board with equal numbers of experts from each country. Board members are chosen for their professional abilities, not as representatives of a particular organization or region (for more, see http://www.ijc.org/en_/Role_of_the_Commission#sthash.XaZ0bEPK.dpuf/).
- Although the GLWQA has a provision allowing for periodic revision (which can be started by an exchange of letters between the parties involved), it has rarely been used. Activating this option should be done more routinely to ensure that current issues and concerns are addressed in the revised agreements.

References

Bakenova, S. (2008). Making a policy problem of water export in Canada: 1960–2002. *Policy Studies Journal, 36*, 279–300. doi:10.1111/j.1541-0072.2008.00266.x

Bertram, P., Neilson, M., L'Italien, S., Glumac, V., & Williams, D. (Nov 1995). Nutrients: Trends and system response. *Background paper for State of the Lakes Ecosystem Conference, Environment Canada and the U.S. Environmental Protection Agency.* Windsor.

Botts, L., & Muldoon, P. (1996). *The Great Lakes Water Quality Agreement: Its past successes and uncertain future.* Hanover: Dartmouth College.

Botts, L., & Muldoon, P. (2005). *Evolution of the Great Lakes Water Quality Agreement.* East Lansing: Michigan State University Press.

Boundary Waters Treaty. (1909). Treaty relating to boundary waters, and auestions arising between the United States and Canada, 11 Jan. 1909, 12 BEVANS 319, 102 BFSP p. 137, LEGISLATIVE TEXTS, Treaty No. 79, p. 260. Retrieved from http://laws.justice.gc.ca/eng/acts/I-17/page-1.html

Canada EaE. (1997). Great Lakes binational toxics strategy: Canada–United States strategy for the virtual elimination of persistent toxic substances in the great lakes. Retrieved from http://www.ec.gc.ca/grandslacs-greatlakes/default.asp?lang=En&n=941BAF79-1

Canada–Ontario agreement respecting the Great Lakes Basin ecosystem. (1994). Retrieved from http://www.ec.gc.ca/grandslacs-greatlakes/default.asp?lang=En&n=B903EE0D-1

Canada–USA. (2002). Great Lakes binational toxics strategy 2002. Chapter One. Retrieved from http://www.ec.gc.ca/grandslacs-greatlakes/default.asp?lang=En&n=941BAF79-1

Christie, G. C., & Goddard, C. I. (2003). Sea Lamprey International Symposium (SLIS II): Advances in the integrated management of sea lamprey in the Great Lakes. *Journal of Great Lakes Research, 29*, 1–14. doi:10.1016/S0380-1330(03)70474-2

Crane, T. R. (2012). History of and lessons in participatory governance. In V. I. Grover & G. Krantzberg (Eds.), *Great Lakes: Lessons in participatory governance*. Boca Raton, FL: CRC.

Force USFIT. (2005). Great Lakes regional collaboration strategy to restore and protect the Great Lakes. US. Retrieved from http://glrc.us/documents/strategy/GLRC_Strategy.pdf

GLSLR Body. (2007). *Great Lakes–St Lawrence River: Water resources regional body*. Retrieved from http://www.ebr.gov.on.ca/ERS-WEB-External/displaynoticecontent.do?noticeId=MTAxNDQy&statusId=MTUxNjUx

Great Lakes Compact. (1968) PUBLIC LAW 90-419-JULY 24, 1968. Retrieved from http://www2.ohiodnr.com/portals/soilwater/pdf/greatlakescompact/Great%20Lakes%20Basin%20Compact%201968.pdf

Great Lakes–St Lawrence river basin sustainable water resources agreement. (2005). Retrieved from http://www.cglg.org/projects/water/docs/12-13-05/great_lakes-st_lawrence_river_basin_sustainable_water_resources_agreement.pdf

Great Lakes–St Lawrence river basin water resources compact. (2005, December 13). Retrieved from http://www.cglg.org/projects/water/docs/12-13-05/great_lakes-st_lawrence_river_basin_sustainable_water_resources_agreement.pdf

Great Lakes Water Quality Board. (1985). 1985 Report on Great Lakes water quality. Windsor, Ontario, June 1985.

Hall, N. (2006). Toward a new horizontal federalism: Interstate water management in the Great Lakes Region. *University of Colorado Law Review, 77*, 405–455.

Hall, N. D. (2009). Introduction: Canada–United States transboundary environmental protection. *Windsor Review of Legal and Social Issues*, *26*, 1–6.

Hansen, M. J., & Holey, M. E. (2002). Ecological factors affecting the sustainability of Chinook and coho salmon populations in the Great Lakes, especially Lake Michigan. In K. D. Lynch, M. L. Jones, & W. W. Taylor (Eds.), *Sustaining North American salmon: Perspectives across regions and disciplines*. Bethesda: American Fisheries Society.

Hansen, M. J., Peck, J. W., Schorfhaar, R. G., Selgeby, J. H., Schreiner, D. R., Schram, S. T., Swanson, B. L., MacCallum, W. R., Burnham-Curtis, M. K., Curtis, G. L., Heinrich, J. W., & Young, R. J. (1995). Lake trout (Salvelinus namaycush) populations in Lake Superior and their restoration in 1959–1993. *Journal of Great Lakes Research, 21(Supplement 1),* 152–175. doi:10.1016/S0380-1330(95)71088-7

Holmes, J. W. (1981). Introduction to The International Joint Commission seventy years on. Toronto: Centre for International Studies, University of Toronto.

IJC. (1990). Fifth biennial report on Great Lakes Water quality. Retrieved from http://www.ijc.org/files/publications/C85.pdf

IJC. (1991). Cleaning up our Great Lakes: A report on toxic substances in the Great Lakes basin ecosystem. Windsor, ON: IJC.

IJC. (1992). Air quality in the Detroit–Windsor/Port Huron–Sarnia Region. IJC report to governments.

IJC. (1994). *Seventh biennial report under the Great Lakes water quality agreement of 1978 to the governments of the United States and Canada and the state and provincial governments of the Great Lakes basin*. International Joint Commission. Retrieved from www.ijc.org/php/publications/html/7bre.html

IJC. (1995–97). Priorities and progress under the Great Lakes water quality agreement: Chapter six: Other priorities: Annex 2 Review, Indicators, Pesticides And Radioactivity. Retrieved from http://www.ijc.org/rel/boards/iitf/pr9597.html

IJC. (2000). Tenth biennial report on Great Lakes water quality. Retrieved from http://www.ijc.org/files/publications/C176.pdf

IJC. (2003). The status of restoration activities in the Great Lakes AOCs report. Retrieved from http://www.ijc.org/files/tinymce/uploaded/documents/reportsAndPublications/aoc_report-e%202003.pdf

IJC. (2004). Twelfth biennial report on Great Lakes water quality. Retrieved from http://www.ijc.org/files/publications/C200.pdf

IJC. (2011). 15th Biennial report. Windsor, Ontario.

Jetoo, S., & Krantzberg, G. (2014). Donning our thinking hats for the development of the Great Lakes nearshore governance framework. *Journal of Great Lakes Research. 40*, 463–465. doi:10.1016/j.jglr.2014.01.015

Jetoo, S., & Krantzberg, G. (in press). A SWOT analysis of the great lakes water quality agreement. *Electronic Green Journal.*

Krantzberg, G. (1999). Deciding When to intervene. Data interpretation tools for making sediment management decisions beyond source control Sediment Priority Action Committee, Great Lakes Water Quality Board. Retrieved from http://ijc.org/php/publications/html/sedwkshp/intro.html

Krantzberg, G., Bratzel, M., & MacDonald, J. (2006). Contribution of the International Joint Commission to Great Lakes renewal. *The Great Lakes Geographer, 13*, 25–37.

Krantzberg, G., & Zarull, M. A. (2001). Sediment management: Ecological and ecotoxicological effects must direct actions. *Water Quality Research Journal Canada, 36*, 367–376.

LeMarquand, D., & Scott, A. (1980). Canada's environment relations. In O. P. Dwivedi (Ed.), *Resources and the environment: Policy perspectives for Canada.* Toronto: McClelland and Stewart.

Madenjian, C. P., O'Gorman, R., Bunnell, D. B., Argyle, R. L., Roseman, E. F., Warner, D. M., Stockwell, J. D., & Stapanian, M. A. (2008). Adverse effects of alewives on Laurentian Great Lakes fish communities. *North American Journal of Fisheries Management, 28*, 263–282.

Muldoon, P. (2012). Governance in the Great Lakes – A regime in transition. In V. I. Grover & G. Krantzberg (Eds.), *Great Lakes: Lessons in Participatory Governance.* Boca Raton: CRC Press, Taylor and Francis Group.

O'Gorman, R., & Stewart, T. J. (1999). Ascent, dominance, and decline of alewife in the Great Lakes: Food web interactions and management strategies. In W. W. Taylor & P. C. Ferreri (Eds.), *Great Lakes fisheries policy and management: A binational perspective.* East Lansing: Michigan State University Press.

Ontario. (2006) Clean Water Act. Retrieved from http://www.e-laws.gov.on.ca/html/statutes/english/elaws_statutes_06c22_e.htm

Ontario Ministry of Environment. (2013). Website. Retrieved from http://www.ontario.ca/environment-and-energy/ontarios-great-lakes-strategy?__utma=1.199231031.1416576784.1416576784.1416576784.1&__utmb=1.6.10.1416576784&__utmc=1&__utmx=-&__utmz=1.1416576784.1.1.utmcsr=google|utmccn=(organic)|utmcmd=organic|utmctr=(not%20provided)&__utmv=1.cab|1=tag_visitor_type=external=1&__utmk=246060164

Palay, S. (2009). Muddy waters: Congressional consent and the Great Lakes–St Lawrence river basin water resources compact. *Hastings Constitutional Law Quarterly, 36*, 717–738.

SedPAC. (1997). Overcoming obstacles to sediment remediation in the Great Lakes Basin. White paper by the Sediment Priority Action Committee Great Lakes Water Quality Board International Joint Commission. Retrieved from http://www.ijc.org/php/publications/html/sedrem.html

Squillace, M. (2006). Rethinking the great lakes compact. *Michigan State Law Review*, 1347–1374. University of Colorado Law Legal Studies Working Paper No. 07-01. Retrieved from SSRN: http://ssrn.com/abstract=960574

Stein, R. A., & Goddard, C. I. (2006). Reconciling fisheries with the rehabilitation of the Laurentian Great Lakes. In J. Neilson, J. J. Dodson, K. Friedland, T. R. Hamon, J. Musick, & E. Verspoor (Eds.), *Proceedings of the fourth world fisheries congress.* American Fisheries Society. Retrieved from https://fisheries.org/shop/54049p

Vollenweider, R. A. (1968). The scientific basis of lake and stream eutrophication with particular reference to phosphorus and nitrogen as factors in eutrophication, DAS/CSI/67-27. Paris: Organization for Economic Cooperation and Development.

Water Quality Board. (1977). *Biological monitoring program data.* Water Quality Board.

Water Quality Board. (2001). *Alien invasive species and biological pollution of the Great Lakes basin ecosystem.* Report to the International Joint Commission.

An analysis of Turkey's water diplomacy and its evolving position vis-à-vis international water law

Aysegül Kibaroglu

This article analyzes Turkey's transboundary water policy by examining its institutional framework and basic principles. It explores the reasons why Turkey voted against the UN Watercourses Convention. Turkey's harmonization with the water law of the European Union is also scrutinized with an aim to assess its implications for transboundary water policy making. Turkish water diplomacy faces new challenges, such as the devastating impacts of prolonged droughts as well as ongoing instability and conflicts in Syria and Iraq. Hence, it is imperative for Turkey to systematically reconcile its water policy objectives in accordance with the global norms that are adopted in this field.

Introduction

The 1997 UN Convention on the Law of the Non-navigational Uses of International Watercourses (UNWC) entered into force on 17 August 2014, and a new era of implementation began with its legally binding nature (Convention, 1997). Pivotal countries like Turkey and China, together with Burundi, had voted against the UNWC back in 1997. As the convention goes into effect, a fundamental question remains as to whether these rejecting states will ever consider becoming party to the convention, and if not, how they will get along with their neighbours without its guidance (Chen, Rieu-Clarke, & Wouters, 2013). Turkey, like China, was seen as a "hegemon" (Zeitoun & Warner, 2006, p. 444) in its riparian settings. Therefore, it is useful to look deeper into the factors that led to Turkey's rejection of the convention in 1997, and how its transboundary water policy discourse and practice have evolved dramatically ever since.

In this context, this article will focus on Turkey's transboundary water policy and its position vis-à-vis international water law. It will begin with an analysis of how Turkey's transboundary water policy has been formulated, and in this context, the institutional framework of transboundary foreign policy making will be described. Subsequently, the fundamental principles of Turkish transboundary water policy as defined by the relevant government institutions will be elaborated. The article will then attempt to provide an insightful analysis on the reasons why Turkey rejected the UNWC even though, in its state practice, Turkey has observed the main principles of international customary water law as codified in the Draft Articles (1994).

Turkey's water policy has evolved since the early 2000s, both at national and regional levels, mainly due to political developments at home and the state of affairs in its relations with its neighbours as well as with the European Union (EU). Major changes have occurred in the Turkish water bureaucracy and in the national legal framework over the last decade. Hence, the article will also look into these changes and analyze their impact on Turkey's transboundary water relations with its downstream neighbours and in particular how they shape legal discourse and practices.

Turkish foreign policy and water

Turkey's national water policy aims at achieving a number of fundamental objectives, such as increasing agricultural production and food security; meeting the increasing demand for drinking water in rural and urban areas, as well as in the industrial sector; reducing dependency on imported energy sources; reducing regional, economic and social discrepancies in the country; and increasing the welfare of the society (Scheumann, Kibaroglu, & Kramer, 2011). Turkey's transboundary water policy is formulated in accordance with national socio-economic development goals and taking into consideration its specific geographical and historical context.

Water became an issue in Turkish foreign policy in the early 1980s with the planning and building of a series of large-scale dams and giant irrigation projects in the Euphrates-Tigris (ET) river basin within the context of the South-Eastern Anatolia Development Project (Turkish acronym, GAP) with a view to achieving social and economic development in the country. The ET basin is one of 25 basins in Turkey, but it accounts for nearly one-third of the country's surface water resources and one-fifth of its irrigable land. Hence, since the late 1970s, consecutive Turkish governments had followed the strategic policy of full development of water and land resources of the ET basin, with the aim of minimizing the national economy's dependency on imported oil as well as expanding irrigation projects with an aim to establish agro-based industries in coping with unemployment and underdevelopment in the region.

In line with the classical school of thinking in international relations, geography and history play a significant role in transboundary foreign policy making (Rice, 2013). In this context, the state of affairs in the relations with the neighbouring states since the first years of the Republic of Turkey (1923) has determined the main frame of transboundary water policy. Hence, from the early 1920s until the late 1950s Turkey had smooth relations in its neighbourhood, when Turkey and its neighbours were all establishing state bureaucracies and all had similar concerns and similar needs for socio-economic development. Throughout that period, Turkey signed various bilateral treaties, which mostly consisted of the delimitation and use of boundary rivers (Kibaroglu, Scheumann, & Kramer, 2011). Also in this period, none of the countries engaged in major development projects which would have resulted in utilization of water by each to the detriment of the others (Kibaroglu, Sumer, & Scheumann, 2012). On the other hand, the Cold War period also had a decisive impact on Turkey's regional and bilateral relations with its neighbours, in its transboundary water policies. During the Cold War it was not possible for Turkey, as a NATO member, to develop fruitful relations with Syria, Iraq or Bulgaria over the waters of the Orontes, Euphrates-Tigris, or Meric River, respectively. Moreover, from the 1950s onward, Turkey, along with its neighbours, had built technical and financial capacity to carry out large-scale water development projects, which put Turkey at odds with its neighbours, particularly in the ET river basin, where uncoordinated and competitive transboundary water policies became the norm.

On the other hand, since the early years of its republican history, Turkey has adopted a foreign policy of building institutional structures with its neighbours, which were in the 'enemy' camp during the Cold War period, by way of conducting negotiations, signing treaties, and establishing temporary or permanent technical committees, using instruments of diplomacy and international law. For instance, the Arpacay Dam, jointly built with Armenia (Protocol, 1963), the 1987 protocol signed with Syria as an (interim) agreement for sharing the waters of the Euphrates River (Protocol, 1987), and a series of protocols and agreements signed with Bulgaria are testaments to Turkey's determination to solve disputes with its neighbours over transboundary water resources in a peaceful manner, as envisaged in the UN Charter as well as in customary international law.

The geographical characteristics of Turkey have also played a key role in the determination of its transboundary water policy. Turkey is in an upstream position with respect to the ET, Coruh and Aras Rivers, and in a downstream position with respect to the Meric and Orontes. The total surface area of the drainage basins of these rivers is 256,000 km^2, and they constitute one-third of Turkey's total surface area. The average water potential of these rivers inside Turkey is 70 km^3/y, equivalent to 30% of the overall water potential of the country. Transboundary rivers comprise 22% of Turkey's borders with its neighbours, namely Greece, Bulgaria, Georgia, Armenia, Azerbaijan, Iran, Iraq and Syria. Moreover, Turkey has a significant amount of arable and irrigable land in these river basins; the ET basin alone constitutes 20% of the arable land. With the growing numbers and the needs of the urban and rural population in these basins, the role of geography stands out in the formulation of the transboundary water policy (Kibaroglu et al., 2011).

Institutional development in the making of transboundary water policy

The fundamental foreign policy principles with respect to transboundary waters have been determined and institutionalized, in bureaucratic circles, especially since the 1980s, within the framework of the conditions that are shaped by the physical as well as human geography under the influence of the global, regional and bilateral relations that evolved in the second half of the twentieth century.

With the increasing profile of the GAP project in the international arena, a bureaucratic structure has evolved since the 1980s, in which the principles and policies with regard to transboundary waters have been determined. A department in charge of regional and transboundary waters has been formed in the Ministry of Foreign Affairs (MFA), under the Directorate General, which is responsible for issues pertaining to energy, water and the environment. Hence, the MFA has become the main official body for formulating as well as executing Turkey's transboundary water policy. MFA authorities have time and again emphasized that Turkey views water as a catalyst for cooperation rather than a source of conflict (Republic of Turkey, Ministry of Foreign Affairs, 2014). Traditionally, Turkey has also stressed the principle of good neighbourliness, which considers other riparians' interests in dealing with 'transboundary' and 'international' rivers. However, Turkish official discourse explicitly distinguishes between these two terms, considering 'international' rivers to be only those that constitute a border between two or more countries, such as the Meric River, which forms the border between Turkey and Greece, and the Arpacay River (Aras Basin), where it forms the border between Turkey and Armenia. While such "boundary" rivers are to be shared "equally" between the riparian countries, the water of "transboundary" rivers should be allocated "equitably" (Ministry of Foreign Affairs, 1996, pp. 2–9).

The relevant institutions within the state mechanism which provide technical information, such as the status of water resources in terms of quality and quantity, as well as those responsible for the development, management and preservation of water resources, have been entrusted to work in conjunction with the MFA in the formulation of the fundamental principles of foreign (water) policy. First and foremost among these institutions is the State Hydraulic Works (Turkish abbreviation DSİ), which has been responsible for development and management of Turkey's water resources since 1954.

The DSİ, has realized, especially in the GAP region, a series of projects for energy, drinking water, irrigation, flood control and drought management. Its mandate encompassed collection of data pertaining to the quality and the quantity of the surface and groundwater resources in 25 river basins in Turkey, as well as planning, constructing and management of the projects that aim at meeting the demand for water for drinking and use in agriculture, industry, energy and the related sectors. DSİ has provided important technical inputs to the MFA in the formulation of the principles of transboundary water policy, despite the fact that its tasks have recently been partially taken over by the private sector, stakeholders (irrigation associations) and other state institutions (Kibaroglu, Baskan, & Alp, 2009).

Moreover, in 2011, within the restructuring process of the government, the mandate of the Ministry of Environment and Forestry was reorganized under two new ministries. Hence, in this context, the Ministry of Forestry and Water Affairs (MoFWA) was established. Under the MoFWA, the Directorate General for Water Management and the affiliated Department of Water Law and Political Development were established as new public institutions to provide new inputs to transboundary water policy. The main responsibilities of the Department of Water Law and Political Development include studying national and international water law and legislation, scrutinizing international conventions, and leading efforts towards harmonization with regional and global water law instruments in coordination with concerned public institutions. In this context, the department is expected to contribute to the making of transboundary water policy. Moreover, the newly established Turkish Water Institute (Turkish abbreviation SUEN) under the MoFWA is entrusted with the objective of conducting and supporting scientific research to strengthen Turkey's national and international water policy.

Impact of the EU Water Framework Directive on Turkey's water policy

The declaration of Turkey as a candidate country for the EU in December 1999 triggered a wave of changes in Turkey's water management policies. Turkey has started to harmonize its domestic legislation with that of the EU in the field of the environment and water resources. Moreover, the harmonization of the Turkish legislation and the restructuring of institutions were accelerated with the accession talks that were launched 3 October 2005. Within the context defined by the Helsinki Summit, beginning from the early 2000s, a number of steps were taken by Turkey to facilitate the process of harmonization with the EU Water Framework Directive (WFD) and its related EU-level legislation (European Communities, 2000). As Reichert (2014) argues, "The EU WFD constitutes a legal interface between EU water law and international water law, thereby forming the centerpiece of an increasingly elaborate multi-level governance regime for the protection and management of transboundary freshwater resources in Europe."

Accordingly, a series of regulations (by-laws) on water quality and environmental protection have been adopted since 2004 (Orhan & Scheumann, 2011). In this context, the main tasks that should be performed by Turkey in the framework of the WFD are summarized as follows: creating a reliable inventory of water data; establishment of a proper monitoring system; setting up pricing systems for all sectors, taking into account the full cost recovery principle; realizing participation of all interested parties in the processes of water management; and designation of river basin management plans, with a view towards implementing the programme of measures to reach the environmental objectives.

In designing river basin management plans, the transboundary nature of the river basin is specifically considered. Thus, according to Article 13 of the WFD, "If transboundary effects occur within a river basin, the EU member states concerned must establish an international River Basin District (RBD) and coordinate the implementation of the EU WFD through a single River Basin Management Plan (RBMP). Where a RBD extends beyond the territory of the EU, the EU member states concerned must seek appropriate coordination with the non-EU riparians in order to achieve the EU WFD objectives" (European Communities, 2000, p. 16).

Through the initiatives of the Directorate General for Water Management, under the MoFWA, with an aim to improve water quality in river basins in line with the environmental objectives of the WFD, river basin protection action plans have been drafted for several river basins of Turkey since 2009 (DG Water Management, 2009). These plans are intended to function as models for RBMPs. Thus, Turkey adopts a stepwise approach to preparing the RBMPs (Republic of Turkey, 2009). Yet, Turkish authorities underlined that Article 13 of the WFD, particularly the clauses related to "transboundary coordination with the non-EU riparians", would be implemented when Turkey became a full member of the EU.

On the other hand, Turkey has made serious efforts to adapt the basic approach of the EU WFD to its specific needs. Hence, the new approaches developed by the concerned ministries within the framework of the relations with the EU have played a key role in the memoranda of understanding (MoUs) signed with Syria and Iraq in 2009 with respect to the use, development and protection of the water resources of the Euphrates, Tigris and Orontes Rivers (Memorandum of Understanding between Turkey-Syria, 2009; Memorandum of Understanding between Turkey-Iraq, 2009).

These bilateral MoUs were signed by the ministers concerned and reflected, in fact, the political will and strong commitment of the highest-level political decision makers, namely the prime ministers and presidents of Turkey, Iraq and Syria. Yet, they did not become legally effective and binding because they could not fulfil the necessary legal procedures in the respective parliaments. Turkey was eager to go ahead and implement the MoUs in line with the basic norm of international law, namely *pacta sunt servanda*. Yet, the eruption of uprisings in Syria and the subsequent deterioration of Turkish–Syrian relations made it impossible to pursue this goal. On the other hand, the Iraqi Parliament constantly criticized Turkey's use of water upstream and became vocal over water rights, and consequently did not approve the bilateral MoU with Turkey.

Although the MoUs did not become legally effective, the concepts used in the texts of these MoUs, such as river-basin-level water management, establishing emission standards, polluter pays, and full cost recovery, reflect the terminology used in the EU WFD. The bureaucracy concerned (then at the Ministry of Environment and Forestry) therefore supported these protocols vigorously because they felt that their implementation would be a useful practice for the implementation and extension of the new water legislation in Turkey, transposed from the EU water legislation (Kibaroglu & Scheumann, 2013). In other words, the bilateral MoUs between Turkey and Syria and between Turkey and Iraq shared the catalogue of aims and principles laid down by the EU WFD. Most importantly, both the Memorandum of Understanding between Turkey-Syria (2009) and the Memorandum of Understanding between Turkey-Iraq (2009) took the river basin as the managerial unit for the protection and management of freshwater resources (Sümer, 2011). In this respect, Turkish authorities considered the WFD the basic template to derive and develop national water policies and practices within the framework of legal and institutional restructuring at the domestic level, including the effort to draft its new (framework) water law. On 8 October 2012, a new draft law was published by the MoFWA for consultations with various stakeholders, including NGOs, universities and water users (Su Kanun Tasarisi Taslagi, 2012). It aims to eliminate the current situation of fragmented water management and to create an efficient governance scheme, in which the MoFWA would be the main authority. The draft law adopts the view that water management should be based on the river basin. Furthermore, the draft enshrines the principle of full cost recovery. In this respect, it is a step forward in complying with some of the basic principles of the WFD.

However, there are limitations to this as well. To illustrate, although the draft law outspokenly targets the development of a participatory framework for water management, it falls short of explaining how this participation is to be realized (Sümer, 2013). Another example is the management of transboundary waters. There is simply no reference to "transboundary waters". The draft law only mentions "bodies of water forming boundaries". This is mainly because the draft water law was formulated with a basic aim of refining and modernizing existing water legislation and bringing it in line with EU water law. It was quite complicated and frustrating to reach some sort of consensus over a draft water law among various stakeholders at the domestic level. Hence, its authors tried to avoid further complications by excluding transboundary water issues, in line with recommendations from officials at the MFA (personal communication, Ministry of Forestry and Water Affairs, Ankara, March 2014). As of October 2014, the draft law is pending to be transferred to Parliament for further consultations and possible amendments.

Basic principles of Turkish water diplomacy and its position vis-à-vis customary international water law

In this section, Turkey's state practice will be analyzed by scrutinizing the official manuals on legal questions and informational notes published on the websites of the concerned ministries, namely the MFA and MoFWA, as well as statements made by the authorities concerned at international conferences and meetings of international organizations. For analyzing Turkey's stance towards international water law, it is necessary to examine, among others, the fundamental principles of Turkish foreign policy making. These principles constitute the primary sources and evidence of state practice. The MFA has published these principles on its website (Republic of Turkey, Ministry of Foreign Affairs), while the MoFWA elaborates on them in relevant official reports (Orman ve Su İsleri Bakanligi, 2013).

By reviewing these principles, one can observe the major influences of the core norms of customary international water law, as well as Turkey's upstream dilemma, specifically as it relates to the ET basin. In fact, the ET basin constitutes a reference point in Turkey's transboundary water policy. Not only the specific principles related to the basin but also the first set of principles concerning Turkey's overall foreign policy posture have mainly been generated from Turkey's transboundary water relations with Syria and Iraq. Thus, the first principle that Turkey adopts, that water is a basic human need, aims at expressing the understanding that Turkey will always intend to release water (as much as is available under the given hydrological and meteorological conditions) to meet the basic needs of the downstream riparians (Eroglu, 2014). Moreover, this principle also reflects the official

Turkish position that 'needs' rather than 'rights' enable the 'equitable use' principle to be operational. Turkey's needs-based approach was expressed in the Three-Stage Plan put forward by the technocrats of its central water agency, the DSİ, back in 1984. Accordingly, inventory studies of the water and land resources of the ET basin throughout the region comprising the territories of the various states would be undertaken and jointly evaluated by Iraq, Syria and Turkey. On the basis of these studies, the means and measures needed to attain the most reasonable, optimum utilization of resources would be defined (Minutes, 1990).

While endorsing the principles of equitable utilization and no significant harm, Turkey intends to reassure that the notions of 'sovereign right to use of water' and 'equitable, reasonable and optimum use' (rather than simply 'equal use') are the defining principles of Turkish transboundary water policy. Yet, Turkey objects to the claim of the downstream countries that they should have the right of co-sovereignty on the waters of the upstream country, or vice versa (Kibaroglu, 2002). To explore the upstream dilemma for (mainly) upstream states like Turkey, Wouters (2013, pp. 374–375) identifies the concept of "sovereignty paradox", which constitutes a challenge posed by the local–international interface in international relations among riparian states. Turkey's strong argument in claiming its sovereign rights over the portions of transboundary rivers that are situated on its territory basically originates from the foundational principles of international law as codified in the UN Charter as the principle of the sovereign equality of all its members as well as territorial integrity of the members. However, as in numerous UN resolutions and international instruments (hard and soft law), including treaties and transboundary water agreements, Turkey attempts to balance the issues arising out of the sovereignty paradox by putting forward a bunch of cooperation initiatives with respect to its transboundary waters—concluding bilateral treaties, establishing joint mechanisms, and initiating joint projects, e.g. joint dams and joint technical trainings on water use and efficiency (Turan, 2011).

Turkish authorities underline the principle that "transboundary water disputes should be solved between the riparian countries, third party involvement and mediation should not be supported" (Orman ve Su İsleri Bakanligi, 2013, pp. 33–34). Through its experience with donor agencies during the construction of the Keban and Karakaya dams on the Euphrates River in the late 1970s, Turkey developed a negative stance towards the possibility of third-party mediation (or what was seen by MFA officials as third-party intervention) in the issue. International donors did not give much weight to the essential Turkish view that a fair distribution of Euphrates waters could be made by taking into consideration the long-term projects and the needs of the three countries, along with the possibility of transferring water from an adjacent river, namely the Tigris. Turkey contended that the donors' intervention was solely in favour of protecting the rights of the downstream riparians and gave slight recognition to Turkish rights to develop and use the river system (Gürün, 1994).

On the other hand, a significant portion of those principles relate to the ET basin, showing its peculiarity among the other five major transboundary basins that Turkey shares with its neighbours. The principles adopted pertaining to the ET basin are, in fact, the accumulation of years of experience and the practices Turkey developed during diplomatic negotiations with its southern neighbours. Since the 1970s, through bilateral and trilateral negotiations with Iraq and Syria, as well as visits to water and land resource development projects in downstream countries, Turkish authorities came to the conclusion that there would be increasing pressure on the Euphrates due to the magnitude of the planned irrigation projects of all three riparians in the Euphrates region. To deal with this pressure, Turkish negotiators kept advocating that the Euphrates and Tigris rivers constitute a single basin, "due to the fact that they are linked by their natural course when merging at the Shatt-al-Arab in the Gulf, and also because of the Thartar Canal built to connect the two rivers inside Iraq" (Ministry of Foreign Affairs, 1996). Therefore, they suggested that "all existing and future agricultural water uses need not necessarily be derived from the Euphrates" and that "irrigation water for areas fed by Euphrates may also be supplied from the Tigris" (Ministry of Foreign Affairs, 1996, p. 13). This principle, in a way, became an established and traditional principle, without any substantial revision since the mid-1990s.

More than a quarter of a century of living with a water sharing agreement (Protocol, 1987) which has put Turkey under a commitment to release certain amount (500 m^3/s) of water from the Euphrates River to Syria has made Turkish decision makers overly cautious regarding hydro-meteorological conditions in the river system. This obligation urged Turkey to provide the promised amount even in dire conditions of prolonged severe droughts. Thus, another basic principle—that variable natural hydrological conditions must be taken into account in the allocation of the waters of the Euphrates and the Tigris Rivers—has been adopted through that state practice. Likewise, Turkey's emphasis on the principles of achieving efficiency in water use, prevention of pollution, and basin-wide data exchange are the products of the perception developed over the years during the futile negotiation processes in the ET basin, which did not pave the way to efficient and equitable use and management in the basin.

Since the early 2000s, Turkish authorities have adopted new perspectives in dealing with transboundary water issues, namely the 'benefit-sharing' approach, thanks to their increasing participation in global water fora such as the World Water Forum organized every three years by the World Water Council and the World Water Week gathered annually by the Stockholm International Water Institute, as well as their growing understanding and perceptions of the evolving global water management paradigms. Rather than sharing the waters by simple arithmetic, as demanded by Iraq and Syria, Turkey suggested in the 1980s sharing the benefits of water-based development projects and water structures by way of conducting joint inventory studies for water and land resources as a basis for a trilateral final allocation agreement in the ET basin (Ministry of Foreign Affairs, 1996). Turkish policy makers argued that the benefit-sharing approach fitted with Turkey's historical position and provided opportunities for win-win solutions. In this respect, Turkey has come up with more concrete proposals, such as the joint dam development projects in the river basins, as initiatives for enhancing mutual benefits related to hydropower and irrigation.

Turkey and the UN Watercourses Convention

Turkey's position towards the UNWC has constituted one significant aspect of its transboundary water policy. In this context, even though the reasons behind Turkey's rejection of the UNWC were officially stated during the Working Groups in New York in 1997, they were not properly elaborated in such a way as to reflect its concerns and dilemmas to date. Hence, this section aims to perform that task. At the outset, Turkish officials argued that the UN has contributed significantly to the codification and progressive development of the rules of international law that apply to transboundary water resources (Ministry of Foreign Affairs, 1996). Indeed, a thorough analysis of Turkey's state practice reveals that, particularly since the second reading of the Draft Articles on the Law of the Non-navigational Uses of International Watercourses (United Nations General Assembly, 1994), drafting of international water law has become a serious matter for Turkey, which is an upstream country with respect to several transboundary watercourses

(Ministry of Foreign Affairs, 1996). That is to say, Turkey was an interested actor in the drafting as well as the making of the convention and paid particular attention to how the scope and substantive rules evolved in drafting the convention.

Turkish authorities were confident that the idea of regarding transboundary water as a shared resource would not be widely supported in the codification exercises on transboundary waters (Yearbook of the International Law Commission, 1987). During the negotiations in the 1980s, the concept of sharing common water resources through a mathematical formula was proposed by Iraq to guarantee the utilization of water for its projects on the Euphrates River. Syria, with a similar approach, supported this idea. However, that concept turned out to be in contradiction with the principle of equitable utilization, which is the core idea of codification exercises in this field (Ministry of Foreign Affairs, 1996).

Moreover, it was also acknowledged by the Turkish authorities that "the historical and acquired rights" claimed by Syria and especially Iraq were inadequate, in the sense that Article 6 of the draft notes that prior uses of water by downstream countries would represent only one of many factors to be taken into account in reaching an equitable utilization of a transboundary river (Ministry of Foreign Affairs, 1996, p. 18). Furthermore, Turkish diplomats as well as legal advisers (personal communication, Ministry of Foreign Affairs, Ankara, October 24, 1997) carefully followed, and considered it a "very important development" (Ministry of Foreign Affairs, 1996, p. 26), when the key concept of "appreciable harm" (Draft Articles, first reading, 1991) was changed to "significant harm" in the second reading of the Draft Articles (1994).

Thus, the threshold of accountability of an upper riparian state which may have caused damage to a lower riparian state was raised from appreciable harm to significant harm. In this respect, Turkish officials expressed the view that a considerable improvement was achieved in the determination of reciprocal responsibilities of all riparians (Ministry of Foreign Affairs, 1996). In line with this reasoning, the Three-Stage Plan mainly used the terminology developed by the International Law Commission (ILC) of the United Nations, which was entrusted with the codification of the law of non-navigational uses of international watercourses. Hence, an overview of the official Turkish stance with respect to the Euphrates-Tigris river basin leads one to the conclusion that it is in conformity with the essence of the Draft Articles (1994), namely with Part II: General Principles (Kibaroglu, 2002).

Even though Turkey endorsed the primacy of the rule of equitable and reasonable use during the discussions in the UN Sixth (Legal) Committee related to the Draft Articles (Statement of Turkey, 1997), representatives from Turkey expressed serious dissatisfaction with the text of the draft convention in New York in 1997 (Turkish Government's Observations, 1997). Most of the comments in regard to the UNWC relate to its key procedural provisions. During the discussions in the Working Group of the Whole (which was convened by the UN to consider the ILC's Draft Articles of 1994), Turkey was particularly active in advancing its views on the governing rules of law applicable to transboundary watercourses, to ensure that the rights of upstream states are better protected.

Turkish officials asserted that the main purpose of the convention should be to achieve an equitable and reasonable arrangement regulating water utilization among the watercourse states: "Any other approach turns the Convention into a document which unilaterally restricts, in terms of both quantity and quality, the utilization rights of states in which watercourses originate; due attention should also be paid to establishing an equitable balance of rights and obligations among all watercourse states" (Statement of Turkey, 1997, p. 5). Yet, according to Turkish government officials, while these requirements were taken into account to a certain extent in the general principles set forth in Part II of the convention, the same cannot be said of Parts III and IV.

Turkey held that Article 5, the "most fundamental principle of the draft", must be understood "in light of the fundamental principle of the sovereign rights of States over their territory". Essentially, Turkey strongly endorsed Article 5 as the "governing principle" and suggested that factors favourable to their particular situations be added to Article 6 (Statement of Turkey, 1997, pp. 36, 37–40). They should also be applied by taking fully into account all the particularities of the watercourses, including the distinction of whether they are transboundary by nature, or international (forming a boundary between states). In relation to the first paragraph of Article 5, Turkish officials suggested that the principle of optimal utilization should aim at both protecting the watercourse and optimizing the interests of riparian states in a way that avoids water waste. From the Turkish standpoint, the first paragraph of Article 5 does not clearly state that optimal utilization should not be restricted to protection only but should also be seen as comprising the concept of efficient use (Turkish Government's Observations, 1997, p. 3).

In Turkey's view, the draft failed to establish the primacy of the principle of equitable utilization over the obligation not to cause significant harm. Turkey claimed that the no-harm rule was unbalanced and favoured downstream states. Thus, Turkey's concern that the contribution of each watercourse state be included in the formula for the assessment of a reasonable use could be met under the principle of equitable utilization.

The most dramatic change that the Turkish government officials proposed for the convention was deleting Article 7 (the no-harm rule) altogether. The Turkish delegation indicated that this deletion would not completely eliminate the no-harm rule but only subordinate it to the principle of equitable utilization as mentioned in the ILA's Helsinki Rules. To that end, the delegation of Turkey requested that it be made explicit in the text that the no-harm rule would be "without prejudice to the principle of equitable and reasonable utilization" (Turkish Government's Observations, 1997, p. 4).

As Wouters (2013) underlines, the duty to cooperate provides the legal foundation at the core of the UNWC. The broad principle is articulated in Article 8 (Convention, 1997). Turkey favours Article 8 for its wording as well as its objectives because of its emphasis on concepts like sovereign equality, territorial integrity, mutual benefit and good faith in order to attain the "optimal utilization" that is embraced by Turkish water diplomacy (Republic of Turkey, Ministry of Foreign Affairs).

Yet, Turkey reserved its position on procedural rules (Articles 12–19) and proposed that they be replaced with provisions referring primarily to consultations and other methods of cooperation. Moreover, the Turkish government officials totally disagreed with the inclusion of Article 33 (rules of dispute settlement) in the convention on the grounds that "it would be more appropriate not to foresee any compulsory rules as regards the settlement of disputes, and to leave this issue to the discretion of the concerned States" (Turkish Government's Observations, 1997, p. 4). Moreover, Turkish officials argued that "a framework agreement should not attempt to set forth detailed rules in this respect, since it is virtually impossible to respond to the pressing needs of specific and more often than not very complex cases of water disputes" (Turkish Government's Observations, 1997, p. 4). Wouters (1999) surveys the state practice in this field, finding that "there are distinctive regional differences on approaches to dispute settlement" and that "what is important is the commitment to peacefully resolve any disputes that arise". In that manner, Turkey is committed to the fundamental tenet of international law that international disputes must be resolved peacefully.

Turkey's cautious attitude towards the ILC's 1994 draft and its rejection of the 1997 UNWC originated from deep concern about the fact that certain articles, especially in Part III of the convention, might constrain its official stance in future negotiations over the allocation and management of the Euphrates-Tigris river basin. Turkish officials were very keen on the point that the convention should keep the original aim of constituting a framework document. Therefore, they went on asserting that the convention clearly includes provisions which go far beyond that scope. In Turkish contention, the convention should be confined to setting forth the conceptual framework and principles regarding international watercourses. As to specific watercourses, bilateral and regional arrangements between watercourse states have to be concluded by taking into account the characteristics of each river system.

The official discourse developed towards the UNWC in fact built on the knowledge and experience that Turkish diplomats and legal advisers had gained by following the developments in international water law through their interactions in multilateral and bilateral platforms. Academics were also consulted at certain stages to comment on the Draft Articles.

Turkey's negative stance towards the UNWC could also be a result of misconceptions and misperceptions regarding the interpretations of the UNWC, especially as they relate to the issue of dispute settlement and notification processes (Salman, 2013). Turkish authorities have time and again emphasized that Turkey is close to some of world's most volatile regions, such as the Middle East. Hence, the complex regional politics have constituted serious obstacles to basin-wide cooperation and treaty making, as is still the case today (personal communication, Ministry of Foreign Affairs, Ankara, 2004).

With these experiences and perceptions in mind, Turkish officials preferred to stay out of the UNWC ratification process, and they have not considered revisiting this position. Instead, in due course, since 1997, the official discourse developed arguments to justify and explain its stance. Time and again, Turkish authorities stated that the convention has lost its credibility because it took so long for it to enter into force (Rende, 2002; personal communication, Ministry of Foreign Affairs, Ankara, June 2002; February 2010; April 2011; 2014).

Against the legal interpretation that the convention represents the customary international water law, concerned Turkish diplomats claim that Turkey has been a 'persistent objector', particularly with regard to the rules on notification and dispute settlement as codified and progressively developed in the UNWC. Even though Turkey has rejected certain provisions of the UNWC, it has also frequently expressed its adherence to the core norms of international water law, such as the obligation to prevent harm, and equitable and reasonable utilization. Hence, as Kirschner & Tiroch (2012) put forward, on the basis of such express commitment it would be far-reaching to regard Turkey as a persistent objector. Moreover, to be able to act as a persistent objector it is also necessary for the members of the international community to accept this claim. In this particular case, there is no satisfactory evidence showing that the Turkish argument is supported by other states at regional or global levels.

On another front, in Turkey, the nongovernmental entities, namely the water- and environment- related associations, foundations and water user organizations, as well as chambers of various professions, have not displayed an informed and systematic attention towards global legal instruments such as the UNWC. Their attention has mostly been directed towards legal developments at home and the process of harmonization with the EU pertaining to water policy and law (Kibaroglu et al., 2009).

With the entry into force of the UNWC, Turkey's position towards it has been slowly returning to the agenda of the concerned circles in Turkey. Experts have examined the convention from the perspectives of prevailing paradigms and urgent issues; they argue that the convention has a narrow scope and objectives, ignoring the complex implications of climate change for freshwater resources. Moreover, in their contention, the increasing linkages of water, food, energy and environmental security have not been dealt with in the convention (Yildiz, 2014).

Conclusion

In its historical and geographical context, Turkey's transboundary water policy is formulated by taking into consideration, on the one hand, the economic and social developmental needs in the country, and on the other hand, the regional political developments in the Middle East, as well as the obligations undertaken by Turkey within the context of the negotiations regarding accession to the EU.

Turkey's transboundary water policy can be understood and explained by examining its institutional framework as well as the basic principles which have been adopted through its experiences and interactions in transboundary water settings. Moreover, global water management approaches have had impacts on transboundary policy making in Turkey, broadening its vision as well as adding new dimensions.

In transboundary water policy making, decision makers have approved and practised international customary water law, particularly its cornerstone principles. Yet, Turkey did not approve the UNWC, and consequently its relations with global legal instruments became problematic. The reasons why Turkey voted against the UNWC, as well as its evolving position towards international water law, are instructive in exploring its upstream dilemma.

The restructuring of Turkey's water policy framework was very much a product of its interfaces with the EU. As a candidate country to the EU, Turkey has started harmonizing its domestic legislation with that of the EU in the field of environment and water resources, which, in turn, had implications for transboundary water policy making in the ET basin. Hence, as Turkey harmonizes its domestic water law and policy with EU water law, it moves closer to the multilevel governance regimes in international water law for the protection and management of transboundary rivers. The 2009 bilateral protocols (MoUs) with Syria and Iraq reflect these changes.

Turkish water diplomacy faces new challenges, namely the negative impacts of climate change in complex transboundary river settings where basin-wide transboundary management is not the rule. This is particularly problematic in the case of the ET basin, where no trilateral accord has been reached among Iraq, Syria and Turkey. Geographical incompleteness might undermine riparian states' ability to advance collective management in the face of climate change. Rising water demands and climate change–induced variability and uncertainty thus threaten the survival of existing water-sharing agreements, while simultaneously making better, collective management more urgent.

Regional political developments in the Middle East have always affected Turkey's transboundary water relations with its downstream neighbours in the ET basin. In the first decade of the 2000s Turkey tried to create a cooperative framework with the contributions of Syria and Iraq. Yet, the implementation of these initiatives, namely the MoUs signed in 2009, have not been put into force due to the eruption of political instability, crisis and violence in the region. Moreover, in addition to the devastating impacts of prolonged droughts and the civil-war conditions in Syria, the nonstate (illegal) actors, namely the so-called Islamic State (or ISIS), have notoriously emerged and started to terrorize millions of people in Iraq and Syria, for example by taking control of major water structures in both countries. Hence, what Turkey says and what Turkey does in its transboundary water relations matter, given the current state of affairs, particularly in its southern neighbourhood. In these critical conditions, one crucial task for the international community would be to establish an effective dialogue with Turkey to systematically reconcile its water policy objectives with the global norms that exist in this field, such as the UNWC, and also with the need for a predictable framework for conflict management in transboundary basins.

References

Chen, H., Rieu-Clarke, A., & Wouters, P. (2013). Exploring China's transboundary water treaty practice through the prism of the UN Watercourses Convention. *Water International, 38*, 217–230. doi:10.1080/02508060.2013.782134

Convention on the Law of the Non-navigational Uses of International Watercourses. (1997). Retrieved from http://untreaty.un.org/ilc/texts/instruments/english/conventions/8_3_1997.pdf

DG Water Management. (2009). Retrieved from http://suyonetimi.ormansu.gov.tr/anasayfa/resimlihaber/13-06-20/HAVZA_KORUMA_EYLEM_PLANI_TASLAKLARI_HAZIRLANDI.aspx?sflang=tr

Draft Articles on the Law of the Non-Navigational Uses of International Watercourses and Commentaries Thereto. (1994). Adopted on the second reading by the International Law Commission at its forty-sixth session. (12 July 1994). A/CN.4/L.493.

Eroglu, V. (2014). *Turkey favors fair share of waters of the Tigris-Euphrates*. Retrieved from http://www.aa.com.tr/en/turkey/89290–turkey-favors-fair-share-of-waters-of-tigris-euphrates, 10 October 2012

European Communities. (2000). Directive 2000/60/EC of the European Parliament and of the Council of 23 October 2000 establishing a framework for community action in the field of water policy, *Official Journal of the European Communities*, L327, 22.12.2000.

Gürün, K. (1994). *Akıntıya Kürek Çekmek: Bir Büyükelçinin Anıları. [Rowing against the flow: Memoirs of an ambassador]*. Istanbul: Milliyet Yayınları.

Kibaroglu, A. (2002). *Building a regime for the waters of the Euphrates-Tigris river basin*. The Hague: Kluwer Law International.

Kibaroglu, A., Baskan, A., & Alp, S. (2009). Neo-Liberal transitions in hydropower and irrigation water management in Turkey: Main actors and opposition groups. In D. Huitema, and S. Meijerink (Eds.), *Water policy entrepeneurs. A research companion to water transitions around the globe* (pp. 287–304). Cheltenham: Edward Elgar.

Kibaroglu, A., & Scheumann, W. (2013). Evolution of transboundary politics in the Euphrates-Tigris river system: New perspectives and political challenges. *Global Governance, 19*, 279–307.

Kibaroglu, A., Scheumann, W., & Kramer, A. (Eds.). (2011). *Turkey's water policy.* Berlin: Springer.

Kibaroglu, A., Sumer, V., & Scheumann, W. (2012). Fundamental shifts in Turkey's water policy. *Journal of Mediterranean Geography (The journal Méditerrané), 119*, 27–34.

Kirschner, A., & Tiroch, K. (2012). The waters of the Euphrates and Tigris: An international law perspective. In A. von Bogdandy and R. Wolfrum (Eds.), *Max Planck UNYB 16*. Leiden: Martinus Nijhoff Publishers.

Memorandum of Understanding between Turkey-Syria. (2009). Memorandum of Understanding between the Government of the Republic of Turkey and the Government of the Syrian Arab Republic in the field of remediation of water quality. 23 December 2009 (published in Turkish on the Turkish Grand National Assembly website). (on file with the author).

Memorandum of Understanding between Turkey-Iraq. (2009). Memorandum of Understanding between the Ministry of the Environment and Forestry of the Republic of Turkey and the Ministry of Water Resources of the Republic of Iraq on Water, 15 October 2009 (published in Turkish on the Turkish Grand National Assembly website). (on file with the author).

Ministry of Foreign Affairs. (1996). *Water issues between Turkey-Syria and Iraq, Manual*. Ankara: Department of Regional and Transboundary Waters.

Minutes of the Fifteenth Meeting of the Joint Technical Committee. (1990). Ankara. On file with the author.

Orhan, G., & Scheumann, W. (2011). Turkey's water policy for combating water pollution. In A. Kibaroglu, W. Scheumann, and A. Kramer (Eds.), *Turkey's water policy* (pp. 117–139). Berlin: Springer.

Orman ve Su İsleri Bakanligi.[Ministry of Forestry and Water Affairs]. (2013). *Havza Yönetimi ve Su Bilgi Sistemi Calisma Grubu Raporu.[Basin management and water information system working group report]*. Ankara. (on file with the author).

Protocol. (1963). Protocol on the meeting of the Turkish-Soviet Joint Commission pertaining to the construction of a joint dam on Arpacay (Ahuryan), 25/04/1963. (on file with the author).

Protocol. (1987). Protocol on matters pertaining to economic cooperation between the Republic of Turkey and the Syrian Arab Republic. United Nations Treaty Series 87/12171, 17/7/1987.

Reichert, G. (2014). *Entry into force of the UN Watercourses Convention – should europe care*. Retrieved August 21, 2014, from http://www.internationalwaterlaw.org/blog/2014/07/21/dr-gotz-reichert-entry-into-force-of-the-un-watercourses-convention-should-europe-care/

Rende, M. (2002). Head of the Department of Regional and Transboundary Waters, MFA, The Looming global water shortage and Turkey's water management in a transboundary context. NATO Parliamentary Assembly 48th Annual Session. (16–19 November 2002). Istanbul.

Republic of Turkey. (2009). Plan for setting up necessary administrative capacities at national, regional and local level and required financial resources for implementing the environmental acquis. Ankara. (on file with the author).

Republic of Turkey, Ministry of Foreign Affairs. (2014). *Turkey's policy on water issues*. Retrieved August 22, 2014, from http://www.mfa.gov.tr/turkey_s-policy-on-water-issues.en.mfa

Rice, D. (2013). *Reinhold Niebuhr and his circle of influence*. New York, NY: Cambridge University Press.

Salman, M. A. S. (2013). Misconceptions regarding the interpretation of the UN Watercourses Convention. In F. R. Loures and A. Rieu-Clarke (Eds.), *The UN watercourses convention in force* (pp. 28–35). London: Routledge.

Scheumann, W., Kibaroglu, A., & Kramer, A. (2011). Trajectories of Turkey's water policy. In A. Kibaroglu, W. Scheumann and A. Kramer (Eds.), *Turkey's water policy* (pp. xxiii–xxx). Berlin: Springer.

Statement of Turkey. (1997). UN GAOR, 51st Session, 99th Plenary Meeting, UN Doc. A/51/PV.99.

Su Kanun Tasarisi Taslagi. [Draft Water Law]. 2012. Retrieved August 21, 2014, from www.ormansu.gov.tr/osb/haberduyuru/duyurular/12-10-08/Su_Kanunu_Tasla%C4%9F%C4%B1_G%C3%B6r%C3%BC%C5%9Flere_a%C3%A7%C4%B1ld%C4%B1.aspx?sflang=tr

Sümer, V. (2011). The European Union water framework directive and Turkey's water management policy: An analysis (Unpublished PhD Dissertation). Middle East Technical University, Ankara.

Sümer, V. (2013). Legal context of water management policy in Turkey. In A. Kibaroglu, A. Kirschner, S. Mehring, and R. Wolfrum (Eds.), *Water law and cooperation in the Euphrates-Tigris region* (pp. 229–257). Leiden: Brill.

Turan, I. (2011). The water dimension in Turkish foreign policy. In A. Kibaroglu, W. Scheumann, and A. Kramer (Eds.), *Turkey's water policy* (pp. 179–197). Berlin: Springer.

Turkish Government's Observations. (1997). Turkish Government's observations on the draft articles on the law of the non-navigational uses of international watercourses. Report sent to the United Nations General Assembly, Sixth Committee, as Member-States Opinions on the Draft Articles, U.N. Doc. A/C.6/51/SR.61.

United Nations General Assembly. (December 9, 1994). Res 49/52 *Draft articles on the non-navigational uses of international watercourses*. UN Doc A/Res/49/52.

Wouters, P. (1999). The legal response to international water conflicts: The UN watercourses convention and beyond. *German Yearbook of International Law, 42*, 293–336.

Wouters, P. (2013). Sovereignty revisited –examining the rules of international law that govern transboundary water resources with a focus on upstream/downstream state practice–possible lessons learned for the Euphrates-Tigris. In A. Kibaroglu, A. Kirschner, S. Mehring, and R. Wolfrum (Eds.), *Water law and cooperation in the Euphrates-Tigris region* (pp. 373–403). Leiden: Brill.

Yearbook of the International Law Commission. 1987, Volume II, Part 2, A/CN.4/SER.A/1987/Add.1 (Part 2).

Yildiz, D. (2014). *Can we do it with 1997 UN Watercourses Convention.* Retrieved from http://www.hidropolitikakademi.org/en/what-will-1997-un-watercourses-convention-bring-in-new.html

Zeitoun, M., & Warner, J. (2006). Hydro-hegemony—a framework for analysis of transboundary water conflicts. *Water Policy, 8,* 435–460. 10.2166/wp.2006.054

Identifying China's transboundary water risks and vulnerabilities – a multidisciplinary analysis using hydrological data and legal/institutional settings

Yan Feng, Daming He and Wenling Wang

This paper identifies the vulnerabilities and risks linked with China's transboundary waters through an analysis of hydrological data and the legal and institutional settings. The risks and vulnerabilities arise in three areas: (1) serious issues of water security arise, both internally and externally, relating to water quantity and water quality; (2) transboundary waters management is hampered by weak capacity across the region; and (3) underdeveloped legal regimes, including procedural requirements for the exchange of information, prior notification and early warning mechanisms, make it difficult to realize effective international transboundary water cooperation.

Introduction

According to a report published in 2003 by United Nations Environment Programme (UNEP) during 'The Third World Water Forum', over 150 of 276 international rivers in the world, with around 60% of the global flow of available freshwater, are liable to cause transboundary disputes. Many of these rivers are located in developing regions, such as Asia and Africa (Duncan et al., 2012; United Nations, 2003). As global freshwater becomes more scarce, key issues relating to transboundary waters, such as reasonable utilization and coordinated management, equitable allocation and benefit sharing, and the maintenance of water eco-security are receiving more attention (Chen, Rieu-Clarke, & Wouters, 2013; Dombrowsky, 2007; Earle, Jagerskog, & Ojendal, 2010; Eleni, Eleni, & Yannis, 2007; Mirumachi, 2008; Mostert & Barraqué, 2006; Wouters & Chen, 2013).

As regards China's transboundary waters, the past 10 years have witnessed events along international rivers that have resulted in a range of adverse impacts, including heavy economic losses, regional water security impacts due to hydropower development of the Lancang River (upstream Mekong) (Goh, 2004; Zhiguo, He, & Feng, 2011), damage to downstream communities due to glacial-lake outburst and floods in Tibet (Cui, Dang, Cheng, & Scott, 2010), the capture of water in the north-west region due to increasing demand (Zhao, 2005), and cross-border water pollution along the Songhua River at the end of 2005 (Zhang et al., 2010).

China's transboundary waters resources management continues to evolve as one of the most important emerging geopolitical issues for the nation and the region. Three areas of vulnerabilities and risks linked to China's transboundary waters can be identified. First, the outflows from the international rivers of China are decreasing due to increasing demand and consumption of water within China and the adverse effects of climate change, which affects downstream countries, including the eco-security of watersheds. Further, the degradation of China's water resources also affects the water security of China and its neighbours. Second, the capacity to deal with these complex problems is limited across the basin, making effective responses difficult. Third, effective transboundary water management is impeded by fragmented water management practices, including under-developed legal regimes at the domestic and regional levels. This paper uses hydrological data and water quality data (including changing water volume over the past 30 years) to analyse water policy and institutional mechanisms with a view to identifying the vulnerabilities and risks associated with China's transboundary waters.

Background on China's transboundary waters

China, located at the centre of Asia with approximately 220,000 km of borders shared with 14 neighbouring countries, is the source of several of the great rivers of the world and is the key water source of the Asian continent. The international rivers in China mainly flow in three regions: first, in the north-east (NE) region the major international rivers are the Helong-Amur (including the Wusuli/Ussuri, Khanka Lake, etc.), Suifen, Tumen and Yalu. The total water volume of this region is 156.5 km^3, 135.2 km^3 of which is from the Helong, 16.2 km^3 from the Yalu and 5.1 km^3 from the Tumen. Second, the north-west (NW) region has five major international rivers: Irtysh-Ob, Ili, Emin, Tarim and Ulungur. The total annual volume of water resources is around 35 km^3, with an outflow of 24 km^3 and an inflow of 8.8 km^3. This region is located in the hinterland of the Eurasian continent, and is the driest area of China. Within the south-west (SW) region there are eight cross-border rivers that cluster together from west to east, including: the Senggecangbu/ Seng-geKambab/Lion (upstream of the Indus), Pengqu (upstream of the Ganges), Brahmaputra, Dulong-Irrawaddy, Nu-Salween, Lancang-Mekong, Yuan-Red and Pearl. In 2011, the annual outflow of this region was 527.9 km^3 (Wang, Yang, Guo, & Mu, 2012). This was over 80% of the total discharge of China's transboundary waters (The Ministry of Water Resources, 2012).

During the period between 1956 and 1979, China's international rivers had a total annual average outflow of nearly 732 km^3, whereas there was an annual average inflow of only 17.2 km^3 (Liu & Chen, 2001). At the same time, the mean annual volume of water availability per capita in China was approximately 2200 m^3, one-quarter the world average. By 2030, water consumption in China will peak at a total volume of 650 km^3. At that time the water utilization ratio will be 79.85% of the total water available, meaning that China will be one of many countries suffering from 'water stress' (Liu, Wang, & Shao, 2009). On the other hand, China's transboundary waters create a wealth of resources including hydro-power, forestry, mineral deposits, biological and cultural diversity. For example, 70% of the total potential hydropower within China is located on the international rivers of the SW region, while three of 13 planned national hydropower bases and three cross-border oil and gas pipelines are located within international river basins. Thus, in order to maintain national water, food, energy and ecological security, the only possible source of reserve water resources large enough for China are its transboundary waters.

Problems intertwining with China's transboundary waters

China faces a very complex situation not only with regards to transboundary waters but also with respect to regional geopolitical issues, including various water resources issues (i.e. demand trends, varying objectives of water resources development, an uneven distribution of water resources, and the degradation of ecosystems along the international rivers within the three regions), both within and outside of China. The major problems include the following. (1) In the NW region, the largest arid zone in the Eurasian continent, riparian states will face water shortage and ecosystem risks because of high levels of water consumption by China and its neighbours (Chen, Wang, & Yu, 2000; Feng & He, 2009), soil salinization caused by improper irrigation, competitive water usage within watersheds which are linked to the safety of Central Asia–China oil and gas pipelines from Kazakhstan, and other Central Asia's countries. (2) In the NE region, due to high levels of water usage and pollution, the issues include: transboundary water pollution control, the maintenance and protection of aquatic ecosystems, increasing the efficiency of water usage and allocation due to the limited availability of water, and the impossibility of water transfer from water abundant areas. (3) In the SW region, transboundary waters are the key strategic reserve water resource for China due to the lower water usage and high levels of water quality. The large potential for hydropower within this region makes it a primary source for China's hydropower construction. However, the construction of hydroelectric dams, the development of international river navigation, cross-border flood control and the lower cross-border runoff have attracted attention in recent years by China's neighbours. There are around 2000 km of territorial boundaries, and approximately 125,000 km^2 of territory between India and China is still disputed (Gleick, 2002). Transboundary water issues intertwine with the territorial and developmental disputes of Tibet and Kashmir.

Water security relating to China's transboundary waters

The most serious situation faced by China is its difficulty to satisfy current domestic water demand. Nearly 400 out of 655 Chinese cities lack an adequate water supply, with approximately 200 of these cities suffering severe water shortages (Ministry of Water Resources, 2012). Since some of China's neighbours are already experiencing a risk of water shortage, China is experiencing, and will experience, increasing 'water pressure' from its neighbours, resulting in growing challenges regarding transboundary water allocation, utilization and the safeguarding of water rights. For example, India needs to feed 17% of the total world population with only 4% of the world's total water resources and constantly pays close attention to the water developments and hydrological changes in Tibet. Transboundary waters in Central Asia are closely related to China's transboundary waters in the NW region.

The upstream reaches of the majority of international rivers within the SW region are fed by melting glaciers, the hydrological regimes of which are changing due to glacial recession. The Himalayan glaciers receded at an average speed of 7.8 m/a throughout 1951–2009 (The Documentation Commission of The Second National Assessment Report on Climate Change, 2011). With this in mind, a number of complex problems arise: how does one balance domestic water demands with the use of shared transboundary waters to satisfy the water demands both of China and its neighbours?; how does one adapt the discharge changes feed by glacier melting in the future of both China and its neighbours?;

and how does one establish warning and alarm mechanisms between China and its neighbours? Each of these topics impacts the water security of China and its neighbours.

Vulnerabilities and risks of China's transboundary waters

Fragmented administrative institution

In China, the responsibilities and duties on water governance among the relevant ministries and official departments are separately endowed by the Chinese Central Government (Table 1), the Ministry of Water Resources (MWR) and the Ministry of Environmental Protection (MEP), which are the principal bodies. Other departments are also involved in general water resources management and are collectively called the Nine Dragons Administrate Water. Amongst the Nine Dragons, the Ministry of Foreign Affairs (MFA) takes the leading position in matters of transboundary waters due to its relation to foreign affairs. The MWR is the technical supporting department of the MFA. Therefore, there is not a single lead agency responsible for all the issues relating to China's transboundary waters. The arrangement of the Nine Dragons means that the tasks of water management overlap and fragment across these departments. This hampers the process of integrated development and coordinated management of China's transboundary waters. Limited

Table 1. Chinese administrative institution on transboundary waters (Ministries under the State Council).

Ministry	Tasks on transboundary waters
Ministry of Foreign Affairs	Develops policies concerning land boundaries, manages land boundary delimitation and demarcation, handles external boundary matters and cases concerning territories, negotiations on international river agreements
Ministry of Water Resources	International water allocation, hydrologic monitoring and management, international water law, transboundary water disasters, international water cooperation, river bank treatment and conservation along frontier rivers' reach
Ministry of Environmental Protection	Transboundary water pollution, international environmental law
Ministry of Housing and Urban–Rural Development	Cities and towns construction, planning, management, drainage system planning and management, and so on, along the border region
Ministry of Agriculture	Fish diversity protection along international rivers, prevention and control of aquatic bio-invasion
State Bureau of Forest	Bio-invasion control, wetland management, development and management of cross-border natural reserves
State Power Group	Hydropower development, planning and management on international rivers
State Development and Reform Committee	Transboundary infrastructure development, regional cooperation, cooperation strategy and action plan of international river basin
Ministry of Transport	International navigation development and management, channel maintenance, port construction, regulations on international watercourse navigation along international rivers
Ministry of Health	Cross-border disease prevention and control, protection and management for drinking water sources in the border region
State Bureau of Tourism	Development and management of cross-border tourism routes, tourist sites construction and management

communication between the relevant departments cause delayed responses in relation to transboundary issues and as a result compromise decision-making.

On the institutional structure of water resources, China adopts 'a system which combines river basin administration with administrative division administration' (Article 12 of China Water Law). Water development planning and the management of transboundary waters are entrusted to four watershed agencies of the MWR: the Chanjiang Water Resources Commission, the Yellow River Conservancy Commission, the Pearl River Water Resources Commission, and the Songliao River Water Resources Commission. In reality, the watershed agencies have some powers regarding water development planning, but have fewer powers relating to water resources management. Local governments, which have fewer powers regarding transboundary waters management, have much stronger powers regarding the development of regional tributaries, even those of international rivers. Therefore, from the view of river basin water resources management, the responsibilities of local governments include water protection, administration and water policy implementation, which will have important impacts on integrated water development and management. The Chinese Central Government will face considerable pressure regarding integrated water resources development and ecosystem conservation of its international rivers when a macro-strategy on water development planning of these rivers is not formulated.

There are currently few inter-governmental coordinating mechanisms that have been developed in order to build capacity regarding transboundary water cooperation such as summit and ministerial meetings. The Division of International Rivers has been established under the MWR, and the Department of Boundary and Ocean Affairs was established under the MFA in 2009. Each is partially responsible for transboundary water management.

Information gap on transboundary waters

Hydrological regime change

Using the China Water Resources Bulletin, covering the period between 1980 and 2012, the total annual water volumes of the three regions were compared with the long-term average discharge (1956–79). The annual changes in the total volume of water resources within the three regions (Figure 1) showed that the following. First, in the NE region, water volumes underwent the biggest change among the three regions, ranging between extreme values from 49% in 1998 to −37% in 2007, indicating a significant overall decrease in water volume during this period. Second, the NW region has the lowest water volume amongst the three regions, but is the only region where the volume of water showed a sustained increase from 1980 to 2012, especially during the period after 1997. The largest change in water volume occurred in 2010, with an increase of 26%. Third, the SW region has the largest share of total water volume amongst the three regions, with water volumes in 1980–2012 decreasing steadily, although the rate of change was the lowest among the three regions.

The correlation between annual precipitation and the volume of water resources in each region in 1995–2012 was analysed (Figure 2). The results show that the following. (1) The values of each correlation coefficient are very high, meaning that rainfall directly affected water volume in each region. In other words, rainfall is the determining factor for water volume. (2) While the value of the correlation coefficient in the NW region is significantly different from those in the SW and NE regions, those in the NE and SW regions are very close and very high, meaning that the water volume of the SW region is the most strongly controlled by rainfall in the three regions, whereas the water volume in

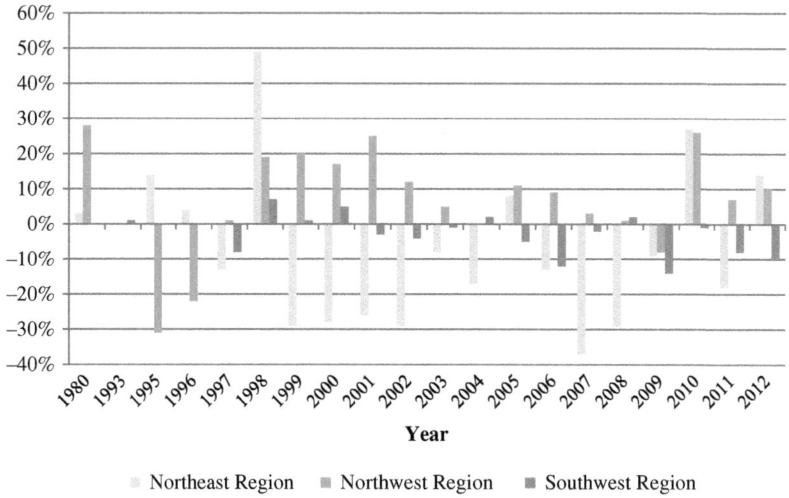

Figure 1. Changes in annual water volumes within the three regions.

the NW region is less controlled by rainfall than in the NE and SW regions. In other words, the water volume in the NW region is impacted not only by rainfall but may also be simultaneously impacted by groundwater or glacial melting.

Outflow change

The ratios of water utilization in the three regions during the period between 1980 and 2012 can be determined based upon data released in the China Water Resources Bulletin. The results, as seen in Figure 3, show that the following. (1) A large percentage of water resources, over 30% and nearly 50% of water resources on average, have been used in the NE and NW regions respectively in the past 20 years, and water utilization within both has shown a tendency to increase slowly. In the NE region, water utilization has grown faster during 1980–2000, from a low of 18% in 1980 to a high of 45% in 1999, while in 2001–12 water utilization decreased from 50% at its highest level in 2007 to 27% at the lowest level in 2010. In the NW region, water utilization increased from 43% in 1980 to 57% at its highest level in 1995, and from 36% in 2001 to 49% in 2012. (2) In the SW region, a limited portion of water resources was used in the past 30 years, rising from 0.7% in 1980 to 2.1% in 2012, and with very slow growth. (3) The total average annual outflow from China's international rivers in 1997–2012 was approximately 680 km^3, 81% of which is contributed by the international rivers of the SW region. Therefore, the large quantity of unused waters in the SW region could be the main source of reserve water resources for China in future, as there is limited available water to support increased demand in the NE and NW regions if water demand for supporting ecosystem functions and structures are taken into consideration. This is especially true in the NW region.

Usually an increase in water utilization results in decreased outflow – however, the reality is not so simple. The China Water Resources Bulletin data from 1997 to 2012 on the annual outflows from China to its neighbours measure this in three ways: as the total outflow, as the outflow from the NE region, and as the outflow from the NW and SW regions. Analysis of these data reveals that: (1) the total outflow, as well as that from the

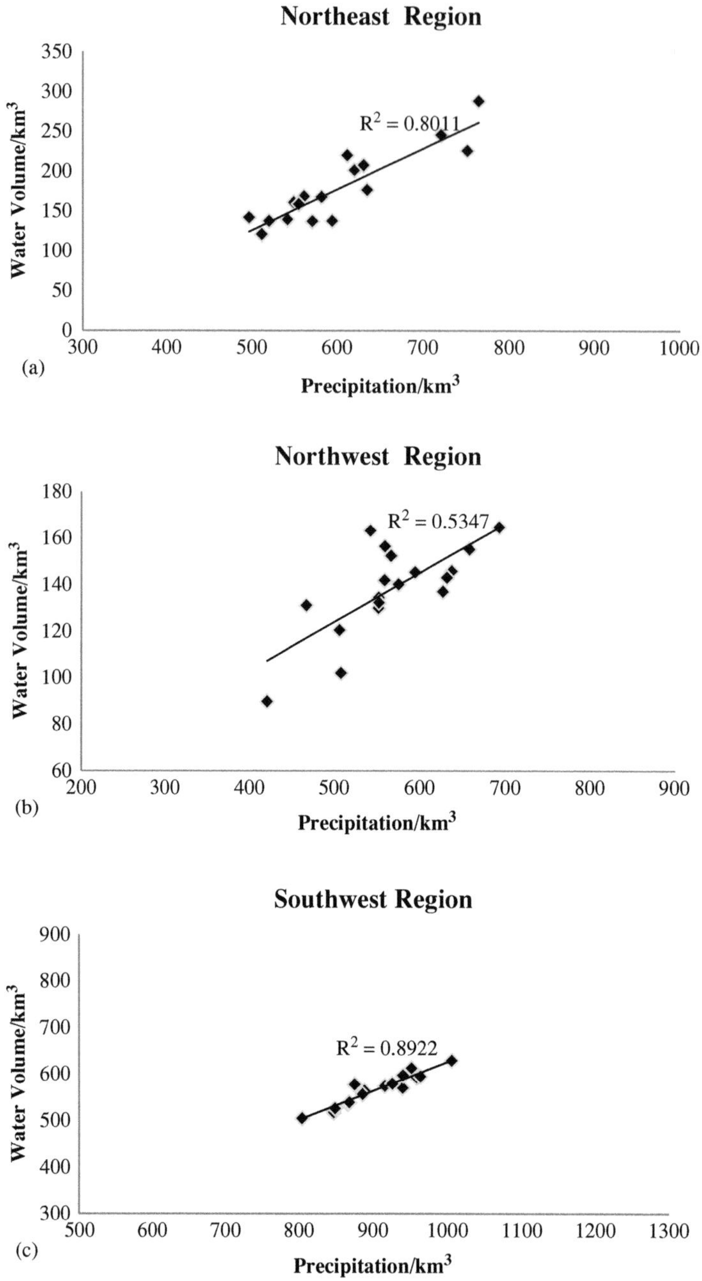

Figure 2. Correlations between annual precipitation and water volume within the three regions.

NW and SW regions, has decreased in the past 15 years, but the outflow from the NE region has shown a tendency to increase slowly (Figure 4); (2) a weak positive correlation and a strong positive correlation between outflow and water utilization, and water volume in the NE region respectively can be determined based upon the values of the correlation coefficients between them (Figure 5); (3) when comparing correlation coefficients

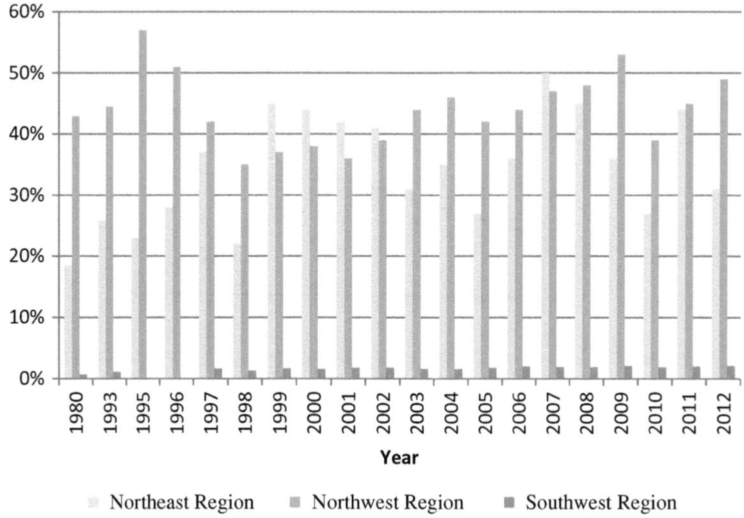

Figure 3. Water utilization ratios in the three regions, 1980–2012.

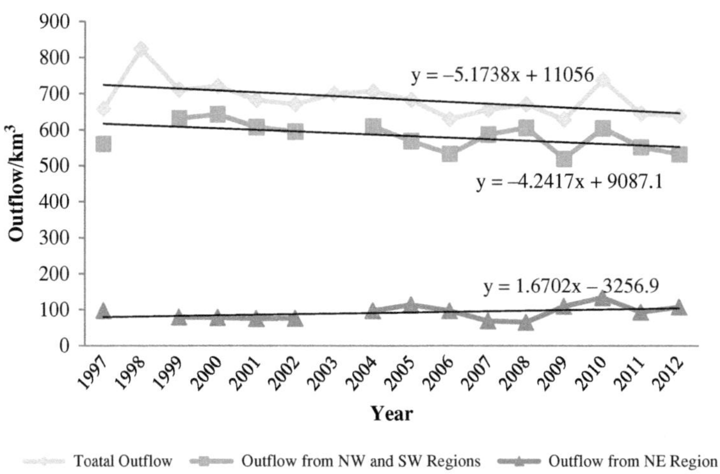

Figure 4. Annual total outflows in the north-east, north-west and south-west regions.

between the outflow and water usage from the NW and SW regions (Figure 6), it can be seen that the outflow has a negative correlation to water use in the NW region, and has little to no relationship with that in the SW region; (4) according to an analysis of the correlations between the outflow from the NW and SW regions with water volume in the NW region and that in the SW region (Figure 7), the outflow has the highest positive correlation with water volume in the SW region, and has an obvious positive correlation with that in the NW region; and (5) comparing the correlative coefficients between the outflow from the NW and the SW regions and water use (Figure 6) and volume (Figure 7) in the NW region shows that outflow has a stronger relationship with water volume.

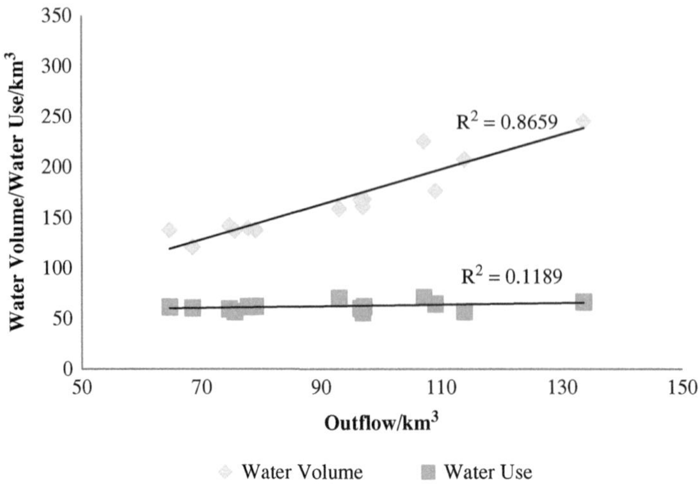

Figure 5. Correlations between water volume, use and outflow in the north-east region.

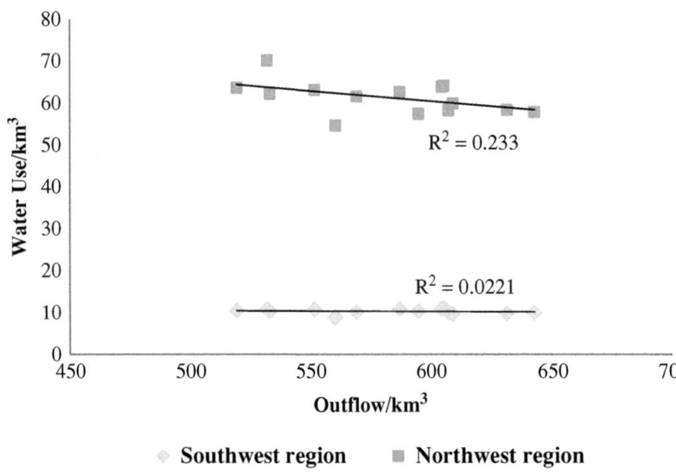

Figure 6. Correlations between the outflow from the north-west and south-west regions with water use in the north-west and south-west regions.

Given these results, in combination with the results of increasing water volume in the NW region, decreasing water volume in the SW region (Figure 1), and increasing water utilization in both the NE and NW regions, it appears that: (1) the change in the total annual outflow is mainly affected by the changes in annual water volume in the SW region; (2) higher and increasing water utilization in the NE region in recent years has not resulted in decreased outflow; and (3) water utilization in the NW region may, to some extent, affect the outflow, but its effect is less than that of water volume on the outflow. In other words, the changes in outflow are controlled by the changes in rainfall, namely controlled by natural factors. The decreasing outflow within the SW and NW regions is

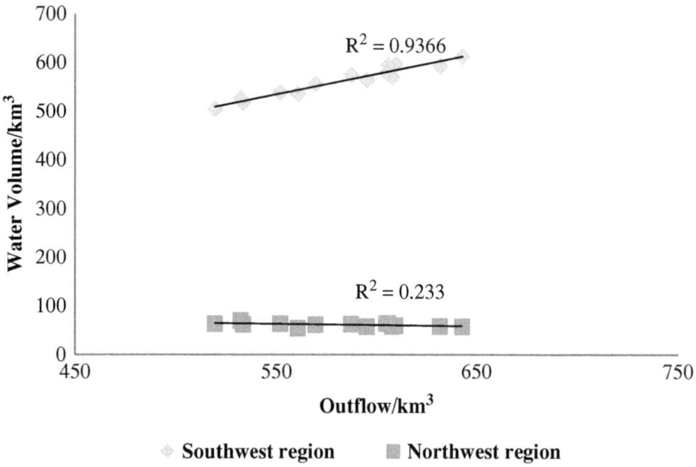

Figure 7. Correlations between the outflow from the north-west and south-west regions and water volumes in the north-west and south-west regions.

primarily affected by decreasing rainfall, especially by decreasing levels of rainfall within the SW region, and partly affected by water use, namely by human activities.

The data analysis reveals that the decreasing outflows within the SW and NW regions of China are affected directly by levels of rainfall as a primary factor. Nonetheless, decreased outflows led Kazakhstan to complain that China's increased withdrawal of regional water resources has resulted in a marked decline in the Ili River's flow (Stone, 2012), without appreciating this other factor. In the SW, China's dam development on the Lancang River (upper Mekong) led downstream states to claim adverse impacts as a direct result of these upstream planned measures (Menniken, 2007; Nickum, 2008). While the lower water flow at the Lower Mekong in spring 2010 was ascribed to the dams' operation within China, in fact China suffered from the most severe drought during the period, which significantly affected the flow. This information gap between China and its neighbouring countries results in these misunderstandings; a lack of formal data-sharing, procedures for notification and lack of academic exchanges all combine to prevent an integrated scientific response on transboundary waters in this region. There is clearly room for improvement for more meaningful transboundary water cooperation across the scientific research domain.

Lack of early warning mechanisms

Transboundary water pollution

Data on surface water quality changes in the three regions can be found in the China Water Resources Bulletin of 1980–2012 and in the China Environmental Bulletin of 2003–12. Combining this information with the Environmental Quality Standards for Surface Water released by the MEP, the following observations can be made (Figure 8). (1) In the NE region, approximately 60% of the monitored reaches of the rivers have varying levels of pollution. This means that China is facing a serious water pollution problem. This includes transboundary water pollution due to the development of China's industrial base since the 1950s, particularly industries such as mining, mechanical

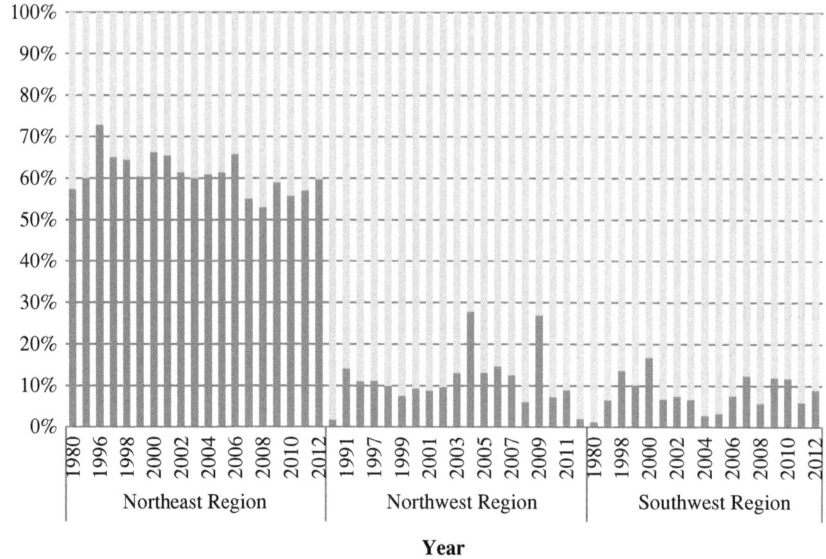

Figure 8. Water quality within the three regions, 1980–2012.

processing, chemical, steel and automotive industries. However, past years have shown a decreasing trend in the levels of pollution. (2) The NW region experienced higher levels of water quality during this period with the exception of 2004 and 2009, but in recent years the situation is changing and keeping at the higher levels of water quality. (3) During these years water quality in the SW region was maintained at a very high level, and the contention that these particular transboundary waters will be the source of China's reserve water resources in future is also supported not only by the quantity of water available but also by its quality.

Transboundary hydro-meteorological disasters

Climate warming is increasing snow and glacier melting in upstream regions in China, directly affecting the transboundary water availability in Central Asia and South Asia (Mukhopadhyay, 2012). Under the influence of climate change, the occurrence of glacial lake outburst events in the Himalayan region has drastically increased in regularity in 1930–2010 (Cui, Chen, Xiang, & Su, 2014), and more severe weather events cause increasingly frequent hydro-meteorological disasters across borders (Li, He, & Feng, 2011). For example, the most severe drought in 80 years hit Yunnan from autumn 2009 to spring 2010, drying up 273 rivers and 413 small reservoirs, and giving rise to a series of transboundary disputes throughout the Mekong Subregion (Lü, Ju, Ren, & Gan, 2012). The largest-ever landslide on the Yigong River tributary of the Yaluzangbu River in Tibet occurred on 9 April 2000 as a result of a sudden temperature increase. The landslide dammed the river and formed the vast reservoir with some 3.0 km^3. On 10 June 2000, the water burst from the reservoir and created a huge flash flood downstream with a maximum velocity 11.0 m/s (i.e. per second) and maximum discharge 135,000 m^3/s as it entered India, where over 30 people died, 100 went missing and more than 50,000 were left homeless; 20 large bridges were also washed away (Shrestha, 2008).

A range of issues affecting China's transboundary waters, from water pollution to natural disasters, continue to challenge water security across the region, potentially resulting in substantial adverse transboundary impacts. There is an urgent and compelling need to establish an early-warning mechanism for any cross-border emergency among China and its neighbouring countries so as to reduce the possible adverse impacts and to enhance regional cooperation.

Weak legal institutions

As of 2011, around 80 agreements, protocols and memorandums of understanding (MOUs) had been signed between China and its neighbouring countries and international organizations, covering eight international rivers (Wouters & Chen, 2013). Over 70% and 40% of these agreements were signed after the years 1990 and 2000 respectively. This pattern shows that issues relating to transboundary waters have received more attention since the opening and reform of China, especially after the year 2000 since international cooperation has been promoted in nearly half of the 15 major international rivers. Looking at the contents of these agreements, it is clear that cooperation has been promoted, especially in technical areas, such as the sharing of information and hydrologic data, navigation, water quality monitoring, hydropower development, power trading, and log floating. Although these agreements cover a variety of areas and some topics, such as academic cooperation, environmental cooperation, the protection and utilization of water resources, international institution building, and so forth, overall these are covered in a rather general legal framework or in single-purpose agreements. The core issues, such as water allocation, benefit sharing, operational reasonable and equitable utilization, integrated water resources management and protection, are not included in any detail in most of these agreements. China needs to promote regional cooperation on transboundary waters and gradually establish a series of legal frameworks, domestically and regionally, that address these outstanding and complex issues.

In recent years, integrated water resource management has been widely accepted and implemented around the world, such as across the European Union. According to Article 14 of China Water Law, 'master plans for development, utilization, saving, and protection of water resources and for prevention and control of water disasters shall be developed in a unified manner based on basin and regional boundaries', and the master plans of major river systems have been gradually prepared. On international rivers, any development strategies and implementation schemes within Chinese boundaries cannot be called an 'integrated watershed plan'. At present, due to variances within political and economic institutions, legal regimes, cultural, religious and diplomatic relationships, water development objectives and shared natural resources, any 'integrated water development and management plan' on the basin by all riparian countries is impossible without an integrated legal system within one river basin. This means that transboundary disputes over unilateral water development are easily raised but remain difficult to resolve, especially without a comprehensive legal framework. China could learn from best practices across Europe and in regional transboundary water contexts from across the world as it moves forward with improving its transboundary waters management practices.

Conclusions

Based upon the research presented here (changes in water volume, utilization and quality within the three regions during 1980–2012, the distribution of China's transboundary waters, the complex regional geopolitical conditions, current Chinese water institutions and the status of international cooperation), the authors identify the following transboundary waters vulnerabilities and risks:

- China is facing, and will continue to face, serious and complex challenges related to water security, both domestically and regionally. This is caused by a number of factors, including decreasing outflow caused by climate change and potentially by increasing water consumption, and deteriorating water quality of the international rivers, especially within the NE region.
- China has to strengthen its capacity in order to respond to present and future challenges regarding transboundary water management, including to clarify relationships among the involved departments and ministries involved in this area, and to develop coordinate master plans for water development and the protection of transboundary waters. The current system of fragmented authority coupled with an underdeveloped legal frameworks for transboundary waters mitigates against progress in this important field. The lack of information exchange, prior notification of planned measures and absence of effective early warning mechanisms preclude China from addressing the risks and vulnerabilities linked with transboundary waters.

Acknowledgments

The authors would like to thank Professor Patricia Wouters and Professor Huiping Chen and the China International Water Law (CIWL) research team at Xiamen Law School for their support for this paper. Thanks also to the anonymous reviewers.

Funding

This work was supported by the Key Project of National Nature Science Foundation of China [grant number U1202232], the Key Project of National Social Science Foundation of China [grant number 11AZD04], and the National Science and Technology Support Program [grant numbers 2013BAB06B03 and 2011BAC09B07]

References

Chen, H., Rieu-Clarke, A., & Wouters, P. (2013). Exploring China's transboundary water treaty practice through the prism of the UN watercourses convention. *Water International, 38*(2), 217–230. doi:10.1080/02508060.2013.782134

Chen, M., Wang, H., & Yu, F. (2000). The general situation on international rivers issues in China. *Water Forum, 28*(4), 24–26 (in Chinese)

Cui, P., Chen, R., Xiang, L., & Su, F. (2014). Risk analysis of mountain hazards in Tibetan plateau under global warming. *ProgressusInquisitiones DE MutationeClimatis, 10*(2), 103–109 (in Chinese)

Cui, P., Dang, C., Cheng, Z., & Scott, K. M. (2010). Debris flows resulting from glacial-lake outburst floods in Tibet, China. *Physical Geography, 31*(6), 508–527. doi:10.2747/0272-3646.31.6.508

Dombrowsky, I. (2007). *Conflict, cooperation and institutions in international water management: An economic analysis*. Cheltenham: Edward Elgar Publishing.

Duncan, J., Stefano, L., Dinar, S., Stahl, K., Strzepek, K., & Wolf, A. T. (2012). Mapping the resilience of international river basins to future climate change-induced water variability. *Journal of Peace Research, 29*(1), 193–209.

Earle, A., Jagerskog, A., & Ojendal, J. (2010). *Transboundary water management: Principles and practice*. Landon: Earthscan.

Eleni, K., Eleni, E., & Yannis, M. (2007). Implementing equitable water allocation in transboundary catchments: The case of river Nestos/Mesta. *Water Resources Management*, *21*(5), 909–918. doi:10.1007/s11269-006-9108-1
Feng, Y., & He, D. (2009). Transboundary water vulnerability and its drivers in China. *Journal of Geographical Sciences*, *19*(2), 189–199. doi:10.1007/s11442-009-0189-7
Gleick, P. (2002). *The world's water* (pp. 219–238). Washington, DC: Island Press.
Goh, E. (2004). *China in the Mekong river basin: The regional security implication of resources development on the Lancang Jiang*. Singapore: Institute of Defence and Strategic Studies.
Li, Z., He, D., & Feng, Y. (2011). Regional hydropolitics of the transboundary impacts of the Lancang Cascade dams. *Water International*, *36*(3), 328–339. doi:10.1080/02508060.2011.585447
Liu, C., & Chen, Z. (2001). *Chinese water resource status assessment and the supply-demand tendency analysis* (pp. 60–68). Beijing: Chinese Water Publisher (in Chinese).
Liu, C., Wang, Y., & Shao, M. (2009). *Water science & technology in China: A roadmap to 2050*. Beijing: Science Press and Springer.
Lü, J., Ju, J., Ren, J., & Gan, W. (2012). The influence of the Madden-Julian oscillation activity anomalies on Yunnan's extreme drought of 2009–2010. *Science China Earth Sciences*, *55*, 98–112. doi:10.1007/s11430-011-4348-1
Menniken, T. (2007). China′ s performance in international resource politics: Lessons from the Mekong. *Contemporary Southeast Asia*, *29*, 97–120. doi:10.1355/CS29-1E
Ministry of Water Resources. (2012). *The strictest water resources control system*. Beijing: China Water Power Press (in Chinese).
Mirumachi, N. 2008. Transboundary river basins and the water politics: Some perspectives for water and earth system governance. http://www.glogov.org/images/doc/Mirumachi.pdf
Mostert, E., & Barraqué, B. 2006. Transboundary river basin management in Europe. UNDP Human Development Report 2006. http://hdr.undp.org/en/reports/global/hdr2006/papers/barraque_bernard%20et%20al.pdf
Mukhopadhyay., B. (2012). Detection of dual effects of degradation of perennial snow and ice covers on the hydrologic regime of a Himalayan river basin by stream water availability modeling. *Journal of Hydrology*, *412–413*, 14–33. doi:10.1016/j.jhydrol.2011.06.005
Nickum, J. (2008). The upstream superpower: China's international rivers. In O. Varis, A. K. Tortaj, & A. K. Biswas (Eds.), *Management of transboundary rivers and lakes* (pp. 227–244). Berlin: Springer.
Shrestha, M. S. (2008). Impacts of floods in South Asia. *Journal of South Asia Disaster Study*, *1*, 85–106.
Stone, R. (2012). Transboundary rivers for China and Kazakhstan, no meeting of the minds on water. *Science*, *337*, 405–407. doi:10.1126/science.337.6093.405
The Documentation Commission of "The Second National Assessment Report on Climate Change". (2011). *The second national assessment report on climate change*. Beijing: Science Press (in Chinese).
The Ministry of Water Resources. 2012. China's water bulletin 2011. Retrieved from http://www.mwr.gov.cn/zwzc/hygb/szygb/qgszygb/201212/t20121217_335297.html (in Chinese).
United Nations. (2003). United Nations: 150 "international" rivers may cause international disputes. Retrieved from http://www.people.com.cn/GB/guoji/22/86/20030321/949273.html (in Chinese)
Wang, X., Yang, Y., Guo, H., & Mu, H. (2012). *Changjiang & southwest rivers water resources bulletin 2011*. Wuhan: Changjiang Press (in Chinese)..
Wouters, P., & Chen, H. (2013). China's "Soft Path" to transboundary water cooperation examined in the light of two UN global water conventions – exploring the "Chinese Way". *The Journal of Water Law*, *22*, 229–247.
Zhang, J., Mauzerall, D. L., Zhu, T., Liang, S., Ezzati, M., & Remais, J. V. (2010). Environmental health in China: Progress towards clean air and safe water. *The Lancet*, *375*, 1110–1119. doi:10.1016/S0140-6736(10)60062-1
Zhao, J. (2005, September 23). Russia exaggerated "China capture water". *International Herald Leader*. Beijing: Xinhua News Agency (in Chinese).

Contemporary legal analysis of China's transboundary water regimes: international law in practice

Yu Su

China's rapid development places growing pressure on its water resources, including its considerable transboundary waters. This paper analyses the current situation of China's transboundary water regimes through a legal analytical framework. The evaluation of treaty and state practice reveals that while China adheres to and follows some fundamental rules of international water law, gaps remain. New trends in China's state practice and in the field of international law offer interesting opportunities for addressing these shortcomings.

Introduction: setting the scene in the context of China transboundary water challenges

Water is essential for human society in a myriad of ways. With social and economic development comes increased demand for water. This is particularly true in the case of China, which has been enjoying an economic growth rate averaging 10% for more than 20 years, with water playing an integral role in sustaining this development (China Water Risk, 2010). Rising water needs at the domestic level puts added pressures on China's transboundary water resources, leading to complex water-related challenges with China's 14 neighbouring riparian nations (Gleick, 2009).

The first major challenge is water *quality*. The heavy agricultural and industrial use of water is having a significant adverse impact on water quality in China. According to statistics provided by the Ministry of Environmental Protection, in 2009 approximately 43% of the water in the main rivers in China was unfit for human consumption. The accumulation and discharge of pollutants leads to pollution of China's international rivers as well (Tyler, 2006). In 2005, for example, the chemical spill on the Songhua River in China caused serious pollution in the city of Khabarovsk, a Russian city situated downstream on the Amur River, as it is called in Russia (UNEP, 2005; Vinogradov & Wouters, 2013, pp. 11ff.). Despite the fact that Russia has not made any compensation claims against China to date, there are important legal implications for environmental responsibility or liability for the transboundary damages caused (Ying, 2008).

The second major challenge is water *quantity*. Water resources, though renewable, are not unlimited. Competing needs for water is likely to lead to conflict over shared water resources both within and possibly beyond China's borders. The Chinese 'South–North Diversion' project, introduced to overcome the uneven water allocation in China, raises

Kazakhstan's concern about China's diversion of water from the Irtysh River under its 10th Five-Year Plan to develop its north-western region. One critic has called this a 'totalitarian' and 'gigomaniac' project, which, in that author's view, might violate rules of customary international law (Sievers, 2002), Moving to the south-west, China has various transboundary water issues across that region with a number of riparian neighbours. China's dam development in the Lancang River, widely known as the Mekong River outside China, is viewed as 'being unilateral in harnessing the water resources in the region' and has become a point of contestation in recent years (UNEP, 2009). China is also seen to be involved in a water rivalry with India over the Yarlung Zangbo (or Brahmaputra) River (Chellaney, 2011).

The issue of *emergency preparedness and response* is another challenge facing China's transboundary water regimes. Apart from transboundary pollution, such as the Songhua incident mentioned above, cross-border floods require robust emergency response procedures. Although China and Russia have conducted 'timely and effective cooperation' in the face of the heavy flooding that hit parts of north-western China and eastern Russia in 2013 (Xinhua News, 2013), China still faces critical challenges in its relationship with its southern neighbours in terms of flood prevention. Thailand, for example, suffered from the worst Mekong River flooding in a century and Chinese dams are suspected to have aggravated this flooding (Khai, 2008).

When there are insufficient water resources to satisfy all competing needs among riparian states, the question of how to turn potential conflict into opportunities of cooperation becomes a key challenge (McCaffrey, 2003b, p. 88). International water law (IWL) plays an important part in addressing transboundary water issues and contributes to enhancing cooperation among riparian states. This article analyses the rules of international law that govern China's transboundary waters, identifying both strengths and weaknesses in current Sino treaty practice in this field.

Sources of IWL governing China's transboundary waters

Treaties and customary law are two of the most important sources of international law (Jia, 2010, p. 82). To date China has concluded close to 110 treaties that govern matters related to international watercourses. These treaties, together with customary international law, which will be dealt with in the following sections, constitute the main body of the applicable law governing China's international watercourses. Thus far there are three water-specific agreements, including the 1994 China–Mongolia Agreement on the Protection and Utilisation of Boundary Waters (hereinafter China–Mongolia Agreement), the 2001 China–Kazakhstan Agreement on the Utilisation and Protection of Transboundary Rivers (hereinafter China–Kazakhstan Agreement), and the 2008 China–Russia Agreement on the Reasonable Utilisation and Protection of Transboundary Waters (hereinafter China–Russia Agreement). Other agreements are not water-specific agreements, but merely water-related agreements, i.e. agreements that contain articles governing international watercourses (Wouters & Chen, 2013, pp. 229–247).

For customary rules, it is widely recognized that certain provisions and principles in the United Nations Convention on the Law of the Non-Navigational Uses of International Watercourses (hereinafter UNWC) and the United Nations Economic Commission for Europe Convention on the Protection and Use of Transboundary Watercourses and International Lakes (hereinafter UNECE Convention) reflect, in part, a codification of rules of customary international law in this field (McCaffrey, 2001; Tanzi & Arcari, 2001, p. 91). China's vote against the adoption of the UNWC does not mean that it rejects the customary rules reflected in these conventions. China, as other states, disagreed with certain provisions of the UNWC and explained its position at the time. However, it has expressly affirmed its adherence to most of the fundamental principles of the UNWC, including rules of customary international law, as revealed in its statement made during the UN General Assembly session for adopting the UNWC (UN, 1997e). Further, while some of the provisions of the conventions have not yet acquired the status of customary international law, analysing China's treaty practice against these framework instruments is an important exercise since they were designed to serve as guidelines for future watercourse agreements (Rieu-Clarke, 2013; UN, 1997a, p. 8).

Apart from treaties and custom, soft law instruments also play a central role in IWL, since most of its principles result from developments in state practice (McIntyre, 2006, pp. 164–166; Zodrow, 2010, p. 40). Therefore, instruments such as the International Law Association (ILA) Helsinki Rules and joint declarations or statements signed between China and its riparian states should also be taken into account when examining China's IWL practice.

Legal analytical framework for IWL: five core elements

Five core elements provide a legal framework for analysing international water regimes: scope; substantive rules; procedural rules; institutional mechanisms; and dispute settlement. Together this framework facilitates a comprehensive overview of the legal system governing transboundary waters (Vinogradov & Wouters, 2003, pp. 45–75). Each of these elements plays a unique role and should be considered individually and then all collectively together as a whole. 'Scope' covers the legal reach of a treaty, in terms of the parties eligible to participate, the subject-matter governed, and the territorial space included within the legal regime (VCLT, 1969, article 29). In treaties governing international transboundary waters, *scope* determines the geographical and hydrological limits of the legal regime; the types of water uses and activities covered; and the parties to which the treaty applies (Rieu-Clarke, Moynihan, & Magsig, 2012, p. 67; Vinogradov & Wouters, 2003, p. 46). In IWL the importance of scope cannot be overemphasized (McIntyre, 2007, p. 40) and the issue of scope often becomes a focus of debate both in cases and treaty-making conferences (River Oder case, 1929, p. 25; UN, 1991, 1997d, p. 12). The category of *substantive rules* refers to rules that set out the substantive, or material, rights of the riparian states (Vinogradov & Wouters, 2003, p. 50). These include, but are not limited to, the following fundamental principles in IWL: equitable and reasonable utilization, due diligence to prevent significant harm, and the environmental protection of international watercourses. *Procedural rules* refer to mechanisms or processes that provide the means though which substantive rules are implemented and how the changing watercourse regime is managed. It is generally believed that procedural rules include the duty to exchange information, to notify, negotiate and consult, and to warn. Concerning the relationship between substantive and procedural rules, it should be noted that even though the International Court of Justice (ICJ) in the *Pulp Mills* case held that procedural and substantive obligations can be examined separately, the court at the same time recognized the existence of a functional link between the two sets of obligations (*Pulp Mills*, 2010, paras 71–79; McIntyre, 2010). In fact, the line between 'substantive' and 'procedural' rules is not always clear and both are *rules* of international law, the breach of which will attract legal remedies (Leb, 2013, pp. 107ff.) As argued by McCaffrey (2003a, p. 397), substantive and procedural rules in IWL are in effect intertwined and inseparable. This line is being drawn here only for analytical purposes.

Institutional mechanisms are essential in ensuring the effective implementation of treaty regimes and serve to enhance and promote functional cooperation between the riparian states dealing with transboundary waters issues. They may be permanent or ad hoc, formal or informal, regional or bilateral cooperative mechanisms (Menon, 1972). These joint bodies 'serve both as forums and tools for enriching dialogue and decision-making' concerning transboundary waters (UNECE, 2009) coordinating states' efforts in managing the shared resources and preventing tension from escalating into disputes that require formal dispute resolution (Vinogradov & Wouters, 2003, p. 57). The legal rules on institutional mechanisms set out, *inter alia*, mandate, composition and capacity of the joint bodies. Lastly, *dispute settlement* is an integral part of any international legal system (Shaw, 2008, p. 1010) and maybe even more so in IWL (McCaffrey, 2003b, pp. 49–90). This element is concerned with the provisions and mechanisms that apply in the event of a disagreement over the treaty or implementation of the treaty regime.

The five elements are an integrated whole. *Scope* delineates the body of waters over which the legal system governs; *substantive rules* set out the principles according to which the waters included within the scope should be utilized or protected; *procedural rules* put forward the means by which the substantive principles are implemented; *institutional mechanisms* provide the forums for enhancing the dialogue and cooperation among the riparian states; and finally *dispute settlement* enables the states, when their diverging uses or needs over the waters cannot be reconciled, to settle disputes peacefully. In summary, the five elements provide a legal approach to determining such key issues as who gets what water, when, why and how; and offers means to avoid and resolve disputes that may arise in this context. This legal analytical framework will be applied to examine China's international water treaty practice, including also relevant comparisons with the UNWC and UNECE global water conventions.

China's treaty practice in light of IWL

Scope

China's transboundary water treaties appear to employ more or less the same terms when defining its geographical scope. A typical example is the 2008 China–Russia Agreement, which provides in article 1: 'The Agreement applies to the protection and utilization of transboundary waters. Transboundary waters refers to any rivers, lakes, streams or swamps that straddle or rest on the boundary line between the two states.' The formulations of the 1994 China–Mongolia Agreement differs slightly from that of the 2008 Agreement. In addition to using similar terms with those abovementioned, it also makes a list of the transboundary waters that fall within the remit of the agreement (China–Mongolia Agreement, 1994, article 1). Upon closer comparison between the three Sino water-specific treaties, an interesting evolution of treaty terminology can be discovered. In the 1994 China–Mongolia Agreement, the term 'boundary' (*bian jie*) waters was used to denote the types of waters the agreement covers. This seems to be a static and narrow scope for a transboundary waters treaty, because the term 'boundary' fails to capture the movement across the boundary line vividly conveyed by the term 'transboundary' (*kua jie*). In the later treaties, such as the 2001 China–Kazakhstan Agreement (2001, article 1) and the 2008 China–Russia Agreement (2008, article 1), China changed its use of terms and adopts the term 'transboundary' (*kua jie*) instead. Drafting techniques seem to improve and the definition of scope in China's water treaties has become broader over the years. Despite the improvement, however, the scope of China's treaties is generally speaking still narrower than that of the UNWC, which governs the utilization of 'International Watercourses', a term defined to cover both surface water and connected groundwater (UNWC, 1997, article 2). China's approach seems to be more consistent with the relatively restricted notion of scope adopted by the UNECE Convention, i.e. 'transboundary waters' (UNECE Convention, 1992, article 1).

As to the functional scope of China's treaties, they are concerned primarily with the 'protection' (*bao hu*) and 'use' (*li yong*) of transboundary waters, while measures of 'use', 'protection, preservation and management' fall within the remit of the UNWC approach (UNWC, 1997, article 1). The difference between the two sets of instruments becomes obvious if one examines the evolution of article 1 in the drafting history of the UNWC. In the 1994 Draft articles on which the UNWC is based, the scope as originally defined covered measures related to 'conservation and management' of the watercourses (UN, 1994, article 1). After lengthy discussion in the Working Group, it was changed to what now appears in article 1 of the convention, i.e. 'protection, preservation and management', terms that the working group of the Sixth Committee believe would set out a broader scope of application of the convention than that would be laid down by the term 'conservation' (Tanzi & Arcari, 2001, pp. 54–56; UN, 1997a, p. 4). This amendment is interpreted to indicate an intention to emphasize the broad scope of the convention, thus covering 'both quantitative and qualitative aspects and, in addition to conservation, measures of "control" such as regulating flow, floods, pollution and erosion, saline intrusion and mitigating drought' (Rieu-Clarke et al., 2012, p. 68). Therefore, it can be seen that the UNWC compared with China's treaties is wider in scope. It should also be noted that the convention, subject to a few exceptions, is quite clear on its non-application to the navigational uses of water. China's treaties, however, do not make clear whether the agreements will apply to navigational uses of the transboundary waters.

Substantive rules

The status of the principle of equitable and reasonable utilization as a customary rule of international law is beyond doubt (McCaffrey, 1986). China accepts the principle and considers it to be 'the cornerstone of the draft articles', as the summary records of the Working Group discussion show (UN, 1996a, p. 7). Its acceptance of the principle is reflected in its treaty practice (China–Mongolia Agreement, 1994, articles 2, 4). Although the 2008 China–Russia Agreement, a more recent watercourse treaty does not have any article expressly referring to the principle of equitable utilization, it nevertheless makes reference to the principle in its title and the preamble (China–Russia Agreement, 2008, preamble) and implicitly refers to it in one provision (article 6). China has also embraced the principle in a variety of other treaties or in joint statements (China–Bangladesh Statement, 2005, para. 11; China–Kazakhstan Statement, 2004, para. 11; 2006, sections 1(4), 5(4); 2012, section 3; China–Russia Treaty, 2001, article 19). It should be noted that the provisions mentioned above are couched in rather vague terms and make no reference to relevant factors other than mere mention of the term 'equitable and reasonable'. While the 2008 China–Russia Agreement could be seen as an exception, which did refer to three relevant factors in its preamble, i.e. 'economy, society and population', the broad formulations of these factors, however, cannot provide much guidance on how to determine whether a specific use is 'equitable and reasonable'. In this regard, UNWC and the ILA Helsinki Rules might be of assistance in better determining the content of this principle, since both instruments provide a list of relevant factors that are to be considered and balanced against each other when determining equitable and reasonable utilization (Helsinki Rules, 1966, article V; UNWC, 1997, article 6).

Given the fact that it is both an upstream and a downstream state, China attaches particular importance to the no-harm principle (Salman, 2007; UN, 1996c), whose status as part of customary international law is well recognized (McCaffrey, 2003a, p. 349; Schwebel, 1981, p. 101). This obligation can be found in many of China's treaties (China–Kazakhstan Agreement, 2001, article 3; China–Mongolia Agreement, 1994, article 4; 2010, article 13; China–Russia Agreement, 2008, article 7; China–Vietnam Agreement, 2009b, article 9). It is widely recognized in treaties and doctrinal statements that the no-harm principle does not impose an absolute obligation upon the states – although a few exceptions can be found in some treaties – meaning that it is a *due diligence* obligation (McIntyre, 2007, p. 209). In any provision setting out the no-harm principle, care is usually taken to qualify the term 'harm' (Schwebel, 1981, p. 98). The formulation of qualitative terms, however, varies from one treaty to another in China's treaty practice. Some treaties adopt the term 'significant' (China–Kazakhstan Agreement, 2001, article 3; China–Russia Agreement, 2008, article 7); some use the term 'minimum' (China–Kazakhstan Agreement, 2011, article 3(3)); and in some others no qualitative terms are inserted before the term 'harm', leaving the reader in doubt whether it is imposing an absolute obligation prohibiting any harm, or just significant harm (China–Mongolia Agreement, 1994, article 4; Wouters & Chen, 2013, p. 237). Compared with the UNWC, China's divergent use of terms might cause confusion in implementation. Since China is upstream on most its transboundary waters, it faces the so-called 'upstream dilemma' where the due diligence requirement not to cause significant harm is an especially important consideration, linked to the procedural rules of prior notification, exchange of information and fully functioning institutional mechanisms. This connectivity between substantive rules, procedural rules and institutional mechanisms will be discussed in more detail below.

Another substantive rule, the duty to protect the environment of international watercourses, was a thorny issue in the UN Working Group debate on the UNWC. The debate revolved around the question of whether to add 'sustainable development' to article 5. While the proponents believed that the reference in article 5 to 'optimal utilization' was misleading and therefore better replaced by the term 'sustainable development', some other states, including China, however, objected to the proposal on the grounds that 'sustainable development' had no place in this provision since it had already been provided for in article 20 dealing specifically with the protection of international watercourses (UN, 1996a). This is interpreted to be a reflection of the assumption that the principle of equitable and reasonable utilization deals primarily with utilization, in disregard of any question concerning environmental protection (Tanzi & Arcari, 2001, p. 114). This interpretation of China's 'assumption', however, is not entirely accurate. In fact, when discussing draft article 20, China stated clearly that it attached great importance to the protection and conservation of ecosystems (UN, 1996b). China's proposal to replace the term 'ecosystem' with that of 'ecological balance' was due to its concern about the ambiguity of the former term, rather than the notion of 'protection of ecosystem' itself (UN, 1996b, 1997b). It should also be noted that the CPC Central Committee and the State Council's Number 1 Document for 2011 declare in the first sentence that 'Water is the origin of life, the essence of production and the basis of ecology'. The link between watercourses and protection of the environment is thus expressly recognized by China in a public document widely considered to be an indication of its priority policy in the coming year. This recognition can also find support in a number of China's other treaties, including some multilateral environmental agreements to which China is party. As one relevant example, in article 19 of the 2001 Treaty of Good-Neighbourliness and Friendly Cooperation Between China and Russia, the parties agree to cooperate in the protection and improvement of the environment, and more specifically in 'the equitable and reasonable use of water resources ...' and 'make joint efforts in protecting rare flora, fauna and the natural ecosystem ...'. The fact that the principle of equitable utilization is placed side by side with the duty to protect natural ecosystems in an article dedicated to the protection of environment demonstrates China's consistent recognition of the close relationship between the utilization of watercourses and protection of the environment/ecosystem. In the 2008 China–Russia Agreement, China affirms that, with respect to the development of water resources, 'utilization' and 'protection' are of equal importance and the link between them is so strong that one is inseparable from the other. Support can also be found in many other agreements China has signed (China–Laos Agreement, 1994, article 4; China–Mongolia Agreement, 1994, article 4).

Procedural rules

As pointed out by McCaffrey, to be effective, procedural rules must at a minimum provide mechanisms and processes capable of dealing with a variety of situations. Broadly speaking there are obligations concerning, *inter alia*, regular cooperation, planned measures and emergency situations.

The obligation to exchange information on a regular basis can be found in all of China's water-specific treaties (China–Russia Agreement, 2008, article 5). It appears that China's treaties tend to draw a distinction between information already prescribed by the parties and that not yet prescribed. For the former, the exchange of information is to be conducted according to the plans agreed upon by the parties, and information is to be exchanged on a regular basis. For the latter, the information requested by the one of the parties has not been previously agreed by the parties. The exchange, therefore, will be conducted on an *ad hoc* and case-by-case basis. Notwithstanding similarities between article 5 of the 2008 China–Russia Agreement and article 9 of the UNWC, there exists an obvious difference between the two provisions. While the latter imposes on the parties an obligation to provide information 'not readily available', the former obligates the states to provide information that has not been agreed upon by the parties, 'if conditions permit'.

Procedural rules are not confined to the requirement for the regular exchange of information only, which is of an ongoing and routine nature, but extends also to the incidence of planned measures. Such a duty can be found in many non-water-specific treaties that China has signed, especially in agreements on border management, all of which contain provisions or even chapters dedicated to boundary waters. Provisions in relation to planned measures *per se*, however, are scant in China's water-specific treaties. While the 2008 China–Russia Agreement enumerates a number of areas of cooperation between the parties (China–Russia Agreement, 2008, article 2(8)), it does not spell out the details of the notification procedures to be followed, but leaves it to the parties to decide in a separate agreement. The 2001 China–Kazakhstan Agreement makes no mention of planned measures. Although the 1994 China–Mongolia Agreement does touch upon the issue of water projects, it only obliges the parties to maintain and utilize equitably the water works and flood prevention and water conservation facilities with no specific mention of rules of procedure to support such measures (China–Mongolia Agreement, 1994, article 2(5)). In sum, China's water-specific agreements either make no provision for the duty of prior notification, or provide only a rather limited duty. This is believed to be influenced by China's adherence to the principle of state sovereignty

(Wouters & Chen, 2013, p. 237). The UNWC, in contrast, is much more detailed and covers the whole process of planned measures starting from the initiation to the implementation of the planned measures, in a comprehensive package of rules of procedure.

The obligation to develop emergency preparedness and early warning systems procedures is a third type of procedural obligation in addition to the two obligations abovementioned and is widely endorsed by global water law agreements (UNECE Convention, 1992, article 3(j); UNWC, 1997, article 28) other relevant agreement (UNCLOS, 1982, article 199) and legal instruments made by authoritative organizations (International Law Association, 2004, article 32; UN, 2001, article 16). In this regard, it should be noted that China is working closely with Russia on transboundary emergency prevention and mitigation. The 2008 China–Russia Agreement contains provisions on preventive obligations related to emergency situations, in which the parties agree to establish notification and information exchange mechanisms necessary for preventing emergencies in transboundary waters (China–Russia Agreement, 2008, article 6). What is most notable is that China and Russia have concluded the 2006 Agreement on Cooperation on the Prevention and Mitigation of Emergency Situations (hereinafter Agreement on Emergency Situation), a legally binding instrument dedicated to emergency management procedures. Under this agreement, the parties, in order to prevent emergency situations (China–Russia Agreement, 2006b, article 1), pledge to keep a close watch on any dangerous human activities or changes in eco-processes, and to exchange information for the purpose of risk assessment of emergency situations and potential threats to the environment and citizens (China–Russia Agreement, 2006b, article 5). Provisions on *prevention* of emergency situations, however, cannot be found in other water-specific agreement; the agreements signed between China and Russia therefore constitute rare exceptions in China's practice. It should nevertheless be noted that the procedural obligations in response to an *actual* emergency situation can be found in a number of non-water-specific agreements (China–Mongolia Agreement, 2010, article 9; China–Russia Agreement, 2006a, article 25; China–Vietnam Agreement, 2009a, article 5; China–Tajikistan Treaty, 2007, article 11).

Last but not least, environmental impact assessment (EIA) is considered to be a rule of customary international law (*Pulp Mills* case, 2010, para. 204) and scholars have also argued that the operation of EIA is an inherent requirement of the procedural obligations in IWL (McIntyre, 2007, p. 239). Such an important obligation, however, cannot be found in any Sino water-specific treaties and is also absent in many water-related treaties. In this regard, a notable exception is the Memorandum of Understanding on the Environmental Principles of Tumen Economic Development Zone and Northeast Asia (Memorandum of Understanding on Tumen Environment) (Vinogradov & Wouters, 2013, pp. 50–51). This instrument contains thus far the most elaborated and detailed provisions on transboundary EIA concerning China's transboundary waters. This instrument, however, is strictly speaking not legally binding. The UNWC briefly refers to EIA as part of the information that is to be exchanged between the riparian states, but does not specifically provide for this notion. While the UNECE Convention is also brief in formulating the duty to perform EIA, the UNECE Espoo Convention on Environmental Impact Assessment in a Transboundary Context is a comprehensive treaty on EIA and complements the UNECE Water Convention, providing detailed provisions in this regard.

Joint bodies

Joint bodies have been established under every water-specific agreement that China has signed. The relevant provisions contained in these agreements are to a certain extent similar, all providing that a joint commission for protection and utilization of a transboundary river is to be established, consisting of representatives from each of the parties. One obvious difference between the agreements is that while in both the China–Mongolia Agreement (1994, article 1) and the China–Kazakhstan Agreement (2001, articles 8–10), broad language is used and the joint commission is entrusted to supervise 'matters related to the implementation of the treaties', the 2008 China–Russia Agreement provides a relatively detailed list of the tasks and functions accorded to the joint commission. Scholars seem to hold positive views of the use of broad language, stating that the 1994 China–Mongolia Agreement 'establishes a Joint Committee with monitoring functions over compliance with the Agreement', a joint body that is 'entrusted with a wide array of tasks' (Tanzi & Arcari, 2001, p. 188). As a result of the compromise struck between opposing views at the time, the UNWC does not oblige the parties to set up joint bodies but only encourages riparian states to establish such mechanisms to facilitate cooperation (UNWC, 1997, articles 8(2), 24 (1)). China has followed this recommendation in establishing joint bilateral commissions with Kazakhstan, Mongolia and Russia.

Dispute settlement

While China firmly supports the peaceful settlement of international disputes, it is averse to third-party dispute settlement (for a number of reasons), preferring instead bilateral negotiations and consultation (Beyer, 2006). While China has accepted the compulsory dispute settlement mechanisms under World Trade Organization (WTO), that was an entry requirement to the General Agreement on Trade and Tariffs (GATT), and so an entirely different regime. In other international engagements, China persistently holds the view that disputes should be settled peacefully through consultations, dialogue and negotiations. Examining China's water treaty practice in this regard, it can be seen that none of the water-specific agreements provides for compulsory dispute settlement. Any dispute arising out of the interpretation and application of the agreements will be resolved through 'friendly consultation' (China–Kazakhstan Agreement, 2001, article 12; China–Mongolia Agreement, 1994, article 14) or 'negotiation' (China–Kazakhstan Agreement, 2011, article 10; China–Russia Agreement, 2008, article 12). As to the agreements on border management, another group of treaties that have provisions on international watercourses, the dispute settlement mechanism contained therein also follows the same pattern, i.e. the dispute is to be resolved by 'boundary representatives' from both parties; should the representatives fail to reach an agreement, the dispute will be referred to officials superior in rank or be resolved through diplomatic channels (China–Mongolia Agreement, 2010, article 42; China–Russia Agreement, 2006a, article 43; China–Vietnam Agreement, 2009a, article 45). China tends to resolve any disputes over transboundary waters through diplomatic means such as negotiation and consultation. In view of this, Julian Ku argues that China still shuns participation in compulsory dispute settlement (Ku, 2012) since 'consultation' or 'negotiation' is a legal procedure, but not a compulsory one (Zhang, 2007). This is echoed by China's remarks on its position on this issue, i.e. that China does not oppose any means of dispute settlement as such, which include international adjudication (UN, 1997c) but is strongly against the compulsory nature of any dispute settlement mechanism, which is contrary to the Charter of the United Nations (UN, 1997d, p. 9). In fact the compulsory nature of the fact-finding procedure is one of the main reasons why China voted against the UN resolution adopting the UNWC in 1997 (UN, 1997e, p. 7).

Evaluating China's treaties: identifying merits and gaps

The legal analysis conducted above offers an overview of China's current transboundary water treaty practice. This provides the platform for evaluating the strengths and weaknesses of China's approach in this field.

As to the matter of *scope*, it should be noted that the geographical scope of China's treaties has been expanding over the years. However, the geographical scope as well as the functional scope of the treaties is still narrower than that of the UNWC. The relatively narrow scope as laid down in the treaties, of course, may possibly be intended to be so by China. Another point to note is China's consistent preference to deal with one party in each agreement, despite the fact that the watercourse might cross more than one state. This bilateral approach to scope tends to result in inconsistency among treaties, especially where a basin extends beyond two watercourse states (basin treaty incoherence). The more bilateral agreements signed, the more inconsistencies might possibly result, leading to the so-called fragmentation of IWL (Tanzi & Arcari, 2001, pp. 24ff.; UN, 2006). As discussed above, the three water-specific agreements contain divergent provisions on the no-harm principle. In the 1994 China–Mongolia Agreement, no qualitative term is inserted before the term 'harm'; while such a qualification can be found in both the 2001 China–Kazakhstan Agreement and 2008 China–Russia Agreement, which stipulate the duty on the part of the parties to refrain from causing 'significant' harm. Although Tanzi believes that 'whatever the adjective qualifying the harm falling within the purview of this obligation, any adjective would provide little guidance for the assessment *in concreto* of the exact nature and extent of the harm to be prevented and, conversely, of that which should be acceptable' (Tanzi & Arcari, 2001, p. 147), it is possible that the distinctions between different agreements signed by the same state regarding similar matters might be construed as legally meaningful. In addition, the bilateral approach, likely to result in 'fragmented governance' is contradictory to the optimal and sustainable approach to manage an international river, which is, as argued by hydrologists and engineers, to 'treat it as an integral unit that respects the ecological interdependence within the ecosystem' (Zawahri & Mitchell, 2011, p. 837). Admittedly, this bilateral approach to transboundary water regimes is not unique to China (Whittington, 2004; Zawahri & Mitchell, 2011, p. 835). It might even be justified considering that every international watercourse has its own uniqueness. However, as we can see using the legal five-element analysis, some rules in IWL are so vital that it is hardly conceivable that any 'uniqueness' of a watercourse can justify them being taken out from any international watercourse agreements. The obligation of notification in emergency situations, for example, widely recognized as part of the corpus of customary international law (Birnie, Boyle, & Redgwell, 2009; Handl, 2008, p. 542; Okowa, 1996), can be found in the agreement concluded between China and Russia but is absent in water-specific agreements China signed with other countries.

As to the substantive and procedural rules, the examination of China's IWL practice shows that China has broadly accepted the fundamental principles of IWL, which include, *inter alia*, the principle of equitable and reasonable utilization of water resources (Weiss, 2007) and the duty to cooperate (Wouters, 2013). By accepting these principles in its treaty practice, China has in effect accepted certain limitations on its sovereign rights. Therefore, China's concern over the omission of the term 'sovereignty' in the text of the UNWC, one of the reasons for China's negative vote, seems to be increasingly misplaced. While on the positive side, it is commendable that China has concluded treaties on a number of its major transboundary waters, important gaps nonetheless remain. In the first instance, most of the substantive and procedural provisions in those instruments are rather broad and vague. While on the one hand this approach might be deliberate, so as to make the agreement flexible enough to address changing circumstances (McCaffrey, 1986, p. 130; McIntyre, 2007, p. 23), recognizing also the difficult if not impossible task of determining an accurate approach to determining water use allocation, on the other hand it might lead to considerable legal indeterminacy, especially with regard to such questions as 'the extent to which each co-riparian State may share in the utilisation of waters' (Lipper, 1967, pp. 41–42). The 2001 China–Kazakhstan Agreement, for example, simply provides that the parties 'undertake to address problems related to the protection and utilisation of transboundary waters in an *equitable* and *reasonable* manner' (China–Kazakhstan Agreement, 2001, preamble; added emphasis). It becomes even more obvious when comparing it with article 6 of the UNWC, which has listed factors that might be taken into account when deciding the equitableness and reasonableness of the planned measures – a helpful indicative list and methodology for determining 'equitable and reasonable use' on a case-by-case basis. In fact, provisions contained in the UNWC, a 'framework agreement', are designed to be general in nature so that it can apply to the widely diverse watercourses around the globe, leaving space for states to work out the details that apply to specific transboundary water regimes (UNWC, 1997, article 3). China's water-specific treaties concluded with its neighbours could build upon this framework and provide more details – and yet, as shown above, Sino water treaty provisions are couched in terms even broader than that of the UNWC.

In terms of the implementation of China's transboundary water treaties, i.e. joint bodies and dispute settlement, there are also some shortcomings worthy of note, especially in the area of dispute settlement. Substantive rules alone are insufficient to ensure the effective implementation of international agreements. Operational rules of procedure and institutional mechanisms have an integral role to play. Few would dispute the indispensability of sharing relevant information and data to achieving effective watercourses management (Schwebel, 1981, p. 119). For example, information is required for deciding the 'equitableness and reasonableness' of all measures. However, not all the information needed for implementing the obligations in IWL can be obtained within a single state's own borders. It requires cooperation between riparian states on such procedural obligations as exchanging and sharing information. The procedural rules or other mechanisms that enhance cooperation between the parties are, therefore, vital for the operation of IWL.

Although the obligation to establish joint bodies for transboundary water management might not have matured into a rule of customary international law (Tanzi & Arcari, 2001, p. 191) the need (UN, 1974) and importance (Bogdanovic, 2001; McCaffrey, 1990; Vinogradov & Wouters, 2003, p. 61) of the establishment of joint bodies for facilitating cooperation between the parties and implementation of the treaties is well recognized and evidenced in 20 years of successful state practice under the UNECE. China has established joint commissions under every water-specific agreement it has signed. While there are still lingering problems concerning the international watercourses between the two countries to be addressed (Stone, 2012), these joint commissions are making progress in terms of enhancing transboundary water cooperation between China and its neighbours. For example, both China and Kazakhstan expressed satisfaction with the work of China–Kazakhstan Joint Commission on the Use and Protection of Transboundary Rivers (Declaration, 2013, section 3). The problem, however, arises mainly from China's selective treaty practice, i.e. the North–South divergence – China has concluded water-specific treaties with most of its northern neighbours, but none with its southern neighbours (Wouters & Chen, 2013, p. 239). In addition, no joint body is established for managing the transboundary waters shared between China and its southern neighbours. While China is a 'dialogue partner' to the Mekong River Commission (MRC) and has promised to promote further cooperation with it (MRC, 2013), the fact that China is still not a member to the Mekong Agreement regime is strongly criticized by commentators as not 'setting a good example' (Kongrut, 2013). Overall, across China's transboundary waters there is a 'lack of adequate normative and institutional frameworks capable of reconciling competing demands' (Wouters & Chen, 2013, p. 240). The UNECE Convention, which places special emphasis on institutional mechanisms for transboundary water management, (UNECE Convention, 1992, article 9), might be a model for China to consider in future.

China prefers diplomatic means for resolving disputes and is averse to third-party involvement and compulsory dispute settlement, a preference that might be deeply embedded in Chinese history and culture (Pan, 2009). The Chinese tend to rely on moral rules (*Li*) instead of legal rules (*Fa*) for regulating social behaviours, since it is believed that only *Li*, but not *Fa* (Pan, 2011) is capable of cultivating a 'harmonious society'. The Confucian concept of such a peaceful society is featured by the minimization of litigation. It therefore should not come as a surprise that contentious court proceedings are considered to be an anomaly for a harmonious society in the Chinese context. Although China did not object to judicial means for settling water disputes during the UN Sixth Committee discussion on the UNWC, it could not accept the compulsory nature of the dispute settlement mechanism, which it considered inappropriate for a framework instrument (which was intended to give guidance only, not prescription that impeded on state sovereignty). It should be noted that none of the water-specific or -related agreements concluded by China provide for judicial dispute settlement and all have opted for diplomatic means, such as negotiation. Admittedly, such preference is not in itself problematic. In fact, in accordance with the fundamental rules of international law and under the UN Charter, states have the sovereign right to choose any peaceful means for settling disputes, and third-party dispute settlement is rare in international practice (Boisson de Chazournes, 2003, p. 108). However, what is problematic in the Sino-treaty practice is the vacuum left where disputes cannot be resolved through diplomatic means. In the field of transboundary water cooperation, institutional mechanisms and dispute settlement provisions are important cornerstones for facilitating dispute avoidance and settlement (Wouters, 2003, pp. 151–154). The China–Mongolia and China–Kazakhstan Agreements do not entrust the joint commission with such tasks; and China is not a member to the MRC and hence is not bound by the dispute settlement mechanism provided for in the 1995 Mekong River Agreement (Mekong Agreement, 1995, article 18). The absence of judicial means, together with the lack of ability of the joint body to entertain the dispute, is likely to lead to 'feelings of frustration and alienation' and might 'breed adversity' among riparian states (McCaffrey, 2003b, p. 88). It does not mean that China has to adopt any specific mode of dispute settlement mechanism, since different regional approaches have been adopted around the world for addressing transboundary water disputes (Wouters, 2003, pp. 111–154). However, China has an important opportunity to devise its own approach in this area, to seek a method for effective dispute settlement.

Assessing China's IWL practice in light of recent developments

An economically and politically emerging China, upstream on almost all of its international watercourses, has inevitably raised concerns among its riparian neighbours. China's upstream hydropower development and dams have fuelled criticisms, leading some commentators to refer to China as a 'hydro-hegemon' (Creak, 2010; Onishi, 2005). Added to these allegations, the fact that China voted against the UNWC and its patchy water treaty practice surveyed above has led another commentator to assert that 'China has a bad record when it comes to water cooperation' (Biba, 2014). Despite these views, however, the legal analysis conducted above through the five-element framework reveals somewhat a different story. China, in its existing water treaty practice has embraced the fundamental principles of IWL, albeit with some shortcomings. And it would appear that things are moving in the right direction, given recent developments both with respect to China's interstate relations and in the field of IWL more generally, as discussed below.

IWL is evolving to meet current challenges. One emerging trend in IWL, referred to as the 'structurization' of IWL, might provide support to address some of China's water-related treaty gaps. Already, the plurality of sources of law in this field are evolving into 'an eclectic regulatory continuum' (Castro, 2000, p. 158). Jing Lee, for example, shows how it is possible, by making use of the interpretive mechanism provided by article 31 (3)(c) of the Vienna Convention on the Law of Treaties, to integrate the relevant rules of water-related multilateral environmental agreements, many of which China is a party to (Convention on Wetlands of International Importance especially as Waterfowl Habitat, Convention on Biological Diversity) so as to give legal meaning to the 'ecosystem of the watercourse' under the UNWC (Lee, 2014). This cross-fertilization across various rules of international law provides important normative connectivity and legal clarity for fundamental rules of IWL. Thus, China's bilateral and sometimes inconsistent water treaty practice might benefit from this regulatory continuum which could serve also to address issues of fragmentation.

In this regard, a notable example may be the interaction between China's treaties and the two global water conventions (UNWC and the UNECE Convention). This interaction is conducive to a better understanding of what the fundamental principles of IWL might mean and how they can be better implemented in the Chinese context. The UNWC, for example, is the result of almost 30 years of study by the UN International Law Commission and had been subject to years of discussion and debate of states; similarly the UNECE Water convention has a rich 20 years of successful state practice on the ground. The understandings drawn out of these two global conventions, coupled with China's obligations under a number of multilateral environmental agreements, could help China flesh out the substantive and procedural aspects of its water-related treaties and state practice, resulting in normative coherence.

A second major trend that is emerging is the *proceduralization* of IWL. With the overarching substantive rule of 'equitable and reasonable use' now universally endorsed, the movement has been towards implementation of that norm through procedural rules and mechanisms. The effective day-to-day functioning of IWL regimes hinges upon the actual interaction and measure of effective cooperation among riparian states (Leb, 2013) and requires more than substantive rules. Thus, processes for the exchange of data, continued dialogue, cooperative joint activities – all lead to the more effective operation of the principle of equitable and reasonable utilization (McCaffrey, 2003a, pp. 341ff.). In fact, the introduction of the 'procedural law of cooperation' into the draft UNWC was observed to be one of the major developments in the entire drafting process (Higgins, 1991). The recent ICJ decision in the *Pulp Mills* case (Argentina v. Uruguay), which considered a water-treaty dispute, is considered by McIntyre to be illustrative of the trend towards the 'proceduralization' of IWL (McIntyre, 2010, pp. 475–497). As demonstrated by that case, institutional mechanisms have an integral role to play in transboundary water cooperation, supported through rules of procedure.

The importance of these two trends – structurization and proceduralization – contribute also to addressing the shortcoming in the area of dispute settlement. China's cultural influence of '*Li*' and its commitment to state sovereignty as a fundamental rule of international law, means that it is unlikely that China's attitude towards water-related disputes will change. It should be noted, however, that China's reluctance towards judicial settlement is not unique: most states prefer the sovereign choice to select the means for addressing disputes, especially when there are water disputes (Caflisch, 2003, p. 236). In light of the fact that judicial means are not available in most cases, cooperative mechanisms between the parties play an important role in dispute avoidance. Undoubtedly legal rules are in themselves important, but their implementation needs to be realized through the process of functional application; in this context transboundary water commissions can play an integral role (Laitta, 2010). There are signs that China is moving in this direction. For example, as noted above, China has established joint commissions overseeing transboundary waters with most of its northern neighbours, which, notwithstanding their limited functions, have made considerable progress over the years. For example, both China and Kazakhstan have spoken highly of the achievement of the China–Kazakhstan Joint Commission and pledge to enhance cooperation on transboundary water issues. As a sign of improved effective cooperation, China and Kazakhstan have recently agreed that the joint commission go forward to determine water allocation under the agreement (Declaration, 2013, section 3). The joint institutional mechanisms set up between China and Russia are also believed to be actively functioning and conducive to fostering continuing cooperation (Vinogradov & Wouters, 2013, pp. 44–49). While no joint institutional mechanisms have been established between China and its southern neighbours, incremental transboundary water cooperation is happening there also; China has signed agreements on information exchange with most of its southern neighbours. Given that information exchange is considered by many to be the first step towards enhanced cooperation (Chenoweth & Feitelson, 2001) it can be concluded that China is taking steps towards building cooperative relationships with these neighbouring countries, albeit step by step. As a dialogue partner with the MRC, China has concluded the Agreement on Provision of Information Regarding the Lancang River to the commission since 2002 and has been updating the agreement every few years. Furthermore, in a speech delivered during his recent visit to India, Premier Li Keqiang reaffirmed China's duty to provide assistance to India in terms of sharing flood-season hydrological information and managing emergency situations, and reassures the Indian side that China is 'ready to enhance cooperation with India in this regard, and to strengthen communication on the utilization and protection of trans-border rivers' (Ministry of Foreign Affairs, 2013). This new wave of bilateral cooperation is significant in that it has either touched upon issues that were not dealt with in the past in China's water treaties (i.e. water allocation with Kazakhstan), or enhanced existing cooperation (i.e. water data exchange with India). China's progressive outreach is not limited to bilateral relations, but also extends to the regional level. Cooperation with regional organizations such as the Association of Southeast Asian Nations (ASEAN), and notably the regular meetings of the Shanghai Cooperation Organisation are considered to be ideal platform for promoting cooperation over transboundary water issues (Mustafina, 2014).

While the phenomena of 'structurization' and 'proceduralization' of IWL are not new to international law, they will have an important impact. This growing current of 'flows', flows across legal normativity, flows across legal and practical understanding, all highlight the over-arching *communicative* dimension of international law:

> Where countries are involved in a disagreement or a dispute, it is handy to have recourse to the rules of international law even if there are conflicting interpretations since at least there is a common frame of reference and one state will be aware of how the other state will develop its argument. They will both be talking a common language and *this factor of communication* vital since misunderstandings occur so easily and often with tragic consequences. (added emphasis; Brunnee & Toope, 2010; Shaw, 2008, p. 7; Venzke, 2012)

The importance of communication and cooperation among legal subjects becomes even more obvious in IWL, as demonstrated by the trends identified above. The analysis of recent developments in both IWL and China's IWL practice supports the view that China is taking part in the *communicative process* of the making of IWL. While China can become more engaged with IWL by drawing insights from the global conventions and participating in joint mechanisms, it can also become engaged by contributing its own understanding of the legal norms to the international community. A good starting point might be that of environmental protection of the transboundary waters, building on the new wave of environmental protection measures being implemented across China. China has launched a 'war on pollution' (Xinhua News, 2014, March 5), introducing weapons such as the amended Environmental Protection Law, which elevates environmental priorities over development (Wee & Stanway, 2014). Considering that the emphasis on harmony between humans and nature has always been innate in Chinese culture, is it not the opportune time for China to contribute more fully to address the pressing environmental issues linked with transboundary water regimes (McIntyre, 2007, pp. 359ff.)? Through informed and increased efforts in this important field, China can truly contribute to its goal of being 'the good neighbour'.

Acknowledgement

Thanks to Professor Patricia Wouters (Xiamen Law School), Dr Owen McIntyre (visiting scholar, Xiamen Law School) and members of the CIWL team at Xiamen Law School – without them, this article could not have been completed.

References

Agreement, China–Laos (1994). Agreement on Transportation of Goods on the Mekong River between the PRC and Laos. Retrieved June 17, 2014, from http://vip.chinalawinfo.com/newlaw2002/eag/index.asp.

Agreement, China–Mongolia (1994). Agreement on Protection and Utilisation of Boundary Waters between PRC and Mongolia. Retrieved June 16, 2014, from http://vip.chinalawinfo.com/newlaw2002/eag/index.asp.

Agreement, China–Kazakhstan (2001). Agreement on the Utilisation and Protection of Transboundary Rivers between the PRC and Kazakhstan. Retrieved June 16, 2014, from http://vip.chinalawinfo.com/newlaw2002/eag/index.asp.

Agreement, China–Russia (2006a). Agreement on Border Management between the PRC and Russia. Retrieved June 17, 2014, from http://vip.chinalawinfo.com/newlaw2002/eag/index.asp.

Agreement, China–Russia (2006b). Agreement on Cooperation on the Prevention and Mitigation of Emergency Situations between the PRC and Russia. Retrieved June 16, 2014, from http://vip.chinalawinfo.com/newlaw2002/eag/index.asp.

Agreement, China–Russia (2008). Agreement on the Reasonable Utilisation and Protection of Transboundary Waters between the PRC and Russia. Retrieved June 16, 2014, from http://vip.chinalawinfo.com/newlaw2002/eag/index.asp.

Agreement, China–Vietnam (2009a). Agreement on Border Management between the PRC and Vietnam. Retrieved June 16, 2014, from http://vip.chinalawinfo.com/newlaw2002/eag/index.asp.

Agreement, China–Vietnam (2009b). Agreement on Border Management between the PRC and Vietnam. Retrieved June 16, 2014, from http://vip.chinalawinfo.com/newlaw2002/eag/index.asp.

Agreement, China–Mongolia (2010). Agreement on Border Management between the PRC and Mongolia. Retrieved June 16, 2014, from http://vip.chinalawinfo.com/newlaw2002/eag/index.asp.

Agreement, China–Kazakhstan (2011). Agreement on the Protection of Water Quality between the PRC and Kazakhstan. Retrieved June 16, 2014, from http://vip.chinalawinfo.com/newlaw2002/eag/index.asp.

Beyer, S. (2006). Environmental law and policy in the People's Republic of China. *Chinese Journal of International Law, 5*(1), 185–211, 190. doi:10.1093/chinesejil/jmk002

Biba, S. (2014). *China cooperates with Central Asia over shared rivers*. Chinadialogue. Retrieved July 9, 2014, from https://www.chinadialogue.net/article/show/single/en/6741-China-cooperates-with-Central-Asia-over-shared-rivers-.

Bogdanovic, S. (Ed.). (2001). *International law of water resources – contribution of the International Law Association (1954–2000)*. The Hague: Kluwer Law International.

Birnie, P. W., Boyle, A., & Redgwell, C. (2009). *International law and the environment*. 3rd ed. Oxford: Oxford University Press.

Brunnee, J., & Toope, S. (2010). *Legitimacy and legality in international law*. Cambridge: Cambridge University Press.

Caflisch, L. (2003). Judicial means for settling water disputes. In International Bureau of Permanent Court of Arbitration (Ed.), *Resolution of international water disputes* (pp. 235–245). The Hague: Kluwer Law International.

Castro, P. C. d. (2000). The future of international water law. In *Shared water systems and transboundary issues*. Lisbon: Luso-American Foundation.

Chazournes, L. B. d. (2003). The role of diplomatic means of solving water disputes: A special emphasis on institutional mechanisms. In International Bureau of Permanent Court of Arbitration (Ed.), *Resolution of international water disputes* (pp. 91–110). The Hague: Kluwer Law International.

Chellaney, B. (2011). *Water: Asia's new battleground*. Washington DC: Georgetown University Press.

Chenoweth, J. L. Ca. E., Feitelson, E. (2001). Analysis of factors influencing data and information exchange in international River basins: Can such exchanges be used to build confidence in cooperative management? *Water International, 26*(4), 499–512. doi:10.1080/02508060108686951

China Water Risk. (2010). *Part I Introduction*. Retrieved June 15, 2014, from China's Water Crisis: http://chinawaterrisk.org/wp-content/uploads/2011/06/Chinas-Water-Crisis-Part-1.pdf

Creak, S. (2010). A Chinese Perspective on the Dammed Mekong. Retrieved June 17, 2014, from http://asiapacific.anu.edu.au/newmandala/2010/07/01/a-chinese-perspective-on-the-damned-mekong/.

Declaration, China–Kazakhstan (2013). China–Kazakhstan Joint Declaration. Retrieved from http://vip.chinalawinfo.com/newlaw2002/eag/index.asp.

Gleick, P. G. (2009). *The world's water 2008–2009* (pp. 79–100). Washington DC: Island Press.

Handl, G. (2008). Transboundary impact. In D. Bodansky, J. Brunnee, & E. Hey (Eds.), *The Oxford handbook of international environmental law* (pp. 532–549). Oxford: Oxford University Press.

Higgins, R. (1991). General course on public international law. *Recueil des cours, 230*, 11–341, 183.

International Court of Justice. (2010) *Case concerning Pulp Mills on the River Uruguay (Argentina v. Uruguay)*. ICJ Reports, Judgment of 20 April 2010. Retrieved July 16, 2014, from http://www.icj-cij.org/docket/files/135/15877.pdf.

International Law Association. (1966). *The Helsinki rules on the uses of the waters of international rivers*. Retrieved June 17, 2014, from http://www.unece.org/fileadmin/DAM/env/water/meetings/legal_board/2010/annexes_groundwater_paper/Annex_II_Helsinki_Rules_ILA.pdf.

International Law Association (2004). The Berlin Rules on Water Resources. International Law Association: Retrieved from http://www.unece.org/fileadmin/DAM/env/water/meetings/legal_board/2010/annexes_groundwater_paper/Annex_IV_Berlin_Rules_on_Water_Resources_ILA.pdf.

Jia, B. (2010). The relations between treaties and custom. *Chinese Journal of International Law, 9* (1), 81–109. doi:10.1093/chinesejil/jmq001

Khai, N. (2008). Thailand: Worst Mekong River Flooding in 100 Years: IRIN Retrieved June 16, 2014, from http://www.irinnews.org/report/79869/thailand-worst-mekong-river-flooding-in-100-years.

Kongrut, A. (2013). China Should Set Good Example on the Mekong River. The Economic Observer. Retrieved June 17, 2014, from http://www.eeo.com.cn/ens/2013/0125/239281.shtml.

Ku, J. (2012). China and the future of international adjudication. *Maryland Journal of International Law, 27*(1), 154–173, 157.

Laitta, M. (2010). *Canada-U.S. Transboundary Hydrographic Data Harmonization Efforts Gain Momentum*. Geobase. Retrieved July 12, 2014, from http://www.ijc.org/rel/boards/watershed/Canada-US_Hydro_Harmonization_e.pdf.

Leb, C. (2013). *Cooperation in the law of transboundary water resources*. Cambridge: Cambridge University Press.

Lee, J. (2014 (forthcoming in August)). *Preservation of ecosystems of international watercourses and the integration of relevant rules*. Leiden: Brill Publishers.

Lipper, J. (1967). Equitable utilisation. In A. H. Garretson, R. D. Hayton, & C. J. Olmstead (Eds.), *The law of international drainage basin* (pp. 15–88). New York: Oceana Publications.

McCaffrey, S. (1986). Second Report on the Law of the Non-Navigational Uses of International Watercourses. (A/CN.4/399). Retrieved June 17, 2014, from http://untreaty.un.org/cod/avl/ha/clnuiw/clnuiw.html.

McCaffrey, S. (1990). Sixth Report on the Law of the Non-Navigational Uses of International Watercourses. (A/CN.4/427). Retrieved June 17, 2014, from http://untreaty.un.org/cod/avl/ha/clnuiw/clnuiw.html.

McCaffrey, S. (2001). The contribution of the UN convention on the law of the non-navigational uses of international watercourses. *International Journal of Global Environmental Issues, 1*(3/4), 250–261. doi:10.1504/IJGENVI.2001.000980

McCaffrey, S. (2003a). *The law of international watercourses*. First ed. Oxford: Oxford University Press.

McCaffrey, S. (2003b). Water disputes defined: Characteristics and trends for resolving them. In International Bureau of Permanent Court of Arbitration (Ed.), *Resolution of international water disputes* (pp. 49–90). Kluwer Law International.

McIntyre, O. (2006). The role of customary rules and principles of international environmental law in the protection of shared international freshwater resources. *Natural Resources Journal, 46*(1), 157–210.

McIntyre, O. (2007). *Environmental protection of international watercourses under international law*. Aldershot: Ashgate.

McIntyre, O. (2010). The proceduralisation and growing maturity of international water law: Case concerning *Pulp Mills* on the River Uruguay (Argentina v Uruguay), International Court of Justice, 20 April 2010. *Journal of Environmental Law, 22*(3), 475–497. doi:10.1093/jel/eqq019

Mekong River Commission. (1995). Agreement on the Cooperation for the Sustainable Development of the Mekong River Basin. Retrieved July 11, 2014, from http://www.mrcmekong.org/assets/Publications/policies/agreement-Apr95.pdf.

Mekong River Commission. (2013). Mekong River Commission and China Boost Water Data Exchange. Retrieved July 11, 2014, from http://www.mrcmekong.org/news-and-events/news/mekong-river-commission-and-china-boost-water-data-exchange/.

Menon, P. K. (1972). Institutional mechanism for the development of international water resources. *Revue belge de droit international, 8*, 81–100, 181.

Ministry of Foreign Affairs. (2013). Speech by Premier Li Keqiang, "Seize the New Opportunities in China-India Strategic Cooperation". Retrieved July 12, 2014, from http://www.fmprc.gov.cn/mfa_eng/wjdt_665385/zyjh_665391/t1051164.shtml.

Mustafina, A. (2014). Transboundary water issues between Kazakhstan and China. *IOSR Journal of Humanities and Social Science, 19*(1), 91–94. doi:10.9790/0837-19139194

Okowa, P. N. (1996). Procedural obligations in international environmental agreements. *British Yearbook of International Law, 67*(1), 275–336, 281. doi:10.1093/bybil/67.1.275

Onishi, K. (2005). Hydropolitics of China and downstream countries in the Mekong river basin. *Proceedings of the International Symposium on Role of Water Sciences in Transboundary River Basin Management*, 10–12 March 2005, Ubon Ratchathani, Thailand.

Pan, J. (2009). *Toward a new framework for peaceful settlement of China's territorial and boundary disputes*. Leiden: Martinus Nijhoff Publishers.

Pan, J. (2011). Chinese philosophy and international law. *Asian Journal of International Law, 1*(2), 233–248, 234. doi:10.1017/S2044251310000354

Permanent Court of International Justice. (1929). Case relating to the territorial Jurisdiction of the International Commission of the River Oder. (Series A, No. 23).
Rieu-Clarke, A. (2013). Will 2013 Be Seen as the International Year of Water Law Cooperation? SIWI. Retrieved June 16, 2014, from http://www.siwi.org/publication/will-2013-be-seen-as-the-international-year-of-water-law-cooperation/.
Rieu-Clarke, A., Moynihan, R., & Magsig, B. -O. (2012). *UN Watercourses convention: User's guide*. Dundee: IHP-HELP Centre for Water Law, Policy and Science.
Salman, S. (2007). The Helsinki rules, the UN watercourses convention and the Berlin rules: Perspectives on international water law. *International Journal of Water Resources Development, 23*(4), 625–640, 633. doi:10.1080/07900620701488562
Schwebel, S. (1981). Third Report on the Law of the Non-Navigational Uses of International Watercourses. (A/CN.4/348). Retrieved June 17, 2014, from http://untreaty.un.org/cod/avl/ha/clnuiw/clnuiw.html.
Shaw, M. (2008). *International law*. Cambridge: Cambridge University Press.
Sievers, E. W. (2002). Transboundary Jurisdiction and Watercourse Law: China, Kazakhstan, and the Irtysh. *Texas International Law Journal, 37*(1), 1–42.
Statement, China–Kazakhstan (2004). China–Kazakhstan Joint Statement. Retrieved June 17, 2014, from http://vip.chinalawinfo.com/newlaw2002/eag/index.asp.
Statement, China–Bangladesh (2005). China–Bangladesh Joint Statement. Retrieved June 17, 2014, from http://vip.chinalawinfo.com/newlaw2002/eag/index.asp.
Statement, China–Kazakhstan (2006). Cooperation Strategy between China and Kazakhstan in the Twenty-First Century. Retrieved June 16, 2014, from http://vip.chinalawinfo.com/newlaw2002/eag/index.asp.
Statement, China–Kazakhstan (2012). Joint Statement between the PRC and Kazakhstan. Retrieved from http://vip.chinalawinfo.com/newlaw2002/eag/index.asp.
Stone, R. (2012). For China and Kazakhstan: No Meeting of the Minds on Water. Science, 337. Retrieved July 12, 2014, from http://211.144.68.84:9998/91keshi/Public/File/41/337-6093/pdf/405.full.pdf.
Tanzi, A., & Arcari, M. (2001). *The United Nations convention on the law of international watercourses: A framework for sharing*. The Hague: Kluwer Law International.
Treaty, China–Russia (2001). Treaty of Good-Neighborliness and Friendly Cooperation between the People's Republic of China and the Russian Federation. Retrieved June 16, 2014, from http://vip.chinalawinfo.com/newlaw2002/eag/index.asp.
Treaty, China–Tajikistan (2007). Treaty on Good-Neighbourliness Friendship and Cooperation between the PRC and Tajikistan, Article 11. Retrieved June 16, 2014, from http://vip.chinalawinfo.com/newlaw2002/eag/index.asp.
Tyler, Z. (2006). Transboundary water pollution in China – an analysis of the failure of the legal framework to protect downstream jurisdictions. *Columbia Journal of Asian Law, 19*(2), 572–613, 573.
UN. (1974). Report of the Secretary-General on the Legal Problems relating to the Utilization and Use of International Rivers. Retrieved June 17, 2014, from http://www.un.org/law/ilc/index.htm.
UN. (1991). Summary Records of the 2214th Meeting, 43rd Session, ILC. (Vol. I (1)). Retrieved June 17, 2014, from http://www.un.org/en/documents/ods/.
UN. (1994). The Law of Non-Navigational Uses of International Watercourses. Draft Articles and Commentaries thereto adopted by the Drafting Committee on Second Reading. (A/CN.4/L.493). Retrieved July 12, 2014, from http://www.un.org/en/documents/ods/.
UN. (1996a). Summary Record of the 15th Meeting, Fifty-First Session, Sixth Committee. (A/C.6/51/SR.15). Retrieved June 16, 2014, from http://www.un.org/en/documents/ods/.
UN. (1996b). Summary Record of the 21st Meeting, Fifty-First Session, Sixth Committee. (A/C.6/51/SR.21). Retrieved June 16, 2014, from http://www.un.org/en/documents/ods/.
UN. (1996c). Summary Records of the 17th Meeting, Fifty-First Session, Sixth Committee. (A/C.6/51/SR.17). Retrieved June 16, 2014, from http://www.un.org/en/documents/ods/.
UN. (1997a). Report of the Sixth Committee Convening as the Working Group of the Whole. (A/51/869). Retrieved June 16, 2014, from http://www.un.org/en/documents/ods/.
UN. (1997b). Summary Record of the 53rd Meeting, Fifty-First Session, Sixth Committee. (A/C.6/51/SR.53). Retrieved June 16, 2014, from http://www.un.org/en/documents/ods/.
UN. (1997c). Summary Record of the 59th Meeting, Fifty-First Session, Sixth Committee. (A/C.6/51/SR.59). Retrieved June 16, 2014, from http://www.un.org/en/documents/ods/.

UN. (1997d). Summary Records of the First Part of the 62nd Meeting, 51st Session, Sixth Committee. (A/C.6/51/SR.62). Retrieved June 16, 2014, from http://www.un.org/en/documents/ods/.

UN. (1997e). Veratim Records of Plenary Meeting No. 99, UN General Assembly, 51st Session. (A/51/PV.99). Retrieved June 16, 2014, from http://untreaty.un.org/cod/avl/ha/clnuiw/clnuiw.html.

UN. (2001). Draft Articles on Prevention of Transboundary Harm from Hazardous Activities. Retrieved July 12, 2014, from http://legal.un.org/ilc/texts/instruments/english/commentaries/9_7_2001.pdf.

UN. (2006). Report of the Study Group of the International Law Commission on "Fragmentation of International Law: Difficulties Arising from the Diversification and Expansion of International Law". (A/CN.4/L.682). Retrieved July 12, 2014, from http://legal.un.org/ilc/guide/1_9.htm.

UNCLOS. (1982). United Nations Convention on the Law of the Sea. Retrieved from http://www.un.org/depts/los/convention_agreements/texts/unclos/unclos_e.pdf.

UNECE Convention. (1992). Convention on the Protection and Use of Transboundary Watercourses and International Lakes. Retrieved June 16, 2014, from http://www.unece.org/fileadmin/DAM/env/water/pdf/watercon.pdf.

UNECE. (2009). *River Basin Commission and Other Institutional for Transboundary Water Cooperation*. (ECE/MP.WAT/32).

UNEP (United Nations Environmental Programme). (December 2005). The Songhua River Spill Retrieved June 16, 2014, from http://www.unep.org/PDF/China_Songhua_River_Spill_draft_7_301205.pdf.

UNEP (United Nations Environmental Programme). (2009). Hydropolitical Vulnerability and Resilience along International Waters. Retrieved June 16, 2014, from http://www.transboundarywaters.orst.edu/research/UNEP_Atlas/hydropolitical_AS.pdf.

UNWC. (1997). United Nations Convention on the Law of the Non-Navigational Uses of International Watercourses. Retrieved June 16, 2014, from http://untreaty.un.org/ilc/texts/instruments/english/conventions/8_3_1997.pdf.

VCLT. (1969). *Vienna Convention on the Law of Treaties*. Retrieved June 17, 2014, from https://treaties.un.org/doc/Publication/UNTS/Volume%201155/volume-1155-I-18232-English.pdf.

Venzke, I. (2012). *How interpretation makes international law: On semantic change and normative twists*. Oxford: Oxford University Press.

Vinogradov, S., & Wouters, P. (2003). *Transforming potential conflict into cooperation potential: The role of IWL*. Paris: UNESCO.

Vinogradov, S., & Wouters, P. (2013). *Sino-Russian Transboundary Waters: A Legal Perspective on Cooperation*. Institute for Security and Development Policy. Retrieved June 16, 2014, from http://www.internationalwaterlaw.org/bibliography/articles/general/Vinogradov-wouters-sino-russian-transboundary-waters-legal-perspective-2013.pdf.

Wee, S.-L., & Stanway, D. (2014). *China Set to Elevate Environment over Development in New Law*. Reuters. Retrieved June 19, 2014, from http://uk.reuters.com/article/2014/04/15/uk-china-environment-idUKBREA3E03A20140415.

Weiss, B. (2007). The evolution of international water law. *Recueil des cours, 331*, 163–404.

Whittington, J. S. D. (2004). Why have some countries on international rivers been successful negotiating treaties. *Water Resources Research, 40*(5), 1–18.

Wouters, P., & Chen, H. P. (2013). China's "soft path" to transboundary water cooperation examined in the light of two UN global water conventions – exploring the "Chinese way". *Journal of Water Law, 22*(6), 229–247.

Wouters, P. (2003). Universal and regional approaches to resolving international water disputes: What lessons learned from state practice. In International Bureau of Permanent Court of Arbitration (Ed.), *Resolution of international water disputes* (pp. 111–154). The Hague: Kluwer Law International.

Wouters, P. (2013). "Dynamic cooperation" in international law and the Shadow of State Sovereignty in the context of Transboundary waters. *Environmental Liability, 21*(3), 88–97.

Xinhua News. (2013, August 22). Chinese, Russian Premiers Stress Cooperation on Flood Control, Disaster Relief. Retrieved June 16, 2014, from http://news.xinhuanet.com/english/china/2013-08/22/c_132654925.htm.

Xinhua News. (2014, March 5). Xinhua Insight: China Declares War against Pollution. Retrieved July 12, 2014, from http://news.xinhuanet.com/english/special/2014-03/05/c_133163557.htm.

Ying, S. (2008). International legal aspect of the Songhua River incident. In M. G. Faure & S. Ying (Eds.), *China and international environmental liability: Legal remedies for transboundary pollution* (pp. 315–332). Cheltenham: Edward Elgar Publishing Limited.

Zawahri, N. A., Mitchell, S.M. (2011). Fragmented governance of international rivers: Negotiating bilateral versus multilateral treaties. *International Studies Quarterly, 55*(3), 835–858. doi:10.1111/j.1468-2478.2011.00673.x

Zhang, Q. (2007). *Consultation within WTO dispute settlement: A Chinese perspective.* Bern: International Academic Publishers.

Zodrow, I. (2010). International aspects of water law reforms. In P. Cullet, A. G. Gualtieri, R. Madhav, & U. Ramanathan (Eds.), *Water law for the twenty-first century: National and international aspects of water law reform in India* (pp. 36–58). Abingdon: Routledge.

China's practice on the non-navigational uses of transboundary waters: transforming diplomacy through rules of international law

Yanmei He

China's abundant transboundary water resources face new challenges as demands increase, internally and externally. As China seeks new ways to cooperate with its riparian neighbours, more attention is being paid to international water law in the evolving diplomatic practice across the region. China's (mostly) bilateral transboundary water cooperation is here characterized as 'responsive diplomacy'. This paper explores China's state practice in this context and concludes by recommending that it moves from 'responsive' towards 'preventive' diplomacy. The author offers strategic and policy responses in order to effect this change.

Introduction

China is blessed with abundant transboundary water resources across its vast nation. In recent years, as China has shifted its economic and energy development to its border reaches, which hold a wealth of transboundary waters, closer attention has been paid to the non-navigational uses of these waters, resulting in varying reactions from downstream states and other stakeholders. This upstream–downstream tension has resulted in diplomatic tensions between China and its riparian neighbours – including as some examples increased uses on the Lancang River (Mekong), the Yarlung Zangbo River (Brahmaputra), the Ili River and the Irtysh River. These developments are linked with many interests: China's energy security, the livelihoods of its citizens in border areas, as well as regional security and stability, and thus have become one of the primary concerns in Chinese foreign affairs.

International law is a system of rules that govern international relations, including transboundary water diplomacy. This trend of 'legalization' of international diplomacy has evolved within contemporary international relations. With the end of the Cold War and rapid globalization, states and their citizens desire a sense of stability and predictability within international relations (Liu, 2010). This article focuses on the challenges faced by China regarding the non-navigational uses of its transboundary waters from the perspective of international law, analysing its state practice and its diplomatic response. This article provides an overview of China's transboundary waters and then examines the status of their usage so as to provide the necessary hydro-geographic and political context as a backdrop for analysing China's diplomatic practice in this field. The paper concludes by recommending policy responses to address the issues identified.

China's transboundary waters

China is home to one-third of the major transboundary waters in the world, with 42 transboundary rivers accounting for 40% of China's total annual runoff (Mehtonen, Keskinen, & Varis, 2008). These rivers run through 3.2 million km^2 of land area, more than one-third of the total land area of China (Gleick, 2000). China ranks fourth in the world for the area covered by transboundary basins after Russia (8 million km^2), the United States (6 million km^2) and Brazil (5 million km^2) (Nickum, 2008). The amount of water leaving Chinese territory is 28 times greater than the amount of water entering, making China an upstream neighbour in a large majority of its transboundary rivers (He, 2013a; He & Feng, 2006).

Among these rivers, 17 are of extreme importance due to their geopolitical and economic influence on China and its co-riparians. These rivers include the Heilong River/Amur River (Mongolia, China and Russia), Suifen River/Razdolnaya River (China and Russia), Tumen River/Tumannaya River (China, North Korea and Russia), and Yalu River/Amrok River (China and North Korea) in the North-east region; Irtysh-Ob River (China, Kazakhstan and Russia), Argun River (China and Russia), Ili River–Lake Balkhash (Kyrgyzstan, China and Kazakhstan), and Tarim River (China and Kyrgyzstan) in the North-west region; Senge Zangbo-Indus (China, India and Pakistan), Yarlung Zangbo/Brahmaputra River (China, Bhutan, India and Bangladesh), Ganges (China, Nepal, India and Bangladesh), Irrawaddy River (China and Myanmar), Nu River/Salween River (China, Myanmar and Thailand), Lancang/Mekong River (China, Myanmar, Laos, Thailand, Cambodia and Vietnam), Yuan River/Red River (China, Vietnam and Laos), Beilun River/Ka Long River (China and Vietnam), and Pearl River/North River (China and Vietnam) in the South-west Region.

The states of these rivers influence the lives and livelihoods of nearly 3 billion people. Low levels of quantity or quality can increase the risk of regional water conflict (He & Feng, 2006; He, 2013a). Competition for these water resources will become increasingly fierce due to growing populations and economic development in and across these basins. According to the 12th Five-Year Plan on Chinese Energy Development released by China's State Council in 2013, due to the over-exploitation and pollution of inland waters, the acceleration of uses of transboundary waters may be one method for the Chinese government to respond to the impending domestic water crisis. This practice will test China's diplomatic practice on its transboundary waters. The next section will examine the status and current trends in China's uses of these waters.

China's non-navigational uses of transboundary waters: status and trends

The non-navigational uses of transboundary rivers have a decisive role to play in China's sustainable development within the context of its broad geographical distribution, vast resources, biological and cultural diversity. However, many of these river basins are ecologically sensitive, covering remote, hard-to-reach areas, and some, such as the Yarlung Zangbo/Brahmaputra, have unresolved border disputes. Opposition to development activities from environmental groups and downstream states make it even more difficult to develop and utilize these transboundary water resources (He, 2013a). As a consequence, the development of China's transboundary waters has proven to be unsatisfactory, with the exception of the Lancang, Ili and Irtysh rivers.

For example, there are 15 hydropower stations planned for the Lancang River in Yunnan province. Thus far four dams (Manwan, Dazhaoshan, Jinghong and Xiaowan) have been built and are in full operation, while the dams at Nuozhadu and Gongguoqiao are under construction, with more dams at the planning stage (Meng, 2014). There are also plans for a cascade of six dams on the Tibetan stretch of the Lancang River at Cege, Yuelong, Kagong, Banda, Rumei and Guxue. Construction is expected to begin in 2015 and finish in 2030 (Meng, 2014). The development of these hydropower stations has been questioned by downstream states and third parties (Magee, 2014).

China's vast Xinjiang region shares 20 rivers with Kazakhstan, among which the Irtysh and the Ili are the largest (Biba, 2014). Since the late 1980s China has, as part of the 'Go West Policy', constructed canals to divert water away from the primary tributary of the Ili for the expansion of the cotton and energy industries. This has resulted in large, unsustainable growth in regional water usage. Water diversion schemes on the Irtysh and Ili are required in order to guarantee the future development of regional oil fields. China has excavated a 300 km canal (the Kara Irtysh-Karamay Canal) in north-west Xinjiang to divert water from the Irtysh to Ulungur Lake, from which water is distributed for local agriculture and to the Karamay oil field (Stone, 2012).

Over the past decade the Chinese government has been cautious when constructing dams in ecologically sensitive and seismically active regions such as the Nu River in Yunnan province and the Yarlung Zangbo River in Tibet. During Wen Jiabao's tenure as Premier of China we have witnessed the delay of a number of projects. Between 2006 and 2010 only one-third of plans identified as priority went ahead (Walker, 2014). A combination of increasing energy demands, intensive lobbying by major dam builders and utility companies, and ambitious plans to meet low-carbon intensity goals within the 12th Five-Year Plan have paved the way for a new era of dam construction. The 2011–15 energy sector blueprint, within the 12th Five-Year Plan on Chinese Energy Development, confirmed plans to 'actively push forward with hydropower development' in the Nu, Lancang and Yarlung Zangbo river basins. According to the plan, construction of approximately 120 gigawatts (GW) of new hydropower stations nationwide will begin by 2015. The plan has caused alarm and opposition amongst downstream neighbours and other stakeholders (Gupta, 2014; Walker, 2014). The construction of a cascade hydropower development, including five dams on the Yarlung Zangbo, is currently underway. As part of the plan, the construction of a large-scale hydropower station had officially begun in September 2010. In November of the same year, the successful closure of the river had been declared and the project entered the main phase of construction (He & Jiang, 2010).

China's utilization of its transboundary waters is not only concerned with its economic, energy and environmental needs, but also is related to geopolitics, regional security and stability, international image, etc. China is simultaneously accelerating the use of its transboundary water resources while paying more attention to transboundary water diplomacy and international water law through the practice of 'responsive diplomacy'. The following two sections will examine this diplomatic practice and its primary challenges.

China's diplomatic practice on the non-navigational uses of transboundary waters

In recent years the Chinese government has attached great importance to transboundary water diplomacy and international water law. This section evaluates China's diplomatic practice on the non-navigational uses of its transboundary waters from the perspective of international law. It examines China's attitude and position towards the 1997 Convention on the Law of the Non-Navigational Uses of International Watercourses (UNWC) and its water treaty practice. I conclude that China's diplomatic practice can be classified as 'responsive diplomacy' and describe the features of this practice.

China's attitude and position on the UNWC

The UNWC, negotiated under the auspices of the United Nations, is a framework global water convention open to all states. It has been accepted, approved, ratified or acceded by 35 states and entered into force on 17 August 2014 (UN, 2014). China has sent diplomatic representatives to take part in the negotiation and conclusion of the UNWC, but ultimately voted against its adoption at the United Nations General Assembly, one of three states to do so.[1] According to the Chinese representative at the meeting, China voted against the convention for the following reasons. First, the convention does not represent or reflect universal international agreement; second, it fails to recognize the principle of territorial sovereignty; and third, the provisions of the UNWC on the rights and obligations of up- and downstream states on international watercourses show a significant imbalance (UNGA, 1997). In voting against the UNWC, the Chinese representative stated the official position of China, in which it reserved the 'right to address the question of the non-navigational uses of international watercourses with its neighbours in a fair and reasonable manner and *in accordance with relevant international practice and with bilateral watercourse agreements*' (UNGA, 1997, emphasis added).

However, as emphasized by Vinogradov and Wouters, China's status as a non-contracting state does not lead to the conclusion that the UNWC is irrelevant to China since its key provisions reflect established state practice and can be viewed as customary international law, binding on all states irrespective of their participation. The authors examine how China's treaty practice compares with the key provisions of the UNWC (Vinogradov & Wouters, 2013). Lingjie Kong suggests that 'key provisions', such as the obligation of equitable and reasonable utilization (Art. 5), the obligation not to cause significant harm (Art. 7), and the general obligation to cooperate (Art. 8), are the codification of principles and rules of customary international law (Kong & Tian, 2011). Notwithstanding the alleged inadequacies of the UNWC, China's water treaty practice demonstrates that, as a whole, it endorses the principal provisions of the convention. A number of the convention's substantive and procedural rules can be found in numerous bilateral water agreements concluded by China (Vinogradov & Wouters, 2013; Wouters & Chen, 2013).

China's water treaty practice

China deals with issues regarding its transboundary waters by concluding and implementing formal and informal water treaties and other legal instruments such as Memorandums of Understanding (MoUs) with its downstream neighbours. At the formal diplomatic level, China practices the principle of 'good neighbourliness', meaning that it desires to live in peace and harmony with its neighbours in accordance with its overall approach to international law and international relations. In order to promote this, China has concluded bilateral water agreements with Russia, Kazakhstan, Mongolia and North Korea. In Southwest China, some informal arrangements and cooperation is under way, including MoUs with the Mekong River Commission (MRC), with India and Bangladesh, related to the provision of hydrological information. These agreements and MoUs are the result of China's diplomatic practice regarding its transboundary waters, which constitute the basis of China's current and future diplomacy and cooperation, reflecting the distinctive characteristic of the 'one river one agreement' practice in transboundary water management (Chen, Rieu-Clarke, & Wouters, 2013; Wouters & Chen, 2013).

China's 'responsive diplomacy' and its features

Diplomacy is the process, means and methods a state uses to achieve its foreign policy objectives and safeguard its interests (Li, 2002). States can passively or actively engage in foreign activities. Almost all of China's water agreements and MoUs are negotiated and concluded in response to requests from its downstream neighbours, not proactively, but rather in response to requests. Its diplomatic behaviour occasionally follows this pattern. For example, when India's former Prime Minister, Manmohan Singh, and China's President, Xi Jinping, met in March 2013, Singh included on the agenda the need for a joint mechanism to look at China's hydropower projects on the Yarlung Zangbo River. Clearly aware of the apprehensions that these projects caused in India, Xi assured Singh that China was aware of its responsibilities towards lower riparian states and that he would ask officials to consider creating a joint mechanism (Gupta, 2014). I refer to this type of diplomacy, that which includes a passive response after a request or dispute, as 'responsive diplomacy'. The next section discusses the key features of this type of diplomacy through the lens of China's transboundary waters.

Theoretical basis of 'responsive diplomacy'

China continues to practise the principle of good neighbourliness (in line with the Five Principles of Peaceful Coexistence; Wouters & Chen, 2013) when dealing with issues on its transboundary waters, to an extent embodying the theory of limited territorial sovereignty. That theory holds that a state shall exercise 'limited' sovereignty over its shared watercourses without harming the interests and rights of other riparian states significantly (He, 2007). Since states are equally sovereign, when one state exercises sovereignty it must respect the sovereignty and interests of other states – correlative duties and entitlements. In essence, this theory guides us to balance the rights and interests of different nations. The obligations of equitable and reasonable utilization, the due diligence duty not to cause significant harm and the general obligation to cooperate, codified within the UNWC, reflect the need to balance the rights and interests of up- and downstream states (He, 2007, 2013a).

Three of China's water agreements – the 1994 Agreement on the Protection and Utilization of Border Waters between China and Mongolia; the 2001 Cooperation Agreement on the Utilization and Protection of Transboundary Waters between China and Kazakhstan; and the 2008 Agreement on Reasonable Utilization and Protection of Transboundary Waters between China and Russia – provide examples of legal frameworks for bilateral water cooperation and embody these three obligations (Chen et al., 2013). The general duty to cooperate is a key theme across China's water treaty practice. Most agreements concluded by China elaborate this through listing mostly technical methods of bilateral cooperation, such as the exchange of information, the monitoring of water quality, joint scientific research, emergency response cooperation, the creation of joint management institutions and the resolution of disputes by means of consultation (Chen et al., 2013).

Under these agreements, China has deepened functional bilateral cooperation through the establishment of bilateral committees with some of its downstream neighbours. For example, China and Russia have established an emergency notification and reporting system and will, in future, commence joint monitoring of water quality and the protection of water biodiversity. China and Kazakhstan have established an emergency notification and reporting system and also cooperate on issues of water quality on transboundary rivers (He, 2013b). China and Kazakhstan are currently in negotiations regarding agreeing a methodology for the allocations of the Ili and Irtysh rivers. However, these negotiations have faced challenges due to their overemphasis on technical cooperation and their failure to afford proper attention to the common interests of stakeholders.

Aim of 'responsive diplomacy'

China has many interests in its transboundary waters, each of which must be protected through diplomatic means. These include economic development, political interests, ecological security, etc. (Li, 2002). China's political interests typically involve territorial sovereignty, good neighbourliness and friendly relations (hereafter referred to as 'traditional political interests'), social stability ('social interests'), regional stability and security ('regional interests'), and its desire to be perceived as a responsible upstream power ('international image'). These interests are interrelated in many ways. For example, ecological harm may lead to ecological migration, social instability and regional conflicts, potentially causing serious damage to China's ecological, social and regional interests as wells as its international image and even economic interests.

China's position towards the UNWC as well as its state practice concerning its transboundary waters leads us to observe that the primary goal of China's 'responsive diplomacy' is to safeguard its economic and traditional political interests. Ecological, social and regional interests as well as its international image are seldom considered in this context. For example, China respects Kazakhstan's interest in its transboundary rivers. Biba suggests that Kazakhstan is an important partner for China, allowing it to ensure border security in its fight against separatism, extremism and terrorism. Additionally, Kazakhstan is a major exporter of oil to China as well as a transit state for importing natural gas from Turkmenistan – both vital assets for China's energy security and its continued economic growth (Biba, 2014).

According to He and Feng, China's water cooperation focuses on economic cooperation for the purpose of national economic gains primarily. China has not yet paid attention to practices of 'harm avoidance', use and cooperation consistent with basin-wide sustainable development, the mitigation of water pollution, the protection of biodiversity and the maintenance of ecological security (He & Feng, 2006). Although China's water agreements relate to the protection of transboundary waters and their respective ecosystems (Chen et al., 2013), they do not provide key measures such as transboundary environmental impact assessment (TEIA) or regulations on ecological flow. In practice, China does not take proactive action on ecosystem protection, which may threaten ecological security and social stability within its own territory as well as regional stability and security, potentially leading to conflict with its downstream neighbours.

Levels of 'responsive diplomacy'

Diplomacy can be bilateral or multilateral. With China's geographically privileged position as an upstream state, and in order to safeguard its own geopolitical and economic interests, China adopts a one-river/one-country approach, preferring to sign bilateral agreements with its neighbours instead of regional agreements with multiple states. For those multinational rivers including the Ili-Lake Balkhash, Irtysh-Ob, Yarlung Zangbo/ Brahmaputra and Lancang/Mekong, China has not yet participated in multilateral negotiations, signed multilateral agreements or established multilateral management institutions specifically as regards its transboundary waters. China has not entered into formal bilateral agreements with co-riparians in the Lancang/Mekong, leaving this basin without an institutional framework to coordinate competing needs. The four downstream states (Thailand, Cambodia, Laos and Vietnam) concluded the Agreement on the Cooperation for the Sustainable Development of the Mekong River Basin in 1995 (1995 Agreement), and established the Mekong River Commission (MRC) to implement the agreement. China is not a party to the 1995 Agreement and is only a dialogue partner of MRC. According to He and Feng, the reason for China's reluctance to join the MRC is caused by its concerns related to the range of issues that would need to be addressed if it were to become a member. Joining the MRC would require China to provide sensitive information and fulfil many obligations, while receiving few benefits (He & Feng, 2006). However, China's ecological, social and regional interests as well as its international image can be best maintained through multilateral cooperation, instead of its current bilateral approach. Some examples of how this might occur include China's already existing engagement in several significant multilateral environmental agreements (Vinogradov & Wouters, 2013).

Methods of 'responsive diplomacy'

There are many methods of cooperation on transboundary waters, including technical cooperation, water allocation, joint development, cooperation on planned measures, and so forth. Although China and its neighbours have, to different degrees, begun cooperating through some of these methods, they are limited mostly to technical activities. China prefers not to conclude water-sharing agreements with its neighbours. China, under significant pressure, has agreed to negotiate water-sharing agreements for the Ili and Irtysh rivers with Kazakhstan, but it has been blamed for stalling the negotiations. Chinese officials have requested more time to create terms for water-sharing agreements on the two rivers; shifting the deadline from the end of 2014 to the end of 2016 (Stone, 2012). With regards to cooperation on planned measures, the related substantive and procedural rules, such as prior notification, are not present in the majority of China's agreements with its neighbours (Wouters & Chen, 2013), and China has few express statements on this. Such limited levels of cooperation have led to water-use conflicts and diplomatic disputes. As noted, responsive diplomacy has its limitations, which is now analysed in more detail.

Primary challenges facing China's diplomatic practice on its transboundary waters

China's practice of responsive diplomacy has experienced unprecedented challenges and, with increased exploitation and utilization of many its transboundary water resources, they are likely to increase. This section explores challenges related to three primary issues: planned measures, pollution control and ecosystem protection, and bilateral transboundary water cooperation.

Planned measures

China's planned measures on its transboundary watercourses have been a recurring source of concern for its downstream neighbours, third parties and other stakeholders over the past decade. Planned measures include projects and activities that any watercourse state proposes to implement, such as the construction of dams and reservoirs, hydropower installations, inter-basin water transfers, and long-distance watercourse installations. The planned measures of one riparian state often adversely affect the hydrology, climate, water allocation, quality and ecological conditions of the entire watercourse, and hence have the potential to cause significant harm to economic, social uses and the ecology of neighbouring riparians. These adverse effects can be considered a primary source of water-use conflicts and diplomatic disputes among co-riparians. The continued water-use conflicts and diplomatic disputes between China and its neighbours may affect China's geopolitical and regional interests as well as its international image.

Among these conflicts and disputes, the most frequent is the alleged 'dams threat'. Those concerned point to issues of the timeliness of notification and the quality of shared information. They are also concerned about unilateral action, which may result in adverse impacts on other riparian nations. For these reasons, China has been paid close attention to by downstream states and stakeholders such as inter-governmental organizations, non-governmental organizations (NGOs), international observers and scholars (Biba, 2014), who sometimes classify the actions as 'upstream hegemony' or 'water hegemony' (Chellaney, 2011).

As an upstream state on almost all of its transboundary watercourses, China faces a great challenge: how to balance its need to use transboundary waters with the needs of its downstream neighbours (Wouters & Chen, 2013) instead of proceeding with unilateral development and utilization. For example, Kazakhstan has criticized China in this regard (He, 2013a; Stone, 2012). Kazakhstan is worried about China's unilateral water diversion and utilization on the Ili River. Hydrologists and environmentalists have warned that the Ili-Balkhash Basin may suffer the same fate as the Aral Sea if water consumption patterns in the basin continue and the ecological needs of Lake Balkhash are ignored (IRIN, 2007; Stone, 2012). Kazakhstan requested the cancellation of China's planned projects on the Ili in 2007, however China did not comply with this request (Greenberg, 2007). In order to save Lake Balkhash, Kazakhstan has actively sought the help and support of the European Union (EU). Being an important part of its global water diplomacy, the EU has instilled the concepts of water governance, provided financial and project support to Kazakhstan's water utilization and governance mechanisms under the framework of the EU–Central Asia Strategic Cooperation Partnership (European Union, 2010), the EU–Central Asia Environment and Water Initiative (European Council and European Commission, 2012), and the EU Water Initiative (European Commission, 2013).

In the Lancang/Mekong River Basin, the four MRC member states have been paying close attention to a range of issues regarding China's planned measures, including: first, the impact of hydropower, shipping, mining and other development activities on water allocation, ecology and biodiversity conservation in the Lancang/Mekong River; and second, downstream states are concerned with the levels of hydrological information exchange with China (He, 2013a). In recent years, the MRC has had frequent contact with China, requesting information concerning its plans for a cascade hydropower project along the river. On the other hand, all riparian states on the Lancang/Mekong vie to develop their economies and forge energy, trade and infrastructure links. Eleven dams are proposed along the mainstream of the Mekong in Southeast Asia, which will have potential impacts on rare fish habitats, fishery resources and shipping (China Dialogue, 2014). Some states in the downstream Mekong have extracted water from the river for farmland irrigation, resulting in low estuary water levels and salt water encroachment (He, 2013a). Some NGOs and scientist have warned that the competition for water from every riparian state of the Lancang/Mekong water resources could cause significant environmental and social impacts (Hirsch, 2014).

The involvement of the United States and Japan has made issues regarding China's planned measures on the Lancang River even more complex. According to Zhifei Li, their involvement has the potential to alter the geopolitical structure of the region and restrict China's dam construction (Li, 2012). The recent expansion of China's influence in Southeast Asia, the Mekong region in particular, has raised concerns in the United States. Ever since 2009 when the Barack Obama administration took office, the United States has initiated a high-profile 'return' to the Asia-Pacific region. It has become actively involved in the development of water resources and water resources management in the Mekong River Basin, and has made capital investments supporting regional projects (Ren, 2013). On the other hand, given the rising influence of China, Association of South East Asian States (ASEAN) have made a strategic choice to involve the United States in order to constrain the power of China, allowing ASEAN states to meet their own security needs. Therefore, cooperation between the United States and the Lower Mekong states is likely to allow each state to take what they need (Ren, 2013), while China's geopolitical, regional and economic interests are hindered.

Pollution control and ecosystem protection

China's responsive diplomacy is challenged to maintain the balance between economic utilization, pollution control and ecosystem protection on transboundary waters. Due to the mutual dependence of transboundary water ecosystems, pollution and ecological harm in one state can have adverse impacts on all co-riparian states. Since China is the upper riparian in the majority of its transboundary waters, protecting this balance has a significant impact on China's neighbours. This is a prominent issue for rivers located in the North-west and North-east regions, such as the Ili, Heilong/Amur, Argun and Ussuri rivers. It is estimated that China contributes 87.5% to the total pollution of the Argun, 75% in the middle section of the Amur, and 97.6% to the pollution of the Ussuri (Vinogradov & Wouters, 2013).

The ecological situation of the Ili River–Lake Balkhash Basin has caught the attention of the EU. In September 2010, according to the Commercial Counselors of the Chinese Embassy in Kazakhstan, the EU noted the deterioration of the status of ecological protection in Lake Balkhash and indicated that the 2001 Agreement between China and Kazakhstan 'failed to establish specific mechanism in preserving the water ecological system in upstream Lake Balkhash'. In order to reach a solution for the protection of water resources in Lake Balkhash, the EU recommended that Kazakhstan and China reconsider their 2001 Agreement, develop more effective programmes and enact ecological standards regarding the quality of upstream water (especially in the Ili) flowing into Lake Balkhash so that upstream water pollution in China does not affect lakes and reservoirs in Kazakhstan (Commercial Counselors of the Chinese Embassy in Kazakhstan, 2010).

Bilateral path of transboundary water cooperation

China's 'bilateral path' of transboundary water cooperation has been challenged by co-riparian states and third parties. For example, in the Lancang/Mekong River Basin, the four MRC member states have put pressure on China to sign the 1995 Agreement and to become an official member of the MRC, a move that has gained support from ASEAN. In the Irtysh-Ob River, China and Kazakhstan have signed a series of bilateral agreements. Kazakhstan is actively soliciting Russia to take part in the process. While Kazakhstan hopes to create a tripartite negotiating committee, China insists on negotiating with each state separately.

Furthermore, for the past decade Kazakhstan has been advocating for the implementation of integrated water resources management (IWRM) in the Ili-Balkhash Basin with the help of the EU. In 2005–07, the European Commission funded the 'Development of the Ili-Balkhash Basin Integrated Management Plan'. At its completion, Kazakhstan requested that China cancel its planned projects on the Ili and that it signs the trilateral agreement concerning the Ili-Balkhash Basin. However, there was no positive response from China. In October 2010, the International Conference on Integrated Water Management in the Ili-Balkhash Basin, organized within the framework of the EU–UNDP project (United Nations Development Programme), was held in Almaty, Kazakhstan. The resolution adopted at the conference recommended that, given that the ecological stability of the region is at stake, the lake should be subject to negotiations on regional water allocation as a separate and independent entity in order to claim its due for sustaining an optimal level of water (Central Asia, 2010). In order to respond to the aforementioned challenges, China is now in a position to make positive changes to its diplomatic practices.

Strategic and policy recommendations for China's diplomatic practice

China is encountering complex challenges in relation to its diplomatic practice on its transboundary waters. Ignoring these challenges may impair China's overall national interests. I suggest that China takes a more active and open approach, transforming its practice of 'responsive diplomacy' into a strategy of 'preventive diplomacy'. 'Preventive diplomacy' occurs when a state proactively interacts with foreign states on issues of concern with a view to anticipating and preventing disputes (dispute avoidance). In situations concerning transboundary waters this would mean that states should proactively anticipate and attempt to address riparian concerns so as to prevent disputes and ecological damage. This section analyses the main features of this strategy and suggests policy measures for its implementation.

Features of 'preventive diplomacy'

The objective of diplomacy is to safeguard the interests of the state. An ideal diplomatic policy is one that maximizes the overall national interest, thus requiring a comprehensive implementation strategy. In order to maximize its overall national interests concerning the uses of transboundary waters, especially in safeguarding its ecological, social, regional interests and its international image, I suggest that China creates and implements a comprehensive strategy of preventive diplomacy, and engages more fully on relevant issues with its neighbours in order to prevent water disputes and ecological harm instead of waiting until a dispute arises. The notion of 'preventive diplomacy' includes the following elements:

- *Preventive diplomacy follows the theory of limited territorial sovereignty, taking into account the common interests of all co-riparian states.* The theory of common interests implies that every international basin is an integrated geographic, economic and ecological unit; every basin state is a member of 'the community of interests' who shares the common interests of the basin, including ecological security, regional peace and stability and the sustainable development of the basin. In order to achieve optimal and sustainable development of the entire basin, every basin state should establish the idea of 'the community of interests', conclude watercourse agreements and undertake joint management of the basin (Yang & Shen, 2009).
- *Preventive diplomacy aims to maximize the nation's overall interests, including economic, political, ecological, social and regional interests, and to protect its international image.* In a globalized world, the preservation of a state's national interests should, within reason, be balanced with the interests of its co-riparians. In order to achieve this, China should take measures to strengthen water cooperation, with the intent of preventing water-use conflicts and ecological harm. Given the legalization of international relations, international law can play an important role since the strategy of preventive diplomacy and international law are intertwined. The formation and implementation of preventive diplomacy involves the evaluation and observation of international law and states may contribute to the development of rules of international law that can safeguard their interests during the formation and implementation of this strategy.
- *Preventive diplomacy adopts a multilevel approach that contains bilateral and multilateral diplomacy.* Through this multilevel approach states can prevent water conflicts and ecological harm, thus maintaining regional stability and security. While signing bilateral water agreements and undertaking bilateral cooperation, China should consider participating in multilateral agreements and management institutions and gradually implement the practices of IWRM.
- *Preventive diplomacy includes various means and areas of cooperation, such as technical cooperation, water allocation measures and methods, joint development and cooperation on planned measures.* While strengthening technical cooperation, consultation and negotiation, China should commence cooperation on planned measures, including prior notification, making arrangements for water allocation or the trading of water rights. The primary goal of undertaking such cooperation is to balance the interests of up- and downstream states, thereby preventing water-use conflicts and diplomatic disputes.

Suggested policy measures for the implementation of preventive diplomacy

Given the characteristics of preventive diplomacy, there are appropriate policy measures that China can implement in order proactively to prevent water disputes and ecological harm through preventive diplomacy, including the following.

Undertake cooperation on planned measures

In order to avoid water-related disputes with downstream neighbours and to maintain its geopolitical, regional and economic interests, China should increase levels of cooperation concerning its planned measures, including observing the duty of prior notification, making arrangements for water allocation or carrying out water rights trading. China should consider observing the duty of prior notification on planned measures, an obligation of customary international law (McCaffrey, 2000), which may have a significant impact upon other watercourse states since, through this policy, China would provide an opportunity for all concerned states to voice their opinions and balance their interests so as to avoid disputes (Wang, 2002).

The allocation of waters shared by China and its co-riparian states is of the greatest concern for every co-riparian state, especially in the cases of the Ili River–Lake Balkhash, the Irtysh-Ob and the Lancang/Mekong rivers. According to the principle of equitable and reasonable utilization, every riparian state within an international watercourse enjoys the equal right to develop and use the water resources within its territory. As an upstream state in the majority of its shared river basins, China may adopt arrangements for water allocation with its co-riparian states. The method of allocation can be determined on a case-by-case basis dependent upon the factors relevant to equitable and reasonable utilization under Art. 6 of the UNWC.

China may consider implementing water rights trading and conduct complementary developments instead of fixed water allocation schemes dealing only with surface water. 'Water rights trading' refers to the trading of water for other commodities including energy, resources, etc. For example, the trading of land in exchange for the control of water, trading water for energy or development aid, or controlling upstream pollution or reducing the share of water resources in exchange for improvements in downstream waterways. If the resolution of water allocation issues solely focuses on issues of water quantity then it may become a zero-sum game, i.e. gains for one party may mean losses for others, however if the involved parties trade water rights for other goods or money, then everyone will benefit. This may assist states in avoiding conflict and to create a win–win situation (Biswas, 2008).

China may undertake negotiations and consultations with co-riparians, trade their respective water rights and conduct development projects which are complementary. For example, in the upper portion of the Lancang/Mekong River Basin, land is scarce and water resources are relatively abundant, while in the lower portion of the basin it is the opposite. If the two issues are tackled simultaneously, it may be possible to achieve significant complementary benefits (He & Feng, 2006). Similarly, in the Ili River–Lake Balkhash Basin, China could trade water for oil with Kazakhstan, as water resources on the Ili within China are relatively abundant, while Kazakhstan has abundant oil resources. However, when conducting water rights trading, the focus should be on long-term benefits, taking the balance of the ecosystem into account.

Control pollution and protect ecosystems within international watercourses

As an upstream state on almost all of its transboundary watercourses, China should take appropriate measures to control pollution and protect local ecosystems. It has become a duty of customary international water law that every watercourse state must prevent, eliminate and mitigate transboundary water pollution (ILA, 2004). This duty has two aspects: first, China should take appropriate measures to control pollution; and, second, it should enact comprehensive measures to protect ecosystems within watercourses. Since transboundary water pollution can cause significant harm to co-riparian states, the principle of no significant harm obligates states to control transboundary water pollution.

The duty to protect transboundary water ecosystems has the potential to become one of the basic principles of international water law (He, 2013b; He & Feng, 2006). Such protection and preservation of ecosystems provides the necessary foundation for sustainable development (Birnie & Boyle, 2002). In order to do so, states ought to adopt the precautionary approach, which contains TEIA, water pollution control and prevention, guarantees of ecological flow, the establishment and implementation of water quality standards, the prevention of invasive species into international watercourses, etc. (He, 2007, 2013b).

Through TEIA China could move forward by addressing a range of water-related problems, including tackling pollution and taking measures to protect the ecosystems of international watercourses. In accordance with domestic law requirements China already conducts EIAs, but it is not obligated to conduct TEIAs. The importance of TEIA has been generally recognized by the international community (Kong, 2011). It is one of the key measures that a state is required to take in order to fulfil its duty not to cause significant harm and to prevent and control water pollution. TEIA is also an embodiment of the precautionary principle (He, 2007). It would appear that TEIA is now emerging as an obligation in customary international law, and certainly a duty for the future (Kong, 2011; Zhang, 2012). In the Pulp Mills on the River Uruguay Case, the International Court of Justice (2010, para. 203) referred to TEIA as 'a practice, which in recent years has gained so much acceptance among States that it may now be considered a requirement under general international law [. . .]'.

Develop rules to balance rights and obligations concerning pollution control and ecosystem protection in upstream states

International law imposes many obligations on upstream states such as pollution control, ecosystem protection and international cooperation, but it offers few corresponding rights. This is neither beneficial in safeguarding the economic interests of upstream states nor is it helpful in mobilizing support. China can take action to correct these imbalances through transboundary water ecological compensation, also called eco-compensation. In a transboundary context, eco-compensation occurs when up- and downstream states take appropriate measures to protect transboundary water ecosystems. Generally, this involves the upstream state taking action to protect the resource, with downstream states, who benefit from these measures, paying some form of compensation for these benefits. In essence, eco-compensation aims to balance the interests of all riparians within an international watercourse, achieving benefits for all (Wang & Wang, 2010).

Gradually implementation of integrated water resources management (IWRM) in response to a bilateral approach

In response to the challenges and pressures arising from bilateral cooperation and unilateral development, China ought to consider cooperating with its co-riparian neighbours in the implementation of IWRM practices. According to the theory of common interests, co-riparian states ought to conclude watercourse agreements and establish joint management institutions. The best way to do this is through the implementation of IWRM, that is, for all riparian states to participate in the negotiation, conclusion and implementation of comprehensive agreements and the establishment of a joint management institution involving the entire watercourse (He, 2013a). While following the theory of limited territorial sovereignty, China's utilization of transboundary waters may borrow ideas from and even actively implement the theory of common interests, and ultimately implement IWRM. The implementation of IWRM is complex, involving interaction between various national jurisdictions over multiple sections of an international watercourse (Wouters & Chen, 2013). Its objective is gradually to strengthen cooperation, a goal which is impossible to accomplish in one step. The path of IWRM can be established through the following gradual approach: first, the exchange of information and data; second, the conclusion of agreements covering the partial watercourse; third, the establishment of joint management institutions for the partial watercourse; fourth, the implementation of water allocation; fifth, joint development of hydropower projects or other projects; sixth, the conclusion of agreements covering the entire watercourse; seventh, the establishment of joint management institution for the entire watercourse; and finally, the achievement of IWRM for the entire watercourse (He, 2012).

Conclusions

While accelerating the uses of its transboundary waters, China is paying more attention to international water law through evolving practices of responsive diplomacy. 'Responsive diplomacy' occurs when a state undertakes foreign activities on the uses of transboundary waters and passively responds to the requests and disputes of other states. Responsive diplomacy, in practice, embodies the theory of limited territorial sovereignty. The main goal of China's responsive diplomacy is to safeguard its economic and traditional political interests. Consistent with this goal, China adopts a bilateral approach to dealing with its transboundary waters, the means of which are limited primarily to technical activities. This approach has faced many challenges, and as water usage increases it is likely that these will grow in strength and number. The primary issues arise from the following issues: planned measures, pollution control and ecosystem protection, and China's bilateral path of water cooperation in the context of multilateral river basins.

In order to maximize its national interests, including economic, political, ecological, social and regional interests and its international image, I suggest that China moves to transform its diplomatic practice from being responsive to preventive. 'Preventive diplomacy' in the context of transboundary waters occurs when a state proactively undertakes a foreign policy strategy aimed at anticipating and addressing relevant transboundary water resources issues with a view to preventing water disputes and ecological damage. It follows the theory of limited territorial sovereignty, while taking into account the theory of common interests, much in line with China's foreign policy approach of 'good neighbourliness'. To implement the strategy of 'preventive diplomacy', China may take appropriate policy measures to control pollution and protect the ecosystems of international watercourses, while also developing new rules to balance the rights and obligations of upstream states. Moreover, China ought to cooperate with its neighbours through various means and on multiple levels, including prior notification of planned measures, exchange of information, water allocation, water rights trading, the conclusion of bilateral or multilateral agreements, the establishment of bilateral or multilateral management institutions, and so forth.

China's transformation from 'responsive diplomacy' to 'preventive diplomacy' will involve significant changes not only in China's diplomatic practices but also in other policy areas, involving combined efforts of many other actors. In order to make this switch, China and its neighbours will require the political will of policy-makers, the adjustment of regional energy and economic development strategies, the improvement of transboundary water management strategies, and the support and engagement of co-riparians.

Acknowledgements

The author would like to give special thanks to Professor Patricia Wouters for her encouragement and the opportunity to present this research; to the anonymous peer reviewers for their constructive comments; to David Devlaeminck for his much welcomed assistance with the English editing; and to Liping Dai for all her logistical support throughout.

Note

1. The other two states that voted against the UNWC are Turkey and Burundi; coincidentally all three are upstream states.

References

Biba, S. (2014). China cooperates with Central Asia over shared rivers. Retrieved from http://www.chinadialogue.org.cn/Art./show/single/en

Birnie, P., & Boyle, A. (2002). *International law and the environment* (2nd ed. p. 54). Oxford: Oxford University Press.

Biswas, A. K. (2008). Management of transboundary waters: An overview. In O. Varis, C. Tortajada, & A. K. Biswas (Eds.), *Management of transboundary rivers and lakes* (p. 39). Berlin: Springer.

Central Asia. (2010). The International Conference "Integrated Water Management in the Ili-Balkhash Basin: Joint Solutions and Perspectives". Retrieved from http://centralasia.iwlearn.org/news-events/news/international-conference

Chellaney, B. (2011). The water hegemony, Project Syndicate. Retrieved from http://Project-Syndicate.org/ commentary/the-water-hegemony

Chen, H., Rieu-Clarke, A., & Wouters, P. (2013). Exploring China's transboundary water treaty practice through the prism of the UN Watercourses Convention. *Water International*, *38*(2), 217–230. doi:10.1080/02508060.2013.782134

China Dialogue. (2014). China and the Mekong: An uncertain future, In China Dialogue (Eds.). *Forging a new course for the Mekong*. Retrieved from https://www.chinadialogue.net

European Commission. (2013). The First Ten Years of the European Union Water Initiative 2002-2012. Retrieved from http://www.euwi.net/files/EUWI_10Anniversary_Web.Pdfeuwi.net/files.

European Union. (2010). European Union Assists Tajikistan in Fighting Polio. Retrieved from http://eeas.europa.eu/delegations/tajikistan/press corner/all news/news/2010/20100604en.htm.

Gleick, P. (2000). *The world's water 2000-2001* (p. 249). Washington, DC: Island Press.

Greenberg. (2007). Talks stall over Central Asia basin. *International Herald Tribune*, *03*, 03.

Gupta, J. (2014). Yarlung Tsangpo River is a living ecosystem, not just a source of hydropower. Retrieved from http://www.chinadialogue.org.cn/Art./show/single/en

He, D., & Feng, Y. (2006). *Rational utilization and coordinated management of transboundary water resources in international watercourses* [in Chinese] (pp. 159–162). Beijing: Science Press.

He, Y. (2007). *The legal theories and practice of the utilization and protection of international water resources* [in Chinese] (p. 188). Beijing: Law press.

He, H., & Jiang, Y. (2010, December 9). Low-key in thirty years but once open: The rough forward pass of hydropower in Yarlung Zangbo River [in Chinese]. *Southern Weekly*, Guangzhou. p.4

He, Y. (2012). The modes and its development of the equitable and reasonable use of international river water resources [in Chinese]. *Resources Science*, Beijing, *2*, 238.

He, Y. (2013a). *Legal issues on China's utilization and protection of transboundary water resources* [in Chinese] (pp. 27–58). Shanghai: Fudan University Press.

He, Y. (2013b). On the protection of transboundary water ecological systems in China: From the point of International Law [in Chinese]. *Hebei Law Science*, *03*, 66–67.

Hirsch, P. (2014). Laos mutes opposition to controversial Mekong dam. In C. Dialogue, Eds. *Forging a new course for the Mekong*. Retrieved from. https://www.chinadialogue.net

International Court of Justice. (2010). Pulp Mills on the River Uruguay, the Judgment of 20 April 2010, Para.203. Retrieved from http://www.icj-cij.org/docket/files/135/15877.pdf

ILA. (2004). Commentary on Art. 59 of Berlin Rules on Water Resources. Retrieved from http:// www.asil.org/ilib/ waterreport2004.pdf

IRIN. (2007). Kazakhstan: Efforts under way to save Lake Balkhash, Almaty, 3 May.

Kong, L. (2011). The International law on TEIA in the development of transboundary water resources [in Chinese]. *Journal of Chongqing University (Social Science Edition)*, *02*, 25.

Kong, L., & Tian, X. (2011). *On the research of international treaties and law involving waters* [in Chinese] (p. 48). Beijing: China Water and Power Press.

Li, S. (2002). *International politics* [in Chinese] (p. 223). Shanghai: Shanghai Peoples Press.

Li, Z. (2012). Utilization of international water resources and the surrounding security of China [in Chinese]. *Teaching and Researching*, *02*, 48–49.

Liu, Z. (2010). *The development of contemporary international law: An analysis from the perspective of international relations theory* [in Chinese] (p. 2). Beijing: Law Press.

Magee, D. (2014). China fails to build trust with Mekong neighbors, In China Dialogue (Eds.), *Forging a new course for the Mekong*. Retrieved from https://www.chinadialogue.net

McCaffrey, S. (2000). An overview of the UN convention on the law of non - navigational uses of international watercourses. *Journal of Land Resources and Environment*, *20*, 69.

Mehtonen, K., Keskinen, M., & Varis, O. (2008). The Mekong: IWRM and institutions. In O. Varis, C. Tortajada, & A. K. Biswas (Eds.), *Management of transboundary rivers and lakes* (p. 228). Berlin: Springer.

Meng, S. (2014). The fate of people displaced by the Mekong dams, In *China Dialogue (Eds.), Forging a new course for the Mekong*. Retrieved from https://www.chinadialogue.net

Nickum, J. (2008). The upstream superpower: China's international rivers. In O. Varis, C. Tortajada, & A. K. Biswas (Eds.), *Management of transboundary rivers and lakes* (pp. 227–230). Berlin: Springer.

Ren, Y. (2013). The Obama administration's policies on the Mekong River and their impact on China [in Chinese]. *Contemporary International Relations*, *02*, 25–27.

Stone, S. (2012). For China and Kazakhstan, no meeting of the minds on water, Science (published by AAAS) (p.406). Volume 337, 27 July.

The Commercial Counselors of Chinese Embassy in Kazakhstan. (2010).The EU proposed Kazakhstan reconsider Sino-Kazakhstan Water Agreement. Retrieved from http://www.mofcom.gov.cn/aArt./i/jyjl/m/201009/20100907161900.html

UN. (2014). United Nations Treaty Collection. Retrieved from https://treaties.un.org/pages/ View Details

UNGA. (1997), Official Records of the General Assembly, Fifty-first Session, 99th plenary meeting, A/51/PV.99.

Vinogradov, S., & Wouters, P. (2013). *Sino-Russian transboundary waters: A legal perspective on cooperation* (pp. 21–27). Stockholm: the Institute for Security and Development Policy.

Walker, B. (2014). China gives green light to new era of mega-dams, In China Dialogue (Eds.), *Forging a new course for the Mekong*. Retrieved from https://www.chinadialogue.net

Wang, K., & Wang, R. (2010). On the progress of the research concerning ecological compensation system and its modes [in Chinese]. *Environmental Science and Management*, *12*, 89.

Wang, X. (2002). On the articles of the liability on the harmful impacts arising from the behaviors not prohibited by international law. In S. Shao & M. Yu (Eds.), *On the problems of international law* [in Chinese] (p. 329). Wuhan: Wuhan University Press.

Wouters, P., & Chen, H. (2013). China's "Soft Path" to transboundary water cooperation examined in the light of two UN global water conventions – exploring the "Chinese Way". *Journal of Water Law*, *22*, 229–247.

Yang, S., & Shen, X. (2009). The international law foundation of resolving the problem of water allocation in international rivers [in Chinese]. *Journal of Lanzhou University (Social Science Edition)*, *04*, 11.

Zhang, W. (2012). EIA as the obligation in international law: Comment on the decision of ICJ. *Quarterly of Taiwan International Law*, *01*, 36.

Benefit-sharing and upstream/downstream cooperation for ecological protection of transboundary waters: opportunities for China as an upstream state

Owen McIntyre

For obvious reasons, upstream states have tended to favour the principle of equitable and reasonable utilization while downstream states invoke the duty to prevent significant transboundary harm. However, while downstream states seek to rely on ecosystem services safeguarded by upstream states, harm can flow in both directions with restrictions on water utilization impacting on upstream interests. Benefit-sharing arrangements which take account of emerging conceptual methodologies for ecosystems protection can play a major role in reconciling the interests of upstream and downstream states.

Introduction

The 1997 United Nations Convention on the Non-Navigational Uses of International Watercourses (UNWC) finally entered into force on 17 August 2014. If we accept that the UNWC is largely a codification of generally applicable customary international law and represents a broad consensus among states about the legal framework required to facilitate inter-state cooperation concerning shared water resources, it is notable that upstream states have yet to endorse the regime set down therein. For example, all three of the states that voted against the Convention in the UN General Assembly (i.e. Burundi, China and Turkey), as well as several of the states which abstained (e.g. Bolivia, Ethiopia, Mali and Tanzania), are predominantly upstream riparians on major shared international watercourses. Many upstream states seem to consider that the convention 'is biased in favour of lower riparians because of its separate and specific mention of the obligation not to cause harm' (Salman, 2014, p. 8), despite the compromise language employed in Article 7, which appears to subordinate this obligation to the broader principle of equitable and reasonable utilization. On the other hand, demonstrating the scope for confusion and disagreement over this point, the perceived subordination of the 'no-harm rule' is the principal reason that a number of predominantly downstream states, including Egypt, France, Pakistan and Peru, also abstained in the General Assembly vote (Salman, 2014). The so-called 'no-harm' rule is generally understood as operating to protect the interests of downstream states, as it is presumed that 'it protects existing uses against impacts resulting from activities undertaken by upstream states', while upstream states in turn tend to favour equitable and reasonable utilization because they often presume that 'it provides more scope for states to utilize their share of the watercourse for activities that might affect downstream riparians' (Salman, 2014, p. 6).

However, the view that the UNWC favours downstream states is based on a misunderstanding of the rules of general international water law. The cardinal rule of international water law under the UNWC and generally the principle of equitable and reasonable utilization is inherently flexible and quite capable of taking account of a very wide range of needs and interests of riparian states, including potential and future uses and the need to protect the entire watercourse ecosystem. In recent years, a number of co-riparian states have sought to give effect to this principle by entering into formal arrangements *inter se* for the equitable sharing of downstream benefits. This approach rests on a broad and innovative understanding of equitable and reasonable utilization, and has received a measure of endorsement from leading commentators recognizing its unique potential to reconcile upstream and downstream interests (Paisley, 2002; Sadoff & Grey, 2005; Tarlock & Wouters, 2007). Similarly, the no-harm rule is capable of addressing not only the more obvious types of harm caused to downstream states by upstream states' utilization of shared waters, typically involving interference with existing uses downstream, but also the harm caused to upstream states by downstream states, 'by foreclosing their future uses of water through the prior use of, and the claiming of rights to such water' (Salman, 2010, p. 350).

It becomes apparent, therefore, that a more sophisticated understanding of the key substantive and procedural rules of generally accepted international water law may be required in order to promote meaningful and effective cooperation between upstream and downstream co-riparian states leading to equitable and environmentally sustainable outcomes. The emphatic emergence and ongoing development in international water law of the obligation to protect ecosystems connected to shared international watercourses and, more specifically, the continuing evolution of the closely related concepts of 'environmental flows', 'ecosystem services' and 'eco-compensation' provide the conceptual and methodological frameworks to help facilitate such cooperation. For example, the 'ecosystem services' concept can help foster a common understanding among co-riparian states regarding the nature and value of socially beneficial services provided by international watercourse ecosystems, while 'eco-compensation' involves systems of payment for ecosystem services, which can provide an incentive for upstream states to forego or significantly modify the planned development or utilization of a watercourses in order to secure ecosystem services throughout the basin (McIntyre, 2014).

China is quite clearly the most important upstream country for transboundary water and ecological security in Asia. It hosts within its territory the 'Asian Water Towers' of the Tibetan-Qinghai Plateau, '[s]haring 110 international rivers and lakes along its southwest, northwest, and northeast borders and being home to most of Asia's great rivers that flow into 18 downstream countries' (He et al., 2014, p. 1). To date, China has not concluded formal agreements with its southern downstream neighbours in what is referred to as the 'Longitudinal Range-Gorge Region' (LRGR), despite the fact that many of the shared basins in this region are deemed to be at risk, including the Yarluzangbu-Brahmaputra, the Lancang-Mekong, the Nu-Salween, the Dulong-Irrawaddy and the Yuan-Red basins (He et al., 2014; Wolf, Yoffe, & Giordano, 2003). China has been criticized for 'its preoccupation with defending a relatively absolutist understanding of sovereignty and the rights flowing from it' and characterized as a recalcitrant actor 'emphatic in its sovereign decision to prioritise economic development over [...] the environment' (Saul, 2011, p. 1). Indeed, Chinese officials have routinely protested against external interference in China's exploitation and utilization of its shared water resources (Ma, 2007; Saul, 2011). Regarding its use of the critically important international watercourses shared with its southern neighbours, and the

Lancang-Mekong more specifically, it has been noted that 'China's upstream development of hydropower dams and other riverine infrastructure has been pursued with relative disregard for the less powerful downstream riparian states of South-East Asia' (Saul, 2011, p. 1). The scope for disputes to emerge is increasing with the adoption by the State Council in early 2013 of China's 12th Five-Year Plan for State Energy Development, which details 50 large dam projects proposed for construction in the LRGR. According to a leading commentator, '[t]hese hydro-electric projects pose major threats to ecosystem services in the LRGR and downstream, with the potential to escalate environmental problems in China into regional conflicts over resources' (He et al., 2014, p. 6).

It is also true, however, that China has demonstrated a willingness to cooperate with its downstream neighbours in order to protect transboundary water ecosystems (Chen, Rieu-Clarke, & Wouters, 2013; Lee, 2014; Wouters, 2014; Wouters & Chen, 2011). For example, partly on the basis of a study identifying the Buyuan River as a key fish migration corridor between the upper and lower Mekong, the Chinese government cancelled the Mengsong Dam project in 2010 (He et al., 2014). Similarly, the Chinese authorities adopt routine measures to reduce the downstream impact of dams situated upstream on the Lancang-Mekong in China, including the introduction of ecologically sensitive operating thresholds for the Xiaowan and Nuozadu dams which seek to minimize interference with natural flow patterns to the greatest extent possible (He et al., 2014). It is precisely the massive reservoirs behind these dams that have caused the greatest degree of concern among downstream states, as 'their multi-seasonal regulating capacity could fundamentally alter flow patterns downstream' so that 'impacts to downstream fisheries, rice cultivation, navigation, and human livelihoods could result' (Magee, 2012, p. 173).

In this context, the 'benefit-sharing', 'e-flows', 'ecosystem services' and 'eco-compensation' concepts may greatly assist China and its downstream neighbours in negotiating and concluding mutually beneficial arrangements. China could then participate in the integrated basin-wide development and optimized utilization of a shared river system, such as the Lancang-Mekong, by sharing equitably in the benefits deriving there from. Such benefits might even include 'eco-compensation' in one form or another in lieu of opportunities to utilize shared waters foregone by China as the upstream state for the purpose of safeguarding downstream benefits for its co-riparian neighbours.

Benefit-sharing

Definition and origins

As the concept of benefit-sharing is not expressly recognized under any formal legal instrument, there exists no authoritative definition. However, commentators have loosely defined the practice to include 'any action designed to change the allocation of costs and benefits associated with cooperation' which, in most cases, 'will require some form of redistribution or compensation' (Sadoff & Grey, 2005, p. 422). Benefit-sharing arrangements would typically involve some form of payments for benefits, or compensation for costs, associated with enhanced stewardship of a shared transboundary watercourse normally undertaken by an upstream state:

> In some instances it might be appropriate to make payments to upstream riparians for watershed management that brings benefits downstream (e.g. reduced flooding and sediment load). Thus stewardship of headwaters and watersheds might entitle upstream riparians to share some portion of the downstream benefits that their stewardship helps to facilitate, and thus share the costs of the stewardship.
>
> (Sadoff & Grey, 2005, p. 423)

The same commentators broadly define benefits 'to include economic, social, environmental, and political gains' (Sadoff & Grey, 2005, p. 421). Therefore, in situations where the simple allocation of a quantum share of water would prove inefficient, benefit-sharing arrangements would permit riparian states to cooperate in taking a basin-wide approach in order to optimize benefits and allocate costs, by providing a framework for the equitable sharing of those benefits and costs. The availability of such arrangements could function to encourage inter-state cooperation, while also greatly enhancing the range and scope of cooperative initiatives in which states might engage, by facilitating broad issue-linkage. In addition to benefits directly connected to water resources utilization, such as irrigated food production or hydropower generation, benefit-sharing arrangements might also take account of non-water use-related benefits, such as increased trade or improved diplomatic relations.

The practice of benefit-sharing with respect to international water resources is widely understood to have its origins in the 1961 Treaty Relating to Cooperative Development of the Water Resources of the Columbia River Basin (Columbia River Treaty) concluded between Canada and the United States. This treaty provided for the construction and operation of three infrastructure projects in Canada designed to maximize benefits associated with hydropower generation capacity, irrigation and flood control in the United States. The treaty also dealt with compensation payments to Canada related to those benefits enjoyed by the United States. Before the treaty's adoption, it was a proposal by the United States to construct the Libby Dam, which would have militated against the development of several viable hydropower sites in Canada and included a reservoir extending into Canada, that provided the impetus to consider inter-state benefit-sharing on the Columbia River. The International Joint Commission (IJC), established under the 1909 Boundary Waters Treaty, facilitated these discussions, lasting many years. In order to inform negotiations, the two riparian states requested the IJC develop a set of principles intended to govern any sharing of benefits. The agreed principles included the central requirements that an agreement 'should result in both the equitable sharing of the downstream benefits attributable to any cooperative undertakings that might take place, and an advantage to each country as compared to any alternatives that might be available to them' (Paisley, 2002, pp. 286–287, citing IJC, 1959). Thus, it appears that benefit-sharing arrangements would essentially require that the project(s) or other forms of cooperation contemplated would benefit both or all riparian states insofar as such cooperation somehow 'enlarges the pie' (Sadoff & Grey, 2005, p. 423).

The Columbia River example also suggests that, due to the novelty of benefit-sharing as a cooperative paradigm and the inherent complexity of the considerations and calculations involved, a sophisticated legal and institutional framework for cooperation would be required for formulating related proposals. In facilitating discussions over the detailed arrangements for the Columbia basin, the IJC established the International Columbia River Engineering Board, which in turn set up an Engineering Committee charged with 'obtaining data and analysing the situation'. This committee continued to carry out extensive technical studies until December 1959 (Paisley, 2002; Pitzer, 1999). In this sense, Tarlock and Wouters compare the relative success of the Columbia River benefit-sharing regime with the problems experienced in attempts to introduce benefit-sharing in the Amu Darya and Syr Darya basins in Central Asia. These authors suggest that the lack of binding legal and competent institutional arrangements are largely to blame, pointing to 'a dysfunctional, ad hoc allocation regime, augmented by endless soft law declarations and agreements' (Tarlock & Wouters, 2007, p. 532). Such elaborate institutional structures are commonly associated with the general obligation of watercourse states to cooperate in the utilization and protection of an international watercourse, one of the generally applicable rules of international water law providing a legal basis for the practice of benefit-sharing, as discussed further below.

Legal basis: permitted not required

The practice of entering into formal inter-state arrangements for the broader sharing of benefits derived from the utilization of the water resources of an international watercourse, rather than the narrower and more simplistic practice of volumetric allocation of the available transboundary waters, has been described variously as 'a shared benefits *strategy*' (Tarlock & Wouters, 2007, p. 523) and as 'a *principle* of equitable sharing of downstream benefits' (Paisley, 2002, p. 288). It has also been argued that this practice 'has become a general principle of both international water law and environmental law' (Tarlock & Wouters, 2007, p. 527). While any lawful practice entered into by sovereign states on the basis of consensus does not require to be legitimated by reference to an established rule or principle of international law, it is still useful to investigate if such a practice is consistent with the key rules of international water resources law and, therefore, actually 'lawful'. Further, it is important to examine the practice's relationship with such rules in order to understand better the objectives, considerations and legal parameters that ought to guide states in crafting benefit-sharing arrangements. In this regard, it appears that benefit-sharing can rely for support upon all three of the key rules of general international water resources law, i.e. the principle of equitable and reasonable utilization, the duty to prevent significant transboundary harm, and the general duty to cooperate in the utilization and protection of an international watercourse.

Equitable and reasonable utilization

However benefit-sharing is characterized, there can be little doubt as to the inherent legality of the practice as a means of giving practical effect to the principle of equitable and reasonable utilization, now almost universally accepted as the cardinal and overarching rule of international water law (ILA, 1966; ILC, 1994). Article 5(1) of the UNWC, which sets out the principle of equitable and reasonable utilization and is, like much of the rest of the convention, widely regarded as indicative of customary international law (McIntyre, 2007, p. 1), provides that

> Watercourse States shall in their respective territories utilize an international watercourse in an equitable and reasonable manner. In particular, an international watercourse shall be used and developed by watercourse States with a view to attaining optimal and sustainable utilization thereof *and benefits therefrom, taking into account the interests of the watercourse States concerned*, consistent with adequate protection of the watercourse.
>
> (added emphasis)

The wording employed in Article 5(1) suggests that 'customary international law anticipates the possibility that co-riparian states may either allocate transboundary water resources volumetrically, or agree to the sharing of benefits arising from shared watercourses (or possibly, both of these in concert)' (Phillips, Daoudy, McCaffrey, Öjendal, & Turton, 2006, pp. 11–12). In other words, benefit-sharing arrangements might serve to substitute for or complement arrangements for allocating a quantum share of water resources. Indeed, even those authors who refer to the practice as a 'strategy' are quick to explain that 'shared benefits' is one of '[t]wo competing models of equitable utilization [to] have emerged'. When employed by co-riparian states, benefit-sharing would typically mean that 'some nations will have to forgo the actual use of wet water but are entitled to monetary compensation for allowing other states to put the water to its most efficient use' (Tarlock & Wouters, 2007, p. 527).

This understanding of benefit-sharing as a means of giving effect to the principle of equitable and reasonable utilization highlights the practice's general objective of achieving an *equitable* apportionment of whatever additional benefits are to be achieved by means of enhanced inter-state cooperation. Any legal understanding of what constitutes an 'equitable' share is necessarily vague, flexible and context-dependent. It is clearly even more so in the case of water- and/or non-water-related benefits than in the more straightforward case of water quantum. Nonetheless, lawyers have some grasp of the function and application of equity in the international water law context (McIntyre, 2013b, 2007). Lawyers increasingly recognize, for example, the significance of the equitable principle of proportionality in ensuring that 'equity as applied in international water law is more distributive in nature' than equity as applied to other contested natural resources (McIntyre, 2007, p.149; see further, Fuentes, 1996).

The flexible and open-textured nature of the principle of equitable and reasonable utilization is confirmed by the non-exhaustive list of factors relevant to its determination, which normally accompanies any detailed articulation of the principle. Though the list set out in Article 6 of the UNWC does not expressly include any reference to benefit-sharing, cost-sharing or the payment of compensation, this list is quite clearly intended to be open-ended and there is nothing to preclude such arrangements being taken into account in reaching or identifying an equitable accommodation of the interests of co-basin states. In support of this contention, the list included in the seminally important Helsinki Rules, the first systematic codification of international water law adopted by the International Law Association in 1966, and on which the UNWC is largely based, appears to contemplate just such arrangements. Article V(II)(10) expressly includes as a factor relevant to the determination of *a reasonable and equitable share in the beneficial uses of the waters* of an international drainage basin, '[t]he practicality of *compensation to one or more of the co-basin States* as a means of adjusting conflicts among uses' (ILA, 1966).

Duty to prevent significant transboundary harm

Benefit-sharing arrangements would also tend to function as a means of giving effect to the second principal substantive rule of general international water law, the duty to prevent significant transboundary harm, or the 'no-harm' rule. Article 7(1) of the UNWC sets out the basic requirement, firmly established in customary international law, that '[w]atercourse States shall, in utilizing an international watercourse in their territories, take all appropriate measures to prevent the causing of significant harm to other watercourse States'. However, Article 7(2) elaborates further on the nature of this obligation, providing that

> Where significant harm nevertheless is caused to another watercourse State, the State whose use causes such harm shall, in the absence of agreement to such use, *take all appropriate measures*, having due regard for the provisions of articles 5 and 6, in *consultation* with the affected State, *to eliminate and mitigate such harm* and, where appropriate, to discuss the question of *compensation*.
>
> (added emphasis)

This authoritative formulation of the no-harm rule as applied to international watercourse, while acknowledging its close interrelationship with the principle of equitable and reasonable utilization, envisages the possibility of a state agreeing to activities within the territory of another state that might adversely impact upon the former's own interests, along with some form of arrangement for compensation in respect of such adverse impact.

Under the no-harm rule, the concept of harm has always been broadly understood to include adverse impact upon a wide range of interests, including both water- and non-water-related interests. For example, one leading commentator notes that the obligation in general international law to prevent transboundary harm is not confined to one state's direct use of a watercourse that causes harm to another state's use thereof, as 'activities in one state not directly related to a watercourse (e.g. deforestation) may have harmful effects in another state (e.g. flooding)' (McCaffrey, 2001, p. 349). Similarly, the commentary to Article X of the ILA's 1966 Helsinki Rules noted that 'an injury in the territory of a state need not be connected with that State's use of the waters' (ILA, 1966, p. 500). Therefore, agreed arrangements concerning the sharing of a wide range of water- and non-water-related benefits, and intended to eliminate, mitigate or compensate for actual or potential harm to a wide range of state interests, can be regarded as effective means for the implementation of the obligation to prevent significant transboundary harm.

Environmental harm, and damage to the ecosystems of an international watercourse in particular, is one of the key types of adverse impact upon the interests of another riparian state envisaged under Article 7 of the UNWC and under customary international law (Birnie & Boyle, 1992; Dupuy, 1991; Wolfrum, 1990). For example, Handl points out that many early treaties relating to the utilization of international watercourses contain 'no-harm' provisions concerning the prevention and abatement of harmful transboundary pollution (Handl, 1975, p. 171). A formulation representative of such instruments is contained in Article X(1) of the Helsinki Rules:

Consistent with the principle of equitable utilization of the waters of an international drainage basin, a state

(a) Must prevent any new form of water pollution or any increase in the degree of existing water pollution in an international drainage basin which would cause substantial injury in the territory of a co-basin State, and
(b) Should take all reasonable measures to abate existing water pollution in an international drainage basin to such an extent that no substantial damage is caused in the territory of a co-basin State.

Any modern articulation of such environmental concerns would be framed sufficiently broadly to include a requirement of ecosystems protection. In this sense, Article 8 of the ILA's 2004 Berlin Rules on Water Resources Law provides, rather inclusively, that 'States shall take all appropriate measures to prevent or minimize environmental harm.' Articles 6 and 7, respectively, of the 1995 Agreement on Cooperation for the Sustainable Development of the Mekong River Basin deal with minimum flow requirements and the prevention of harmful effects. Article 2 of the 1992 United Nations Economic Commission for Europe (UNECE) Convention on the Protection and Use of Transboundary Watercourses and International Lakes (UNECE Water Convention), which has recently been opened for accession to non-UNECE states, contains obligations to 'prevent, control and reduce pollution of waters causing or likely to cause transboundary impact' and to 'ensure conservation and, where necessary, restoration of ecosystems' (McIntyre, 2007, pp. 87–89).

The UNWC itself contains explicit ecosystems obligations, with Article 20 providing that '[w]atercourse States shall, individually and, where appropriate, jointly, protect and preserve the ecosystems of international watercourses' (McIntyre, 2004, p. 6). Article 20 is supplemented by Article 21, requiring watercourse states to prevent, reduce and control pollution of an international watercourse; Article 22, requiring watercourse states to prevent the introduction of alien or new species detrimental to the watercourse ecosystem; and Article 23, requiring watercourse states to protect and preserve the marine environment. In its articulation of ecosystem protection requirements, the UNWC reflects wider state practice and, almost certainly, the position in customary international law (McIntyre, 2004). The international community's shared understanding of what ecosystems protection involves, based on sophisticated analytical and methodological concepts, such as those regarding environmental flow requirements and the evaluation of ecosystem services, is continually evolving. As a result, the normative nature and implications of ecosystem obligations under international law are becoming increasing clear (McIntyre, 2014).

With regard to benefit-sharing, Articles 20, 21 and 23 of the UNWC expressly require states to act jointly, where necessary, to achieve the objectives under each provision. The stipulation that states ought to act jointly, where appropriate, could be understood to support the use of benefit-sharing arrangements under which downstream states agree to compensate upstream states in lieu of forgone water resources development opportunities or for the adoption of measures aimed at protecting and preserving international watercourse ecosystems to the benefit of such downstream states.

Harm to the interests of watercourse states, however, does not only 'travel downstream'. Commentators are increasingly conscious of the fact that water utilization by downstream states can cause harm upstream (McCaffrey, 2007a). This is true despite the fact that '[i]t is much less obvious, and generally not realized, that the upstream riparians can be affected, or even harmed by the foreclosure of their future uses of water, caused by the prior use, and the claiming of rights to such water by the downstream riparians' (Salman, 2010, p. 351). In the specific context of an examination of benefit-sharing arrangements, Sadoff and Grey acknowledge that 'downstream extraction generates externalities upstream by diminishing future flows available for abstraction upstream, by virtue of perceptions of acquired rights to that water downstream' (Sadoff & Grey, 2005, p. 424). Salman outlines a relatively rich and diverse body of international practice that shows that states, judicial bodies and other international actors have long understood that downstream uses can establish a *fait accompli*, thereby pre-empting future upstream use and thus causing harm to the interests of upstream riparians (Salman, 2010).

Arguments citing the 'prior use' or 'acquired rights' doctrines as a basis for favouring existing uses over potential future uses as a matter of legal principle have largely been discredited (McIntyre, 2007). This is reflected in the formulation of factors relevant to equitable and reasonable utilization under Articles 6 and 10 of the UNWC, which attributes no inherent priority to existing uses over future uses (ILC, 1994). Nonetheless, the practice of international water law demonstrates that existing uses would tend to enjoy careful consideration on the basis of other relevant factors, such as economic and social dependence, vital human needs, or the obligation to avoid causing significant harm to other watercourse states (Tanzi & Arcari, 2001). Generally, for the purposes of determining a reasonable and equitable regime of utilization of shared water resources, the difficulties inherent in reliably considering the beneficial character or adverse impacts of a future use are manifest (McIntyre, 2007).

In sum, such 'foreclosure' of water utilization opportunities for upstream states can amount to harm for the purposes of the no-harm rule. This means that benefit-sharing arrangements involving some form of compensation for upstream states which forgo such opportunities for the sake of ecosystems protection would be entirely consistent with the principle articulated in Article 7 of the UNWC, especially in light of the express reference to the possibility of compensation in Article 7(2).

Duty to cooperate

Benefit-sharing arrangements also find legal support in the third key rule of general international water resources law, the general duty of watercourse states to cooperate in the utilization and environmental protection of an international watercourse. This obligation is largely comprised of procedural elements, including the well-established duties to notify co-riparian states of planned projects; to consult and, where necessary, negotiate with co-riparian states where disagreements arise; and to engage in the regular exchange of relevant water-related information. The duty to cooperate is generally regarded as essential in facilitating meaningful implementation of the two key substantive obligations outlined above (Leb, 2012, 2014; McIntyre & Leb, 2013a).

Article 8 of the UNWC, on the '[g]eneral obligation to cooperate', can be considered an authoritative indication of that norm's position in customary international law. That provision establishes that '[w]atercourse States shall cooperate on the basis of sovereign equality, territorial integrity, *mutual benefit* and good faith in order to attain *optimal utilization* and *adequate protection* of an international watercourse' (added emphasis). The emphasis on 'mutual benefit', 'optimal utilization' and 'adequate protection' suggests that benefit-sharing arrangements under which upstream states are rewarded for maintaining the ecosystem of an international watercourse fall within the scope of the cooperative initiatives envisaged. In elaborating upon the precise legal origins and meaning of Article 8, the ILC's Commentary to its 1994 Draft Articles explains that inter-state watercourse cooperation 'is an important basis for the attainment and maintenance of an equitable allocation of the *uses and benefits of the watercourse*' (ILC, 1994, p. 105, added emphasis). In providing an example of an instrument incorporating typical provisions on transboundary water cooperation, the ILC cites the 1964 Agreement between Poland and the USSR concerning the use of water resources in frontier waters, pointing out that

> Paragraph 3 of article 3 states that the purpose of the Agreement is to ensure cooperation between the parties in *economic, scientific and technical activities* relating to the use of water resources in frontier waters. Articles 7 and 8 of the Agreement provide for cooperation with regard, *inter alia*, to *water projects* and the regular exchange of data and information.
>
> (ILC, 1994, p. 106, added emphasis)

This illustration of the obligation to cooperate seems to corroborate the understanding that cooperation under Article 8 would cover joint arrangements incorporating elements of benefit-sharing.

Consistent with the experience of the riparian states on the Columbia River basin in crafting a benefit-sharing regime, Article 8(2) of the UNWC proceeds to suggest that:

> In determining the manner of such cooperation, watercourse states may consider the establishment of joint mechanisms or commissions, as deemed necessary by them, to facilitate cooperation on relevant measures and procedures in the light of experience gained through cooperation in existing joint mechanisms and commissions in various regions.

The key function of any joint watercourse mechanism or commission is the regular and structured exchange of data and information. Such exchange is acknowledged as an absolutely critical prerequisite for the conclusion of any benefit-sharing arrangements. In this regard, Sadoff and Grey highlight the key role of Cooperative Regional Assessments (CRAs), which would ideally function 'to reach common understanding, change perceptions, and achieve information symmetry, in order to build trust and catalyse cooperation' (Sadoff & Grey, 2005, p. 425). Ultimately CRAs can 'provide a common point of departure' as they serve to 'provide riparian states with the information needed to reach consensus regarding the way forward in projects of mutual interest' (Sadoff & Grey, 2005, p. 426).

With the three norms examined above serving as legal foundations for benefit-sharing, this practice may be viewed as a means of ensuring their harmonized and coherent application. There is, in this sense, a close connection between the duties to cooperate and to prevent significant transboundary harm, including environmental harm, from which a need arises for states to work together to develop more sophisticated legal arrangements to compensate for harm. Adopting a broad understanding of environmental damage to include the denial of water-related ecosystem services, it is worth noting that Principle 22 of the 1972 Stockholm Declaration has long provided that

> States shall cooperate to develop further the international law regarding liability and compensation for the victims of pollution and other environmental damage caused by activities within the jurisdiction or control of such States to areas beyond their jurisdiction.
>
> (UNCHE, 1972)

This exhortation is repeated as Principle 13 of the 1992 Rio Declaration, which further proposes that states act 'in an expeditious and more determined manner' in developing international law in this area (UNCED, 1992a). More generally, Stockholm Principle 24 states that

> International matters concerning the protection and improvement of the environment should be handled in a cooperative spirit by all countries. [...] Cooperation through multilateral or bilateral arrangements or other appropriate means is essential to effectively control, prevent, reduce and eliminate adverse environmental effects resulting from activities conducted in all spheres.
>
> (UNCHE, 1972)

Such universally adopted soft-law principles can also be understood as an acknowledgement by the international community of the need for more sophisticated arrangements, within the framework of international law, for ensuring that environmental costs are fully and equitably internalized, including by means of benefit-sharing and eco-compensation arrangements. In an exhaustive study categorizing the various types of provisions relating to a prohibition on transboundary harm found in freshwater treaties, Fuentes includes that of 'treaties that establish mechanisms of cooperation to deal with the problem of the environmental degradation of international rivers'. She refers to numerous examples of such provisions, including Article 2(2) of the 1994 Agreements on the Protection of the Rivers Meuse and Scheldt and Article 4 of the 1994 Convention on Cooperation for the Protection and Sustainable Use of the Danube River (Fuentes, 1998, pp. 145–163).

State practice

As regards support for benefit-sharing in state practice, there exist several high-profile examples of basins illustrating benefit-sharing arrangements, whether in practice, most notably the Columbia shared between Canada and the United States, or in theory, such as the Amu Darya and Syr Darya basins in Central Asia (Tarlock & Wouters, 2007). Paisley lists eight examples of 'international agreements which provide for the return, either in kind or in monetary form, of a share of the benefits received in a state or states as a result of acts done in another state or states' (Paisley, 2002, p. 288). On this basis, he concludes that

> These examples confirm that state practice can be invoked in support of an emerging principle of customary international law regarding the equitable sharing of downstream benefits where the act that confers the benefit on one state appears to have been done, or not done, at the request of another state.
>
> (Paisley, 2002, p. 289)

It would be overstating the case, however, to argue that the practice of benefit-sharing is firmly established in international water law. Benefit-sharing as a means of achieving equitable and reasonable utilization has been actively discussed at a number of international fora, including the 2001 Bonn International Conference on Freshwater, the 2003 3rd World Water Forum and Ministerial Conference, and the 2005 Stockholm World Water Symposium. Yet, Phillips *et al* consider that the topic is somewhat neglected by the academic community and, further, that 'when reading the literature [...] little substance is discernible beyond the catch-phrase level' (Phillips et al., 2006, p. 29). McCaffrey (2001, p. 264) acknowledges that 'it is not uncommon for some form of compensation (e.g. sharing electric power) to be part of an overall package of equitable apportionment of the uses and benefits of an international watercourse'. But he cautions that this practice does not imply that modern international law recognizes states as having any property in shared waters which could underlie a firm right to such compensation:

> On the contrary, upper riparians are under an obligation not to prevent such waters from flowing to a lower riparian country. The only interference with such flow that would be permissible are those that would be equitable and reasonable in the context of the states' fluvial relations.
>
> (McCaffrey, 2001; Paisley, 2002)

Thus, while benefit-sharing arrangements are permissible under international water law, no obligation or entitlement to enter into such arrangements arises by virtue of customary international law. Any such compulsion or right, arising otherwise than by the clear and explicit consent of watercourse states, would be an intolerable interference with state sovereignty. However, in its 1957 deliberation upon the true nature of the duties imposed upon a watercourse state under international law, consistent with respect for the principle of state sovereignty, the arbitral tribunal in the *Lac Lanoux* arbitration found that any such state

> under the rules of good faith, has an obligation to take into consideration the various interests concerned, to seek to give them every satisfaction compatible with the pursuit of its own interests and to show that it has, in this matter, a real desire to reconcile the interests of the other riparian with its own.
>
> (*Lac Lanoux* 1957; McCaffrey, 2007b)

This suggests that watercourse states are at least obliged to consider seriously any overture made to them regarding possible benefit-sharing arrangements.

Challenges to benefit-sharing

Critiquing the arguments for a shift in emphasis from water allocation to the sharing of benefits, Tarlock and Wouters articulate 'three possible objections to this shift: unequal bargaining among states; the premature "sale" of future use opportunities; and the increased risk of aquatic ecosystem degradation' (Tarlock & Wouters, 2007, p. 524). The challenge of inequality of bargaining power among co-riparian states will always be present whether states seek to allocate a share of water resources or of water-related or other benefits, even though the growing emphasis in international water law on procedural equity should function to level the playing field to a certain extent (McIntyre, 2013a, 2013b). Concluding generally that often 'headwaters states cannot bargain equally with wealthier downstream states', Tarlock and Wouters describe China as a notable exception on the Upper Mekong (Tarlock & Wouters, 2007, p. 528). Indeed, China is unlikely to negotiate 'premature' benefit-sharing arrangements to its longer-term detriment. As a comparatively developed Mekong state, it is improbable in China's case that '[s]hort-term monetary gains may come at the expense of foregone future uses' (Tarlock & Wouters, 2007, p. 528). Rather, it is fair to assume that China possesses the resources and capacity to identify and safeguard its long-term strategic interests. Furthermore, the considerable weight given within the framework of equitable and reasonable utilization to environmental considerations generally, and ecosystems protection in particular, should help to address concerns that benefit-sharing may present a risk to aquatic ecosystem integrity (McIntyre, 2007). Such concerns should recede further with the increasing normative sophistication represented by the continuing, rapid evolution of technical concepts and methodologies to give practical effect to the so-called 'ecosystems approach' to the protection of international watercourses stipulated under international water law (McIntyre, 2007, 2014).

Practical arrangements

As regards the kind of practical arrangements required to facilitate benefit-sharing, Sadoff and Grey identify the aforementioned CRAs as playing a central role. These are described as

> essentially practical tools for identifying opportunities for regional actions, for promoting the recognition and optimization of all four types of benefits of cooperation [economic, social, environmental and political], for analyzing the distribution of costs and benefits associated with cooperative regional programs, and for identifying benefit-sharing and institutional mechanisms for realizing these benefits.
>
> (Sadoff & Grey, 2005, pp. 425)

These are high-level, strategic exercises that 'bring riparian specialists together in a process that builds common understanding and relationships' and would normally include: a 'Transboundary Analysis', providing a basin-wide view of the best possible water management and development opportunities; a 'Distributive Analysis', exploring the various options for equitably sharing the costs and benefits of cooperation; and an 'Institutional Analysis', examining the possible modes of cooperation necessary to generate the greatest net benefits (Sadoff & Grey, 2005, pp. 425–426). Cooperation by means of benefit-sharing arrangements is an 'iterative, adaptive and dynamic' process, but '[t]he CRA can also provide riparian states with the information needed to reach consensus regarding the way forward in projects of mutual interest' (Sadoff & Grey, 2005, p. 426).

Stressing the need for accurate, detailed and relevant information to support cooperative initiatives concerning transboundary waters, He et al. focus on the potential role of the discipline of applied ecology. They suggest that 'applied ecology can play an important role, not only in diagnosing problems and designing solutions but also in providing a firm scientific basis on which constructive dialogue can take place' (He et al., 2014, p. 8).

Emergence and evolution of international watercourse ecosystems protection obligations

It is now beyond debate that the rules of modern international water law include a requirement for all watercourse states to protect ecosystems connected to international watercourses. Many commentators have detailed the origins and evolution of such a legal requirement (Brunnée & Toope, 1994, 1997; McIntyre, 2004, 2007; Tarlock, 1996). For example, whilst conceding, perhaps somewhat conservatively, that '[t]here is no *specific* duty to protect aquatic ecosystems in customary international law', Tarlock cites the earlier work of other leading authorities in support of the conclusion that '[t]he duty can be derived from the principle of international environmental law of state responsibility for environmental harm' (Caponera, 1987; Tarlock, 2014). To this end, watercourse states are expected to adopt an 'ecosystems approach' to the use and protection of international basins, which requires careful consideration of possible impacts upon ecosystem functions and services in basin development planning. According to Chapter 18 of Agenda 21,

> water resources have to be protected, taking into account the functioning of aquatic ecosystems and the perenniality of the resource, in order to satisfy and reconcile needs for water in human activities. In developing and using water resources, priority has to be given to the satisfaction of basic needs and the safeguarding of ecosystems.
>
> (UNCED, 1992b, p. 167)

The UNECE Water Convention provides an example of a conventional instrument requiring parties, under Article 3(1)(i), to ensure that 'sustainable water-resources management, including the application of the ecosystems approach, is promoted'. The practical requirements for such an approach, however, have until recently been largely uncertain. Other than some detailed early guidance (UNECE, 1993), international practice had tended to place emphasis on procedural rules, concerning mechanisms for information exchange, early warning, and scientific and technical cooperation, rather than substantive rules. The Latter norms are notoriously difficult to elaborate (McIntyre, 2004).

Analysing the 1994 ILC Draft Articles upon which the UNWC was based, leading commentators have concluded that the ILC had aimed to create frameworks 'establishing a process for the subsequent development of substantive obligations, rather than seeking to elaborate an immediate set of specific substantive rules' (Brunnée & Toope, 1994, p. 57). Presciently, the same commentators argued that procedural rules act as 'catalysts' for the development of substantive rules:

> The exchange and cooperative practice they generate can contribute significantly to the convergence of positions, thus allowing gradual agreement on substantive obligations and standards. At the same time, cooperative practice serves important trust-building and conflict-prevention functions.
>
> (Brunnée & Toope, 1994, p. 57)

This may help to explain why recent developments in this particular area of international water law have been characterized by a proliferation of technical guidance elaborating on innovative concepts and methodologies which can aid in the effective implementation of ecosystems protection requirements. The international community has devoted considerable energy to this issue over the past 10 years or so. Convention secretariats, international non-governmental organizations, academic researchers and others active in the discourse of international environmental law have been busy producing such technical guidance (McIntyre, 2014). Key conceptual methodological approaches set out in such guidance include those of environmental or ecological flows (e-flows), ecosystem services, and compensation for ecosystem services or 'eco-compensation'.

e-flows

The concept of e-flows provides 'a methodological approach that incorporates environmental concerns into the process of allocating water rights among different users' (Brels, Coates, & Loures, 2008, p. 13), and is absolutely central to the implementation of an ecosystems approach. This concept has been elaborated upon by bodies such as the Secretariats of the Ramsar and Biodiversity Conventions (Adams, 2012), and the Global Environmental Flows Network (Forslund et al., 2009), a collaborative network involving such organizations as International Union for Conservation of Nature (IUCN), World Wide Fund for Nature (WWF) and United Nations Environment Programme (UNEP). The Scientific and Technical Review Panel (STRP) established under the Ramsar Convention has also been actively engaged in the preparation of technical guidance on a range of aspects of the concept, including the ecological vulnerability of wetlands and reviews of environmental flow methodologies (Brels et al., 2008).

The concept has been endorsed in international soft law instruments, notably the 2007 Brisbane Declaration, which defined the concept as 'the quantity, timing and quantity of water flows required to sustain freshwater and estuarine ecosystems and the human livelihoods and well-being that depend on these ecosystems' (McIntyre, 2014, p. 90). Commentators explain that '[t]he overriding objective of e-flows is to modify the magnitude and timing of flow releases from water infrastructure (e.g. dams) to restore natural or normative flow regimes that benefit downstream river reaches and their riparian ecosystems' (LeRoy Poff & Matthews, 2013, p. 667). Of course, natural flow regimes might equally be maintained by the decision of an upstream state to refrain from developing water infrastructure altogether.

It seems clear that e-flow obligations are now recognized in general international water law. The adoption in 2012 by the United States–Mexico International and Boundary Water Commission (IBWC) of Minute 319, amending by interpretation the 1944 United States–Mexico Water Treaty, marks the recognition by both the United States and Mexico of the need to ensure certain minimum ecological flows to the Colorado Delta in Mexico, parts of which have been designated as protected wetlands under the Ramsar Convention (IBWC, 2012). Minute 319 establishes a pilot programme to deliver 158,088 acre-feet (195,000 million m^3) of annual water flow to the Colorado Delta in two ways: a one-time pulse flow of 105,392 and a base flow of 52,696 acre-feet per year (Buono & Eckstein, 2014; Tarlock, 2014). Tarlock recognizes the intrinsic link between e-flows and ecosystems obligations and views the restoration of ecological flows essential for maintenance of the Delta as 'a de facto implementation of the ecosystem conservation mandates of the United Nations Convention on the Law of the Non-navigational Uses of International Watercourses' (Tarlock, 2014). Indeed, he regards the advocacy work of non-governmental organizations (NGOs) regarding the Colorado Delta after its chance revival due to 'two wet decades in the 1980s and 1990s' and the recent implementation of a number of ad hoc river restoration programmes in the United States as having 'helped to legitimate the idea that environmental flows are an integral part of many allocation regimes'

(Tarlock, 2014). Describing Minute 319 as an important precedent in international environmental and water law, Tarlock explains that it provides an example of 'a more modest, bottom-up approach to the development of law' whereby

> A 'common law' of international environmental and water law can develop from *de facto* rather than *de jure* examples of the adoption and implementation of proposed foundational legal principles by a range of international and national or bi-national legal regimes. Minute 319 is an important example of the *de facto* development of a 'common law' of international environmental and water law.
>
> (Tarlock, 2014)

Minute 319 might also be regarded as an example of the noted modern phenomenon whereby the evolution of international environmental and water law is facilitated by institutionalized treaty regimes which, '[u]pon their establishment, they become machineries for the making of new law and for the development of existing law in their respective areas of competence' (Gehring, 2007).

A 2013 report on basin water allocation planning, developed by a broad coalition of international actors and published by the United Nations Educational, Scientific and Cultural Organization (UNESCO), concludes that '[t]here is now wide recognition of importance of maintaining an appropriate flow regime to maintain the ecological health of river basins' and, consequently, that 'water allocation plans are increasing allocating water to meet instream ecological requirements' (Speed, Ji, Le Quesne, Pegram, & Zhiwei, 2013, p. 58). Even prior to such arbitral and judicial support, commentators had observed that '[t]he need to provide environmental flows in order to conserve ecological integrity of water basins is becoming more and more important' (Aguilar & Iza, 2011, p. 99). Other commentators note that the e-flows concept has evolved very significantly over the last 25 years and 'has unfolded to encompass a wide range of ecological, social and economic goals' (LeRoy Poff & Matthews, 2013, p. 671). The same authors point out that the utility of the e-flows framework has already been tested in China (LeRoy Poff & Matthews, 2013; Zhang et al. 2012).

Regarding the status of an obligation to maintain e-flows under international water law, it is worth referring to recent case law. In February 2013, a tribunal established under the auspices of the Permanent Court of Arbitration (PCA) to settle a dispute between Pakistan and India over shared waters in the Indus basin held that 'hydro-electric projects [...] must be planned, built and operated with environmental sustainability [and minimum environmental flow in particular] in mind' (PCA, 2013a, para. 454). Having found that '[i]t is established that principles of international environmental law must be taken into account even when [...] interpreting treaties concluded before the development of that body of law', the tribunal has established that an obligation to maintain minimum environmental flows can be read into existing water treaties (PCA, 2013a, para. 452). Beyond such existing treaties, the tribunal has suggested that such an obligation might also arise under 'the "principle of general international law" that states have "a duty to prevent, or at least mitigate" significant harm to the environment when pursuing large-scale construction activities' (PCA, 2013a, para. 451). In its Final Award in this case handed down in December 2013, the tribunal specified the precise quantum of water required to satisfy India's obligation to maintain the minimum environmental flow in the Kishenganga River (PCA, 2013b). Having determined that there were no existing agricultural uses and that Pakistan had failed to provide sufficient evidence to allow future agricultural uses to be taken into account, the tribunal proceeded to consider the impact of the project on the downstream environment. It found that, in order to determine an appropriate minimum environmental flow, some form of in-depth assessment of the downstream environmental impacts was be required. While the tribunal was not prepared to be overly prescriptive about the nature of such an assessment, it advised that 'for any given river project, the correct approach will depend upon the existing state of the river, the magnitude of anticipated changes, the importance of the proposed project, and the availability of time, funding and local expertise' (PCA, 2013b). On this basis the tribunal concluded that a minimum flow of 12 m^3 per second (cumecs) would be appropriate to preserve the ecological condition of the river. Nevertheless, taking account of additional factors, in particular the economic efficiency of the overall project, it lowered the mandated minimum flow to 9 cumecs, a level that the tribunal estimated would prevent significant environmental harm to the river whilst also providing enough water to generate sufficient power at the plant (Harrison, 2014). This approach was based less on any perceived inherent legal requirement to balance ecological and economic interests, however, that on a rather rigid interpretation of the restrictive wording of the applicable provisions of the 1960 Indus Waters Treaty, so that '[t]he Court's authority is more limited and extends only to mitigating significant harm' (Harrison, 2014; PCA, 2013b). In an acknowledgement of the inherent variability of the ecological situation in a river basin, and of the resulting need for adaptive ecosystems management, the tribunal noted that factors may arise in the future justifying a modification of this minimum flow and held that either party has the right to request a reconsideration of the tribunal's decision by the Permanent Indus Commission after a period of seven years (Harrison, 2014; PCA, 2013b).

The International Court of Justice (ICJ) has also acknowledged the importance of environmental flows in a recent case concerning the San Juan River, shared between Costa Rica and Nicaragua (ICJ, 2011). For example, in respect of Costa Rica's concerns that dredging of the river and diversion of water 'was creating a risk of irreparable prejudice to Costa Rica's environment or to the flow of the San Juan River', Judge Sepulveda-Amor noted in his Separate Opinion 'the risks of loss of habitat and progressive erosion and changes in the groundwater aquifer beneath the wetlands' (ICJ, 2011, para. 25).

The emergence of this obligation in international water law, coupled with the development of sophisticated methodologies for calculating ecological flows (Beaton & Bradford, 2013), provides upstream states with an incentive to cooperate with their downstream neighbours on the ecological protection of transboundary watercourses, while also providing clarity as to the adequate level of protection expected. Such technical clarity contributes towards achieving the 'information symmetry' likely to be vital for an agreement to be reached on benefit-sharing arrangements (Sadoff & Grey, 2005).

Ecosystem services

The concept of ecosystem services is now very closely linked to the ecosystems approach to the protection of international watercourses (McIntyre, 2014). The concept will be central to future benefit-sharing arrangements intended to secure the protection of watercourse ecosystems. Ecosystems services are the benefits that people obtain from healthy ecosystems (Pardy, 2014). The concept provides a methodology for the economic and social valuation of such benefits, including non-marketable benefits, natural ecosystems and can thus permit their integration into benefit-sharing arrangements. Common understanding of the value of ecosystems was greatly assisted by the 2005 Millennium Ecosystem Assessment, which provided an essential typology of four categories of ecosystem services (Millennium Ecosystem Assessment, 2005):

- Supporting services, such as soil formation and pollination.
- Provisioning services, such as food, water supply and timber.
- Regulating services, such as climate control, flood control and water quality.
- Cultural services, such as recreational uses and spiritual values.

The ecosystem services concept can play a particularly significant role in transboundary water cooperation involving benefit-sharing. Co-riparian states must communicate and agree upon equitable measures for the utilization and protection of shared waters based upon a common understanding of their cost and benefit for each state. Generally accepted methodologies for the valuation of benefits provided by watercourse ecosystems can help to provide 'a common point of departure' for negotiations over benefit-sharing (Sadoff & Grey, 2005).

A great deal of highly specific guidance has emerged in recent years regarding the use of the ecosystem services concept in policy-making. This includes, for example, the World Resources Institute (WRI) guidance on methodologies for review of ecosystem services in the context of a process of impact assessment (Landsberg et al., 2011) and the recent report by the Institute for European Environmental Policy and the Ramsar Secretariat on the economics of ecosystems and biodiversity for water and wetlands (Russi et al., 2013). Further technical guidance has recently been issued by such bodies as the United Nations Development Programme (UNDP) (Alpizar & Bovarnick, 2013), the United Nations Environment Programme (UNEP) (Brander, 2013), and the European Commission (European Commission, 2013). A 2008 report published by the Secretariat of the Convention on Biological Diversity includes an Annex I that examines in detail 'ecosystem services by inland waters which can be affected by inappropriate water allocations and unsustainable water use' (Brels et al., 2008).

This level of interest has caused Rieu-Clarke and Spray to suggest that '[i]n this regard, water and wetland ecosystems are perhaps among the best studied of habitats in terms of ecosystems services', and to note further that '[t]his linkage between the upstream provision of services [...] and the downstream utilisation of services thus provided (often water-related) has now become widely recognised and can be seen to operate on very large, often transboundary scales' (Rieu-Clarke & Spray, 2013, p. 25). Such ongoing elaboration of the ecosystem services concept in the context of shared waters can only serve to clarify the value to downstream states of such services provided by international watercourse ecosystems maintained by upstream states, and can thus serve to inform negotiations over possible benefit-sharing arrangements.

Eco-compensation

The current active discourse on ecosystem services is usually accompanied by discussion of the potential role of systems of payment for ecosystem services, or 'eco-compensation'. Key international actors, such as the IUCN (IUCN, 2006) and the UNECE (UNECE, 2009), have provided guidance on how such payment systems might work, with the latter explaining that

> Such financing mechanisms operate at many levels, between and within countries, from and to governments, the private sector and local communities. Payment for ecosystem services (PES) is an innovative tool for rewarding ecosystem managers for their sustainable management practices, which increase ecosystem resilience.
>
> (UNECE, 2009, p. 101)

However, the concept of eco-compensation is not yet highly developed in the practice of international water law. One authority notes that this area 'is still emerging and frankly not well explored by the legal scholarship', before identifying, among the key issues remaining to be addressed by policy-makers, the basis for payments, the parties to whom payments should be made, the services for which payments should be available and the amounts to be paid (Benjamin, 2013, p. 2). Similarly, Rieu-Clarke and Spray caution that 'pitfalls may be identified when considering the linkages between ecosystem services, law and transboundary freshwaters' and recognize 'the need for further research through case studies at the transboundary basin level' (Rieu-Clarke & Spray, 2013, p. 29). Having regard to existing experience of benefit-sharing arrangements, Paisley readily concedes that '[t]he identification of these benefits can be difficult, and precise calculations complex' (Paisley, 2002, p. 299).

Despite the availability of a wide variety of possible modes of cooperation arranged along a 'continuum' of benefit-sharing, ranging from unilateral action through coordination and collaboration to joint action (Sadoff & Grey, 2005), commentators agree that '[i]n most cases, benefit sharing will require some sort of redistribution or compensation, which will be highly situation specific' (Paisley, 2002; Sadoff & Grey, 2005; Tarlock & Wouters, 2007).

At the domestic level, China is one of very few states with considerable experience in implementing working eco-compensation mechanisms in conformity with a domestic legal framework and thus should have practical experience of the operation of such payment systems which, potentially, could be transferred to the international level. A number of large-scale eco-compensation programmes have been implemented in China to protect and restore natural capital and ecosystem services, including the Natural Forest Protection Programme, the Forest Eco-Compensation Programme, the Grassland Eco-Compensation Programme, the Wetland Restoration Programme and the Grain-to-Green Programme (He et al., 2014; Liu, Ouyang, Yang, Xu, & Li, 2013). Eco-compensation schemes have also been implemented with the specific aim of protecting and restoring major watercourse ecosystems (Dai, 2014).

Eco-compensation in lieu of ecological benefits

In the context of enhanced cooperation over shared international water resources, Sadoff and Grey characterize the potential benefits to be shared as falling into four categories: environmental, economic, political and catalytic (Sadoff & Grey, 2002; Phillips et al., 2006). Similarly, in a later comprehensive study of the practice of benefit-sharing, Phillips et al. identify three specific categories of benefits, i.e. those arising in the security sphere, in the economic sphere and in the environmental sphere. As regards the environmental sphere, they point out that 'a strong case can be made for the maintenance of some form of ecosystem integrity, either to sustain essential ecological goods and services on which human livelihoods are dependent, or to enable those ecosystems to function as sinks in a sustainable manner' (Phillips et al., 2006, p. 40). Indeed, these same authorities explain that 'issue-linkage can take place to the extent that a common currency is developed, through which benefit-sharing can start to emerge' (Phillips et al., 2006, p. 40). In the environmental sphere, the emergence of the ecosystems approach to the utilization and management of international watercourses, and the continuing development of related conceptual tools, help to provide just such a common currency. Such conceptual tools include, *inter alia*, the ecosystem services concept as 'a methodology for the economic and social valuation of natural ecosystems' (McIntyre, 2014, p. 92), and the e-flows concept as 'a methodological approach that incorporates environmental concerns into the process of allocating water rights among different users' (Brels et al., 2008, p. 13). As noted above, Tarlock and Wouters highlight three challenges presented by what they regard as a shift in emphasis in recent years, intended to 'temper unilateral use', from quantum allocation of water to the sharing of benefits among riparian states, including: 'unequal bargaining power among states; the premature "sale" of future use options and the increased risk of aquatic ecosystem degradation' (Tarlock & Wouters, 2007, p. 524). These authors elaborate further on their third concern, pointing out that inter-state cooperation on both the Columbia basin and the Amu and Syr Darya basins has resulted in environmental disasters of one kind or another. Tarlock and Wouters go on to note that benefit-sharing appears to function largely to achieve the inter-state consensus required to facilitate the development of hydropower projects, despite the undisputed environmental impacts of such projects and the prospect that climate change will make many major international watercourses 'even less reliable for economically feasible large hydropower projects, and their waters more precious for other uses' (Tarlock & Wouters, 2007, pp. 534–535).

However, where the benefit to be optimised, and in lieu of which compensation would be owed to an upstream state, is that of ecological integrity, and/or the ecosystem services accruing from a functioning riverine ecosystem, such problems could be avoided or significantly ameliorated. The same authors identify three possible benefits that can flow from benefit-sharing: '(1) better ecosystem management, (2) rivers services such as hydroelectric power, and (3) the achievement of regional water security through cost-sharing rather than inefficient duplicate development'. They then conclude, on the basis of state practice to date, that 'only the second claimed benefit, hydroelectric power production, defines the current practice of benefit sharing' (Tarlock & Wouters, 2007, p. 533). Nevertheless, there is no legal impediment preventing co-riparian states from adopting a more expansive and eco-centric view of the shared benefits accruing from development opportunities forgone or environmental measures undertaken by an upstream state. In fact, such an approach is ever more likely to be considered consistent with the rapidly emerging ecosystems obligations of basin states, such as the obligation to maintain environmental flows identified in the recent Kishenganga Arbitration. Moreover, due to profound changes in our understanding of ecosystem benefits, exemplified by the 2005 Millennium Ecosystem Assessment, it is today ever more likely that riparian states might consider the maintenance of a watercourse's ecological integrity as the 'most efficient use' of the shared waters in question, a key aim and justification of benefit-sharing arrangements (Tarlock & Wouters, 2007, p. 527).

At the same time, the difficult task of crafting complex benefit-sharing arrangements focused on ecosystem protection, including arrangements to compensate the upstream states concerned, might now be more possible than before due to the sophisticated tools and methodologies emerging under the 'ecosystems approach' to transboundary water management. Without common understanding fostered by widely accepted methodologies for identifying minimum environmental flows, evaluating ecosystems services and calculating appropriate compensation payments for those ensuring the continued provision of such services, the inter-state agreement required for effective ecological benefit-sharing would simply not be possible.

Conclusions

The entire history of international water resources law, or at least that of modern attempts to codify and clarify this field of law dating back to the 1960s, can be understood as a search for means of reconciling the sharp differences of opinion between, and thus of ameliorating the potential for water-related conflict among, upstream and downstream co-riparian states

(Bourne, 2001; Tarlock & Wouters, 2007; Wouters, Vinogradov, Allan, Jones, & Rieu-Clarke, 2005). The possibility of entering into benefit-sharing arrangements, along with the recent emergence and continuing development of such methodological concepts as e-flows, ecosystem services and eco-compensation, offer an unprecedented opportunity to achieve mutually beneficial and ecologically sound cooperation between upstream and downstream states. As a predominantly upstream state, China would appear to have little incentive to cooperate with its downstream neighbours for the basin-wide ecological protection of shared international watercourses, in the absence of benefit-sharing arrangements, most likely including eco-compensation payments in one form or another. However, as regards water resources, China's Peaceful Rise strategy charges the Chinese authorities with acting 'to reach informed, long-term understanding of the complexity of the problems and to find meaningful and amicable solutions to water security issues' (He et al., 2014, p. 6). In addition, a 2013 policy statement issued by China's Ministry of Foreign Affairs stipulates that 'new projects in transboundary rivers must go through scientific planning and study, with the consideration of the interests of both downstream and upstream riparian countries' (He et al., 2014, p. 8). Such exhortations strongly suggest that China should engage fully in regional cooperative initiatives where the feasibility of broad benefit-sharing arrangements could be explored with its southern neighbours, especially as 'China and its Mekong neighbors may be in agreement on broad development goals such as subregional poverty alleviation, improved transportation and more reliable access to energy resources' (Magee, 2012, p. 191; see also Lee, 2014).

Acknowledgements

The author would like to thank all the academic staff at Xiamen University Law School for their kind hospitality and assistance, in particular Professor Chen, Huiping, Professor Han, Xiuli and Professor Patricia Wouters.

Funding

This article was prepared during a research visit to Xiamen University Law School, China, generously supported by the Chinese Government under the 2014 High-End Foreign Experts Programme of the State Administration of Foreign Affairs.

References

Adams, J. (2012). *Determination and implementation of environmental water requirements for estuaries* (Ramsar Technical Report No. 9/CBD Technical Series No. 69). Gland/Quebec: Ramsar Convention Secretariat/Secretariat of the CBD.

Aguilar, G., & Iza, A. (2011). *Governance of shared waters: Legal and institutional issues*. Gland: IUCN.

Alpizar, F., & Bovarnick, A. (2013). *Targeted scenario analysis: A new approach to capturing and presenting ecosystem service values for decision making*. New York, NY: UNDP.

Beaton, A. D., & Bradford, A. (2013). Demonstration of a methodology for setting ecological flow and water level targets. *Canadian Water Resources Journal, 38*(4), 296–310. doi:10.1080/07011784.2013.830371

Benjamin, A. H. (2013). Payment for ecosystem services. *Potchefstroom Electronic Law Journal, 16* (2), 2–5.

Birnie, P., & Boyle, A. (1992). *International law and the environment*. Oxford: Oxford University Press.

Bourne, C. (2001). The international law association's contribution to international water resources law. In S. Bogdanovic (Ed.), *International law of water resources: Contribution of the international law association (1954–2000)*. The Hague: Kluwer Law International.

Brander, L. (2013). *Guidance manual on value transfer methods for ecosystems services*. Nairobi: UNEP.

Brels, S., Coates, D., & Loures, F. (2008). *Transboundary water resources management: The role of international watercourse agreements in implementation of the CBD*. Quebec: Secretariat of the CBD.

Brunnée, J., & Toope, S. J. (1994). Environmental security and freshwater resources: A case for international ecosystems law. *Yearbook of International Environmental Law, 5*, 41–76. doi:10.1093/yiel/5.1.41

Brunnée, J., & Toope, S. J. (1997). Environmental security and freshwater resources: Ecosystem regime building. *The American Journal of International Law, 91*(1), 26–59. doi:10.2307/2954139

Buono, R. M., & Eckstein, G. (2014). Minute 319: A cooperative approach to Mexico–US hydro-relations on the Colorado River. *Water International, 39*(3), 263–276. doi:10.1080/02508060.2014.906879

Caponera, D. (1987). The role of customary international water law. In M. Ali, A. Khan, & G. Radosewich (Eds.), *Water resources policy for Asia* (pp. 365). London: Taylor & Francis.

Chen, H., Rieu-Clarke, A., & Wouters, P. (2013). Exploring China's transboundary water treaty practice through the prism of the UN watercourses convention. *Water International, 38*(2), 217–230. doi:10.1080/02508060.2013.782134

Dai, L. (2014). Exploring China's approach to implementing 'eco-compensation' schemes: The Lake Tai Watershed as case study considered through a legal lens. *Water International, 39*(5), 755–773. doi:10.1080/02508060.2014.950860

Dupuy, P.-M. (1991). Overview of the existing customary legal regime regarding international pollution. In D. B. McGraw (Ed.), *International law and pollution* (pp. 61). Philadelphia: University of Pennsylvania Press.

European Commission. (2013). *Mapping and assessment of ecosystems and their services*. Brussels: European Union.

Forslund, A., Malm Renöfält, B., Barchiesi, S., Cross, K., Davidson, S., Farrell, T., ... Smith, M. (2009). *Securing water for ecosystems and human well-being: The importance of environmental flows* (Swedish Water House Report 24). Stockholm: Swedish International Water Institute.

Fuentes, X. (1996). The criteria for the equitable utilization of international rivers. *British Yearbook of International Law, 67*, 337–412. doi:10.1093/bybil/67.1.337

Fuentes, X. (1998). Sustainable development and the equitable utilization of international watercourses. *British Yearbook of International Law, 69*, 119–200. doi:10.1093/bybil/69.1.119

Gehring, T. (2007). Treaty making and treaty evolution. In D. Bodansky, J. Brunnée, & E. Hey (Eds.), *The oxford handbook of international environmental law* (pp. 467–497). Oxford: Oxford University Press.

Handl, G. (1975). Balancing of interests and international liability for the pollution of international watercourses: Customary principles of law revisited. *Canadian Yearbook of International Law, 13*, 156–194.

Harrison, J. (2014). Significant international environmental law cases: 2012–14. *Journal of Environmental Law, 26*, 519–540. doi:10.1093/jel/equ028

He, D., Wu, R., Feng, Y., Li, Y., Ding, C., Wang, W., & Yu, D. W. (2014). Review: China's transboundary waters: new paradigms for water and ecological security through applied ecology. *Journal of Applied Ecology, 51*(5), 1159–1168. doi:10.1111/1365-2664.12298

IBWC. (2012). Interim International Cooperative Measures in the Colorado Basin through 2017 and Extension of Minute 318 Cooperative measures to Address the Continued Effects of the April 2010 Earthquake in the Mexicali Valley, Baja California (Minute 319, Coronado, 20 November 2012), interpreting the US–Mexico Treaty Regarding Utilization of the Waters of the Colorado and the Tijuana Rivers and of the Rio Grande (Washington DC, 3 February 1944, entered into force 8 November 1945).

ICJ. (2011, March 8). *Certain activities carried out by Nicaragua in the Border area (Costa Rica v. Nicaragua)*, Request for Provisional Measures Order.

IJC. (1959). *Report of the international joint commission on principles for determining and apportioning benefits from cooperative use of storage waters and electrical interconnection within the Columbia River System*.

ILA. (1966). Report of the Fifty-Second Conference. London: International Law Association.

ILA. (2004). *Berlin rules on water resources law*. London: International Law Association.

ILC. (1994). *Report of the international law commission on the work of its forty-sixth session*, UN Doc. A/49/10 (1994).

IUCN. (2006). *PAY: Establishing payments for watershed services*. Gland: IUCN.

Lac Lanoux Arbitral Tribunal. (1957). *Lac Lanoux Arbitration (France v. Spain)*, award of 16 November 1957. *International Law Reports*, *24*, 101.

Landsberg, F., Ozment, S., Stickler, M., Henninger, N., Treweek, J., Venn, O., & Mock, G. (2011). *Ecosystem services review for impact assessment: Introduction and guide to scoping*. Washington, DC: World Resources Institute.

Leb, C. (2012). *Cooperation in the law of transboundary water resources*. Cambridge: Cambridge University Press.

Leb, C. (2014). One step at a time: International law and the duty to cooperate in the management of shared water resources. *Water International*, *40*(1). doi:10.1080/02508060.2014.978972

Lee, S. (2014). Benefit sharing in the Mekong river basin. *Water International*, *40*(1). doi:10.1080/02508060.2014.978974

LeRoy Poff, N., & Matthews, J. H. (2013). Environmental flows in the Anthropocence: past progress and future prospects. *Current Opinion in Environmental Sustainability*, *5*(6), 667–675. doi:10.1016/j.cosust.2013.11.006

Liu, J. G., Ouyang, Z. Y., Yang, W., Xu, W. H., & Li, S. X. (2013). Evaluation of ecosystem services policies from biophysical and social perspectives: The case of China. In S. A. Levin (Ed.), *Encyclopedia of biodiversity* (2nd ed., Vol. *3*, pp. 372–384). Waltham: Academic Press.

Ma, X. (2007). Statement on the ILC's work on shared natural resources and responsibility of international organizations. *Chinese Journal of International Law*, *6*(3), 769–772. doi:10.1093/chinesejil/jmm045

Magee, D. (2012). The dragon upstream: China's role in Lancang-Mekong development. In J. Öjendal, S. Hansson, & S. Hellberg (Eds.), *Politics and development in a transboundary watershed* (pp. 171–193). New York: Springer.

McCaffrey, S. (2001). *The law of international watercourses: Non-navigational uses*. Oxford: Oxford University Press.

McCaffrey, S. C. (2007a). Some developments in the law of international watercourses. In M. Kohen (Ed.), *Promoting justice, human rights, and conflict resolution through international law: Liber Amicorum Lucius Caflisch*. Leiden: Martinus Nijhoff.

McCaffrey, S. C. (2007b). *The law of international watercourses* (2nd ed.). Oxford: Oxford University Press.

McIntyre, O. (2004). The emergence of an 'ecosystem approach' to the protection of international watercourses under international law. *Review of European Community and International Environmental Law*, *13*(1), 1–14. doi:10.1111/j.1467-9388.2004.00379.x

McIntyre, O. (2007). *The environmental protection of international watercourses under international law*. Aldershot: Ashgate Publishing.

McIntyre, O. (2013b). Utilization of shared international freshwater resources – the meaning and role of 'equity' in international water law. *Water International*, *38*(2), 112–129. doi:10.1080/02508060.2013.779199

McIntyre, O. (2014). The protection of freshwater ecosystems revisited: Towards a common understanding of the 'ecosystems approach' to the protection of transboundary water resources. *Review of European, Comparative & International Environmental Law*, *23*(1), 88–95. doi:10.1111/reel.12068

McIntyre, O. (2013a). The contribution of procedural rules to environmental protection of transboundary rivers. In L. Boisson de Chazournes, C. Leb and M. Tignino (Eds.), *Freshwater and international law: The multiple challenges* (pp. 239–265). Cheltenham: Edward Elgar.

Millennium Ecosystem Assessment. (2005). *Ecosystems and human well-being: Synthesis*. Washington, DC: Island Press.

Paisley, R. (2002). Adversaries into partners: International water law and the equitable sharing of downstream benefits. *Melbourne Journal of International Law*, *3*, 280–300.

Pardy, B. (2014). The logic of ecosystems: Capitalism, rights and the law of 'ecosystem services'. *Journal of Human Rights and the Environment*, *5*(2), 136–152. doi:10.4337/jhre.2014.03.02

PCA (2013a, February 18). *Indus waters Kishenganga Arbitration (Pakistan v. India)*, Partial Award.

PCA (2013b, December 20). *Indus waters Kishenganga Arbitration (Pakistan v. India)*, Final Award.

Phillips, D., Daoudy, M., McCaffrey, S., Öjendal, J., & Turton, A. (2006). *Trans-boundary water cooperation as a tool for conflict prevention and for broader benefit-sharing*. Stockholm: Ministry of Foreign Affairs Sweden.

Pitzer, P. (1999). *Annex 11: Negotiating the Columbia basin treaty, draft Grand Coulee dam and Columbia basin project case study* (Working Paper, World Commission on Dams).

Rieu-Clarke, A., & Spray, C. (2013). Ecosystem services and international water law: Towards a more effective determination and implementation of equity. *Potchefstroom Electronic Law Journal, 16*(2), 12–65.

Russi, D., Ten Brink, P., Farmer, A., Badura, T., Coates, D., Förster, J. ... Davidson, N. (2013). *The economics of ecosystems and biodiversity (TEEB) for water and wetlands*. London and Brussels/Gland: IEEP/Ramsar Secretariat.

Sadoff, C. W., & Grey, D. (2002). Beyond the river: the benefits of cooperation on international rivers. *Water Policy, 4*(5), 389–403.

Sadoff, C. W., & Grey, D. (2005). Cooperation on international rivers: A continuum for securing and sharing benefits. *Water International, 30*(4), 420–427. doi:10.1080/02508060508691886

Salman, S. M. A. (2010). Downstream riparians can also harm upstream riparians: The concept of foreclosure of future uses. *Water International, 35*(4), 350–364. doi:10.1080/02508060.2010.508160

Salman, S. M. A. (2014). Entry into force of the UN Watercourses Convention: why should it matter? *International Journal of Water Resources Development*, 1–13. doi:10.1080/07900627.2014.952072

Saul, B. (2011, November). China, Resources and International Law. *Sydney Law School, Legal Studies Research Paper* No. 11/82. Retrieved from: http://ssrn.com/abstract=1954180

Speed, R., Ji, Y., Le Quesne, T., Pegram, G., & Zhiwei, Z. (2013). *Basin water allocation planning: principles, procedures and approaches for basin allocation planning*. Paris: UNESCO.

Tanzi, A., & Arcari, M. (2001). *The United Nations convention on the law of international watercourses*. The Hague: Kluwer Law International.

Tarlock, A. D. (1996). International water law and the protection of river system ecosystem integrity. *Brigham Young University Journal of Public Law, 10*(2), 181–211.

Tarlock, A. D. (2014). Mexico and the United States assume a legal duty to provide Colorado River delta restoration flows: An important international environmental and water law precedent. *Review of European, Comparative and International Environmental Law, 23*(1), 76–87. doi:10.1111/reel.12066

Tarlock, A. D., & Wouters, P. (2007). Are shared benefits of international waters and equitable apportionment? *Colorado Journal of International Environmental Law and Policy, 18*(3), 523–536.

UNCED. (1992a), *Rio declaration on environment and development*. UN Doc. A/CONF/151/5 Rev.1. *International Legal Materials*, 31, 876.

UNCED. (1992b), *Agenda 21*. UN Doc. A/CONF/151/26, Vol. 1.

UNCHE. (1972). *Stockholm declaration on the human environment*. Report of the United Nations conference on the human environment, Part I, Chapter I. Reprinted in *International Legal Materials* 11, 1416.

UNECE. (1993). *Guidelines on the ecosystem approach in water management*, UN Doc. ECE/ENVWA/31, 1993.

UNECE. (2009). *Guidance on water and adaptation to climate change*. Geneva: UNECE.

Wolf, A. T., Yoffe, S. B., & Giordano, M. (2003). International waters: Identifying basins at risk. *Water Policy, 5*, 29–60.

Wolfrum, R. (1990). Purposes and principles of international environmental law. *German Yearbook of International Law, 33*, 308.

Wouters, P., Vinogradov, S., Allan, A., Jones, P., & Rieu-Clarke, A. (2005). *Sharing transboundary waters – An integrated assessment of equitable entitlement: The legal assessment model* (IHP-VI Technical Documents in Hydrology, No. 74).

Wouters, P. (2014). The Yin and Yang of international water law: China's transboundary water practice and the changing contours of state sovereignty. *Review of European, Comparative & International Environmental Law, 23*(1), 67–75. doi:10.1111/reel.12069

Wouters, P., & Chen, H. (2011). China's 'Soft-Path' to transboundary water cooperation examined in the light of two un global water conventions – exploring the 'Chinese Way'. *Water Law, 22*(6), 229–247.

Zhang, Y., Arthington, A. H., Bunn, S. E., Mackay, S., Xia, J., & Kennard, M. (2012). Classification of flow regimes for environmental flow assessment in regulated rivers: The Huai River Basin, China. *River Research and Applications, 28*(7), 989–1005. doi:10.1002/rra.1483

Approaches to investment in Chinese transboundary waters

Xiuli Han

This article identifies three methods of Chinese investment in transboundary waters: (1) investment in waters upstream from its own state; (2) cooperative investment in a transboundary body of water in a neighbouring state; and (3) cooperative investment in a transboundary body of water which forms the border between two states. On the basis of this introduction, the article explores and analyzes the challenges faced by Chinese investments in transboundary waters and identifies approaches to respond to these challenges, including multilateral dispute settlement mechanisms, data sharing and communication, as well as following the general principles of international law.

Introduction

'Transboundary water', as the name suggests, not only refers to transboundary rivers, but also includes lakes, inland waters and aquifers. It has been defined as "shared water resources, including a river, lake and its tributaries, or estuaries (commonly referred to as the drainage area), and underground water systems under the jurisdiction of two or more states" (Wang, 2002, p. 272). Within Chinese transboundary water treaties, transboundary rivers include all rivers and river flows that cross state borders or are located on the border (China-Kazakhstan Agreement, 2001, Article 1); and correspondingly, 'transboundary waters' refers to any river, lake, stream or swamp that crosses state borders or is situated on the border (China-Russia Agreement, 2008, Article 1). Although transboundary waters may also be referred to as 'transnational', 'trans-state' or 'international' water or watercourses (UN Water, thematic paper, 2008, n. 2), China's relevant treaties prefer the phrase 'transboundary water'. Therefore, transboundary water is water which either constitutes the boundary or crosses the boundary. In other words, transboundary waters are contiguous or successive, and make up a system of surface and underground waters (the latter being largely disregarded by China's transboundary water treaties). Although this definition is used in the mentioned treaties, the scope of the definition of transboundary water may differ in others.

In total there are 23 significant transboundary water bodies in China. They are primarily distributed in three geographical areas: the North-East, North-West and South-West. China is a state with rich transboundary water resources. With the advancement of the Chinese economy and the increasing needs of Chinese economic development and livelihoods of its peoples, China is utilizing these transboundary water resources through individual, cooperative, public and private investment projects. One method to promote reasonable use of these transboundary waters is the creation of legitimate investment projects. In turn, the reasonable use of transboundary waters depends upon legitimate investment, and here international water law and international investment law intersect. At this intersection, investments fund hydropower dams, reservoirs, fishery projects, mineral resource development, and tourism. Among these investments, the construction of hydropower dams and reservoirs is typically the most controversial, often causing tension and disputes among riparian states. For China, dealing with these disputes is of considerable importance, as they may affect smooth economic development and friendly relations with other states. With this in mind, it is essential to fully understand the relevant investments and associated issues in order to prevent and settle such disputes and to eliminate misunderstanding and controversy.

Survey of investments and relevant laws relating to Chinese transboundary waters

Investments in Chinese transboundary waters

In North America and Europe, most transboundary waters have already been fully developed, and relevant legal and management regimes have matured over time. The level of investment in the development of China's transboundary waters, however, has been relatively low (Tan & Kong, 2012). Thus, with China's rise, it is now seeking to increase investment in the development of its transboundary water resources.

China has detailed plans to develop its transboundary waters. Under the 12th Five-Year Plan for National Economic and Social Development (2011–2015), China will actively develop hydropower, with a special focus on the construction of large hydropower stations in the South-West (NDRC, 2011). Under the 12th Five-Year Energy Development Plan (2011–2015), China will construct a comprehensive national energy base, including developments in South-Western and North-Western China. Specifically, China will promote the construction of hydropower facilities on the middle and lower reaches of the Lancang and Yarlung Zangbo Rivers, and will initiate the systematic construction of hydropower bases on the Nu River and the upper reaches of the Lancang River (State Council, 2013). In addition, China plans to steadily promote international cooperation on energy, suggesting that China is ready to cooperate with its neighbours regarding its transboundary water investments.

In South-West China in particular, the Nu River (known as the Salween outside China) and the Lancang River (known as the Mekong outside China) serve as important resources for riparian states, including downstream states in South-East Asia. These two rivers are especially attractive for development because of their combined hydropower potential, steep terrain and high volume. There are 13 proposed cascade hydropower stations on the main tributary of the Nu River, but most of them are still in the stages of planning or site preparation. On the Lancang River, China has invested in the construction of dams and cascade reservoirs to generate hydropower and in navigation development. On the whole, however, cascade hydropower development is still very limited.

In contrast, the Mekong River has been utilized more fully with the assistance of foreign aid. For example, Japan is very active in investing in the Mekong River for a variety of reasons, and besides actual investment, Japan concluded investment agreements with Laos and Cambodia (Fan, 2008). On 22 April 2012, Japan decided to provide approximately USD 7.4 billion in economic aid to the five lower riparians of the Mekong river basin (Laos, Myanmar, Thailand, Cambodia and Vietnam) to take place from 2013, and exempted Myanmar from USD 3.7 billion in debt (Han & Ji, 2012).

In 2010, China began construction on the Zangmu Hydropower Project, the first on the midstream of the Yarlung Zangbo River; and in 2013, China approved the construction of three more hydropower projects on this river. In North-West China, the Ili and Irtysh-Ob Rivers are the largest, with the Irtysh-Ob River also being the longest river, flowing across China, Russia, Kazakhstan and Mongolia. To hasten economic development and improve the local water supply of the North-West region, China's investment since the 1990s has primarily focused on these two rivers (Wang & Hu, 2011). For example, construction on the No. 635 water diversion project on the Irtysh-Ob River began in 1998. It includes a channel, reservoirs, and power plants, with total investment reaching CNY 14.4 billion. In addition, the first China–Kazakhstan cooperative investment, the joint construction of the Horgos River Friendship Joint Water Diversion Control Project, started in April 2011 and was successfully completed on 5 July 2013.

Another option for transboundary water development is to invest in Chinese transboundary waters in neighbouring riparian states, which may also promote the principles of cooperation found within international water law. Although the duty to cooperate is typically understood as comprising a number of procedural obligations—for example duties to notify, exchange information, consult, negotiate, etc.—other forms of cooperation, including investment, should not be excluded. Chinese investment in the Hatgyi Dam, in Hlaing Bei, Myanmar, on the downstream portion of the Salween River, is a typical example. In June 2006, Sinohydro Corp. Ltd and EGAT (Thailand) signed a formal memorandum of understanding on the cooperative development of the Hatgyi Dam Hydropower Project. On 24 April 2010, Sinohydro, Myanmar's Ministry of Electric Power No. 1, EGAT (Thailand) and IGOEC (Myanmar) signed a further memorandum of understanding on the cooperative development of the Hatgyi Dam Hydropower Project, to make the project one of trilateral cooperation. The investment takes the form of a build-operate-transfer (BOT) project, with Sinohydro as the controlling shareholder with an investment of USD 1 billion (SOE News, 2014). At present, studies are being conducted regarding the construction of the proposed dam.

In January 2011, China Southern Power Grid declared that the Salween Meng Dong Hydropower Project on the Salween River will be implemented by a consortium of investors composed of Three Gorges Group, Southern Power Grid Company and Sinohydro, with foreign parties including EGAT (Thailand), the Myanmar government and IGE (Myanmar). Similar to the project on the Hatgyi River, the Meng Dong Hydropower Project will also take the form of a BOT investment. At present, the parties are conducting feasibility studies for the project (Suidianzhan, 2014).

In summary, Chinese investments in transboundary waters take three forms: individual investment in a body of transboundary water upstream from its own territory; cooperative investment in a body of transboundary water downstream, in a neighbouring state's territory; and cooperative investment in a body of transboundary water which acts as a border between China and its neighbour, or crosses the boundary. This may give the impression that China has ambitious investment plans for Chinese transboundary waters, but in actuality the majority of these investments are in a preliminary stage.

Relevant treaty laws on Chinese transboundary waters

Although some Chinese transboundary waters are multinational, China's approach is to conclude bilateral treaties with each riparian state, and to seldom sign or accede to any multilateral water treaty or institution aimed at integrated development of the entire drainage area of a shared river. In recent years, an increasing body of transboundary water law has been developed, and this law is becoming increasingly specific and practical. In addition, the Water Law of the People's Republic of China (2002, article 78) states that "where any international treaty or agreement relating to international or border rivers or lakes, concluded or acceded to by the People's Republic of China, contains provisions differing from those in the laws of the People's Republic of China, the provisions of the international treaty or agreement shall apply, unless the provisions are ones on which the People's Republic of China has made reservations". However, generally speaking, there have been few treaties and other instruments concluded between China and its neighbouring states, and the operability of such treaties remains poor, so that they cannot satisfy the practical requirements that arise. China also has a national source of law to regulate investment in its transboundary waters. The Regulatory Documents of the State Council are typical regulations. For example, the Notice of the State Council on Issuing the Catalogue of Investment Projects Subject to the Approval of Government of the State Council was issued on 2 December 2013. This document stipulates that reservoir and other water projects concerning transboundary waters, hydropower stations in the main river and overseas investment projects in which Chinese investment reaches USD 1 billion and above, and overseas investment projects concerning sensitive countries, regions and industries, shall be approved by the Department of the State Council in charge of investment, i.e., the National Development and Reform Commission.

The existing treaties related to investment in Chinese transboundary waters can be classified as follows. First, there are boundary treaties which touch upon transboundary waters and investment. These treaties principally include the China-Myanmar Border Management and Cooperation Agreement (concluded 25 March 1997), the China-Russia Border Management System Agreement (concluded 9 November 2006), the China-Kazakhstan Border Management System Agreement (concluded 20 December 2006), the China-Vietnam Land Border Management System Agreement (concluded 18 November 2009), the China-Mongolia Border Management System Agreement (concluded 1 June 2010), and the China-Laos Border Management System Agreement (concluded 23 August 2011).

These agreements are evolving and become more complex as they mature; however, only a few provisions within them touch upon the issue of transboundary waters and investment. For example, Article 1.10 of the latest China-Laos Border Management System Agreement defines 'transboundary water' in such a way as to refer only to rivers running across the border between two states. Under Article 1.11, 'cross-border facilities', which cover artificial facilities such as bridges and dams, are enumerated. Two other articles relate to investment in transboundary water. Article 9 embodies the fundamental principles of protection and utilization of transboundary water, i.e., equality and mutual benefit, mutual respect for each other's interests, avoidance of losses to other parties, and methods to deal with issues related to transboundary waters through consultation. Article 10 also provides that consultation is the preferred approach to dispute settlement, i.e., if one party causes a significant impact upon transboundary waters through constructing or dismantling dams or other hydropower projects, or through the use of water, changing the water flow, or influencing the river regime, the two parties should settle the dispute through consultation, in the spirit of equality, mutual convenience and respect for each other's interests.

Second, there are agreements which typically focus on the protection and use of transboundary waters, or more specifically the reasonable utilization and protection of transboundary waters. Examples of such treaties include the China-Mongolia Agreement on the Protection and Utilization of Transboundary Waters (concluded 29 April 1994), the

China-Kazakhstan Cooperation Agreement on Utilization and Protection of Transboundary Rivers (concluded 12 September 2001), and the China-Russia Agreement on Reasonable Utilization and Protection of Transboundary Waters (concluded 29 January 2008).

A survey and comparison of these three treaties shows many commonalities. First, each treaty simultaneously emphasizes protection and utilization. In this respect, the titles of these agreements are self-evident, and this idea is reflected in the preambles and specific provisions of these treaties (China-Mongolia Agreement, 1994, Preamble, Articles 2, 4, 5; China-Kazakhstan Agreement, 2001, Preamble, Articles 2, 4, 5, 8, 9; China-Russia Agreement, 2008, Preamble, Articles 1, 2, 4). Second, each emphasizes the principles of equity, equality, rationality and cooperation. Generally, these principles are discussed in the preamble of each agreement. Each agreement also articulates the broad areas of cooperation (China-Mongolia Agreement, 1994, Preamble, Articles 2, 3, 5, 9; China-Kazakhstan Agreement, 2001, Preamble, Articles 2, 5; China-Russia Agreement, 2008, Preamble, Articles 2, 3, 6) and the creation of a cooperative agency, commonly a joint commission. This commission is responsible for the development of a statute which regulates its activities and decides on issues relevant to the effectuation of the respective agreement (China-Mongolia Agreement, 1994, Articles 4.2, 10, 11; China-Kazakhstan Agreement, 2001, Articles 8, 9, 10; China-Russia Agreement, 2008, Articles 2, 4). Third, each treaty emphasizes the resolution of disagreements concerning interpretation and implementation through friendly consultation instead of arbitration or litigation (China-Mongolia Agreement, 1994, Article 14; China-Kazakhstan Agreement, 2001, Preamble, Article 12; China-Russia Agreement, 2008, Preamble, Article 7).

Each treaty has its own unique features, as they were concluded at different times and with different states. The China-Mongolia Agreement on the Protection and Utilization of Transboundary Waters is simple in comparison to other agreements. In contrast, the China-Russia Agreement on Reasonable Utilization and Protection of Transboundary Waters lists 16 different aspects of cooperation. When necessary, it is possible for the two parties to utilize transboundary waters through joint investment, under Article 7 of the treaty. Under Article 8, only those hydraulic engineering projects that are being constructed or are proposed for construction, and which might cause serious impacts, require notification of the other party, and the implementing party shall take necessary measures to prevent, control and mitigate such impacts. Therefore, in this case, only notification, instead of prior permission or assent, is needed for specific investments. In addition, Article 4 of the China-Kazakhstan Cooperation Agreement on Utilization and Protection of Transboundary Rivers states: "Taking into account mutual interests, no Party shall limit the other Party in the rational use and protection of the water resources of transboundary rivers." This reflects the shared preference not to place limitations on the rational use and protection of transboundary water.

Third, certain investment agreements involve specific transboundary waters. This kind of agreement ensures that specific projects are governed by international law rather than the domestic law of state parties. These primarily include the Agreement on Joint Construction, Management and Maintenance of Ji'an-Manpu Border-River Road Bridge on Yalu River (concluded 10 May 2012), the China-Kazakhstan Cooperation Agreement on the joint construction of the Horgos River Friendship Joint Diversion Control Project (concluded in November 2010), and the Agreement on Joint Construction, Management and Maintenance of the Border-River Road Bridge on Yalu River (concluded 25 February 2010). These agreements define all aspects related to the specific project. For example, under the agreement on the

Horgos River Project, the total investment reaches USD 9.56 million, with each party contributing half.

Fourth, international investment agreements (IIAs) also cover investment in transboundary waters. China has concluded IIAs, including bilateral investment treaties (BITs), with India, Kazakhstan, Kyrgyzstan, Laos, Mongolia, Myanmar, North Korea and Russia (MOC, Bilateral Investment Treaties, 2014). According to these BITs, Chinese companies as well as overseas investments in China's transboundary waters can be considered as qualified investors and investments which can achieve legal protection from these treaties. Although these IIAs do not specifically mention investments regarding transboundary waters, it can be inferred that such investments are qualified investments. For example, the China-Myanmar BIT (2001) says that "the term 'investment' means every kind of asset invested in by investors of one Contracting Party in accordance with the laws and regulations of the other Contracting party in the territory of the latter" (Article 1) and that "an economic entity, including a company, association, partnership and other organization, incorporated or constituted under the laws and regulations of either Contracting Party, and having its seat in that Contracting Party, is the 'investor'" (Article 2(b)). Therefore, the investor and their investment shall enjoy the treatment and protection afforded by the treaty. Under Articles 8 and 9, any dispute concerning the interpretation or application of the agreement between the two states, and any investment dispute between the investor and the host state, can be submitted for arbitration. Finally, as regards compensation for damages and losses caused by war, a state of national emergency, insurrection, riot or other similar events in the territory of the contracting party, the investor can enjoy national treatment or most-favoured-nation treatment under Article 5.

Under Article 1 of the Investment Agreement found within the China-ASEAN Agreement (2009), both private and public investments in transboundary waters are considered to be qualified investments and investors. Under Articles 13 and 14, any dispute between states or any specific investment dispute between an investor and the state can be settled through arbitration. Under Article 9, the investor will receive most-favoured-nation treatment or national treatment, whichever is more favourable, for the compensation of damages and losses caused by war, a state of national emergency, insurrection, riot or other similar events in the territory of the contracting party.

Finally, with certain states, such as India and Bangladesh, treaties directly involving their transboundary waters are limited to the Memorandum on Exchange of Flood Information on the Yarlung Zangbo River-Brahmaputra River, which does not mention investment in transboundary waters. Therefore, the scope of cooperation is limited, since the content of exchange does not cover data and information on the condition of transboundary waters or the possible effects of planned measures on the state of transboundary water bodies. In some instances, states such as Kyrgyzstan, Nepal, Bhutan and Afghanistan actively share such information with China even though there is no formal agreement obligating them to do so.

In conclusion, it is clear that the body of international law relevant to both investment and Chinese transboundary waters is developing. Regarding the latter, it should be noted that treaties transforming general legal principles into specific actionable rules are more helpful. In addition to transboundary water treaties wherein "the international rights and obligations respecting water in its natural state are contained in separate treaties and agreements negotiated for that purposes" (Kibel & Schutz, 2011, p. 757), IIAs should also be considered, allowing China to take a holistic approach to investment in transboundary waters.

Risks and controversies of Chinese investment in transboundary waters

Generally speaking, the utilization of transboundary waters requiring investment can bring many benefits to riparian states and stakeholders, and can attract support from the UN, the World Bank and developed countries. The benefits brought about by the optimal utilization of shared waters may include hydropower generation, high-quality water for domestic uses, tourism, prevention of floods, and so on. Adverse effects, however, may also arise. Therefore, the key to investment in transboundary waters is determining how to obtain these advantages while avoiding or reducing the disadvantages. When discussing Chinese investments in transboundary waters, concerns relating to environmental protection, ecological destruction, the resettlement of displaced persons, and even the threat of armed conflict arise; however, although these are possible, such concerns are typically overstated.

As illustrated above, investment in transboundary waters can be unilateral or cooperative, and within China's borders or in neighbouring riparian states; but no matter what kind of investment is involved, risk and controversy may arise. Generally speaking, when the border between riparian states comprises shared transboundary waters, cooperative investment is likely to be much easier; however, this is not always the case. For example, the Gabčíkovo-Nagymaros Project is a failure, as will be illustrated later. However, when China lies upstream on a transboundary water body, Chinese investment in hydropower projects or other facilities has proven to be controversial, and in the case of investments in downstream waters in neighbouring riparian states, Chinese companies face greater risks and uncertainty.

Regarding the investment in 13 cascade hydropower stations on the Nu River, the majority of proposed projects have been unable to advance past this stage due to social controversy, especially criticism from Western scholars and international nongovernmental organizations (Xinhuanet report, 2013). On the Nu River, 9 of the 13 proposed dams are in an area which, as of July 2003, was recognized by the United Nations Educational, Scientific, and Cultural Organization (UNESCO) for its unique value to global biodiversity and designated a UNESCO World Heritage Site. This designation exposed these projects to further risks.

The investment in hydropower projects on the Lancang River has also been controversial, and because of this, these investments are ineligible for financing from the Asian Development Bank. These proposed dams, as well as the Xiaowan Hydropower Project upstream on the Lancang River, have been blamed for potential drought in the future and the serious drought experienced, in 2010, by downstream states on the Mekong River (Li, 2010).

Downstream states, primarily India and Bangladesh, have expressed concern regarding Chinese investment in dams on the Yarlung Zangbo River since 2003. At present, moderate Chinese investment in Sino-Indian transboundary waters has led Indian and global media to warn of the "Theory of the China Water Threat" (Hao, 2012, p. 4). India countered China's construction of a hydropower project on the Yarlung Zangbo River by building a hydropower project in southern Tibet (Geng, 2013). China and India are now competing to push their respective hydropower projects on the Yarlung Zangbo River (Jha, 2014).

Among Chinese investments in dams in Myanmar, the dispute concerning the Myitsone Dam is the most striking, and the project has been suspended. In the same vein, investment in dams on the Salween River is facing a variety of risks, especially political bias and armed conflict. For example, in relation to the Hatgyi Dam Hydropower Project on the Salween River, Karen rebels have warned that if the government of Myanmar proceeds with the project, conflict will be unavoidable (Eleven, 2014).

In the past, China's large-scale development and utilization of the Ili and Irtysh-Ob Rivers have been an area of concern and criticism for Kazakhstan, causing Kazakhstan to demand negotiations with China. China's related investments have also caught the world's attention (Wang & Hu, 2011). At present all Chinese investments within Vietnam are adversely affected by violence aimed at Chinese citizens, and it seems that the Vietnamese government has failed to take effective measures to stop such violence (Xinhua report, 2014).

In summary, Chinese investment in its transboundary waters is facing a variety of difficulties involving environmental protection, ecological diversity, the resettlement of displaced peoples, and adverse impacts upon the interests of downstream states. Most seriously, some projects even face political risks and armed conflict.

The way forward regarding investments in transboundary waters

Adhering to fundamental principles to prevent investment controversy

Since the majority of investments on transboundary waters will touch upon the relationship between two or more states, we ought to comply with the fundamental principles of international law, such as sovereign equality, *pacta sunt servanda*, peaceful settlement of international disputes, refraining from the threat or use of force, etc. In fact, the Convention on the Law of Non-navigational Uses of International Watercourses (UNWC) and the Convention on the Protection and Use of Transboundary Watercourses and International Lakes reflect these principles, in addition to the other specific principles of international water law, such as the principle of cooperation, the principle of equitable and reasonable utilization, and the principle of non-harm. Furthermore, although China has not signed or ratified these treaties, it recognizes and practises these core principles, as reflected in Chinese treaties and other practices regarding the utilization and protection of transboundary waters. Through these principles we can promote the interests of riparian states while preventing disputes.

China generally emphasizes the legal principle of good neighbourliness, the importance of which is affirmed by the UNWC. For example, China has paid constant attention to the national interests of downstream states during Chinese investment in the hydropower projects on the Lancang River, thus ensuring that investment in hydropower on the Lancang River will play a mitigating role instead of aggravating the risks of flood and drought on the lower reaches of the river. In addition, China took the initiative to forgo construction of the Mengshong Hydropower Station, as well as 30 hydropower stations on the Yili River, so as to protect fish migration routes and local ecosystems (Dizhou News, 2012; Tan & Kong, 2012).

China has also complied with principles of sustainable development in the process of its hydropower investments. The 12th Five-Year Energy Development Plan guides China to develop its hydropower in an efficient manner. It aims "to harmonize hydropower exploitation, enrich displaced persons, protect the environment, optimally utilize the water resource, and develop the local economy; to strengthen hydropower planning of the drainage area, and develop hydropower on the precondition of proper ecological conservation and resettlement of displaced persons" (State Council, 2013, Subchapter 1, Chapter 3). In fact, China's transboundary water treaties routinely include environmental and sustainable development provisions, requiring prevention, reduction and control of pollution on transboundary waters, similar to Article 21 of the UNWC (1997).

Compared with general investments, investments in transboundary waters universally involve foreign benefits, have the potential for adverse effects on society, the environment, biological diversity and culture, and may even complicate relations between riparian states. Thus, both stakeholder theory and the fundamental principles of international law are significant to investment on China's transboundary waters.

Expanding communication to eliminate misunderstanding and controversy

Due to historical, cultural, and ideological concerns, China has been regarded as an opaque, mysterious and unusual state. Although China is becoming increasingly open, and taking great leaps in development, bias still exists. With the increase of Chinese political power, some states even bear hostile sentiments towards China according to the outdated logic that a strong country is bound to seek hegemony. Therefore, it is necessary for China to actively expand communications in order to eliminate any misunderstanding. However, this communication must go both ways. The global community also needs to better understand China's rise and should trust China's good faith.

A public investment policy specifically for transboundary waters is contained in China's 12th Five-Year Plan (2011–2015). However, nearly all foreign scholars stress only one aspect of this policy: that it (allegedly) calls for accelerated construction of large-scale hydropower in South-West China. These scholars forget to mention the premise of hydropower development in China illustrated in the same subchapter: that such development occurs "on the precondition of proper ecological conservation and resettlement of displaced persons" (Subchapter 1, Chapter 11). Under the guidance of this policy, in the specific context of investments in the Yarlung Zangbo River, China has insisted that it has not invested in the construction of hydropower projects with a large storage capacity, and also emphasizes its consideration, as a responsible state, of the protection and development of the water body as well as the interests of downstream states (Foreign Ministry Spokesperson Hong Lei's Regular Press Conference, 2013).

With this in mind, China should not be regarded as a threat to its riparian neighbours and their equitable and reasonable use of their shared transboundary waters. President Jinping Xi has repeatedly emphasized that the Thucydides Trap, that is "the phenomenon of a rising power provoking so much fear in a status quo power that it ultimately leads to conflict between the two" (Eikenberry, 2014), can be avoided, and that China does not have a historical or cultural hegemonic instinct (Xi, 2014). In fact, Confucianism, which emphasizes benevolence and empathy in the relationship between human beings and nature, is deeply rooted in Chinese culture.

Investing in transboundary water through cooperation

The cooperative principle in international water law is typically understood to involve a suite of procedural and information-sharing duties, for example notification, (ongoing) exchange of information, consultation, negotiation, etc. However, the cooperative principle ought to have more extensive implications. To avert further disputes over transboundary water, the cooperative principle through cooperative investment should be followed more closely. For example, the Itaipu Hydropower Project on the Paraná River, constructed through cooperative investment from Brazil and Paraguay, has been able to settle historical conflicts over the Paraná River (Jia et al., 2009; Salman, 2010). At the request of the United States, the downstream riparian, Canada has built three upstream dams, which were paid for by the United States for the benefit of flood control, and Canada was granted rights to divert water for hydropower purposes (Giordano & Wolf, 2003). Thus, cooperative investment was able to resolve disputes related to the Columbia River, which had been ongoing from the mid-1940s through the 1960s.

We must consider adopting a similar strategy for dealing with controversies over China's transboundary water. Cooperation between China and Kazakhstan provides an excellent example of successful cooperation through joint investment. The Horgos River Dam, which was under construction from April 2011 to July 2013, was designed by Kazakhstan, constructed by China, and carried out on the basis of equal investment, while the flow of the river was divided equally (Economics, 2013). Thus, the project eased China's worries concerning the use of the waters from Kazakhstan on the middle and lower reaches of the Horgos River, and was highly praised by both states (Economics, 2013). This project was successful for two reasons. First, taking into consideration that the Horgos River forms the Kazakhstan–China border, this project provides an equal distribution and helps mitigate water supply issues for the border regions of both countries. Second, the Joint Commission played an important role in the cooperative investment of the project. In contrast, a cooperative investment project on the Danube River, which forms the border between Hungary and Slovakia, was unsuccessful due to the political environment at that time and uncooperative actions between the two sides; the resulting dispute has lasted for more than two decades (Gabčíkovo-Nagymaros Case, 1997).

With this in mind, cooperative investment in transboundary waters, even the Yarlung Zangbo/Brahmaputra River shared by China, India and Bangladesh, may be the best choice for dealing with potential conflict.

Using a multilateral mechanism to prevent and settle disputes

The cross-border nature of transboundary waters requires a holistic approach to their use even though each individual state unilaterally controls the waters within its territory. Thus, a regional multilateral mechanism is necessary so as to allow all basin states to cooperatively invest in order to maintain peace and avoid conflict.

First, if a transboundary river runs through three or more states, a multilateral treaty should be created. Thus the community of interests within a transboundary basin can be taken as a single whole, limiting the adverse affects on the other states within the basin. Common examples include the Irtysh-Ob River, which runs through China, Mongolia, Kazakhstan and Russia; the Heilong River, which runs through China, Mongolia, North Korea and Russia; and the Yarlung Zangbo River, which runs through China, India and Bangladesh. The Agreement on the Establishment of the Consultative Commission for the Development of the Tumen River Economic Development Area and North-East Asia is an example of such an agreement; it is a rare example of a multilateral mechanism covering Chinese transboundary waters (Tumen River Commission Agreement, 1995).

Second, China ought to, in due time, consider acceding to such an existing regional mechanism. On 6 April 2014, a Chinese delegation attended the summit of the second session of the Mekong River Commission and put forward four suggestions to strengthen cooperation with the commission and its members, without acceding to the commission. These suggestions included: strengthening the harmonization of regional development strategies; strengthening the planning processes for cooperation; further developing cooperation regarding hydropower development; and jointly creating an atmosphere of cooperation (Xinhua report, 2014).

Third, China may, in future, consider agreeing to international conventions such as the UNWC. For China, in comparison to its existing transboundary water treaties, Article 14(b) of the UNWC is too demanding because it provides that, for planned measures with possible "adverse effects", a state has an obligation to notify relevant parties. In these instances the notifying state "shall not implement or permit the implementation of the planned measures without the consent of the notified States". Surely this obligation only applies during the period of notification, which is generally six months, and neither the UNWC nor general international water law requires express consent from an affected riparian state (see e.g. *Lac Lanoux Arbitration*, 1957). However, the UNWC is unclear about the result if the notified states do not consent to the planned measures. Certainly Article 16.1 of the UNWC creates a procedural limitation on the notified states; that is, in the absence of a reply to such notification, the notifying state may proceed with the implementation of the planned measures, "subject to its obligations under articles 5 and 7", which refer to "equitable and reasonable utilization and participation" as well as the obligation "not to cause significant harm", and "in accordance with the notification and any other data and information provided to the notified States". Furthermore, under Article 17, consultations and negotiations concerning planned measures are necessary if the notified state considers that the implementation of planned measures would be inconsistent with the provisions of Article 5 or 7. This limitation seems to balance bilateral benefits; however, it may also impede the planned measures of the notifying state. Studies have shown that downstream states are also capable of causing harm to the benefits of upstream states (Salman, 2010); therefore it is incorrect to assume that the UNWC only protects downstream interests. In addition, one reason that China has not ratified the UNWC is Article 33, which is related to arbitration and tribunal, possibly through the International Court of Justice. However, it can be argued that the UNWC does not grant compulsory jurisdiction to the International Court of Justice or any international tribunal.

Determining appropriate methods for dispute settlement

When resolving transboundary investment disputes, we can employ all the means of international dispute resolution, including consultation, negotiation, good offices, mediation, conciliation, inquiry/fact-finding, arbitration and judicial adjudication. However, nonjudicial means are always the principal and optimal means of dispute settlement.

International courts and tribunals can play a role in resolving transboundary water problems to a certain extent. Historical cases, such as the case of the diversion of water from the Meuse (*Netherland v. Belgium*, 1937) before the Permanent Court of International Justice, laid the foundation for the principle of non-harm. The Lake Lanoux arbitration case (*Spain v. France*, 1957), decided before an ad hoc arbitral tribunal, found that in the case of a water diversion and hydropower project, France was obligated to "notify" Spain; however, they were not obligated to "get approval". Recent examples include cases such as the Indus Waters Kishenganga arbitration before the Permanent Court of Arbitration (*Pakistan v. India*, 2013). Thus far, the International Court of Justice (ICJ) has also heard and decided on several cases.

Three common cases relate to investments on transboundary waters. A first example is the Gabčíkovo-Nagymaros case (*Hungary v. Slovakia*), which arose from investment cooperation in the construction of a system of dams and locks along the Danube River. Hungary and Slovakia decided to jointly invest in building the Gabčíkovo-Nagymaros project on the basis of the Treaty Concerning the Construction and Operation of the

Gabčíkovo-Nagymaros System of Locks, concluded 16 September 1977. However, a dispute arose on 7 April 1993 during the process of cooperative investment. At this time the two states concluded a special agreement to file the case before the ICJ. On 25 September 1997, the court decided that Hungary had no right to terminate the project but that Slovakia, in applying an alternative investment scheme and unilaterally taking control of a shared resource, had also violated international law, as well as the 1977 treaty. The court ruled that Slovakia should have received consent from Hungary. Ultimately, the court ordered them to "re-establish co-operative administration of what remains of the Project". On 3 September 1998, Hungary requested that the ICJ make an additional judgment on the grounds that Slovakia had failed to follow the order. At present, the application for an additional decision is pending, and the decision of the court has yet to be implemented. Thus, this case shows that international adjudication is not always a feasible and effective way to settle a transboundary water dispute. Cases with political, historical and cultural factors may prove to be unsuitable for settlement before international courts.

The case of *Argentina v. Uruguay* arose when, in 2003, Uruguay authorized the construction of pulp mills on the Uruguay River, the border between Argentina and Uruguay. In 2006, Argentina initiated proceedings before the ICJ to forbid Uruguay from constructing the mills. Argentina charged that Uruguay had violated treaty provisions requiring prior notification and consultation before taking actions that could affect river water quality. Argentina's request for a preliminary injunction was rejected because the ICJ believed that Uruguay had intended to and might still comply with its international obligation. As a result, protesters in Argentina blocked roads to prevent construction, and Argentina's government appealed to the ICJ a second time. On 20 April 2010, the court concluded that while Uruguay had breached its international procedural obligations to notify and consult with Argentina before authorizing and commencing construction on the pulp mills, the court's declaration of Uruguay's breach constituted a sufficient remedy for Argentina's claim. In the author's opinion, this decision will not permanently resolve the problem, and because of this, the ICJ may not always be the right forum for such disputes.

In the case of *Nicaragua v. Costa Rica* (2011), Nicaragua claimed that Costa Rica's unilateral action of road construction parallel to the San Juan River was threatening to destroy the fragile ecosystem found in the river. In response, Nicaragua requested restoration to the *status quo ante*, damages, and the production and presentation of an appropriate transboundary environmental impact assessment. Although the case was filed on 22 December 2011, it is still pending before the ICJ.

Arbitration, as a method of pacific dispute settlement, has the potential to resolve transboundary water disputes. For example, the inter-state dispute between India and Pakistan under the 1960 Indus Waters Treaty arose from India's proposal to build a 330 MW hydroelectric project on the Kishenganga River. At the request of Pakistan, an arbitral tribunal under the Permanent Court of Arbitration was constituted on 17 December 2010. The final decision, 20 December 2013, imposed no further restrictions on the operation of the Kishenganga Hydro-Electric Project investment, subject to certain conditions (*Pakistan v. India*, 2013). Investments in transboundary waters may be based on a state–state agreement; however, they may also be based on an investor–investor agreement or an investor–state agreement (for example, a BOT project). The type of agreement will affect the choice of dispute settlement method. For example, the case of *Bayview Irrigation District et al. v. United Mexican States* was concluded in 2007 under the International Centre for the Settlement of Investment Disputes (ICSID) Additional Facility Rules. In this case, related to the Rio Grande River, the tribunal rejected the claimants' arguments, leading it to determine that Texan farmers did not possess investments in Mexico, a determination that deprived the tribunal of jurisdiction to further consider the claim.

A Chinese investor, even if it is a state-owned-enterprise such as the Huadian Corporation, Guodian Corporation or China Huaneng Group, falls within the scope of the definition of 'investor' under most Chinese BITs. Although China's transboundary water treaties do not include an arbitration clause, all of China's BITs with riparian states include such a clause, to different degrees. Therefore, a dispute arising from a transboundary water investment between China and other riparian states or between a Chinese investor and another riparian state can be arbitrated.

In China's culture and traditions, the Chinese prefer other methods over judicial approaches to dispute settlement. As a colonial and semi-colonial state for more than a century, it is understandable that China cherishes its sovereignty. Therefore, it is hard to imagine that China would opt to use ICJ adjudication to settle transboundary water disputes in the near future. However, considering that sensitive issues of sovereignty are not present in a dispute concerning investment in transboundary waters, but are confined to the utilization and protection of water resources, such a case may be a candidate for settlement by international arbitration. In addition, China has accepted arbitration in relation to investment disputes falling within the scope of IIAs. In fact, the diplomatic approach has proven to be effective. For example, a dispute between Mexico and the United States, arising from the allocation of water rights, was settled by consultation between the two parties at the International Boundary and Water Commission (Kibel & Schutz, 2011).

Conclusion

China's investments in transboundary waters have become an important issue, attracting the attention of the international community, causing fear among neighbouring riparian states, and even affecting the relationships between China and its neighbours. When they met at the BRICS summit in Durban, the Indian prime minister, Manmohan Singh, suggested to Chinese president Jinping Xi that India wanted to monitor China's dams on the Yarlung Zangbo River (New Indian Express, 2013). This suggests that China is being carefully watched by the international community for so-called hydrohegemonic tendencies. After examining the status of China's investments in its transboundary waters and the applicable laws, this article argues that restrained investment with due consideration for the interests of riparian states, environmental protection, and the resettlement of displaced persons will allow China to maintain peaceful relations with its neighbours, even though its policy on the utilization of transboundary waters seems 'ambitious'. China has paid close attention to cooperative investment when utilizing its transboundary waters, and thus far cooperation has seen some success. China should extend this kind of cooperation to other riparian states so as to settle ongoing controversies in other shared river basins. Although this will help settle these disputes, we must explore other approaches in order to respond to the risks and controversies related to investment in transboundary waters, including adhering to fundamental principles to prevent disputes, actively using multilateral mechanisms, eliminating misunderstanding by improving external communication, and settling investment disputes on transboundary waters through diplomatic means and arbitration. Given the uncertain political risks faced by China, it should strive for investor protection in its transboundary water treaties and IIAs.

Acknowledgements

Heartfelt thanks are due Professor Patricia Wouters (Xiamen University Law School), Dr Owen McIntyre (visiting scholar, Xiamen University Law School), David Devlaeminck and members of the CIWL team at Xiamen Law School for their great encouragement and help in preparing and revising this article, and also to the two anonymous reviewers for their excellent comments.

Funding

This research is supported by the Fundamental Research Funds for the Central Universities (grant 20720140010), and one of the outcomes of the 2011 Program for New Century Outstanding Talents in University (grant NCET-11-0292).

References

Bayview Irrigation District, *et al.* v. United Mexican States Case. (2007). ICSID Case No. ARB(AF)/05/1, Award rendered on 19 June 2007.

China-ASEAN Agreement. (2009). Agreement on Investment of the Framework Agreement on Comprehensive Economic Cooperation between The People's Republic of China and the Association of Southeast Asian Nations. Retrieved from http://fta.mofcom.gov.cn/inforimages/200908/20090817113007764.pdf

China-Kazakhstan Agreement. (2001). Agreement between the Government of the People's Republic of China and the Government of the Republic of Kazakhstan on Cooperation in the Use and Protection of Transboundary Rivers. Retrieved from http://vip.chinalawinfo.com/newlaw2002/SLC/SLC.asp?Db=eag&Gid=100669255

China-Mongolia Agreement. (1994). Agreement between the Government of the People's Republic of China and the Government of the Republic of Mongolia on Protection and Utilisation of Transboundary Waters. Retrieved from http://www.chinainternationalwaterlaw.org/pdf/treaties/mongolia/1994_en.pdf

China-Myanmar BIT Agreement. (2001). Agreement between the Government of the People's Republic of China and the Government of the Union of Myanmar on the Promotion and Protection of Investments. Retrieved from http://investmentpolicyhub.unctad.org/Download/TreatyFile/762

China-Russia Agreement. (2008). Agreement between the Government of the People's Republic of China and the Government of the Republic of Russia on Cooperation in the Reasonable Use and Protection of Transboundary Rivers. Retrieved from http://vip.chinalawinfo.com/newlaw2002/eag/index.asp

Dizhou News (2012, November 4). Protect the Ecological Environment, More than 30 Hydroelectric Stations Suspended in Yili River Basin. Retrieved from http://news.ts.cn/content/2012-11/04/content_7392373.htm

Economics (2013, July 10). Dam of Friendship on the Horgos River. Retrieved from http://newsite.kazpravda.kz/c/1373451504

Eikenberry, K. (2014). Thucydides Trap. *American Review.* Retrieved from http://americanreview-mag.com/stories/Thucydides-Trap

Eleven (2014, March 9). Ethnic armed groups oppose Hatgyi dam project. Retrieved from http://elevenmyanmar.com/index.php?option=com_content&view=article&id=5319:ethnic-armed-groups-oppose-hatgyi-dam-project&catid=44:national&Itemid=384

Fan, X. J. (2008). Japan's High-profile Mekong Diplomacy. *International Data, 4*, 12–14.

Foreign Ministry Spokesperson Hong Lei's Regular Press Conference. (2013, January 30). Retrieved from http://www.fmprc.gov.cn/mfa_eng/xwfw_665399/s2510_665401/2511_665403/t1009844.shtml

Gabcikovo-Nagymaros Case. (1997). Case Concerning the Gabcikovo-Nagymaros Project (Hungary v. Slovakia), Judgment of 25 September 1997, ICJ Reports 1997, and the subsequent development also can be Retrieved from http://www.icj-cij.org/docket/index.php?p1=3&p2=1&code=hs&case=92&k=8d

Geng, S. W. (2013). *Growing Sino-Indian transboundary water resources problems and its influence on China* (Master Dissertation). Shijiazhuang: Hebei Normal University.

Giordano, M. A., & Wolf, A. T. (2003). Sharing waters: Post-Rio International Water Management. *Natural Resources Forum, 27*, 163–171. doi:10.1111/1477-8947.00051

Han, S., & Ji, P. J. (2012, April 23). International perspective: Japan expands influence in Asia by making use of ASEAN. *Renmin Daily.*

Hao, S. Y. (2012). *The legal study on the harmonious exploitation of cross-border water* (Doctor Dissertation). Xi'an: Changan University.

International Commission on Large Dams. (2009). *The principles and practices on the exploitation and utilization of international shared river.* (J. S. Jia, et al., Trans.). Beijing: China Water Power Press.

Jha, P. S. (2014, March 5). Why India and China should leave the Yarlung Tsangpo alone. Retrieved from https://www.chinadialogue.net/article/show/single/ch/6753-Why-India-and-China-should-leave-the-Yarlung-Tsangpo-alone

Kibel, P. S., & Schutz, J. (2011). Two rivers meet: At the confluence of cross-border water law and foreign investment law. In M.-C. C. Segger,M. W. Gehring, & A. Newcombe (Eds.), *Sustainable development in world investment law* (pp. 749–767). The Netherlands: Kluwer Law International.

Li, H. Y. (2010, May 10). Fight back the theory of responsibility for drought by dam diplomacy. Retrieved from http://www.ennweekly.com/2010/0510/2231.html

MOC, Bilateral Investment Treaties. (2014) Ministry of Commerce of the People's Republic of China, Department of Treaty and Law. Retrieved from http://tfs.mofcom.gov.cn/aarticle/Nocategory/201111/20111107819474.html

NDRC. (2011). The 12th Five-year Plan for National Economic and Social Development. Retrieved from http://news.xinhuanet.com/politics/2011-03/16/c_121193916_2.htm

Netherland v. Belgium Case. (1937). Diversion of Water from the Meuse case, [1937] P.C.I.J. (Ser. A/B) No.70.

New Indian Express. (2013, April 17). Monitor Brahmaputra to rule out Chinese dam. Retrieved from http://www.newindianexpress.com/editorials/Monitor-Brahmaputra-to-rule-out-Chinese-dam/2013/04/17/article1547919.ece

Nicaragua v. Costa Rica Case. (2011). Application Instituting Proceedings, Construction of a Road in Costa Rica along the San Juan River, ICJ, filed in the Registry of the Court on 22 December 2011.

Pakistan v. India (2013). Case of Indus Waters Kishenganga Arbitration, PCA, Final Award. 20 December 2013.

Salman, M. A. S. (2010). Downstream riparians can also harm upstream riparians: The concept of foreclosure of future uses. *Water International, 35*(4), 350–364. doi:10.1080/02508060.2010.508160

SOE News. (2014, April 30). Sinohydro Group Ltd.: The memorandum of Agreement on Burma Haji hydropower station was signed. Retrieved from http://www.sasac.gov.cn/n1180/n1226/n2410/n314289/11740540.html

Spain v. France Case. (1957). Lake Lanoux Arbitration, 24 Int'l L. Rep. 101(1957).

State Council. (2013). The 12th Five Year Plan for Energy Development. Retrieved from http://www.gov.cn/zwgk/2013-01/23/content_2318554.htm

Suidianzhan. (2014, January 23). Three Gorges Group: Inviting Consultations and Comments Conference on the Feasibility Study of Meng Dong hydropower station. Retrieved from http://www.shuidianzhan.com/shijiao/guonei/22679.html

Tan, G. M., & Kong, L. J. (2012). *International law and practices of transboundary water resources*. Beijing: Social Sciences Academic Press.

Tumen River Commission Agreement. (1995). Agreement on the establishment of the consultative commission for the development of the Tumen River Economic Development Area and Northeast Asia. Retrieved from http://www.pkulaw.cn/fulltext_form.aspx?Db=eagn&Gid=100668800&keyword=%e5%9b%be%e4%bb%ac%e6%b1%9f&EncodingName=&Search_Mode=like

UN Water, thematic paper. (2008). Transboundary waters: Sharing benefits, sharing responsibilities. Retrieved from http://www.unwater.org/downloads/UNW_TRANSBOUNDARY.pdf

UNWC. (1997). United Nations Water Courses Convention (adopted by UN General Assembly in 1997). Retrieved from http://www.unwatercoursesconvention.org/the-convention/part-iv-protection-preservation-and-management/article-21-prevention-reduction-and-control-of–pollution/

Wang, X. (2002). *UNEP environmental law training manual*. Beijing: Law Press China.

Wang, J. F., & Hu, Y. (2011). Sino-Kazakhstan trans-boundary water disputes: Origin, progresses and China's countermeasures. *Journal of Xinjiang University (Philosophy, Humanities & Social Science)*, *39*(5), 99–102.

Water Law of the People's Republic of China. (2002 Revision). Retrieved from http://en.pkulaw.cn/display.aspx?cgid=41759&lib=law

Xi, J. P. (2014, May 16). There is No Gene for Invasion in our Blood. Retrieved from http://usa.chinadaily.com.cn/china/2014-05/16/content_17511170.htm

Xinhua report. (2014, April 9). Chinese delegation attended the Summit of the second session of the Mekong River Commission. Retrieved from http://www.mwr.gov.cn/slzx/mtzs/xhsxhw/201404/t20140409_555050.html

Xinhua report. (2014, May 16). Chinese ministry condemns violence in Vietnam. Retrieved from: http://www.ecns.cn/voices/2014/05-16/114419.shtml

Xinhuanet report. (2013, April 28). Prejudice against Hydropower Development is the Biggest Obstacle to Ecological Civilization. Retrieved from http://www.mwr.gov.cn/slzx/mtzs/xhsxhw/201304/t20130428_434997.html.

Mapping Asia's trans-boundary waters, with a focus on China

Vivian Louis Forbes

National interest, enhancing the nation's wealth and welfare, and fostering development are factors cited for the awareness of the need of accurate and detailed topographic maps. These aspects are clearly spelt out in China's Law No. 75 of 2002. An overview of the status of mapping in China in the context of overall management of its trans-boundary rivers is presented. Public access to these graphics is restricted in this regional context; they are available only to bona fide users. The importance of maps in the context of the trans-boundary water discourse is discussed.

When the Emperor did not see to it that there was water for irrigation, that canal barges could transport rice, that rivers did not flood, and that roads were safe for traveling, the people suffered. (Confucius, 551–479 BCE)

Introduction

Questioning the observations, as in the above quotation, of the eminent scholar of China of many centuries ago, one may ask: Did the Emperor see if there were maps depicting where the rivers flowed? Perhaps there was no need for maps in those early days. Did these rivers cross administrative and political divisions? In researching the literature and examining maps dating back centuries, an appreciation is gained of the cartographical skills employed to depict the landscape on a variety of media – paper, silk, stone and wood. The focus of this study is on the twenty-first century and the issue of trans-boundary rivers and the concepts of hydro-politics (that is, the politics of water sharing amongst nations) and of political expediency in negotiating agreements between riparian States (Leb, 2013). The aim of this narrative is to discuss the status of the mapping in the vicinity of trans-boundary waters, with a particular focus on China. International financial aid and expertise have been forthcoming, such as in the European-China River Basin Management Programme of 2006–12 and the Australia-China Environment Development Partnership Project on the Yellow River, and numerous reports have been published. However, they are not necessarily in the public domain for in-depth analysis by academic researchers. This fact is accepted; however, such limitations hinder discussion on the true state of surveying and mapping and compilation of data-sets and availability of geospatial data, especially along international political boundaries and where sovereignty over a parcel of land and territorial limits are disputed by the parties.

Following this introduction a brief comment is offered on the concept of borders and its derivations, and continued by discussions of the historical mapping of China's geography and the rivers of the country. An analysis of China's Law No. 75 of 2002 is presented after a brief discourse on China's international political boundaries and the rivers they cross.

Borders, boundaries and frontiers

Borders reflect a nation's concept of its own territorial and sovereign limits, and thus boundary studies are a topic of political geography (Fravel, 2008; Prescott, 1975; Elleman, Kotkin and Schofield, 2013). However, in the twenty-first century, more than ever, other disciplines are required to have a basic understanding of geography and politics in the context of political boundary studies. Borders, boundaries and frontiers are the physical manifestation of the abstract concept of the separation between nations and states. The way that a nation marks, monitors, and defends its borders is a direct result of its interactions with its neighbours. Walls and guards reflect uneasiness about a permeable border; large border cities and economic zones reflect a high level of trade. Trans-boundary waters (rivers and lakes) offer opportunities for local communities to barter their goods, but these waterways can also become focal points of regional geopolitical conflict.

While boundaries may seem clearly delineated on maps, the reality is that borders remain difficult to define. Even when 'natural boundaries' such as rivers and other topographical features are used, borders are frequently an arbitrary separation of territories, resources and population. Many of the major rivers do, in sections, function as natural riverine boundaries or are cast as trans-boundary in nature by political decisions made over the centuries. Some maps are geographical in context, but cartography is culture. Maps and charts are utilized to delineate the international and provincial political boundaries, as well as natural boundaries, along watersheds and rivers. This was also evident in historical cartography.

Historical mapping in China

Inspection of early mapping conveys to the reader the culture of the day through cartographic lenses. The earliest extent map of China, on silk, dated to the second century BC (in the early Han Dynasty), came from the Mawangdui excavation in 1973. During archaeological excavation of a Qin State tomb in Fangmatan, near Tianshui City, Gansu Province, seven ancient Chinese maps dating back to about the second century BC were discovered. The territories depicted on the seven Qin maps overlap, suggesting that they formed a series (Hsu, 2009). The Qin recognized the importance of documents in accordance with the ideas expressed in classical texts. The maps portrayed tributary river systems of the Jialing River in Sichuan Province. They were drawn in black ink on wooden blocks.

The oldest reference to a map in China comes from the third century BC (Hsu, 2009). The district map of Qin was painted on silk. An early western Han Dynasty (202 BC–9 AD) silk map depicting the kingdoms of Changsha and Nanyue in southern China was found in Mawangdui.

During the Song Dynasty (960–1279), Yuan Xie was director-general of the governmental grain stores. In pursuance of his schemes for the relief of famines he supposedly ordered that each village prepare a map showing the fields and mountains, the rivers and roads in fullest detail (Yee, 1987). Provincial officials could then readily carry out their duties in the collection of taxes and census taking with the aid of these maps (Behr, 2007).

There were large and intricately carved stone stele maps of the Song period – for example, the 0.9 m^2 stone stele map by an anonymous artist in 1137, on a scale of 100 li for each grid square (Yi & Liu, 2010). The map gives incredibly precise detail of coastal outlines and river systems in China (Needham, 1956). It shows 500 settlements and a dozen rivers in China, and extends as far as north as the Korean Peninsula and south to India. On the reverse, a copy of a more ancient map uses grid coordinates on a scale of 1:1,500,000 and shows the coastline of China with great accuracy (Needham, 1956).

The accuracy and importance of maps

The accuracy and importance of maps in the context of the trans-boundary water discourse will now be discussed, thereby bringing to the fore the questions: Why do maps matter? How do maps matter? What are some of the complications with maps? And, how might these be overcome or addressed?

By the mid-to-late twentieth century, states attempted to assert control over how their territories were surveyed and mapped – with 'incorrect' maps seen as a threat to the very identity of the nation. These efforts to control cartography extend from issues of map content to attempts to dictate who is allowed to map the state's territory. Indeed, the last point is increasingly important to present-day governments in many nation-states as a result of the ways in which map creation and distribution have changed. Earlier, state-sponsored mapping projects (for example, the Australian National Topographic Series and the Survey of India, to name just two) provided the most detailed maps with extensive coverage of the country, often at a variety of scales, enabling states' direct control over how their territories were mapped. By the 1980s, the rapid adoption of digital mapping technologies coincided with a decline in state expenditures on general-purpose civilian cartography (and the adoption of the user-pays principle, which citizens perceived as financial double-dipping by the state) and resulted in a significant and continuing shift toward private map production.

An obvious explanation of why maps are such a vital tool for human and physical geographers, town planners and commercial and military navigation – air, land and sea – is their uses in the contextualization of data, specifically for locating features and places and analyzing demographic data such as age structures and employment status. The key component of a map that is so important is that of map scale. By mapping to scale, a study area of demographic information can be localized and expressed in terms of distance relationships. These spatial relationships are critical to any study that demands research based on spatial phenomena. An example of how this could be used is in mapping river streams, road alignments and census data with the information grouped by locale and these groups displayed with maps to show proximity to each other by map symbology and/or colour coding.

Bathymetrical and topographical survey-style maps and hydrographical charts are best suited for this purpose. They are designed – customized – for the role they play in providing services and infrastructure, which often complements the demographic data being studied. A map provides a miniature 'image' (electronic or print) of a very large space. A map is a guide to the landscape, the infrastructure upon it and the resources on the surface and in the substratum. Maps depict distance, mountains, rivers, and shapes of places or destinations. With a map, one does not have to depend on local directions. For a small price, it is a direction finder and a dependable way to take a journey.

Maps are very useful and important to the user in many ways – for example: to calculate and/or measure direction and distance between two places; to obtain information about mountain ranges, alignment of rivers, valleys and route planning; and to deduce other geographical information like the elevation of a place or the incline of railway and road alignments. Thematic maps such as geology, soil and vegetation offer a pictorial image of the natural environment. Geopolitical maps delineate local, national and international jurisdictional and political boundaries, while cadastral maps portray land ownership and allotments which may be used for land tax assessments and town planning and more importantly assist the government administrators of a state or province in keeping a record of land ownership. Weather maps portray forecast and current weather reports, direction of wind and rainfall distribution, which eventually become a historical record of past climate and weather data.

Geographical information systems (GIS) are computer-assisted cartographical software, which is also crucial to geographical research and analysis. GIS gives human geographers the ability to transform basic maps into more specialized data visualizations, whether it be creating a 3-D model of topographical relief and adding layers of infrastructure and data, or grouping data and creating choropleth maps or data distributions. These types of data plotting have introduced novel mapping and offer innovative concepts. Some of the most important features of a useful map are the accuracy of the graphic, its relevance and the authoritative nature of the publication. News of inaccuracies of maps, such as misplaced landmarks, warped images and inaccurate route information, which are misleading for the user, are available from many websites. Maps were once static in nature, but since about the 1980s they are dynamic and interactive and the mapped data can be manipulated. An inherent problem may arise in this interactive concept which the average map user may not be aware of, especially when there is scale change (zoom factor) involved in the process or changes made – accidental or intentional – during the manipulation processes. Digital maps and the representations they make possible could be undermining of the hegemonic ideal of the territorial state. For example, in 2011, there were disagreements between the government of China and commercial companies providing online maps. Such issues revealed that web-based expansion of coverage and accessibility has merely transformed, not eliminated, such conflicts. The most concerning factors are jurisdictional conflict and territorial jurisdiction, because criminals realize that boundaries continue to matter and thus place themselves (consciously) within the physical borders of nonprosecuting states.

The available cartographic techniques have made territorialization legitimate, desirable and possible. Mapping shapes the conditions of possibility of how people conceive of space, territory and political authority. So, is every map a theory, or a legal instrument, or a factual and scientific document? Nautical charts – aeronautical and marine – are explicitly practical, intended as shipboard navigational graphics; they are also legal documents. Topographical maps are ideal for studies relating to landscape planning and riverine and stream flows and morphological change and may be challenged as legal documents in the event of a dispute between stakeholders over a parcel of land or a water body. However, if a map is endorsed with a statement, for example "This graphic is not to be considered authoritative," then its validity as a legal document may be questioned in a court of law.

China's geography and landscape

The melting snow from the mountains in the south-west of the country creates the headwaters for two of China's most important rivers, the Yellow River and the Yangtze River, and is the source of many other rivers that flow to the south into the countries of South and South-East Asia as trans-boundary rivers. The Yangtze River is the longest in China and third-longest in the world. It irrigates the agricultural regions in the south of China as it meanders across many provinces.

Conservatively speaking, China has more than 50,000 rivers. Almost all large rivers belong to the exterior river system, directly or indirectly emptying into the adjacent seas. Because China's terrain is elevated in the west (for example, the Tibetan Plateau) and is relatively lower in the eastern coastal region, most of its rivers flow east and discharge into the Yellow, East and South China Seas. This includes the Hai, Heilong, Liao, the Xi and Pearl Delta, and the Yangtze and Yellow (Huang) Rivers. The extent to which these rivers retain volumes of water throughout the year is dependent on snow melt, precipitation, climate change and utilization. Figure 1 portrays China's 14 neighbours and the major trans-boundary rivers.

Figure 2, compiled by the present author based on a large-scale topographic map, illustrates the numerous streams and rivulets in a surface area of about 2,600 km^2 of western China that is surrounded on three sides by international neighbours, namely Afghanistan, Pakistan and Tajikistan. The elevation of the terrain in Figure 2 may be inferred from the spot heights, which are in metres; mountain passes are indicated, as well as two tri-points (common points where political boundaries meet and established by agreements), which are also trigonometric stations, surveyed points whose geographical locations were determined by astronomical sightings. A true account of the number of rivers that flow throughout the mainland of China is debatable until such time as an inventory is conducted and an official statement is published.

Figure 1. China's 14 neighbours and major rivers. (Maps drawn by the present author except where noted.)

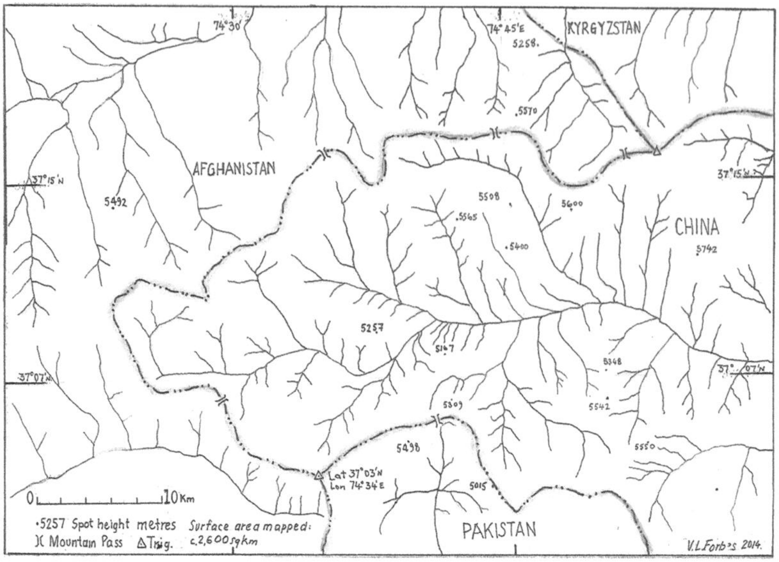

Figure 2. Numerous rivulets in an area of some 2,600 km^2 in western China.

Rivers as China's international political boundaries

The Tumen and Yalu Rivers, flowing north-east and south-west, respectively, from Changbai Mountain (Paektu in Korean), form a natural boundary between China and North Korea (the Democratic People's Republic of Korea). Although there is no current public boundary dispute, there have been a number of issues regarding control of the waterways, ports, islands in the river, and mountain territories in the border region over the past seven decades. The boundary extends for about 1,400 km (Armstrong, 2013; Pinilla, 2004).

Trans-boundary rivers

Along China's southern and western frontiers, the major rivers with their source in the Tibetan Plateau meander mainly in a southerly direction through the territories of China's neighbours and discharge into the Arabian Sea, the Bay of Bengal, or the South China Sea. The delimitation, demarcation and delineation of the China–Laos boundary were basically based on watercourses and watersheds, where possible. The boundary extends eastward from the tri-junction of Laos, China and Myanmar in the Mekong River to the tri-junction point of China, Laos and Vietnam (Figure 3). To the north of the boundary is Yunnan Province of the People's Republic of China, while to the south are the northern provinces of Laos, including Luang Namatha to the west (bordering Myanmar) and Oudomxay and Pongsaly to the east, bordering Vietnam (Townsend-Gault, 2013).

The land boundary agreement between China and Vietnam was ratified and took effect on 6 July 2000 following an exchange of letters of ratification between the parties in Beijing. The demarcation process was initiated on 27 December 2001 and officially concluded towards the end of 2008. The demarcation of the boundary was an essential component for the long-term stability in bilateral relations, as well as for the continued expansion of multifaceted cooperation between the border provinces of China and Vietnam and especially amongst the riverine communities, for whom political issues

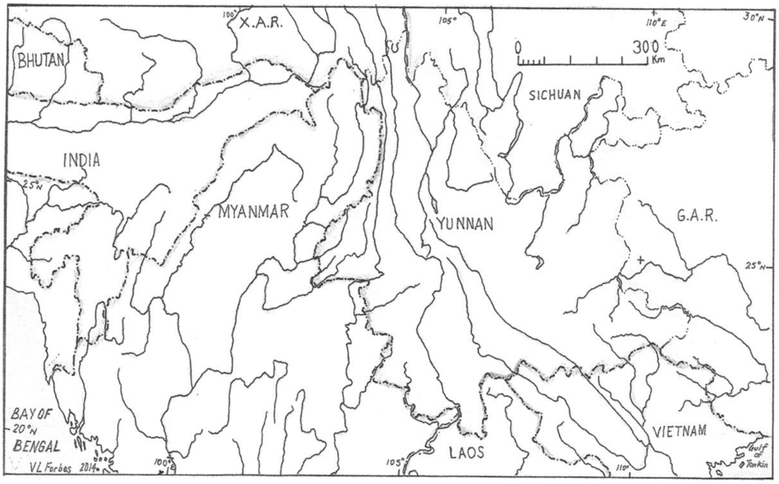

Figure 3. The land boundary of China with its neighbours, Myanmar, Laos and Vietnam.

take a back seat as daily living and bartering of merchandise and natural products are of greater importance (Amer, 2013).

The Bhutan–China boundary negotiations have never been solely about real estate or territory; they also reflect the larger geopolitical dynamics of South Asia. Once the two countries achieve their aim of demarcating their borders (this also involves India), the parties can focus on outstanding trans-boundary problems, including poaching, smuggling and other encroachments. A river, called by the Bhutanese the Drangeme Chhu, at about latitude 27°50′ N, is the only major watercourse that crosses the perceived Bhutan–China boundary (Smith, 2013).

Sharpened geopolitical rivalry is a result of strains in China's bilateral relations with India. The unsettled boundary along the Himalayas and the Hindu Kush remains at the centre of the bilateral tensions, extending even to the 2009 efforts to block approval of the Asian Development Bank's USD 2.9 billion loan assistance plan for India's initiative of a flood-management project in Arunachal Pradesh – just another example of hydro-politics. When the bank overrode China's objections, the latter expressed "strong dissatisfaction," stating that approval could not alter the "immense territorial disputes" with India (Chellaney, 2013: 54).

China's bilateral relations with Pakistan were developed to accommodate many shared concerns; however, the friendship begins and ends with India. The China–Pakistan boundary settlement of March 1963 came hot on the heels of the urgency that was initiated by the China–India border conflict of 1962. The waters of the Indus River and its tributaries have their sources just inside Chinese territory, or in the vicinity of the boundary, depending on which maps are consulted. Herein rest the dynamics of hydro-politics. Afghanistan's boundary with China is delineated, and where demarcated, along the watershed of the mountain ranges; as a consequence there is no trans-boundary river between these two countries (see Figure 2).

Surveying and mapping in China

The State Bureau of Surveying and Mapping (SBSM) was established in 1956 as an administrative agency of the government of China for the functions of surveying and mapping. The SBSM has two main functions: the organization and management of the national basic surveying and mapping work; and the overall administration of the surveying and mapping profession. In the restructuring of the Chinese government in 1998, the SBSM remained a state bureau but moved from the Ministry of Construction to the Ministry of Land and Resources. State-sponsored mapping projects provide the most detailed maps, with extensive coverage of the country, often at a variety of scales. However, this concept enables the state's direct control over not only how its territories are mapped but also who has access to the mapped material, which can create delays and problems for planners and researchers. The rapid adoption of digital mapping technologies does not necessarily entail instant information at hand, because often mapping authorities – as in most countries – insist on 'print on demand' service but with stringent notice and form-filling requirements.

China's Law No. 75 of 2002

The Surveying and Mapping Law of the People's Republic of China was amended and adopted at the 29th Meeting of the Standing Committee of the Ninth National People's Congress of the People's Republic of China on 29 August 2002, and promulgated and entered into effect on 1 December 2002. The law contains 54 articles, each very detailed and outlining the specifics of surveying and mapping the country. A brief analysis is offered here.

The law of 2002 was enacted to strengthen the administration of the surveying and mapping undertaking, to promote its development and to ensure that it renders service to the development of the Chinese national economy, the building-up of national defence, and the progress of the society (Government of China, 2002). All surveying and mapping activities within China's territorial air, land and waters, as well as other sea areas under its jurisdiction, must be conducted in compliance with this new legislation. The activities of surveying and mapping include the conduct to determine, collect and formulate – in essence geospatial data – the key elements of physical geography as well as to process and provide the data, information and results.

Foreign organizations or individuals who wish to conduct surveying and mapping within the territory and under the jurisdiction of the People's Republic of China are subject to approval by the administrative department for surveying and mapping under the State Council and the competent department for surveying and mapping of the Army, and they are required to observe the provisions of relevant laws and administrative rules and regulations of the People's Republic of China. Foreign organizations or individuals are required by law to cooperate and to work in association with the relevant departments or units of the People's Republic of China in the form of Chinese–foreign equity joint ventures or Chinese–foreign contractual joint ventures, and such surveying and mapping may not involve state secrets or endanger state security.

Data and systems for mapping and surveying

In Law No. 75 the state established and adopted nationwide unified geodetic, vertical, depth and gravimetric data, and the data in this respect must be examined and verified by the administrative department for surveying and mapping under the State Council, which, after consulting with the relevant departments under the State Council and the competent department for surveying and mapping of the Army, are required to submit the data to the State Council for approval.

The state has established a nationwide unified geodetic coordinate system, plane coordinate system, vertical system, geocentric coordinate system and gravimetric system, and defined the classification of orders and classes and precision of the national geodetic survey, as well as the national basic scale map series and their basic precision. Specific norms and requirements are formulated by the administrative department for surveying and mapping under the State Council in conjunction with the relevant departments under the State Council and the competent department for surveying and mapping of the Army. Where it is really necessary to adopt and does not endanger state security, the matter is subject to approval by the administrative department for surveying and mapping under the State Council and the competent department of survey and mapping of the Army.

Where it becomes really necessary to establish relatively independent plane coordinate systems for major construction projects of large cities and for the state to meet the needs of construction, urban planning and scientific research, the matter is subject to approval by the administrative department for surveying and mapping under the State Council; and where it is really necessary to establish relatively independent plane coordinate systems for other purposes, the matter is subject to approval by the administrative department for surveying and mapping under the people's government of a province, autonomous region, or municipality directly under the Central Government. The relatively independent plane coordinate systems to be established will be connected with the national coordinate system.

Surveying and mapping of boundaries: international and provincial

The national boundaries of the People's Republic of China must be surveyed and mapped in accordance with the boundary treaties or agreements concluded between the People's Republic of China and its 14 neighbouring states. Standard sample maps defining the international boundaries of the People's Republic of China are drafted by the Ministry of Foreign Affairs and the administrative department for surveying and mapping under the State Council and submitted to the State Council for approval and publication.

The boundary lines between the administrative areas are surveyed and mapped in accordance with the relevant regulations of the State Council. Standard maps delineating the administrative boundaries between provinces, autonomous regions and municipalities directly under the Central Government, and between autonomous prefectures, counties, autonomous counties and cities, must be drafted by the Department for Civil Affairs under the State Council and the administrative department for surveying and mapping under the State Council and submitted to the State Council for approval and publication.

The administrative department for surveying and mapping under the State Council, in conjunction with the department for land administration under the State Council, draws up plans for cadastral surveying and mapping nationwide. The administrative department for surveying and mapping under the local people's governments at or above the county level, in conjunction with the departments for land administration at the corresponding level, draws up plans for cadastral surveying and mapping of their own administrative areas. The administrative department for surveying and mapping under the local people's governments at or above the county level, in accordance with the plans for cadastral surveying and mapping, takes charge of the administration of cadastral surveying and mapping.

Summary

China's vast landscape is graced by its many rivers. Indeed, the country's paper currency carries images of some of these landscapes of gorges, rivers and mountains. The waters of the rivers are vital for the daily living conditions of over 1.3 billion inhabitants; for major construction projects; and, most importantly, for the environment and sustainable development of the nation. For these reasons alone, every care must be taken to ensure that antipollution legislation and administration of the river systems are enforced and supervised. Correct procedures for the surveying and mapping of the country are contained in Law No. 75 of 2002, as analyzed above.

China is both an upstream and downstream state in terms of the river systems that flow through the country; thus, management in all aspects of the rivers, especially transboundary watercourses, can be a true measure of good neighbourliness, friendship and cooperation. These are acknowledged in many of the treaties and arrangements that the government of China has in place with its 14 land neighbours.

Acknowledgements

The author thanks the referees for their valued critical comments that enhanced this submission; thanks also to professor Patricia Wouters and the members of CIWL team at Xiamen University who extended a warm welcome and embrace during the conference in May 2014.

References

Amer, R. (2013). Sino-Vietnamese border disputes. In B. A. Elleman, S. Kotkin and C. Schofield (Eds.), *Beijing's power and China's borders* (p. 300). Armonk, NY: M.E. Sharpe.

Armstrong, C. K. (2013). Sino-Korean border relations. In B. A. Elleman et al. (Eds.), *Beijing's power and China's borders* (p. 111). Armonk, NY: ME Sharpe.

Behr, W. (2007). Placed into the right position—Etymological notes in *Tu* and Congeners. In F. Bray, V. Dorofeeva-Lichtmann and G. Métailié (Eds.), *Graphics and text in the production of technical knowledge in China: The warp and the weft* (p. 113). Leiden: Brill.

Elleman, B., Kotkin, S., & Schofield, C. (Eds.). (2013). *Beijing's power and China's borders.* Armonk, NY: M.E. Sharpe.

Chellaney, B. (2013). India's intractable border dispute with China. In B. A. Elleman et al. (Eds.), *Beijing's Power and China's Borders* (p. 54). Armonk, NY: ME Sharpe.

Confucius. (1979). *The Analects.* A Book of Quotations of Confucius., (D.C. Lau, Compiled and Trans.). London: Penguin Classics.

Fravel, M., Taylor. (2008). *Strong borders secure nations: Cooperation and conflict in China's territorial disputes.* Princeton, NJ: Princeton University Press.

Government of China. (2002). Surveying and Mapping Law of the People's Republic of China, (Order No.) 75 of 2002.

Hsu, H. M. A. (2009). Structured perceptions of real and imagined landscapes in early China. In K. A. Raaflaub, R. J. A. Talbert, (Eds.), *Geography and ethnography: Perceptions of the world in pre-modern societies* (pp. 44–45). New York: John Wiley & Sons.

Leb, C. (2013). *Cooperation in the law of trans-boundary water resources.* Cambridge: Cambridge University Press.

Needham, J. (1956). *Science and civilisation, mathematics and the sciences of the heavens and the earth.* Cambridge: University of Cambridge Press. Volume 3. p. 350 and Plate LXXXI.

Pinilla, G. D. (2004). Border Disputes between China and North Korea. *China Perspectives.* [Online] 52, 25 March to 24 April 2004. Retrieved from http://www.cefc.com.hk/china-perspectives/issues/last-issue/

Prescott, J. R. V. (1975). *Map of mainland Asia by treaty.* Melbourne: University of Melbourne Press.

Smith, P. J. (2013). Bhutan-China Border Disputes. In Elleman, B. A. et al. (Eds.), *Beijing's Power and China's Borders* (p. 32). Armonk, NY: M.E. Sharpe.

Townsend-Gault, I. (2013). The China-Laos Boundary: Lan Xang meets the Middle Kingdom. In B. A. Elleman and et al. (Eds.), *Beijing's Power and China's Borders* (p. 147). Armonk, NY: M.E. Sharpe.

Yee, C. D. K. (1987). Chinese maps in political culture. In J. B. Harley and D. Woodward (Eds.), *The history of cartography* (pp. 71–95). Chicago: University of Chicago Press.

Yi, X., & Liu, X. (2010). The calligraphy and printing cultural heritage of Gansu—the development of the engraved printing process and papermaking: An archaeological approach. In S. M. Allen, L. Zuzao, C. Xiaolan (Eds.), *The history and cultural heritage of Chinese calligraphy, printing and library work*. Munich: De Gruyter Saur.

Appendix

Topographical maps are based on topographical surveys. Performed at large scales, these surveys are called 'topographical' in the old sense of topography: showing a variety of elevations and landforms. The first multi-sheet topographic map series of an entire country, the *Carte Géométrique de la France*, was completed in 1789. The Great Trigonometric Survey of India, commenced by the East India Company in 1802 and taken over by the British Raj (the administration in India) after 1857, was notable as a successful effort on a larger scale and for accurately determining the heights of Himalayan peaks from viewpoints over 100 miles away. During the next 100 years, topographical mapping of a nation generally came under the jurisdiction of a national mapping agency or was the sole function of the military. In India, for example, the large-scale topographical maps produced by them were of high quality in information and cartographic aesthetics. They were printed in colour on linen-backed paper. The scales, as a representative fraction, included the 1:253,440 (one inch to four miles) series, the 1:63,360 (one inch to one mile) series, 1:31,180 and 1:15,590. These have been replaced with metrification values of 1:25,000, 1:50,000, 1:250,000 and other smaller-scale series.

By the 1980s, centralized printing of standardized topographic maps began to be eroded by establishment of databases of coordinates that could be used on computers by moderately skilled end users to view or print maps with arbitrary contents, coverage and scale. By 2014, the standard phrases are 'spatial data', 'geographical information systems', 'geomatics', 'satellite-derived imagery', and 'global positioning systems'. Topographic maps have multiple uses in the present day: any type of geographical planning or large-scale architecture; earth sciences and many other geographic disciplines; mining and other earth-based endeavours; civil engineering; and recreational uses such as hiking and orienteering. Select examples of regional states' practices in topographical mapping, in the context of this paper, are offered below.

China

The State Bureau of Surveying and Cartography compiles topographic maps at 1:25,000 and 1:50,000 scales. It is reported that these maps are accurate and attractively printed in seven colours, and that successive editions show progressive improvement in accuracy. These large-scale maps are the basis for maps at smaller scales. Maps at scales of 1:4,000,000 or smaller are exported by the Cartographic Publishing House, Beijing, while larger-scale maps are restricted as state secrets, their publishing prohibited by legislation, except for Hong Kong and Macau. China's topographic maps follow the international standard map sheet format with 1:100,000 maps spanning 30 minutes longitude by 20 minutes latitude.

India

The Survey of India is responsible for all topographical control, surveys and mapping of India. The Survey Department inherited the medium- and large-scale series produced from 1882 to 1930s by British-administered military survey and mapping units. The series also included maps of Baluchistan, Burma, the Malay Peninsula and Singapore.

Nepal

From 1992 to 2002 a definitive series of large-scale topographic maps were surveyed and published through a joint project of the Survey Department of the government of Nepal and Finland's Ministry for Foreign Affairs, contracting through the consulting firm FinnMap. The Japan International Cooperation Agency replaced FinnMap in the Lumbini Zone. Topographic sheets at 1:25,000 scale covering 7.5 minutes (of arc value) latitude and longitude mapped the densely populated Terai and Middle Mountain regions. Less populated high mountain regions are on 15-minute sheets at 1:50,000. Scanned images of some of these can be accessed by the general public.

Pakistan

The responsibility for topographic mapping and aerial photography lies with the Surveyor General of Pakistan (SGP). Established in 1947, the Survey of Pakistan (SOP) is based in Rawalpindi, with a number of regional offices distributed at urban centres throughout Pakistan. The SOP is a civil organization which, for security reasons, is headed by a Surveyor General and works under the strict control of Army General Headquarters. All departments which require topographic maps make their request to the SGP and many are permanently registered with it for mapping and aerial photograph procurement. The SOP performs these functions under the auspices of the Ministry of Defence. Organizationally, the SOP is overseen by the Surveyor General (SG), who is a direct military appointee and a senior uniformed officer. The SG reports directly to the Secretary of Defence. These departments are divided into Regional Directorates for Topographic Mapping, including the northern region (centred in Peshawar), the eastern region (Lahore), the western region (Quetta), and the southern region (Karachi). Responsibilities for field surveys and the maintenance/update of topographic maps are subdivided according to these geographic areas.

Russia

Detailed, accurate topographic maps of Russia (and the former Soviet Union) have long been a military priority. They are currently produced by the Military-Topographic Service of the armed forces of the Russian Federation. Military topographic mapping departments held other titles in the Russian Empire since 1793 and in the Soviet Union, where these maps also came to be used for internal control and economic development. When Germany invaded in 1941, detailed maps from the USSR's western borders to the Volga River became an urgent need; the task was accomplished in less than a year. After the war, the entire Soviet Union was mapped at scales down to 1:25,000—even 1:10,000 for the agriculturally productive fraction. The rest of the world, except Antarctica, is believed to have been mapped by the Russians at scales down to 1:200,000, with regions of special interest down to 1:50,000 and many urban areas to 1:10,000. In all, there may have been over a million map sheets of high quality and detail. Soviet maps were also notable for their consistent global indexing system.

Thailand

Thailand set up a mapping database worth USD 4.7 billion using a 1:4000 scale in a bid to centralize the use of standardized maps in government agencies. The system supports key projects such as the water management scheme and large-scale infrastructure project. In addition, the system improves efforts in precision farming, forest and land management. It was essential to standardize map usage in government agencies, which had used a variety of maps at different scales to plan development projects. Now that the system is in place, government agencies will be equipped with a mapping database which enables them to leverage aerial photographs and high-resolution satellite images to efficiently and effectively plan long-term development projects.

International Map Series

Special mention is made here of the international map series which have topographical information. Many agencies and national and university libraries have taken the initiative to scan the series and make them available online. Many of the series may be reproduced and are free of copyright, provided that due acknowledgment of the source is given. To the knowledge of the present author, some of the series available that could be consulted for their topographical detail (in the absence of, or inaccessibility of, nationally produced maps) include:

- Operational Navigation Chart series, scale 1:1,000,000
- Tactical Pilotage Chart series, scale 1:500,000
- American Map Series, scale 1:250,000
- American Map Series at scales of 1:100,000 and 1:50,000, generally available for Indochina

Comparable series were produced by the French national map agency, IGN.

China's southbound transboundary river basins: a case of asymmetry

Mirja Kattelus, Matti Kummu, Marko Keskinen, Aura Salmivaara and Olli Varis

An overview is presented of the contemporary societal and environmental development situation in the six major transboundary river basins that drain south from China: the Red River, Mekong, Salween, Irrawaddy, Ganges-Brahmaputra-Meghna and Indus. The overall societal and environmental vulnerability of the basins is assessed using multidimensional river basin vulnerability analysis. The analysis shows that while China has a fairly low level of vulnerability in these basins, its downstream influence is substantial. This setting offers a plethora of opportunities for transboundary cooperation and calls for a high level of responsibility from the upstream riparian countries.

Introduction

The six transboundary rivers originating in South-West China – Red River, Mekong, Salween, Irrawaddy, Ganges-Brahmaputra-Meghna (GBM) and Indus – are amongst the largest rivers in the world. The river basins produce ample discharge to the oceans and provide water and livelihoods for over one billion people. The basins also have outstanding diversity in terms of ecology, biodiversity, environment, culture, ethnicity and governance (e.g. Biodiversity and Nature Conservation Association, 2009; Brown, Tullos, Tilt, Magee, & Wolf, 2009; Grumbine & Pandit, 2013; Grumbine & Xu, 2011; Varis, Kummu, & Salmivaara, 2012; Ziv, Baran, Nam, Rodriguez-Iturbe, & Levin, 2012).

The 6 basins are shared by 12 riparian countries: China, Vietnam, Laos, Myanmar, Cambodia, Thailand, Bangladesh, Bhutan, India, Nepal, Pakistan and Afghanistan (Table 1). They have large and rapidly growing populations, as well as a massive pace of urbanization, particularly in South Asia (Varis et al., 2012; World Bank, 2014). The 12 riparian countries account for 46% of the world's total population and 35% of the world's urban population, as well as 39% of the total population growth and 50% of the total urban population growth during 2005–2010 (United Nations, 2014). In South and South-East Asia over 90% of the population either live in poverty (less than \$2.50 a day) or are vulnerable to it (less than \$10 a day) (World Bank, 2013). The riparian countries account for over half of the global occurrences of malnutrition (FAO, WFP, & IFAD, 2012). There is thus a combined effect of high concentration of people alongside a high level of poverty, especially in the Indus and GBM basins (Sharma et al., 2010), as well as of malnutrition and uncontrolled urbanization (FAO, 2013).

High population and the increasing water use that may follow rising urbanization are fundamental reasons contributing to growing pressure on water resources (Varis, 2005).

Table 1. Population (in millions) in the investigated six transboundary river basins, across countries. Data sources: Grübler et al. (2007) and Klein Goldewijk et al. (2010).

	Afghanistan	Bangladesh	Bhutan	Cambodia	China	India	Laos	Myanmar	Nepal	Pakistan	Thailand	Vietnam	Total
GBM		137.4	0.9		1.7	536.0		0.0	32.9				709.0
Indus	10.2				0.1	87.4			0.0	151.1			248.8
Irrawaddy					1.9	3.0		33.3					38.1
Mekong				14.3	6.5		6.2	0.7			22.5	20.7	70.9
Red river					11.1		0.0					19.0	30.1
Salween					3.5			4.6			0.2		8.3
Total	10.2	137.4	0.9	14.3	24.7	626.4	6.2	38.7	32.9	151.1	22.7	39.7	1105.2

Furthermore, climate change is expected to increase the stress in large areas and particularly in the river deltas through sea level rise, which threatens the GBM, Irrawaddy and Mekong deltas (IPCC, 2007). According to the IPCC (2012), extreme events, including floods and droughts, are projected to increase in the region during the twenty-first century as a result of climate change. The target region is particularly prone to natural disasters, having faced the highest number of natural disasters in the world, with an increasing trend between 1981 and 2010 (World Bank, 2013).

Meanwhile, a significant and growing share of the world's economic growth takes place in this region (Varis et al., 2012). The recent two decades have witnessed major economic growth and development, which has led and is still leading to massive changes and development pressures in the basins. The region is facing serious challenges related to the over-exploitation of water resources combined with the impacts of various human activities, such as deforestation, agricultural land expansion, industrialization, and intensification and construction of water-related infrastructure (Lu, 2004). Gigantic hydropower construction in the Lancang-Mekong (Grumbine & Xu, 2011; Keskinen, Kummu, Käkönen, & Varis, 2012; Magee, 2006; Tilt, Braun, & He, 2009), as well as planned dams across the Indian Himalaya and upstream Brahmaputra (Grumbine & Pandit, 2013; Rahaman, 2012), Irrawaddy (Kattelus, Rahaman, & Varis, 2014) and Salween (Magee & Kelley, 2009; Magee, 2006) river basins are examples of this. The current plans in the Mekong basin, for instance, aim at a 10-fold increase in the active storage capacity of the reservoirs (Kummu, Lu, Wang, & Varis, 2010), while all the proposed Indian dams together would create the highest density of dams in the world in the targeted area of the Indian Himalaya (Grumbine & Pandit, 2013).

The soaring water resource exploitation imposes major stress on the region's riverine ecosystems (Grumbine & Pandit, 2013; Grumbine & Xu, 2011). These ecosystem changes and threats to biodiversity will have implications for traditional livelihoods and even food security in the six study basins. Meanwhile, the increasing water resource exploitation creates geopolitical tensions associated with dams, ranging from the unequal and unfair distribution of costs and benefits, to public participation and governance issues (Moore, Dore, & Gyawali, 2010; Tullos, Tilt, & Liermann, 2009). For instance, tension is increasing in the Mekong region between the proponents of hydropower development and those who rely on the aquatic system (Kirby et al., 2010). The governance systems of the 12 countries under study are in too precarious a state, for instance as described by the State Fragility Index or Corruption Perception Index, to respond to the vast and rapidly changing challenges that these countries and river basins are facing (Varis et al., 2012). At the same time, appropriate institutional set-up for river-basin-level cooperation is lacking in the majority of the transboundary basins.

The various water-related challenges outlined above require broad, multidisciplinary knowledge, as well as extension of the water resources discussion beyond the conventional water-sector-centred discourse (Biswas & Tortajada, 2010). This article assesses these multifaceted challenges by investigating the water-related vulnerabilities of the six transboundary river basins using publicly available global spatial data-sets including data on nature and environment (such as climate, anthropogenic impacts on environment, and water stress), demography, and political stability, as well as social and economic development. The methodology established in Varis et al. (2012) is used to assess basin-level socio-economic and environmental vulnerability.

Geographically, we focus on the six major southbound transboundary river basins that originate from South-West China, as well as on the 12 countries that share these basins. This article uses an intercomparable approach in assessing vulnerability by river basin and by country (or province) to summarize their water-related economic, societal and environmental characteristics and related vulnerabilities. We also discuss how such vulnerabilities link with the development pressures as well as transboundary dynamics in these river basins.

Study area: six transboundary river basins in South and South-East Asia

The study area includes the 6 basins, Red River, Mekong, Salween, Irrawaddy, GBM and Indus, and their 12 riparian countries of China, Vietnam, Laos, Myanmar, Cambodia, Thailand, Bangladesh, Bhutan, India, Nepal, Pakistan and Afghanistan. China plays a unique role as the upstream country in all of the basins (Figure 1). The six river basins have their unique characteristics, combined with different sets of challenges. The population is highly unevenly distributed, with the Indus and GBM being two of the most densely populated areas in the world (Sharma et al., 2010). These two basins have a population of 240 million and 647 million, respectively, while the Salween only has 8 million people (Varis, Kummu, Lehr, & Shen, 2014). It is also worth pointing out that only a miniscule proportion (around 2%) of China's total population of over 1.3 billion is within these six transboundary river basins (Varis et al., 2014).

The Indus and GBM have diverse agro-climatic, social and economic conditions in four countries (India, Pakistan, Nepal and Bangladesh), making them extremely complex (Sharma et al., 2010). The GBM river system is the third-largest freshwater outlet to the world's oceans, exceeded only by the Amazon and Congo river systems (Chowdhury & Ward, 2004). The GBM and Indus basins also contain key agricultural areas, with high irrigation levels that provide regional food security (Biemans et al., 2013). However, the

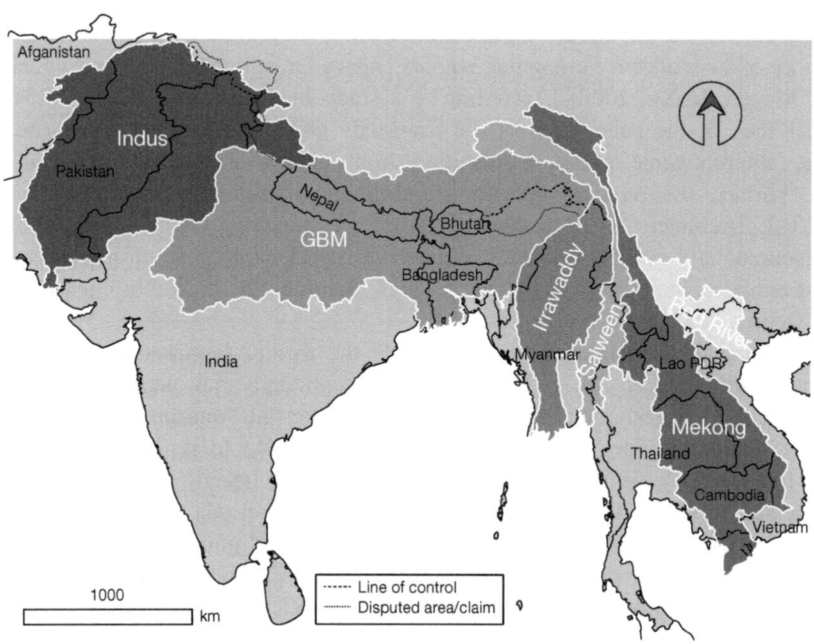

Figure 1. Study area.

Brahmaputra part of the GBM basin has little agriculture, because a large part of it is at high elevation and is unsuitable for crops (Biemans et al., 2013).

Water scarcity is a major concern for agricultural economies such as Pakistan and India that have developed their dry-climate agriculture through intensive use of irrigation (Cook, Fisher, Tiemann, & Vidal, 2011). The groundwater is being exploited unsustainably and at an increasing rate (Sharma et al., 2010), resulting in falling water tables, especially in Rajasthan, Haryana and Punjab in India (Rodell, Velicogna, & Famiglietti, 2009). The Indus basin is now practically closed, with near-zero environmental flows in most years (Sharma et al., 2010). The Ganges is, in general, only moderately water-scarce but contains areas of extreme or increasing scarcity (Cook et al., 2011). Thus, pressures to improve productivity, combined with expanding demands from non-agricultural sectors, create a problematic situation (Cook et al., 2011). Meanwhile, Bangladesh, being the downstream country of both the Ganges and the Brahmaputra basins, without any control of upstream water flows, suffers from flooding during monsoon months and water shortage during summer months (Rahaman, 2012). A particular bottleneck to political collaboration in both the Indus and the GBM is the ongoing border disputes between China and India.

The two largest rivers in Myanmar are the Irrawaddy and the Salween. Both rivers have their headwaters in mountainous areas, broaden into a vast deltaic area, and then empty into the Andaman Sea (Kattelus et al., 2014). The Irrawaddy basin covers the central plains and the vast southern delta area, which represent the most important croplands and the most intensively populated areas in the country (Salmivaara, Kummu, Keskinen, & Varis, 2013). Ninety per cent of the total drainage area is situated in Myanmar, covering about three-fifths of Myanmar's surface area, with a population of around 37.2 million (Varis et al., 2012).

The Salween is shared by China (52.4%), Myanmar (43.9%) and Thailand (3.7%) (Affeltranger, 2008). With its 5516 m head fall within the Chinese part of the basin, it provides very little flat riverine land for cultivation but has tremendous hydropower potential (Feng, He, & Li, 2010; Magee & Kelley, 2009). Despite the difficult conditions, the majority of the workforce along the river are engaged in agriculture for their livelihood (Brown, Magee, & Xu, 2008). According to a study by Salmivaara et al. (2013), the ecological state of the delta areas of the Irrawaddy and the Salween is threatened by aggregate anthropogenic pressures resulting from population growth and land use changes. Further, the ongoing economic and political reforms (Asian Development Bank, 2012; Bremmer, 2012; Robinson, 2012) are creating numerous pressures on water resources and competitive uses through an accelerating rate of investments in hydropower and agriculture (Kattelus et al., 2014; Schmidt, 2012; Webb, Phelps, Friess, Rao, & Ziegler, 2012).

The Mekong River was until recently one of the world's largest rivers still mostly undammed (Grumbine & Xu, 2011). It enjoys exceptionally rich aquatic biodiversity, including the third-largest inland fishery in the world (after the Amazon and Bangladesh), with many highly migratory fish populations (Grumbine & Xu, 2011; Kirby et al., 2010). Peoples' livelihoods – fishing, farming and grazing – rely heavily on the river and its environmental services (Grumbine & Xu, 2011; Kirby et al., 2010; Orr, Pittock, Chapagain, & Dumaresq, 2012). Both economic wealth and population, particularly in the urban centres, have grown remarkably, accompanied by a growing demand for electricity, first and foremost in China, Thailand, and Vietnam (Kuenzer et al., 2013). However, there are large variations within the basin in terms of socio-economic development. Cambodia and Lao PDR are among the poorest in Asia, while Thailand and

Vietnam are more industrial (Cook et al., 2011; Kirby et al., 2010). The basin's population is about 70 million, most of whom are rural poor with livelihoods directly dependent on the availability of water for the production of food (Kirby et al., 2010).

The 1000 km Red River flows from north-east to south-west, from the south-eastern corner of Tibet to the Gulf of Tonkin and the South China Sea (Searle, 2006). A total of 50.3% of the Red River basin is located in Vietnam, 48.8% in China, and 0.9% in Laos (Quynh et al., 2005), with a total population of 30 million (Varis et al., 2014). Land use differs markedly between the upstream and the delta of the Red River. It has a large alluvial delta in Vietnam (Berg et al., 2007), where cultivated land (mainly rice fields) holds the largest share of the land use (63%), while forest occupies the largest part of the upstream (54%) (Quynh et al., 2005).

Major water infrastructure development

There is renewed enthusiasm for hydropower development in the region. This is driven by energy demand growth resulting from rapid economic development across South and South-East Asia and China, as well as increased interest in economic integration. As a result, numerous plans are underway to develop the vast hydropower potential of a number of these transboundary rivers.

Because of the (so far) limited exploitation of the Mekong's hydropower potential, the Mekong countries' governments are fostering large-scale hydropower projects within their territories (Kuenzer et al., 2013). Eleven mainstream dams have been proposed in Lao PDR and Cambodia, while China alone has planned a cascade of eight mainstream dams, of which six are already built, in the upper Mekong basin, in addition to several more further upstream (Grumbine & Xu, 2011; Kummu et al., 2010; Räsänen, Koponen, Lauri, & Kummu, 2012). If all the development plans are implemented, the total number of hydropower dams in the Mekong basin will be at least 165, though it is uncertain how many will eventually be realized (Räsänen et al., 2012). Based on several studies conducted in the Mekong basin, hydropower development can be particularly harmful to fish, by trapping nutrient-rich sediments (Arias et al., 2014; Kummu & Varis, 2007; Kummu et al., 2010), flattening out the flood pulse (Arias et al., 2014; Kummu & Sarkkula, 2008; Lauri et al., 2012; Räsänen et al., 2012) and blocking the migration routes of migratory fish (Ziv et al., 2012). In addition, there are multifaceted social impacts, including displacement, insufficient compensation, effects on rural economies (such as decline of agricultural productivity), and cultural changes (Hall & Bouapao, 2010; Tilt et al., 2009; Wang, Lassoie, Dong, & Morreale, 2013).

The next river basin to the west is the Salween, where a series of dams is under planning in Yunnan Province in China. If completed, the 13-dam cascade would have greater power-generating potential than the Three Gorges Dam (Brown et al., 2008). Chinese companies have also been interested in investing heavily in the vast hydropower potential in the Irrawaddy and Salween Rivers (Kattelus et al., 2014; Sovacool, 2013). According to McDonald, Bosshard, and Brewer (2009) and Urban, Nordensvärd, Khatri, and Wang (2013b), plans for dams on the Salween and Irrawaddy Rivers with financial and construction support from China alone add up to almost 40,000 MW. The estimates of future capacity vary, as it remains uncertain also in the Salween which of the planned dams will eventually be commissioned (Kattelus et al., 2014).

In the GBM and Indus, massive plans are underway in Pakistan, India, Nepal and Bhutan to build several hundred dams in the region (Dharmadikhary, 2008; Rahaman,

2012). India aims to construct 292 dams throughout the Indian Himalaya, doubling their current hydropower capacity (Grumbine & Pandit, 2013). Almost half the total hydropower potential of India lies in the Brahmaputra basin (Rahaman, 2012). Furthermore, Pakistan, Bhutan and Nepal, with numerous undeveloped hydropower sites, are building or planning to build multiple dams, a minimum of 129 projects all together (Grumbine & Pandit, 2013). Meanwhile, China is also planning to develop its hydropower potential in the northern stem of the Brahmaputra, although data on these plans are scarcely available (Bawa et al., 2010; Dharmadikhary, 2008; Rahaman, 2012). The hydropower potential of the main stem and five main tributaries of the Yarlung Zangbo River (this is the Tibetan name for the Chinese part of the Brahmaputra) fosters the second-highest potential in China, after the Yangtze River basin (Rahaman, 2012). A dam proposed there would be the world's largest hydropower station, with an installed capacity of 38,000 MW, which is over twice the generation capacity of the Three Gorges Hydropower Station (Rahaman, 2012). India strongly opposes these upstream projects because its own hydropower and water transfer projects to alleviate water scarcity in the Ganges would be adversely affected (Rahaman, 2012). Bangladesh is downstream of all the Himalayan rivers of Nepal and India, but has little say in the dam-building programme that is planned in these areas (Dharmadikhary, 2008).

China is playing a major part in the dam building in the study basins, having plans for 750 projects in Tibet alone (Bawa et al., 2010). In addition to the upstream dams, China is seeking to electrify South-East Asia by investing in 85 dams in Lao PDR, Myanmar and Cambodia (Grumbine & Xu, 2009). Further, when scaling up beyond the region, China is financing some 200 dams in some 50 countries across four continents (Grumbine & Xu, 2009). Because many of the hydropower deals are sensitive, information is confidential, and many of the dams remain unresearched or under-researched, making it particularly difficult to assess the exact impacts of the projects (Urban, Mohan, & Cook, 2013a).

Needless to say, these proposed projects across South and South-East Asia will have severe environmental and social impacts, causing drastic changes to the economies of the riparian countries and potentially exacerbating regional inequality. Furthermore, considering the already existing water-related vulnerabilities within the basins, they will add new pressures and create new geopolitical and transboundary management challenges. Considering the current institutional and governance status of the riparian countries, it will be difficult to respond to these challenges in a way that ensures equitable sharing of benefits and costs. These implications will be discussed further from the perspective of upstream–downstream dynamics in the Discussion section.

Materials and methods

A spatial vulnerability approach was used to analyze the six transboundary river basins with regard to the multifaceted challenges they are currently exposed to. Because river basins are combined with administrative units, a simultaneous operation at two geographical divisions is essential, and the analysis was thus done both by river basin and by administrative unit (jurisdiction). Both of these divisions are crucial in water management – the river-basin level in the management of water with regard to allocation of water in time and space, and the national, provincial, or other jurisdictional level in policy making and administration.

Materials

Consequently, two types of data were used in this study: spatially gridded data for environment and water resources; and administrative-scale data for policy, governance, and macroeconomy (see Table 2). Detailed documentation of data and their selection rationale can be found in Varis et al. (2012) and the modifications made can be found in Varis et al. (2014). These modifications were made to incorporate subnational data for China in the analysis and to obtain the most recent data. For the present article, state-level data for India were used to replace national averages for several of the components used to calculate the Political Instability Index (PSI) (EIU, 2011). In addition, minor changes were made to the calculation procedure of the PSI. While this change has a relatively small impact on the overall vulnerability maps, it should be taken into consideration if comparing our results to those of Varis et al. (2014). Furthermore, the national-level data for India were replaced by subnational-level values for the Multidimensional Poverty Index (MPI) (Alkire, Conconi, & Seth, 2014). For details on the data and their sources, see Table 2.

River-Basin Vulnerability Index approach

The river-basin vulnerability (RBV) analysis approach of Varis et al. (2012) was used in this study. The approach was developed within the context of analysis of the vulnerability of 10 major river basins in the Asia-Pacific, and used thereafter to investigate the vulnerability of Central Asian river basins (Varis & Kummu, 2012) and Chinese river basins (Varis et al., 2014). This approach was used because it was tailored especially for purposes such as this one. Moreover, comparable river-basin vulnerability results are available from the above-mentioned studies covering an area with a population well over one-quarter of humankind. The approach is only briefly introduced here; details are given by Varis et al. (2012).

The RBV approach is designed to be applicable to analyzing river basins as highly complex societal-environmental systems. Six sources of vulnerability are included: governance, economy, social issues, environment, hazards, and water stress. These were selected to maximally cover the common key dimensions of the general context where water resources management is being performed. They are also aligned with the philosophy of sustainable development and consequently with common policy frameworks in the water sector, most importantly integrated water resources management (Agarwal et al., 2000; Rahaman & Varis, 2005).

The results produced with the RBV approach are basically comparable with any other geographical area in the world. The approach relies on already published composite indicators, because they have been thoroughly investigated globally and are easily available for most locations, allowing comparison.

The implementation of the approach consisted of the following steps.

(1) A geospatial mesh of administrative regions and basins was created using ArcGIS software (as in the maps in Figure 2).
(2) Using the gridded population data for the year 2010 – which was a combination of two data-sets, from Grübler et al. (2007) and Klein Goldewijk, Beusen, and Janssen (2010) – the total population and its proportion of the total basin population were calculated for each mesh unit (see Table 2).
(3) Each indicator value of the administrative-scale data was weighted by the population of the mesh unit.

Table 2. Data used for six vulnerability indices, including data description, resolution, sources, and acronym (used in the text and equations for calculating the vulnerability index).

Source of vulnerability (*indicator*)	Data description	Resolution	Source	Acronym
Data by administrative region				
Governance (Political Instability Index)	The Political Instability Index consists of 12 indicators for underlying vulnerability and three for economic distress (EIU, 2011). Data were not directly available for provinces in China or states in India. To obtain subnational estimates, we used national averages in the case of China for all indicators except inequality, ethnic fragmentation and income per head, from China Statistical Yearbook (2011) and in case of India for all indicators except inequality (from India Census, 2011), growth in income (India Census, 2011), unemployment (Labour Bureau, 2012) and income per head (Ministry of Statistics and Programme Implementation of India, 2014).	Country/province (China)/state (India)	EIU (2011); China Statistical Yearbook (2011); India Census (2011); Labour Bureau (2012); Ministry of Statistics and Programme Implementation of India (2014)	PSI
Economy (GNI (PPP) per capita)	Gross national income per capita (adjusted with purchasing power parity to international dollars) National-level data from World Bank (2014); provincial-level data for China from China Statistical Yearbook (2011); state-level data for India from Ministry of Statistics and Programme Implementation of India (2014).	Country/province (China)/state (India)	World Bank (2014); China Statistical Yearbook (2011); Ministry of Statistics and Programme Implementation of India (2014)	GNI_{pop}
Social (Multidimensional Poverty Index)	The index (originally developed by Alkire and Santos, 2010) represents the nature and intensity of poverty at the individual level in education, health outcomes and standard of living. MPI data were not available for provinces. Provincial values were obtained by scaling China's national value by province-specific Human Development Index components of education and health (UNDP, 2010) as well as level of water services (urban and rural separately – China Statistical Yearbook, 2011). MPI was readily available for Indian states (Alkire et al., 2014).	Country/province (China)/state (India)	Alkire and Santos (2010); Alkire et al. (2014); UNDP (2010); China Statistical Yearbook (2011)	MPI

(*Continued*)

Table 2. (Continued).

Source of vulnerability (*indicator*)	Data description	Resolution	Source	Acronym
Spatially gridded data				
Environment (human footprint)	Global Human Footprint Dataset from the Last of Wild v 2 project covering human population pressure, human land use and human access.	Grid: 30 arcsec. (≈ 1 km × 1 km)	WCS/CIESIN (2005)	HF
Hazards (multihazards)	Natural disaster hotspots: Global Multihazard Frequency and Distribution, classified.	Grid: 2.5 arcmin. (≈ 5 km × 5 km)	Dilley et al. (2005)	MH
Water stress (water stress)	Net water demand divided by amount of renewable blue water resources.	Grid: 0.5 arc degree (≈ 50 km × 50 km)	Wada, van Beek, and Bierkens (2011); Wada, van Beek, Wanders, and Bierkens (2013)	WS

Table 3. Equations for calculating the indicators for the River Basin Vulnerability Index. The final index is calculated as the average of the six components shown in the table. (See Table 2 for the abbreviations and sources of the components.)

Component	Integrated index
Governance	PSI'_{pop}
Economy	$(1 - (\log(GNI_{pop}) - \log(GNI_{pop\ max}))/(\log(GNI_{pop\ min}) - \log(GNI_{pop\ max})))'$
Social	MPI'_{pop}
Environment	HF'
Hazards	MH'
Water scarcity	min(WS,1)

Note. The scaled value for each index is calculated as follows: $x' = (x - x_{min})/(x_{max} - x_{min})$, where x_{max} is the maximum 95% fractal of the global data-set in question, and x_{min} is the minimum 5% fractal of the same set. pop = population-weighted basin average value.

(4) The river-basin value was calculated by dividing the sum of the weighted indicator values by the total basin population. The spatially gridded data were aggregated directly to the basin areas.
(5) The data were scaled to between 0 and 1 based on the global distribution of each analyzed vulnerability component (Table 3). The lower and upper 5% were pinned to 0 and 1 to avoid a few outliers' having a dramatic impact on the scaling of all other data points. The exception was water stress, which was calculated as the ratio of net water demand and the available renewable water resources. If this ratio exceeds 1, the indicator is set to 1.
(6) The six components were combined to give the River-Basin Vulnerability Index as the average value over the components.

Results

This section presents the results from the vulnerability analyses. First, the indicator results are presented according to the six categories of vulnerability – governance, economy, social issues, environment, hazards, and water stress (Figure 2) – followed by the overall vulnerability results at basin and sub-basin levels (Figure 3). Secondly, the results are demonstrated from an upstream–downstream perspective (Figure 4) as well as through population-weighted proportions of vulnerability (Figures A1 and A2 in the appendix).

Indicator values and overall vulnerability

Governance is described by the PSI (EIU, 2011). Higher governance-related vulnerability occurs in Pakistan, Bangladesh, Myanmar and Cambodia; India, China and Vietnam show lower values (Figure 2a). The range relative to the global values (with global minimum being 0 and global maximum 1) is 0.43–0.80, demonstrating that some parts of the study area have very high vulnerability in terms of political stability. Interestingly, India does not show large variability within the country, despite its being a very heterogeneous area. Though China does not stand out in our results as highly vulnerable in terms of governance and economy, the Tibet Autonomous Region and Yunnan Province were identified by Varis et al. (2014) as two of the most vulnerable areas in terms of governance within China.

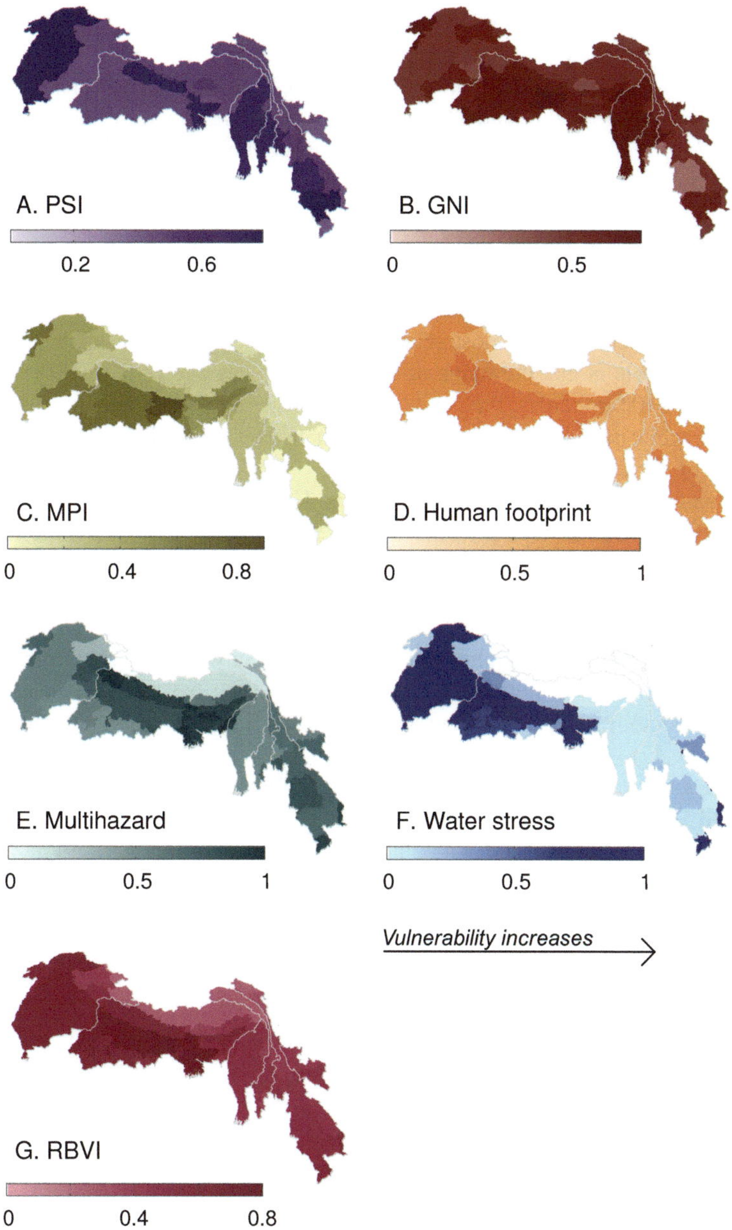

Figure 2. Scaled indicator values mapped for the six river basin systems and their administrative boundaries (countries/provinces/states). a. Governance (Political Instability Index). b. Economy (gross national income per capita adjusted with purchasing power parity). c. Social issues (Multidimensional Poverty Index). d. Environment (human footprint). e. Hazards (Multihazard Index). f. Water stress. g. General River Basin Vulnerability Index, derived as a combination of the six indicators. Higher values indicate higher vulnerability.

Economic vulnerability is described by gross national income (GNI), assuming that vulnerability is reduced with increasing wealth (Figure 2b). This indicator presents major variations across the region, with the highest vulnerabilities in South Asia and Myanmar. The GBM, Indus and Mekong all show a large range relative to the global values: 0.37–0.70, 0.36–0.66 and 0.29–0.66, respectively. In the GBM, both the most and the least economically vulnerable states are in India: Bihar and Delhi. In the Indus basin, the least vulnerable state is Chandigarh in India, and the most vulnerable is Afghanistan. In the Mekong, the least economically vulnerable country is Thailand, and the most vulnerable is Myanmar.

Social issues are described by the MPI (Alkire & Santos, 2010; Alkire et al., 2014). In this category, vulnerability is most severe in Bihar, India, and the least so in Thailand (Figure 2c). Notable is that the range of this indicator was one of the widest amongst the indicators: 0.01–0.90 across the study region. The largest range was in the GBM (0.10–0.90), while in the Mekong it was much narrower (0.01–0.40), demonstrating a lower level of social vulnerability.

Environmental vulnerability is indicated by Human Footprint Index (WCS/CIESIN 2005) (Figure 2d). The studied river basins vary widely in these respects; the values range relative to the global values between 0.11 and 1.00. The highest environmental vulnerability (1.00) occurs in Bangladesh and in various states in India, while the lowest human footprint is in the northern parts of the basins in China and Arunachal Pradesh in the northern Indian parts of the Brahmaputra basin. The Salween and the Irrawaddy basins stand out as having lower human footprints. The Mekong basin experiences high environmental vulnerability, with Thailand and Vietnam being the most vulnerable (0.93 and 0.90, respectively).

Multihazards in terms of environmental and climate hazards are described by the Multihazard Index, combining the total estimated impacts of droughts, floods, volcanoes, storms, earthquakes and landslides and taking into account frequency and hazard-specific mortalities (Dilley et al., 2005) (Figure 2e). In terms of these multiple hazards, the GBM and the Mekong have the highest vulnerability, with basin-level values of 0.66 and 0.68, respectively. In the GBM, the values range between 0.16 in the upper reaches of the basins and around 1 in the central and deltaic regions of the basin. Also in the Mekong basin, the upstream has the lowest vulnerability to multihazards, while downstream Vietnam has the highest, the values ranging between 0.07 and 0.91 throughout the basin. Generally lower vulnerability to multihazards is experienced by the Irrawaddy and the Salween rivers.

Water stress across the study region also ranges remarkably, i.e. between 0.00 and 1.00 – as wide as the global range (Figure 2f). Clearly, the highest level of water stress is experienced by the GBM and Indus rivers, with the vulnerability to water stress being the highest (1.0) in Bangladesh, India (states of Bihar, Chandigarh, Delhi, Haryana, Punjab, Rajasthan, Uttar Pradesh and West Bengal) and Pakistan. The lower reaches of the Mekong and Red River also experience some water stress. The lowest water stress occurs in the northern parts of the river basins and across the Irrawaddy and Salween Rivers.

Overall vulnerability, i.e. the River Basin Vulnerability Index, is highest in the GBM and the Indus, with global-adjusted values of 0.28–0.81 and 0.22–0.68, respectively. This is largely due to high population density and the resulting high human footprint and water stress, combined with higher poverty rates (Figure 2g). Generally lower overall vulnerability is experienced in the Irrawaddy, Salween, Mekong and Red River, despite lower affluence and higher political instabilities, especially in Myanmar. The Mekong River also shows large heterogeneity for all the different vulnerability sources. The northern parts of the basins demonstrate lower overall vulnerabilities and score generally lower in all the vulnerability sources.

Sub-basin vulnerability results

Figure 3a presents the results grouping the vulnerability into societal (PSI, GNI and MPI), environmental (human footprint and multihazards) and water-stress sources. Overall, as demonstrated in Figure 2, the Indus and GBM stand out as the most vulnerable river

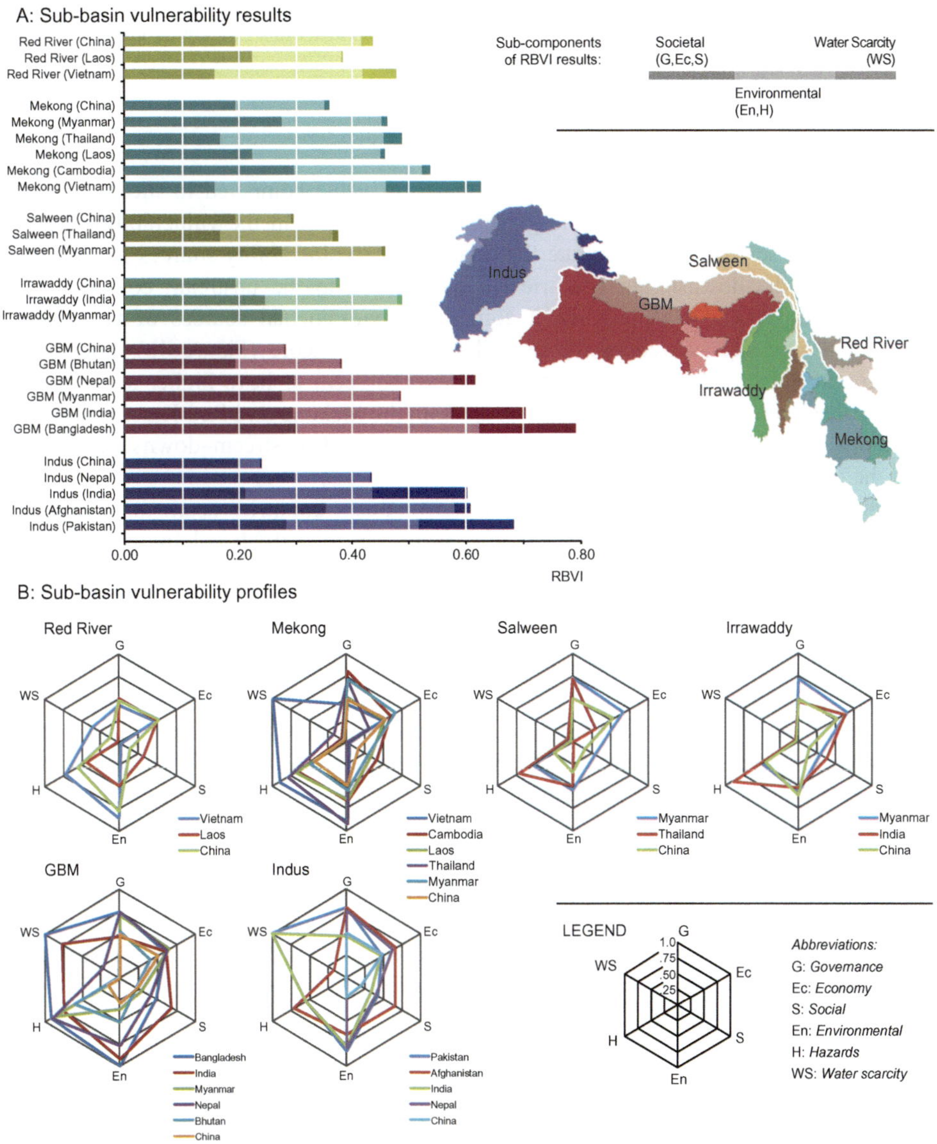

Figure 3. Vulnerability profiles of the river systems.

basins, while the Irrawaddy and Salween have the lowest vulnerability of the six studied basins. In the majority of the basins, the societal sources contribute to around half of the vulnerability score, and in the Irrawaddy and Salween, societal sources comprise the majority of the vulnerability.

When looking at the vulnerabilities according to the basin countries, some interesting differences emerge. For China, as the less populated parts of the basins, their vulnerability comes largely from societal sources, while in the more populated parts of the river basins the environmental vulnerability is also very significant. The river systems show a very interesting geographical pattern (Figure 3b), where Vietnam contributes largely to the vulnerability of the Red River and Mekong in terms of water stress and multihazards, while the other riparian countries are more vulnerable in terms of societal sources, though environmental sources and multihazards also contribute. In the most vulnerable basins, the Indus and GBM, the lower reaches of the rivers contribute to the majority of the vulnerability profiles in all the vulnerability groups.

Upstream–downstream dynamics

Figure 4 presents the results in terms of upstream value compared to the downstream countries' value to demonstrate the geographical dynamics in the vulnerability profiles. Generally, these results demonstrate that many of the vulnerability sources impose more pressure on the downstream countries in comparison to the upstream countries. This pattern is especially evident when it comes to environmental sources of vulnerability (Figure 4d, 4e, 4f) as well as the overall vulnerability score (Figure 4g). For societal sources of vulnerability, downstream countries also seem to have higher social vulnerability than their upstream neighbours, apart from the Mekong and Red River (Figure 4c). In terms of political stability and wealth, no clear overall upstream–downstream pattern can be detected (Figure 4a, 4b).

For political stability (Figure 4a), the downstream countries seem to have higher vulnerability in the Salween and Mekong basins. In the GBM and Indus, the lowest downstream country, Bangladesh or Pakistan, respectively, is the most vulnerable. For the Mekong and Red River no clear pattern is detectable. In economic terms there seems to be no clear pattern in the vulnerability scores, with the Mekong showing the most variance between the riparian countries (Figure 4b). In the Indus and GBM, social vulnerability in terms of poverty seems to increase towards the downstream reaches of the rivers; India especially is much more vulnerable in the GBM than its upstream co-riparians (Figure 4c). Only Bangladesh seems to perform slightly better than its upstream countries in terms of multidimensional poverty. For human footprint (Figure 4d), all the densely populated basins (Indus, GBM and Mekong) share a pattern: vulnerability is always higher in the downstream country than in the upstream countries, on average. The same dynamic prevails for multihazards (Figure 4e) and water stress (Figure 4f) in all the basins. For overall vulnerability (Figure 4g), the downstream country in all the basins is always more vulnerable than its upstream countries, on average. This is especially evident for densely populated India, Bangladesh and Pakistan.

Population-weighted proportions of vulnerability

Figures presented in the appendix provide an overall view of the six categories of vulnerability by presenting them with population-weighted proportions by river basin

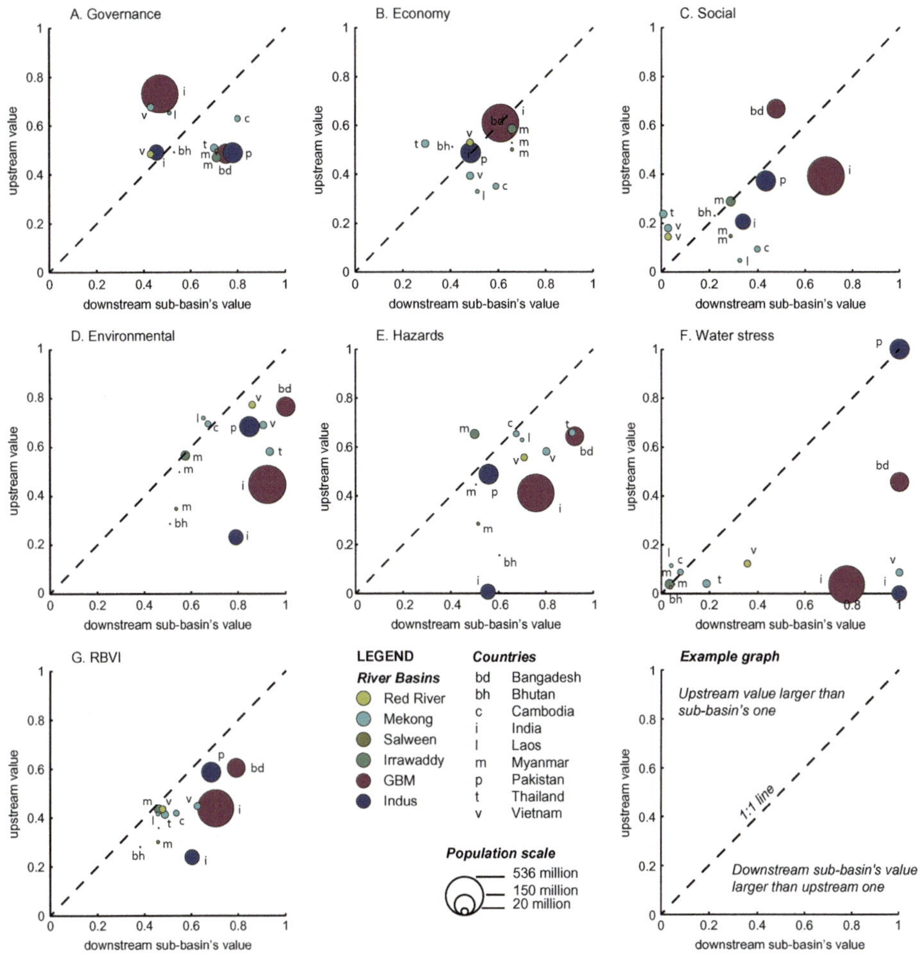

Figure 4. Upstream versus downstream vulnerability values. The *y* axis is the average value for the upstream countries within the basin in question; the *x* axis is the value for each downstream country separately. China is not presented because it is the most upstream country in all the basins. The colours indicate the different river basins, and circle size corresponds to population. a. Governance (Political Instability Index) b. Economy (gross national income per capita adjusted with purchasing power parity) c. Social issues (Multidimensional Poverty Index). d. Environment (human footprint). e. Hazards (Multihazard Index). f. Water stress. g. General River Basin Vulnerability Index, derived as a combination of the six indicators.

(Figure A1) and by country (Figure A2). As can be seen, the GBM and the Indus – and consequently India, Pakistan and Bangladesh – clearly dominate all six vulnerability indices when weighted by population.

Discussion

Water-related vulnerability in China's southbound transboundary basins

Our analysis indicates that economic income level, political stability, poverty, population density, proneness to hazards, and stress on the environment all have large spatial differences within and between the six studied basins. The spatial distribution of population is one of the key factors determining the level of vulnerability in the basins. In many studies, the Indus and the GBM basins have been identified as having serious water-related challenges (Cook et al., 2011; Mustafa & Wrathall, 2011; Rahaman & Varis, 2009; Rahaman, 2009; Sharma et al., 2010), and our study agrees with this (strongly highlighted in Figure A1 in the appendix).

Based on this assessment, the GBM and Mekong appear very heterogeneous in terms of vulnerability. The GBM, which consists of three basins very different from each other, also has very different sources of vulnerabilities in each river basin. Of these three, the Ganges is clearly the most vulnerable due to its high population density and high human footprint, while also providing food for millions of people as the 'breadbasket of the world'. In addition, the GBM shows the largest variability in vulnerability between the different riparians, with India and Bangladesh having the highest and China having the lowest levels of vulnerability.

Our analysis of the upstream–downstream dynamics makes it very clear that vulnerability in each basin, and regarding almost all the sources of vulnerability, increases in the downstream direction. This is partially explained by the commonly higher population densities towards the delta areas of basins. This strongly highlights the precarious position of downstream countries with respect to upstream impacts, especially in the GBM, Salween and Mekong basins. Related to this, China experiences low levels of vulnerability within all of the basins, despite its influential role and political and economic leverage in them. This creates an acute imbalance between China and its neighbours, as also identified by Turner, Shifflett, & Batten (2013). It should be noted that, as shown by Brown et al. (2008) and Varis et al. (2012), the Chinese provinces within the study basins are economically and socially much more vulnerable than other provinces in China, largely due to geographic, political and cultural marginalization. However, such local sensitivities were not captured in our macro-scale analysis; our result was largely determined by the fact that China's parts of the basins are not densely populated and account for quite a minimal part of the basins' total population.

Extremely interesting is the contrast between India and China in their vulnerability profiles within the study basins, which is strongly highlighted when looking at population-weighted proportions of vulnerability (Figure A2 in the appendix). For India, another factor affecting its vulnerability not captured in this assessment is the ongoing border disputes with Pakistan and China. China and India are two geopolitical and economic giants that share numerous environmental challenges and should seek to tackle them bilaterally (Bawa et al., 2010). However, both India's and China's hydropower development and water diversion plans concern areas that are already volatile due to ongoing territorial disputes by these two nations (Rahaman, 2012), hindering cooperation on transboundary water issues and regional economic development (Feng & He, 2009). Despite some hopes, the recent visit of Chinese president Xi Jinping to India to meet Indian prime minister Narendra Modi did not result in any progress towards a border agreement, reportedly due to the simultaneous occurrence of a border incursion (BBC News, 2014; Meyer, 2014). Considering India's position downstream of China with its high level of vulnerability and dependency on the river, lack of cooperation on the shared resources could become a major source of water conflict.

Transboundary cooperation

The strong asymmetries and remarkable downstream vulnerabilities in the six study basins suggest that the ongoing hydropower development – the majority of it by China and other upstream countries – will require careful planning and cumulative impact assessment at the regional (i.e. transboundary) level. Yet, currently the Mekong River Commission (MRC) is the only actively operating river basin commission in the study basins, and even it has only the four Lower Mekong countries of Laos, Thailand, Cambodia and Vietnam as its members (Grumbine & Xu, 2011). The Mekong region also has the Asian Development Bank–facilitated and more economically focused Greater Mekong Subregion (GMS) programme, in which China is also engaged. In the other five river basins, current transboundary cooperation is less institutionalized and driven largely by bilateral agreements and negotiations (McNally, Magee, & Wolf, 2009; Rahaman & Varis, 2009; Rahaman, 2009).

At the same time, the present study indicates how the significance of the six study basins differs for the riparian countries in terms of population, economy and many other dimensions. The combination of the significance of the river for a country and the country's location also partly translates into the interest of the country in transboundary cooperation. At first glance, shared basins should rank high on China's agenda, as one-third of the country's land lies in transboundary basins (Nickum, 2008). However, amongst many reasons identified by Nickum (2008), the lack of salience of international river basins in China's decision making can be partially explained by geography. Firstly, nearly three-quarters of China's runoff crosses no international borders. Secondly, China is at the upstream end of nearly all the international flows. Thirdly, most border areas are relatively unpopulated and are habited by minorities which lack economic or political clout.

In the Mekong River basin, downstream riparian Cambodia is highly dependent on the Mekong for food security and livelihoods, and as a result Cambodia sees the MRC as an extremely important regional forum in which to discuss sustainable development of the Mekong's water resources (Keskinen, Mehtonen, & Varis, 2008). More upstream riparian Thailand has little to gain from the Mekong mainstream that the MRC focuses its actions on, and as a result the MRC is less important for Thailand (e.g. Keskinen et al., 2008). Similarly, Bangladesh, as a downstream country in the GBM basin, seeks multilateral cooperation with other riparian countries, without achieving it as of yet (Rahaman, 2012). In other words, while transboundary negotiations and international river commissions commonly include equal representation of the riparian countries, in practice the member countries' interest or stake in participating and promoting transboundary cooperation differs greatly.

The critical question is thus how transboundary cooperation can be facilitated in the study basins, and how to ensure that China – which is often the least vulnerable but at the same time a major player in water infrastructure development – could also find an incentive to participate in it. As outlined by Chen, Rieu-Clarke, and Wouters (2013), China's ambition to secure regional and global geopolitical leadership, as well as their policy of 'good-neighbourliness', could motivate China to seek enhanced regional cooperation. This could further be supported by China's economic opening, its drive for more transparency and its increasingly important role as a member of the international community (Keskinen et al., 2008). In addition, Turner et al. (2013) mention that broadening the engagement with Chinese stakeholders, such as researchers, hydropower companies and NGOs, to promoting sustainable development could demonstrate the economic benefits of responsible dam building. However, it seems that China is currently more willing to operate through bilateral channels (Chen et al., 2013), where it can more fluently leverage its geopolitical and economic powers and reach those agreements it sees as most beneficial for it. Meanwhile, shared resources such as water cannot be comprehensively and holistically managed through mere bilateral cooperation. However, participating in regional discussions – or even being a member of a regional organization such as a river basin commission – would most probably provide a less powerful position and could even lead to situations where downstream countries create alliances against China.

Yet, the Chinese government seems to be increasingly aware of the necessity of promoting environmentally and socially sound investments, including those related to hydropower (McDonald et al., 2009). Also, Chinese domestic law clearly lays out procedures for environmental and social impact assessment of the existing projects (McDonald et al., 2009). However, as these principles are not directly applicable to transboundary settings, the practices employed in investments largely depend on the financiers, builders or regulators themselves (Urban et al., 2013a).

In addition, generally recognized principles of transboundary cooperation – presented most importantly by the UN Watercourses Convention of 1997, which finally entered into force in August 2014 – call for equitable and reasonable utilization of transboundary water resources as well as the due-diligence obligation of no harm (for more, see *Water International* 38(2), e.g. Rieu-Clarke, Kinna, & Loures, 2013). While no doubt contested and prone to different interpretations, both principles can be called for in the six study basins. China's transboundary water agreements remain, however, relatively unsophisticated, although there is a growing watercourse treaty practice in China (Chen et al., 2013). As a result, Chen et al. (2013) conclude that China's transboundary water treaty practice would benefit from the guidelines set forth in the Watercourses Convention.

Strengths and weaknesses of the vulnerability approach

Water-related vulnerability maps were created for the six transboundary river basins south of China based on an approach introduced by Varis et al. (2012). The analysis demonstrates the complexity of water-related vulnerability beyond mere water-stress aspects. Water stress in this method is considered together with the coping capacity of the society to tackle water challenges as well as stress factors to the environment (Varis et al., 2014).

The analysis also provides new insight into the upstream–downstream dynamics in these basins, thanks to the finer analytic resolution. For India and China, subnational-scale data were used in view of the large heterogeneity within these enormous countries. Because much of the administrative unit–based data are at a fairly coarse scale (commonly only at the national level), finding corresponding data for local jurisdictions can be challenging. This was experienced by Varis et al. (2014) in the case of China's provinces, and the same problem was faced for Indian states in our study. Though the MPI was readily available for India at the state level, the PSI created certain challenges. Only some of the indicators that the PSI includes (EIU, 2011) were available at the state level. As a result, a large part of the internal variability within countries for the PSI comes from the economic variables, which were more readily available at finer scales. Therefore, the PSI indicator does not capture the political nuances within India, such as the border disputes that are a major contributor to unrest in North India and near the China–India border.

Other sources of inaccuracy in the data include the data representing different years, as well as certain assumptions made to enable using different indicators as metrics for vulnerability (see Varis et al., 2012, 2014). Meanwhile, because the analysis method has a certain simplicity, many opportunities for refinement exist (Varis et al., 2012).

Conclusions

The multidimensional vulnerability of the six transboundary rivers originating in South-West China (Red River, Mekong, Salween, Irrawaddy, Ganges-Brahmaputra-Meghna (GBM) and Indus) was analyzed. The importance of these river basins in terms of environmental, economic and social aspects is gigantic, and together they provide water, food and livelihoods for over one billion people. At the same time, the six basins are targeted for some of the most extensive water infrastructure development in the world, led by plans for a massive number of large-scale hydropower dams, rapid urban development and irrigation expansion.

Such complex challenges call for broad, multidisciplinary knowledge. This article employed an intercomparable spatial vulnerability assessment approach that assessed the river basins regarding the multifaceted challenges they are exposed to. The approach incorporated data on nature and environment, demography, and political stability, as well as social and economic development.

According to the analysis, China's six southbound transboundary river basins present notable spatial heterogeneity in terms of their socio-economic and environmental vulnerability, particularly in relation to the environmental aspects (water stress, hazards and human footprint). The range in these vulnerabilities equalled that of the entire globe. Societal aspects described by poverty and political stability between and within the basins also showed large variation, especially in the Mekong and GBM basins.

The GBM and the Indus basins in particular showed very high levels of vulnerability, due to their extreme population densities, consequent high human footprint, and limited societal capacity to address various development- and environment-related pressures. A key finding in the study is that in each of the transboundary basins, the downstream parts are more vulnerable than the upstream parts, and thus are highly sensitive to upstream pressures. In particular, the GBM downstream countries of India and Bangladesh were highly vulnerable regarding all the sources of vulnerability (figures A1, A2). Also, in the Mekong basin, the downstream countries, Vietnam and Cambodia, experienced higher levels of vulnerability than their upstream neighbours.

In sum, the analysis clearly indicates that while China's role is very influential as the upstream country in all six river basins, its own vulnerability in those basins is generally much lower than in its downstream neighbours. This is largely thanks to a much smaller population in the Chinese parts of the studied basins. At the same time, China's various operations and policies within these basins – including plans for intensive hydropower development – are likely to have pronounced downstream effects and thus create novel sensitivities such as governance and geopolitical challenges and increasing proneness to natural hazards in the much more vulnerable downstream countries. This is the case particularly in the extremely vulnerable GBM system, where the other riparians, particularly India, are also putting increasing pressure on the waters, thus creating major stress for its downstream neighbour, Bangladesh. The GBM's extraordinary level of challenge is exacerbated by the perennial border dispute between India and China, which makes the political scene behind transboundary collaboration quite demanding. The Mekong's situation is also quite complicated, because particularly Laos and China are modifying the catchment and the river in a very pronounced manner.

Such findings call for a high level of responsibility from all the riparian countries of these six major rivers. China is in a particular position given that despite its geopolitical strength it has participated fairly weakly in transboundary water cooperation. China has not signed the major international conventions on transboundary waters, nor really participated in a transboundary river commission, preferring bilateral cooperation with its neighbouring countries. To a large extent, India's past and current policies in the GBM basin merit similar concern. Hence, room for improvement exists, and the findings from this vulnerability analysis emphasize the importance of this in a very distinct manner.

Acknowledgements

The authors thank the members of the Water and Development Research Group for support, insightful comments and fruitful discussions.

Funding

This study was partly done within the Academy of Finland–funded Nexus Asia project (no. 269901). Work by Mirja Kattelus was supported by the Doctoral Programme in the Built Environment (RYM-TO), Finnish University Network for Asian Studies and Academy of Finland project (no. 133748). Aura Salmivaara received funding from the VALUE Doctoral School, Academy of Finland project (no. 133748) and Aalto doctoral funds. Matti Kummu received funding from the Academy of Finland's SCART project (grant no. 267463). Olli Varis received funding from the Cultural Foundation of Finland.

References

Affeltranger, B. (2008). Inter-basin water transfers as a technico-political option: Thai-Burmese projects on the Salween River. In N. I. Pachova, M. Nakayama, & L. Jansky (Eds.), *International water security, domestic threats and opportunities* (pp. 161–179). Tokyo, Japan: United Nations University Press.

Agarwal, A., Delos Angeles, M. S., Bhatia, R., Chéret, I., Davila-Poblete, S., Falkenmark, M., ... Kindler, J. (2000). *Integrated water resources management.* Denmark: Global Water Partnership.

Alkire, S., & Santos, M. E. (2010). Multidimensional poverty index: 2010 data [Data set]. Retrieved from www.ophi.org.uk/policy/multidimensional-poverty-index/

Alkire, S., Conconi, A., & Seth, S. (2014). Multidimensional Poverty Index 2014: Brief Methodological Note and Results [Data set]. Retrieved from ophi.qeh.ox.ac.uk

Arias, M. E., Cochrane, T. A., Kummu, M., Lauri, H., Holtgrieve, G. W., Koponen, J., & Piman, T. (2014). Impacts of hydropower and climate change on drivers of ecological productivity of Southeast Asia's most important wetland. *Ecological Modelling, 272*, 252–263. doi:10.1016/j.ecolmodel.2013.10.015

Asian Development Bank (2012). *Myanmar in transition—opportunities and challenges.* Mandaluyong City, Philippines: Asian Development Bank.

Bawa, K. S., Koh, L. P., Lee, T. M., Liu, J., Ramakrishnan, P. S., Yu, D. W., ... Raven, P. H. (2010). China, India, and the environment. *Science, 327*, 1457–1459. doi:10.1126/science.1185164

BBC News. (2014, 18 September). China's Xi Jinping signs landmark deals on India visit. Retrieved from http://www.bbc.com/news/world-asia-india-29249268

Berg, M., Stengel, C., Trang, P. T. K., Hung Viet, P., Sampson, M. L., Leng, M., ... Fredericks, D. (2007). Magnitude of arsenic pollution in the Mekong and Red River Deltas—Cambodia and Vietnam. *Science of the Total Environment, 372*, 413–425. doi:10.1016/j.scitotenv.2006.09.010

Biemans, H., Speelman, L. H., Ludwig, F., Moors, E. J., Wiltshire, A. J., Kumar, P., ... Kabat, P. (2013). Future water resources for food production in five South Asian river basins and potential for adaptation—A modeling study. *Science of the Total Environment, 468–469*, S117-S131. doi:10.1016/j.scitotenv.2013.05.092

Biodiversity and Nature Conservation Association. (2009). *Environmental impact assessment (special investigation) on hydropower development of Ayeyawady river basin above Myitkyina, Kachin state, Myanmar.* Yangon, Myanmar: Biodiversity and Nature Conservation Association (BANCA).

Biswas, A. K., & Tortajada, C. (2010). Future water governance: Problems and perspectives. *International Journal of Water Resources Development, 26*, 129–139. doi:10.1080/07900627.2010.488853

Bremmer, I. (2012, 20 March). Economic reforms solidify the political opening in Burma. *Foreign Policy.* Retrieved from http://blog.foreignpolicy.com/posts/2012/03/20/economic_reforms_to_solidify_political_opening_in_burma

Brown, P. H., Magee, D., & Xu, Y. (2008). Socioeconomic vulnerability in China's hydropower development. *China Economic Review, 19*, 614–627. doi:10.1016/j.chieco.2008.06.002

Brown, P. H., Tullos, D., Tilt, B., Magee, D., & Wolf, A. T. (2009). Modeling the costs and benefits of dam construction from a multidisciplinary perspective. *Journal of Environmental Management, 90*(Supplement 3), S303-S311. doi:10.1016/j.jenvman.2008.07.025

Chen, H., Rieu-Clarke, A., & Wouters, P. (2013). Exploring China's transboundary water treaty practice through the prism of the UN Watercourses Convention. *Water International, 38*, 217–230. doi:10.1080/02508060.2013.782134

China Statistical Yearbook. (2011). *China statistical yearbook 2011.* Beijing: National Bureau of Statistics of China.

Chowdhury, M R., & Ward, N. (2004). Hydro-meteorological variability in the greater Ganges–Brahmaputra–Meghna basins. *International Journal of Climatology, 24*, 1495–1508. doi:10.1002/joc.1076

Cook, S.,Fisher, M., Tiemann, T., & Vidal, A. (2011). Water, food and poverty: Global- and basin-scale analysis. *Water International, 36*, 1–16. doi:10.1080/02508060.2011.541018

Dharmadikhary, S. (2008). *Mountains of concrete: Dam building in the Himalayas.* Berkeley, CA: International Rivers.

Dilley, M., Chen, R. S., Deichmann, U., Lerner-Lam, A. L., Arnold, M., Agwe, J., ... Yetman, G. (2005). *Natural disaster hotspots: A global risk analysis.* Washington, DC: The World Bank.

EIU. (2011). *ViewsWire: Business Intelligence on 203 Countries.* New York, NY: Economist Intelligence Unit.

FAO. (2013). *The state of food insecurity in the world 2013: The multiple dimensions of food security.* Rome, Italy: United Nations Food and Agriculture Organization.

FAO, WFP, & IFAD. (2012). *The state of food insecurity in the world 2012: Economic growth is necessary but not sufficient to accelerate reduction of hunger and malnutrition.* Rome, Italy: United Nations Food and Agriculture Organization.

Feng, Y., & He, D. (2009). Transboundary water vulnerability and its drivers in China. *Journal of Geographical Science, 19*, 189–199. doi:10.1007/s11442-009-0189-7

Feng, Y., He, D., & Li, Y. (2010). Ecological changes and the drivers in the Nu River basin (upper Salween). *Water International, 35*, 786–799. doi:10.1080/02508060.2010.534619

Grübler, A., O'Neill, B., Riahi, K., Chirkov, V., Goujon, A., Kolp, P., ... Slentoe, E. (2007). Regional, national, and spatially explicit scenarios of demographic and economic change based on SRES. *Technological Forecasting and Social Change, 74*, 980–1029. doi:10.1016/j.techfore.2006.05.023

Grumbine, R. E., & Pandit, M. K. (2013). Threats from India's Himalaya dams. *Science, 339*, 36–37. doi:10.1126/science.1227211

Grumbine, R. E., & Xu, J. (2009). China shakes the world—and then what? *Conservation Biology, 23*, 513–515. doi:10.1111/j.1523-1739.2009.01239.x

Grumbine, R. E., & Xu, J. (2011). Mekong hydropower development. *Science, 332*, 178–179. doi:10.1126/science.1200990

Hall, D., & Bouapao, L. (2010). *Social impact monitoring and vulnerability assessment: Regional report.* Vientiane, Laos: Mekong River Commission.

India Census. (2011). India Census Data [Data set]. Retrieved 10 September, 2014 from http://planningcommission.nic.in/data/datatable/

IPCC. (2007). *Climate change 2007: The physical science basis.* Cambridge, UK, and New York, USA: Cambridge University Press.

IPCC. (2012). *Managing the risks of extreme events and disasters to advance climate change adaptation. A special report of working groups I and II of the intergovernmental panel on climate change.* Cambridge: Cambridge University Press.

Kattelus, M., Rahaman, M. M., & Varis, O. (2014). Myanmar under reform: Emerging pressures on water, energy and food security. *Natural Resources Forum, 38*, 85–98. doi:10.1111/1477-8947.12032

Keskinen, M., Kummu, M., Käkönen, M., & Varis, O. (2012). Mekong at the crossroads: Next steps for impact assessment of large dams. *AMBIO, 41*, 319–324. doi:10.1007/s13280-012-0261-x

Keskinen, M., Mehtonen, K., & Varis, O. (2008). Transboundary cooperation vs. internal ambitions: The role of China and Cambodia in the Mekong region. In N. I. Pachova, M. Nakayama, & L. Jansky (Eds.), *International water security: Domestic threats and opportunities* (pp. 79–109). Tokyo, Japan: United Nations University Press.

Kirby, M., Krittasudthacheewa, C., Mainuddin, M., Kemp-Benedict, E., Swartz, C., & de la Rosa, E. (2010). The Mekong: A diverse basin facing the tensions of development. *Water International, 35*, 573–593. doi:10.1080/02508060.2010.514094

Klein Goldewijk, K., Beusen, A., & Janssen, P. (2010). Long-term dynamic modeling of global population and built-up area in a spatially explicit way: HYDE 3.1. *The Holocene, 20*, 565–573. doi:10.1177/0959683609356587

Kuenzer, C., Campbell, I., Roch, M., Leinenkugel, P., Tuan, V., & Dech, S. (2013). Understanding the impact of hydropower developments in the context of upstream–downstream relations in the Mekong river basin. *Sustainability Science, 8*, 565–584. doi:10.1007/s11625-012-0195-z

Kummu, M., Lu, X. X., Wang, J. J., & Varis, O. (2010). Basin-wide sediment trapping efficiency of emerging reservoirs along the Mekong. *Geomorphology, 119*, 181–197. doi:10.1016/j.geomorph.2010.03.018

Kummu, M., & Sarkkula, J. (2008). Impact of the Mekong river flow alteration on the Tonle Sap flood pulse. *AMBIO: A Journal of the Human Environment, 37*, 185–192. doi:10.1579/0044-7447(2008)37[185:IOTMRF]2.0.CO;2

Kummu, M., & Varis, O. (2007). Sediment-related impacts due to upstream reservoir trapping, the Lower Mekong River. *Geomorphology, 85*, 275–293. doi:10.1016/j.geomorph.2006.03.024

Labour Bureau. (2012). *Report on employment and unemployment survey 2011-2012*. Government of India: Ministry of Labour and Employment.

Lauri, H., de Moel, H., Ward, P. J., Räsänen, T. A., Keskinen, M., & Kummu, M. (2012). Future changes in Mekong River hydrology: Impact of climate change and reservoir operation on discharge. *Hydrology and Earth System Sciences, 16*, 4603–4619. doi:10.5194/hess-16-4603-2012

Lu, X. X. (2004). Vulnerability of water discharge of large Chinese rivers to environmental changes: An overview. *Regional Environmental Change, 4*, 182–191. doi:10.1007/s10113-004-0080-0

Magee, D. (2006). Powershed politics: Yunnan hydropower under great western development. *The China Quarterly, 185*, 23–41. doi:10.1017/S0305741006000038

Magee, D., & Kelley, S. (2009). Damming the Salween River. In F. Molle, T. Foran, & M. Käkönen (Eds.), *Contested waterscapes in the Mekong Region: Hydropower, livelihoods and governance* (pp. 115–142). London, UK: Earthscan.

McDonald, K., Bosshard, P., & Brewer, N. (2009). Exporting dams: China's hydropower industry goes global. *Journal of Environmental Management, 90*(Supplement 3), S294-S302. doi:10.1016/j.jenvman.2008.07.023

McNally, A., Magee, D., & Wolf, A. T. (2009). Hydropower and sustainability: Resilience and vulnerability in China's powersheds. *Journal of Environmental Management, 90*(Supplement 3), S286–S293. doi:10.1016/j.jenvman.2008.07.029

Meyer, E. (2014, 23 September). Who Sabotaged Chinese President Xi Jinping's India Visit? *Forbes*. Retrieved from http://www.forbes.com/sites/ericrmeyer/2014/09/23/who-sabotaged-xi-jinpings-india-visit/

Ministry of Statistics and Programme Implementation of India. (2014). Official website. Retrieved 10 September, 2014 from http://mospi.nic.in/Mospi_New/site/home.aspx

Moore, D., Dore, J., & Gyawali, D. (2010). The World Commission on Dams+ 10: Revisiting the large dam controversy. *Water Alternatives, 3*, 3–13.

Mustafa, D., & Wrathall, D. (2011). Indus basin floods of 2010: Souring of a Faustian bargain? *Water Alternatives, 4*, 72–85.

Nickum, J. E. (2008). The upstream superpower: China's international rivers. In O. Varis, C. Tortajada, & A. K. Biswas (Eds.), *Management of transboundary rivers and lakes* (pp. 227–244). Berlin, Heidelberg: Springer-Verlag.

Orr, S., Pittock, J., Chapagain, A., & Dumaresq, D. (2012). Dams on the Mekong River: Lost fish protein and the implications for land and water resources. *Global Environmental Change, 22*, 925–932. doi:10.1016/j.gloenvcha.2012.06.002

Quynh, L. T. P., Billen, G., Garnier, J., Théry, S., Fézard, C., & Minh, C. V. (2005). Nutrient (N, P) budgets for the Red River basin (Vietnam and China). *Global Biogeochemical Cycles, 19*, GB2022. doi:10.1029/2004GB002405

Rahaman, M. M. (2009). Principles of transboundary water resources management and Ganges treaties: An analysis. *International Journal of Water Resources Development, 25*, 159–173. doi:10.1080/07900620802517574

Rahaman, M. M. (2012). Hydropower ambitions of South Asian nations and China: Ganges and Brahmaputra Rivers basins. *International Journal of Sustainable Society, 4*, 131–157. doi:10.1504/IJSSOC.2012.044670

Rahaman, M. M., & Varis, O. (2005). Integrated water resources management: Evolution, prospects and future challenges. *Sustainability: Science, Practice & Policy, 1*, 15–21.

Rahaman, M. M., & Varis, O. (2009). Integrated water management of the Brahmaputra basin: Perspectives and hope for regional development. *Natural Resources Forum, 33*, 60–75. doi:10.1111/j.1477-8947.2009.01209.x

Räsänen, T. A., Koponen, J., Lauri, H., & Kummu, M. (2012). Downstream hydrological impacts of hydropower development in the Upper Mekong Basin. *Water Resour Manage, 26*, 3495–3513. doi:10.1007/s11269-012-0087-0

Rieu-Clarke, A., Kinna, R., & Loures, F. (2013). Guest editorial. *Water International, 38*, 109–111. doi:10.1080/02508060.2013.780230

Robinson, G. (2012, 19 March). Myanmar pushes reform to woo investors. *Financial Times*. Retrieved from http://www.ft.com/intl/cms/s/0/168baaee-71b6-11e1-b853-00144feab49a.html#axzz3D0QjVAsF

Rodell, M., Velicogna, I., & Famiglietti, J. S. (2009). Satellite-based estimates of groundwater depletion in India. *Nature, 460*, 999–1002. doi:10.1038/nature08238

Salmivaara, A., Kummu, M., Keskinen, M., & Varis, O. (2013). Using global datasets to create environmental profiles for data-poor regions: A case from the Irrawaddy and Salween River Basins. *Environmental Management, 51*, 897–911. doi:10.1007/s00267-013-0016-x

Schmidt, C. (2012). As isolation ends, Myanmar faces new ecological risks. *Science, 337*, 796–797. doi:10.1126/science.337.6096.796

Searle, M. P. (2006). Role of the Red River Shear zone, Yunnan and Vietnam, in the continental extrusion of SE Asia. *Journal of the Geological Society, 163*, 1025–1036. doi:10.1144/0016-76492005-144

Sharma, B., Amarasinghe, U., Xueliang, C., de Condappa, D., Shah, T., Mukherji, A., . . . Smakhtin, V. (2010). The Indus and the Ganges: River basins under extreme pressure. *Water International, 35*, 493–521. doi:10.1080/02508060.2010.512996

Sovacool, B. K. (2013). Confronting energy poverty behind the bamboo curtain: A review of challenges and solutions for Myanmar (Burma). *Energy for Sustainable Development, 17*, 305–314. doi:10.1016/j.esd.2013.03.010

Tilt, B., Braun, Y., & He, D. (2009). Social impacts of large dam projects: A comparison of international case studies and implications for best practice. *Journal of Environmental Management, 90*(Supplement 3), S249–S257. doi:10.1016/j.jenvman.2008.07.030

Tullos, D., Tilt, B., & Liermann, C. R. (2009). Introduction to the special issue: Understanding and linking the biophysical, socioeconomic and geopolitical effects of dams. *Journal of Environmental Management, 90*(Supplement 3), S203–S207. doi:10.1016/j.jenvman.2008.08.018

Turner, J. L., Shifflett, S. C., & Batten, R. (2013). China's upstream advantage in the Great Himalayan Watershed. *Asia Policy, 16*, 11–18. doi:10.1353/asp.2013.0026

UNDP. (2010). *China and a sustainable future: Towards a low carbon economy and society. China human development report 2009/2010*. Beijing: United Nations Development Program, Renmin University and Beijing and China Translation and Publishing Corporation.

United Nations, (2014). World urbanization prospects: The 2014 Revision, CD-ROM Edition. [Data set]. Retrieved from http://esa.un.org/unpd/wup/

Urban, F., Mohan, G., & Cook, S. (2013a). China as a new shaper of international development: The environmental implications. *Environment, Development and Sustainability, 15*, 257–263. doi:10.1007/s10668-012-9411-3

Urban, F., Nordensvärd, J., Khatri, D., & Wang, Y. (2013b). An analysis of China's investment in the hydropower sector in the Greater Mekong Sub-Region. *Environment, Development and Sustainability, 15*, 301–324. doi:10.1007/s10668-012-9415-z

Varis, O. (2005). Externalities of Integrated Water Resources Management (IWRM) in South and South-East Asia. In A. K. Biswas, O. Varis, & C. Tortajada (Eds.), *Integrated water resources management in south and southeast Asia* (pp. 1–38). New Delhi, India: Oxford University Press.

Varis, O., & Kummu, M. (2012). The major Central Asian River Basins: An assessment of vulnerability. *International Journal of Water Resources Development, 28*, 433–452. doi:10.1080/07900627.2012.684309

Varis, O., Kummu, M., & Salmivaara, A. (2012). Ten major rivers in monsoon Asia-Pacific: An assessment of vulnerability. *Applied Geography, 32*, 441–454. doi:10.1016/j.apgeog.2011.05.003

Varis, O., Kummu, M., Lehr, C., & Shen, D. (2014). China's stressed waters: Societal and environmental vulnerability in China's internal and transboundary river systems. *Applied Geography, 53*, 105–116. doi:10.1016/j.apgeog.2014.05.012

Wada, Y., van Beek, L.P.H., & Bierkens, M.F.P. (2011). Modelling global water stress of the recent past: on the relative importance of trends in water demand and climate variability. *Hydrology and Earth System Sciences*, *15*, 3785–3808. doi:10.5194/hess-15-3785-2011

Wada, Y., van Beek, L.P.H., Wanders, N., & Bierkens, M.F.P. (2013). Human water consumption intensifies hydrological drought worldwide. *Environmental Research Letters*, 8. doi:10.1088/1748-9326/8/3/034036

Wang, P., Lassoie, J. P., Dong, S., & Morreale, S. J. (2013). A framework for social impact analysis of large dams: A case study of cascading dams on the Upper-Mekong River, China. *Journal of Environmental Management, 117*, 131–140. doi:10.1016/j.jenvman.2012.12.045

WCS/CIESIN, (2005). Last of the Wild Project, v2.0 (LWP-2): Global Human Footprint Dataset (Geographic). [Data set]. Retrieved from http://sedac.ciesin.columbia.edu/data/set/wildareas-v2-human-footprint-geographic

Webb, E. L., Phelps, J., Friess, D. A., Rao, M., & Ziegler, A. D. (2012). Environment-friendly reform in Myanmar. *Science, 336*, 295. doi:10.1126/science.336.6079.295-a

World Bank. (2013). *World development report 2014: Risk and opportunity: Managing risk for development*. Washington, DC: The World Bank.

World Bank. (2014). The data catalog. [Data set]. Retrieved from http://data.worldbank.org/data-catalog

Ziv, G., Baran, E., Nam, S., Rodriguez-Iturbe, I., & Levin, S. A. (2012). Trading-off fish biodiversity, food security, and hydropower in the Mekong River Basin. *Proceedings of the National Academy of Sciences, 109*, 5609–5614. doi:10.1073/pnas.1201423109

Annex

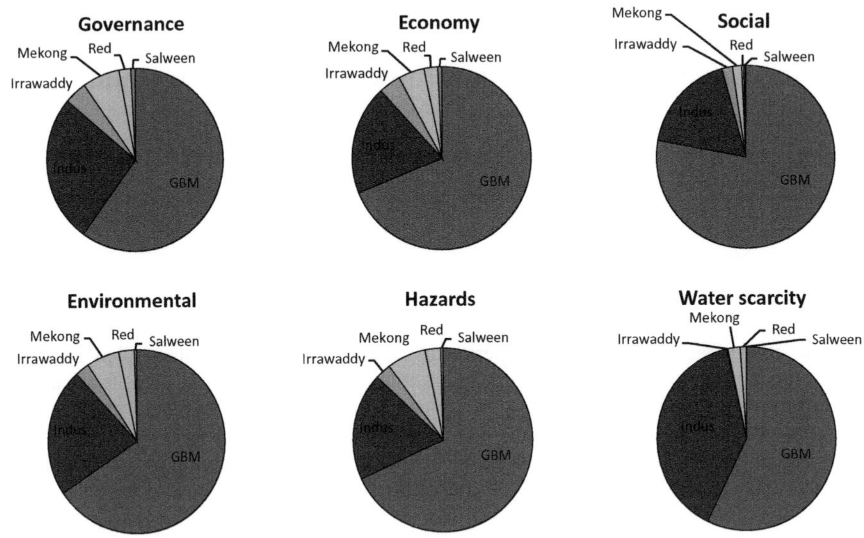

Figure A1. Population-weighted proportions of vulnerability in the entire study area, by river basin.

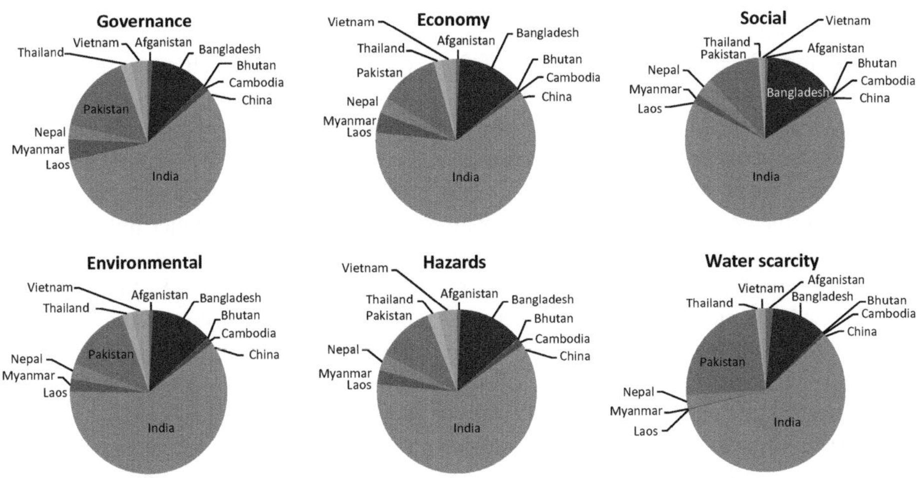

Figure A2. Population-weighted proportions of vulnerability in the entire study area, by country.

Water security in Himalayan Asia: first stirrings of regional cooperation?

Bjørn-Oliver Magsig

The challenges of freshwater cooperation and the links between the global policy agenda and national water management are more obvious and easier to spell out at a regional level than in the global arena. By analysing how the existing regional regime within Himalayan Asia contributes to the development of international water law, this paper identifies gaps in our current understanding of regional approaches in international law and outlines pathways for addressing them.

Introduction

Only a small number of challenges have the potential to create as much friction between states as the allocation and utilization of freshwater resources which cross international boundaries. A surge of water-related crises can be witnessed around the world – like Egypt's disagreement with Ethiopia on the Nile (Abseno, 2013), the Israeli-Palestinian dispute over a more equitable distribution of water (Fröhlich, 2012), Central Asia's intensifying water dispute (Ziganshina, 2014), or the highly contentious issue of dam projects on the Mekong (Cronin, 2013). This is why water, the gossamer linking various other security concerns (like the environment, food and energy), has acquired an independent status within the security discourse (Wouters, Vinogradov, & Magsig, 2009, p. 103). Just as the concept of security has gone through a widening and deepening process, so has the perception of water security. The approach of addressing water security merely as a nation's internal affair of securing its access to a continuous supply of freshwater is untenable in today's interrelated world. A contemporary understanding of water security goes beyond the narrow state-centred zero-sum game debate and provides a platform for various disciplines to engage in strengthening cooperation over shared waters. Hence, this paper follows a broader understanding of security and regards a community to be water secure when it has sustainable access to freshwater of sufficient quantity and quality, or to the benefits derived there from; and the ability to minimize water-related risk and its various repercussions to an acceptable level – without compromising the supporting ecosystems (Magsig, 2014).

This perception of water security has also brought about renewed interest in regionalism in international water law. How can regional approaches to transboundary water cooperation be developed in order to be conducive to a broader regional governance framework aiming at the development of concrete forms of integrated cooperation on a whole range of transboundary environmental issues? As this level of integration is more challenging to achieve at the global level, the normative influence of this cross-sectoral approach represents a significant potential contribution of regional approaches to strengthening international water law (Moynihan & Magsig, 2014, p. 57).

Himalayan Asia is especially interesting for studies on regional water security due to the multitude of water challenges it faces, its complex international relations and the fact that some basins are governed by joint institutions while most are not. Since 'Himalayan Asia' is a rather loosely defined region, it is important to point out that in this paper it is being understood as encompassing Afghanistan, Bangladesh, Bhutan, Burma, Cambodia, China, India, Laos, Nepal, Pakistan, Thailand and Vietnam – all dependent on the transboundary river systems originating in the Himalayas. Here, where more than 1.5 billion people (22% of the global population) are directly dependent on the international rivers originating in the Himalayas (Grey & Connors, 2009, p. 60), the water crisis is aggravating the already tense relations between riparians. Since the outlook for this region in terms of impacts from climate change, economic development and population growth is especially worrisome, one can expect that governments will increasingly look beyond their borders to secure freshwater resources – leading to even more geopolitical tensions.

The main objective of this paper is to examine the current affairs of regional water cooperation in Himalayan Asia and how the novel understanding of common water security and its legal foundations could help to strengthen international cooperation. The hypothesis is that even in such a complex and volatile region, looking at international law governing transboundary freshwater resources through the lens of common water security can help to break with the prevalence of zero-sum games, and thus serve as a change agent towards stronger regional cooperation. It is hoped that this paper will contribute to the renewed international legal discourse about the further development of the architecture governing transboundary freshwater resources.

Regional water security and the general obligation to cooperate

General obligation to cooperate in international law

Just like the general discourse on international environmental law, the debate about the principles and rules governing transboundary water cooperation has long been trapped between the either/or debate on sovereignty versus the joint management of natural resources. Even though today most states follow a more nuanced interpretation of sovereignty, the debate about how sovereignty over freshwater resources is to be interpreted is still highly controversial. This is also due to the fact that, far from being absolute, the notion of sovereignty carries a responsibility to cooperate with it. Rather than being a clear-cut concept, sovereignty has to be regarded as a dynamic fabric which remains fundamental to the existence of states, the international system based on these states (to the extent that we can speak of the system) and, consequently, international relations (Ţuţuianu, 2013, p. 44). Article 1 of the UN Charter reads: '[t]he purposes of the United Nations are: [...] (3) [t]o achieve international co-operation in solving international problems of an economic, social, cultural, or humanitarian character [...]'. This rather vague duty to cooperate was described further by the 1970 Declaration on Principles of International Law concerning Friendly Relations and Co-operation among States in accordance with the Charter of the United Nations (UN General Assembly, 1970):

> States have the duty to co-operate with one another, irrespective of the differences in their political, economic and social systems, in the various spheres of international relations, in order to maintain international peace and security and to promote international economic stability and progress, the general welfare of nations and international co-operation free from discrimination based on such differences.

Even though the declaration does not constitute binding international law, it nevertheless carries considerable legal clout, as it has been universally recognized as a standard of conduct and is being perceived as an elaboration of a key principle of international law. The MOX case before the International Tribunal for the Law of the Sea (ITLOS) also prominently discusses the general obligation to cooperate. Here, Ireland claimed that the UK failed to cooperate as demanded by the rules of the United Nations Convention on the Law of the Sea (UNCLOS) regarding the obligations to communicate and exchange information (Sands & Peel, 2012, p. 251). The tribunal endorsed the duty to cooperate as a general principle of international law, and hence ordered Ireland and the UK to cooperate and enter into consultations. According to Judge Wolfrum, the duty to cooperate

> balances the principle of sovereignty of States and thus ensures that community of interests are taken into account vis-à-vis individualistic State interests. It is the matter of prudence and caution as well in keeping with the overriding nature of the obligation to cooperate that the parties should engage therein [...]. (The MOX Plant Case (Ireland v. United Kingdom), Provisional Measures, 2001)

In interpreting the order of the tribunal, one comes to the conclusion that there is a need ultimately to arrive at a stage where the concept of state sovereignty is being perceived as one of 'cooperative sovereignty' (Perrez, 2000). Inevitability, this becomes even more obvious when addressing the challenges of managing common-pool resources, where the collective action problem leads to unsatisfactory outcomes (Hardin, 2009). Rather than treating sovereignty as an unshakable stumbling block in international environmental negotiations – due to its apparent incompatibility with relinquishing freedoms and making concessions – acknowledging the fact that the responsibility to cooperate is one of the key elements of sovereignty itself seems to be a more promising strategy in addressing the tragedy of the commons (Perrez, 2000, p. 343). In essence, international law has the obligation to provide a path for moving from 'sovereignty as independence' to 'sovereignty as interdependence' (Moynihan & Magsig, 2014, p. 45). Simply acknowledging this need for change, however, does not automatically facilitate the implementation of the general obligation to cooperate. While progress can be witnessed where benefits for cooperation are easily quantifiable in economic terms and occur within a short time frame (e.g., in international trade negotiations), international environmental law is still struggling with perception of cooperative sovereignty.

It was here where the constant political push and pull of state sovereignty and environmental cooperation provoked an element of uncertainty amongst international legal scholars as to whether the duty to cooperate is indeed a binding legal obligation; or whether it is merely a guideline for conduct (Delbrück, 2012, p. 3). Hence, the question remains, whether states actually must (rather than should) cooperate; and if yes, how this obligation can be imposed on states and legally enforced? The consequential debate reveals that while it succeeds in balancing the principle of 'sovereignty over natural resources' and the 'no significant harm' rule, cooperation is too broad to qualify as an international legal rule (Rieu-Clarke, Moynihan, & Magsig, 2012, p. 123). However, its classification as a general principle of international law, rather than a customary rule, does not diminish its important role in international law and policy. This is even more the case since an increasing number of states realizes that cooperation on transboundary challenges is ultimately in their self-interest.

Both global frameworks of international water law – the 1997 UN Watercourses Convention and the 2008 Draft Articles on the Law of Transboundary Aquifers – promote the general obligation to cooperate. Article 8(1) of the UN Watercourses Convention provides that '[w]atercourse states shall cooperate on the basis of sovereign equality, territorial integrity, mutual benefit and good faith in order to attain optimal utilisation and adequate protection of an international watercourse'. This general provision contains the procedural duties of prior information and of prior consultation (Articles 11–19 of the UN Watercourses Convention), which aim at operationalizing the rather vague principle (Rieu-Clarke et al., 2012, p. 125). While the UN Watercourses Convention surely had an impact on the implementation of the general obligation to cooperate regarding transboundary water management (Leb, 2013), it falls short of fundamentally influencing the way states perceive their national sovereignty over shared freshwater resources, as states still have considerable discretion with regard to the particular means of cooperation. For instance, the setting up of joint institutions was not made compulsory in the Convention, even though their tremendous benefit for transboundary freshwater cooperation has long been recognized (Schmeier, 2013). Merely recommending the establishment of joint river basin organizations (Article 8(2) of the UN Watercourses Convention) was, apparently, as far as the drafters of the convention could go in order to still receive widespread support. While this kind of hitch in negotiating global legal frameworks is quite prevalent, it could be overcome in a regional setting, where the urgency to act jointly on a common challenge is usually more obvious. Hence, the general obligation to cooperate could be strengthened by following a regional approach in international water law.

Regional common concern as the conceptual framework for the future development of international water law

Earlier work has addressed the question as to how communality has been treated in international environmental law and what lessons can be learned for international water cooperation (Magsig, 2014). While some approaches are too limited as they only apply to a certain geographical area beyond national jurisdictions and its resources (common area, common heritage), the notion of common concern appears to be the most promising in tackling issues of water security.

At the 1992 UN Conference on Environment and Development, a global framework for environmental responsibilities was designed which, for the first time, was based on a common concern, rather than the concept of good neighbourliness (Birnie, Boyle, & Redgwell, 2009, p. 128). This concern is based on the understanding that some kind of harm to the environment (like the changing climate) has the potential to adversely affect humanity as a whole; and thus mitigating those impacts can only be achieved effectively if the international community in its entirety is involved. Acknowledging this position carries with it both a right and an obligation of the international community as a whole to have concern for the global environment (IUCN Commission on Environmental Law, & International Council of Environmental Law, 2010, p. 39). Both treaties negotiated 1992 in Rio – the UN Framework Convention on Climate Change (UNFCCC) and the Convention on Biological Diversity (CBD) – follow this approach.

In its preamble, the CBD states that 'the conservation of biological diversity is a common concern of humankind [...]'. While it lacks strong procedural support, the inclusion of common concern does ensure that the loss of biodiversity is being acknowledged as a major issue which the international community has to address on a common basis. However, this does not compensate for the shortcomings of the convention – for instance, not extending state responsibility for extra-territorial harm to damage caused to the global commons (Guruswamy, 1999, p. 79). The convention also fails to provide a balancing tool for the conflict between the protection of biological diversity and the national sovereignty to exploit natural resources, evident in the fact that the agreed target to achieve significant reduction of the current rate of biodiversity loss by 2010 has been missed considerably (Secretariat of the Convention on Biological Diversity, 2010, p. 9). While the UNFCCC, which also features the concept of common concern in its preamble, kick-started a process of legal and political engagement to address global climate change, it also failed to overcome the state-centrism of the governance system, which ultimately lead to watered-down obligations in weak agreements which were 'designed to mask the political failure of the international community to create a global climate treaty' (Dimitrov, 2010, p. 795).

While it seems like the two conventions did bring the concept of common concern into bad repute, the Draft International Covenant on Environment and Development tried to reanimate it. The covenant, a collaborative effort by the IUCN Commission on Environmental Law and the International Council of Environmental Law, was designed in 1995 as a 'living blueprint' for an international framework agreement which consolidates and advances existing legal principles related to environment and development (IUCN Commission on Environmental Law, & International Council of Environmental Law, 2010). Article 3 of the covenant stipulates that '[t]he global environment is a common concern of humanity and under the protection of the principles of international law, the dictates of the public conscience and the fundamental values of humanity'. According to the commentary, the concept should to be interpreted as 'the basis upon which the international community at all levels can and must take joint and separate action to protect the environment' (IUCN Commission on Environmental Law, & International Council of Environmental Law, 2010, p. 39). In urging that not only single issues, like climate change or the loss of biodiversity, should be treated as being of common concern to the international community, but also the environment as a whole, it constitutes a departure from previous approaches to common concern. Furthermore, the notion of common concern in the covenant has to be understood not only as a spatial principle, but also as a temporal one, as it recognizes long-term impacts of environmental harm and the responsibility of the international community owed to future generations (IUCN Commission on Environmental Law, & International Council of Environmental Law, 2010, p. 43). Regrettably, while the covenant breathed new life into the debate about communality in international law, it has attracted only little attention amongst international legal scholars due to the lack of legitimacy.

However, even in its embryonic stage, the concept of common concern is of particular interest to the advancement of international water law. Although its focus lies again on (common) benefits, it considers the benefits from common action rather than those derived from the mere exploitation of a resource (Brunnée, 2007, p. 564). Moreover, it fixes its attention on what renders a concern as being common, rather than targeting one particular area or resource, and thus avoids discussions about common property and territorial sovereignty. One of its main advantage over other approaches to communality is that it triggers a shift from the orthodox reciprocity and material benefit sharing we often find in treaties to joint action in the long-term interest of the community (IUCN Commission on Environmental Law, & International Council of Environmental Law,

2010, p. 40). Yet, the weaknesses of the notion are evident in the UNFCCC and the CBD, both facing difficulties in achieving strong legal impact with regard to the common concern at the level the conventions are intended to have an impact on – the global one. This is mainly due to the difficulties in phrasing a global set of values and interests which is detached from the individual interests of states. Given the magnitude of the global water crisis, it should not be impossible to construct an analogous mindset to the loss of biodiversity or climatic changes for transboundary freshwater cooperation. However, given the complexity of water security, it would be an impossible endeavour to agree on a perception of water security as a common concern of humankind. This is why scaling down one level by looking at the regional layer would be a fruitful middle road which can sufficiently accommodate the national interests and the global challenge. Here, the political and economic influence of regional institutions can be utilized in the push for change – change which is home-grown, rather than being perceived as imperialistic. Since, contrary to a river basin, a region is not a narrowly defined geographical area, framing water security as a regional common concern also opens up the enormous potential of including (non-state) actors and interests beyond the basin.

Thus, it may add a new dimension to international freshwater cooperation, which is still being perceived as a zero-sum game, by including non-riparian interests in the design and performance of international water cooperation (Brunnée & Toope, 1994, p. 75). On these grounds, the notion of a regional common concern is the most suitable conceptual vehicle for the endeavour to achieve common water security (Magsig, 2015, p. 104). While this does not necessary require reinventing the wheel of transboundary freshwater cooperation, like the principle of equitable and reasonable utilization, it expects states to subscribe to certain basic rules and minimum managing standards as cornerstones for their transboundary water relations; and thus triggers a process of rethinking focused on the respective regional approach.

Current approaches to regional freshwater cooperation in Himalayan Asia

The question now arises how this regional understanding of water security could shape cooperation over transboundary water resources in Himalayan Asia. How is the general obligation to cooperate being interpreted and implemented here? What are possible pathways for strengthening water diplomacy in the region?

At the same time as most countries in Himalayan Asia have seen their renewable freshwater resources and water availability drop continuously over the last decades (Asian Development Bank, 2013, p. 29), regional cooperation is being hampered by political tensions between several states – e.g., the conflicts in the volatile parts of Kashmir and Tibet. The immense pressures of decreasing water quality and increasing competition for freshwater not only affects the states in their respective national development, but also transforms these domestic challenges into regional ones. The glaciers of the Himalayas feed the headwaters of the mighty rivers Yellow, Yangtze, Mekong, Salween, Irrawaddy, Ganges-Brahmaputra and Indus, on which more than 1.5 billion people directly depend (Grey & Connors, 2009, p. 60). China's and India's emphasis on large-scale infrastructure in addressing their water issues – the South–North Water Transfer Project (Berkoff, 2003) and the River Linking Project (Khalid, 2004), respectively – have huge implications for their downstream neighbours. The governments' ambitious plans to step up hydropower capacity and push forward with inter-basin water transfers will certainly increase the geopolitical risks of international freshwater cooperation in Himalayan Asia.

Insights on how countries in Himalayan Asia perceive the general obligation to cooperate in international water law can be drawn by looking at the negotiation and ratification process of the UN Watercourses Convention. One of this region's countries – China – voted against the adoption of the convention, five of them – Afghanistan, Bhutan, Burma, India and Pakistan – abstained or were absent, and six of them – Bangladesh, Cambodia, Laos, Nepal, Thailand and Vietnam – voted in favour; with only one of the latter, Vietnam, having ratified the convention until now. Given China's key role in Himalayan Asia's water politics, its statement on the convention is of particular interest here. Since the region's hegemon emphasized 'indisputable territorial sovereignty' and bilateral agreements in commenting on why it voted against the convention (United Nations General Assembly, 1997), one has to wonder what the future prospects of a more regional approach actually are. Furthermore, treaty practice does not allow for excessive optimism, either, as all freshwater agreements China has entered are bilateral, despite the fact that many of them govern multi-state watercourses (Wouters & Chen, 2013, p. 232). While there are some success stories of international water law in Himalayan Asia – e.g. (at least to some extent) in the Mekong basin (Rieu-Clarke & Gooch, 2010) – the region remains rather hostile towards the idea of a more common approach to water security. Which role, then, can international law play in moving towards regional water cooperation in Himalayan Asia? How could a regime based on regional common concern emerge and evolve in a climate where the political willingness among the various riparian countries to change course still seems to be lacking? In trying to address these questions, the following section will cast for developments which can be interpreted as potential pathways towards a new approach based on regional common concern.

While the chances for agreeing on a comprehensive regional freshwater agreement anytime soon are rather slim, a gradual development towards closer regional cooperation on freshwater issues is by no means illusive. For instance, few pundits would have considered it possible for China and India to ever agree on sharing information regarding the state of their glaciers (Lamont for the *Financial Times*, 2009). While this kind of trust-building is a valid step towards a more regional approach to freshwater interaction, it needs to be tied into an institutional mechanism in order to be of long-term value. Here, regional organizations can take on an important role as they allow for gradual strengthening and deepening of relations on a diverse range of topics.

An interesting starting point in this regard is to take a closer look at the Association of Southeast Asian Nations (ASEAN), which was founded in 1967 by Indonesia, Malaysia, Philippines, Singapore and Thailand and now includes Brunei, Burma, Cambodia, Laos and Vietnam. The reason for choosing the framework of ASEAN for this study is that it possesses the necessary political clout to involve both China and India – and thus become the central player of Asian efforts towards regionalism (Bower for CSIS, 2010). While neither of the two hydro-hegemons of Himalayan Asia are members of ASEAN, they are nonetheless actively involved – mainly through the ASEAN+3 (China, Japan and South Korea) and the East Asia Summit (AEAN+3 together with Australia, India, New Zealand, Russia and the United States) initiatives which have been established to create an integrated market in the Asia-Pacific region. One of ASEAN's near-term goals is setting up the ASEAN Economic Community (AEC) by 2015 which envisages (1) a single market; (2) a highly competitive economic region; (3) a region of equitable economic development; and (4) a region which is fully integrated into the global economy (*Jakarta Post*, 2013). While historically the regional integration has been driven by economic considerations, cooperation amongst ASEAN countries has continually broadened and deepened to include other topics as well.

In 1976, the Treaty of Amity and Cooperation in Southeast Asia was signed by the founding members of ASEAN and adopted fundamental principles of their international relations: (1) mutual respect for the independence, sovereignty, equality, territorial integrity and national identity of all nations; (2) the right of every state to lead its national existence free from external interference, subversion or coercion; (3) non-interference in internal affairs; (4) settlement of differences or disputes in a peaceful manner; (5) renunciation of the threat or use of force; and (6) effective regional cooperation (Treaty of Amity and Cooperation in Southeast Asia, 1976). Notably, China and India were the first nations outside ASEAN to sign the treaty in 2003. While one could argue that this so-called 'ASEAN way' seems to constitute a stumbling block for overcoming the prevailing focus on national interests and arriving at a truly regional approach to transboundary water management, several initiatives within ASEAN point into another direction.

In 1985, for instance, several member states signed the ASEAN Agreement on the Conservation of Nature and Natural Resources, which stipulates in Article 19(3)(a) that

> Contracting Parties shall especially co-operate together and, where appropriate, shall endeavour to co-operate with other Contracting Parties, with a view to: (a) the conservation and management of (1) border or contiguous protected areas; (2) shared habitats of species listed in Appendix 1; (3) shared habitats of any other species of common concern [...].

While this wording suggests close cooperation between ASEAN countries in the area of nature conservation, it has not entered into force, as only three of the six signatory states (Indonesia, Philippines and Thailand) have ratified it. In 2002, the ASEAN Ministerial Meeting on Environment (AMME) agreed to further synergize the regional environmental cooperation by focusing on 10 areas – freshwater resources being one of them (ASEAN, 2002). Other agreements and initiatives on environmental cooperation under the roof of ASEAN include the 1995 Cooperation Plan on Transboundary Pollution, the 2002 Agreement on Transboundary Haze Pollution, the ASEAN Wildlife Enforcement Network launched in 2005 and the 2007 ASEAN Declaration on Environmental Sustainability. Regarding freshwater resources, the environmental ministers of ASEAN adopted the 2003 ASEAN Strategic Plan on Water Resources Management, which is aimed at promoting 'the attainment of sustainability of water resources to ensure sufficient water quantity of acceptable quality to meet the needs of the people of Southeast Asia in terms of health, food security, economy and environment'. In a recent meeting of AMME, the ministers further agreed to 'enhance existing transboundary water cooperation within ASEAN which aims to reduce poverty, protect natural resources, prevent crisis and resolve conflicts between countries'.

However, these initiatives aiming to strengthen water cooperation have not yet been thoroughly implemented, as the current conflicts over the region's freshwater resources show. The reasons for bilateral arrangements still being the norm are twofold. At the international level, states still see transboundary water interaction as a zero-sum game, as it is the limiting factor of their respective economic growth (see, for example, the Kishenganga Arbitration between India and Pakistan; Magsig, 2015, p. 187); while at the various national levels the power interplay is preventing a joint approach due to the prevailing fragmentation caused by bureaucratic competition over functional responsibilities (Moynihan & Magsig, 2014, p. 56). There is, however, a ray of hope for a more coherent and regional regime for transboundary freshwaters. During the 2nd Asia-Pacific

Water Summit on the topic of 'Water Security and Water-Related Disaster Challenges: Leadership and Commitment', Thailand's Prime Minister, Yingluck Shinawatra, stressed the importance of regional forums for the peaceful resolution of water conflicts, as '[n]o country in this region can handle these challenges alone' (AFP, 2013). The heads of state and government declared in the outcome document of the summit the Chiang Mai Declaration to 'enhance regional and international cooperation on sharing, exchange and dissemination of scientific and technical knowledge, as well as best practices, related to integrated water resources management' (Chiang Mai Declaration, 2013).

Yet, despite these promising soft-law developments, a legally binding regional framework for governing Himalayan Asia's shared freshwater resources still seems to be a long way away. Potentially, China's and India's growing role in the international arena will be a chance to lay stress on their moral obligation to move into the direction of regional common concern. While there can be no doubt that the two regional hydro-hegemons are among the rising powers in world politics, do they also accept new responsibilities? So far, China has shown little evidence that it is seeking international leadership in global governance (Wang & French, 2013, p. 89). With respect to his foreign policy, the current President of China, Xi Jinping, has made it clear that his nation will continue to 'never compromise its core interests' or swallow the 'bitter fruit' of accepting harm to its sovereignty, security and development (Economist Intelligence Unit, 2013). India's strategy in becoming a 'great power' seems even more obscured, as its hands are tied by domestic constraints (Narlikar, 2013, p. 613). The contemporary perception of water security as regional common concern attempts to persuade states that a regional approach to their water security challenges not only benefits the region as a whole, but also, when taking a long-term perspective, it constitutes the best option for the respective national security interests (Magsig, 2015, p. 198). While both regional hegemons, China and India, seem to prefer a stable Himalayan Asia which will allow them to sustain their political clout and economic growth, they still perceive threats to their security differently, and thus employ different strategies (Tellis & Mirski, 2013). To unfold fully, however, the concept of regional common concern needs to be implemented through concerted action at the regional level, even if it is merely through a geographical confined form of special international law (Pulkowski, 2012, p. 3). Considering the current political tensions between China and India, the recently signed accord on sharing hydrological information between the two states is a welcome sign of a more friendly relationship between the two hydro-hegemons (*Daily News* 2013).

Observations

While there is some scepticism about the capacity of ASEAN to be a leading force in regional integration of Himalayan Asia, more subtle processes are often overlooked. Yet, it is through these apparently small steps that perceptions of transboundary water cooperation within the region are being incrementally transformed. Indeed, while ASEAN may have originally focused on economic issues, its very endurance has created an increasing sense of regional identity in what has been a rather arbitrary geographical area. The slowly emerging interest of both hydro-hegemons, China and India, to engage in this process increases the importance of ASEAN as a key player in the region's cooperation over freshwater resources. However, Himalayan Asia will not eagerly implement a perception of cooperative sovereignty which comes from 'outside'. It is much more likely that states within the region will develop their own concepts – based on their regional identity and specific political and cultural environment. International legal scholars bear the responsibility to support this process of nurturing regional norms and institutions by drawing lessons from other regional regimes, as well as developing novel pathways towards an understanding of cooperative sovereignty (Moynihan & Magsig, 2014, p. 56). It is hoped that the initiated discussion about perceiving water security as a regional common concern will trigger fundamental change in how states cooperate on the highly complex and controversial issues concerning their shared freshwater resources in the future.

References

1985 ASEAN Agreement on the Conservation of Nature and Natural Resources (9 July 1985; not entered into force) Retrieved from http://www.jus.uio.no/english/services/library/treaties/06/6-01/asean-conservation-nature.xml

1992 Rio Declaration on Environment and Development (14 June 1992) UN Doc A/CONF.151/26 (Vol. I); Retrieved from http://www.un.org/documents/ga/conf151/aconf15126-1annex1.htm

1995 ASEAN Cooperation Plan on Transboundary Pollution; reprinted in 12 ASEAN Economic Bulletin 89.

2002 ASEAN Agreement on Transboundary Haze Pollution; Retrieved from http://haze.asean.org/docs/1128506236/ASEANAgreementonTransboundaryHazePollution.pdf/view

Abseno, M. M. (2013). The influence of the UN Watercourses Convention on the development of a treaty regime in the Nile river basin. *Water International*, *38*(2), 192–203. doi:10.1080/02508060.2013.782798

AFP, 'Asia-Pacific Leaders Warn of Water Conflict Threat'. (2013, May 20). Retrieved from http://www.asiaone.com/News/AsiaOne+News/Asia/Story/A1Story20130520-424064.html

ASEAN Declaration on Environmental Sustainability. (2007, November 20). Retrieved from http://environment.asean.org/index.php?page=agreements:aseandeclarationenvsus

ASEAN Strategic Plan on Water Resources Management. (2003, December 18). Retrieved from http://cil.nus.edu.sg/rp/pdf/2003%20ASEAN%20Strategic%20Plan%20on%20Water%20Resource%20Management-pdf.pdf

Asian Development Bank. (2013). *Asian Water Development Outlook 2013: Measuring water security in Asia and the Pacific*. Manila: Asian Development Bank.

Berkoff, J. (2003). China: The south-north water transfer project: Is it justified? *Water Policy*, *5*(1), 1–28.

Birnie, P. W., Boyle, A. E., & Redgwell, C. (2009). *International law and the environment* (3rd ed.). Oxford: Oxford University Press.

Bower, E.Z. for CSIS, 'ASEAN's Opportunity to Become the Core of Asian Regionalism'. (2010, April 2). Retrieved from http://csis.org/publication/aseans-opportunity-become-core-asian-regionalism

Brunnée, J. (2007). Common areas, common heritage, and common concern. In D. Bodansky, J. Brunnée, & E. Hey (Eds.), *The oxford handbook of international environmental law* (pp. 550–573). Oxford: Oxford University Press.

Brunnée, J., & Toope, S. J. (1994). Environmental security and freshwater resources: A case for international ecosystem law. *Yearbook of International Environmental Law*, *5*, 41–76. doi:10.1093/yiel/5.1.41

Convention on Biological Diversity (5 June 1992; entered into force 29 December 1993) 1760 UNTS 79; reprinted in 31 ILM 818 (1992).

Cronin, R. P. (2013). Hydropower dams on the Mekong: Old dreams, new dangers. *Asia Policy*, *16*, 32–38. doi:10.1353/asp.2013.0035

Daily News, 'India, China Ink Key Accord on River Information'. (2013, October 23). Retrieved from http://india.nydailynews.com/politicsarticle/cea1292d874ee01b807b9971c2ad1768/india-china-ink-key-accord-on-river-information

Delbrück, J. (2012). The international obligation to cooperate: An empty shell or a hard law principle of international law? A critical look at a much debated paradigm of modern international law. In H. P. Hestermeyer, D. König, N. Matz-Lück, V. Röben, A. Seibert-Fohr, T.-P. Stoll, & S. Vöneky (Eds.), *Coexistence, cooperation and solidarity* (Vol. 1, pp. 3–16). Leiden: Martinus Nijhoff.

Dimitrov, R. S. (2010). Inside UN climate change negotiations: The Copenhagen conference. *Review of Policy Research*, *27*(6), 795–821. doi:10.1111/j.1541-1338.2010.00472.x

Economist Intelligence Unit, 'China: New Leader Hints at Tough Foreign Policy'. (2013). Retrieved from http://country.eiu.com.proxy1.athensams.net/article.aspx?articleid=60102990&Country=China&topic=Politics&subtopic=Forecast&subsubtopic=International+relations&u=1&pid=1510105735&oid=1510105735&uid=1.

Fröhlich, C. J. (2012). Security and discourse: The Israeli-Palestinian water conflict. *Conflict, Security & Development*, *12*(2), 123–148. doi:10.1080/14678802.2012.688290

Grey, D., & Connors, G. (2009). The water security imperative: We must and can do more. In *5th World Water Forum Istanbul 2009* (pp. 58–62). Istanbul: World Water Council.

Guruswamy, L. D. (1999). The convention on biological diversity: Exposing the flawed foundations. *Environmental Conservation*, *26*(2), 79–82. doi:10.1017/S0376892999000120

Hardin, G. (2009). The tragedy of the commons. *Journal of Natural Resources Policy Research*, *1*(3), 243–253. doi:10.1080/19390450903037302

IUCN Commission on Environmental Law, & International Council of Environmental Law. (2010). *Draft international covenant on environment and development* (4th ed.). Gland: IUCN.

Khalid, A. R. M. (2004). The interlinking of rivers project in India and international water law: An overview. *Chinese Journal of International Law*, *3*(2), 553–570. doi:10.1093/oxfordjournals.cjilaw.a000534

Lamont, J. for the Financial Times, 'India and China Co-operate over Himalayan Glaciers'. (2009, August 2). Retrieved from http://www.ft.com/cms/s/0/5e729abe-7fa8-11de-85dc-00144feabdc0.html

Leb, C. (2013). The UN watercourses convention: The éminence grise behind cooperation on transboundary water resources. *Water International*, *38*(2), 146–155. doi:10.1080/02508060.2013.781474

Magsig, B.-O. (2014). Pushing the boundaries: Rethinking international law in light of the common concern for water security. In V. Sancin & M. Kovič Dine (Eds.), *International environmental law: Contemporary concerns and challenges in 2014* (pp. 441–452). Ljubljana: GZ Založba.

Magsig, B.-O. (2015). *International water law and the quest for common security*. London: Routledge.

Moynihan, R., & Magsig, B.-O. (2014). The rising role of regional approaches in international water law: Lessons from the UNECE water regime and Himalayan Asia for strengthening transboundary water cooperation. *Review of European Community & International Environmental Law*, *23*(1), 43–58. doi:10.1111/reel.12065

Narlikar, A. (2013). India rising: Responsible to whom? *International Affairs*, *89*(3), 595–614. doi:10.1111/1468-2346.12035

Perrez, F. X. (2000). *Cooperative sovereignty: From independence to interdependence in the structure of international environmental law*. The Hague; Boston: Kluwer Law International.

Pulkowski, D. (2012). Theoretical premises of regionalism and the unity of international law. *ESIL Conferenc Paper Series*, *2*(16), 1–10.

Rieu-Clarke, A., & Gooch, G. (2010). Governing the tributaries of the Mekong: The contribution of international law and institutions to enhancing equitable cooperation over the Sesan. *Pacific McGeorge Global Business & Development Law Journal*, *22*(2), 193–224.

Rieu-Clarke, A., Moynihan, R., & Magsig, B.-O. (2012). *UN watercourses convention: User's guide*. Dundee: IHP-HELP Centre for Water Law, Policy and Science.

Sands, P., & Peel, J. (2012). *Principles of international environmental law* (3rd ed.). Cambridge: Cambridge University Press.

Schmeier, S. (2013). *Governing international watercourses: River basin organizations and the sustainable governance of internationally shared rivers and lakes*. New York, NY: Routledge.

Secretariat of the Convention on Biological Diversity. (2010). *Global biodiversity outlook 3*. Montreal: Secretariat of the Convention on Biological Diversity.

Tellis, A. J., & Mirski, S. (Eds.). (2013). *Crux of Asia: China, India, and the emerging global order*. Washington, DC: Carnegie Endowment for International Peace.

Jakarta Post, 'ASEAN+6 start regional FTA Talks'. (2013, Mai 10). Retrieved from http://www.thejakartapost.com/news/2013/05/10/asean6-start-regional-fta-talks.html

The MOX Plant Case (Ireland v. United Kingdom), Provisional Measures, Order of 3 December 2001, ITLOS, Separate Opinion of Judge Wolfrum; Retrieved from http://www.itlos.org/fileadmin/itlos/documents/cases/case_no_10/sep.op.Wolfrum.E.orig.pdf

The Second Asia-Pacific Water Summit, Chiang Mai Declaration. (2013, May 20). Retrieved from http://apws2013.files.wordpress.com/2013/05/chiang-mai-declaration.pdf

Treaty of Amity and Cooperation in Southeast Asia, as amended by the First Protocol amending the Treaty of Amity and Cooperation in Southeast Asia, 1987, the Second Protocol amending the Treaty of Amity and Cooperation in Southeast Asia, 1998 and the Third Protocol amending the Treaty of Amity and Cooperation in Southeast Asia, 2010. (1976, February 24). Retrieved from http://www.official-documents.gov.uk/document/cm84/8472/8472.pdf

Țuțuianu, S. (2013). *Towards global justice: Sovereignty in an interdependent world*. The Hague: T.M.C. Asser Press.

UN General Assembly, Declaration of Principles of International Law Concerning Friendly Relations and Co-operation Among States in Accordance with the Charter of the United Nations. (1970, October 24). UNGA Res 2625, 25 UN GAOR Supp 18, UN Doc A/5217 at 121; reprinted in 65 AJIL 243 (1971).

United Nations Framework Convention on Climate Change. (9 May 1992; entered into force 21 March 1994) 1771 UNTS 107; UN Doc A/AC.237/18 (Part II)/Add.1; reprinted in 31 ILM 849 (1992).

United Nations General Assembly, Fifty-first Session, 99th Plenary Meeting. (1997, May 21) UN Doc A/51/PV.99; Retrieved from http://www.un.org/ga/search/view_doc.asp?symbol=A/51/PV.99

Wang, H., & French, E. (2013). China's participation in global governance from a comparative perspective. *Asia Policy, 15*, 89–114. doi:10.1353/asp.2013.0016

Wouters, P., & Chen, H. (2013). China's 'soft-path' to transboundary water cooperation examined in the light of two UN global water conventions - Exploring the 'Chinese way'. *Journal of Water Law, 22*(2/3), 229–247.

Wouters, P., Vinogradov, S., & Magsig, B.-O. (2009). Water security, hydrosolidarity and international law: A river runs through it. *Yearbook of International Environmental Law, 19*(1), 97–134. doi:10.1093/yiel/19.1.97

Ziganshina, D. (2014). *Promoting transboundary water security in the Aral Sea Basin through international law*. Leiden: Brill Nijhoff.

Transboundary water cooperation on the Yarlung Zangbo/ Brahmaputra – a legal analysis of riparian state practice

Yang Liu

The Yarlung Zangbo/Brahmaputra, shared by China, India, Bhutan and Bangladesh, lacks a comprehensive legal framework. Following an overview of the basin, the paper analyses state and treaty practice through a legal lens, comparing the riparian state positions, based on their approaches to sovereignty. Finding fragmented transboundary water cooperation across the basin, two possible approaches that might help are explored. Taking inspiration from the hydropower projects on parts of the watercourse, and from existing multilateral environmental agreements, it is queried whether future cooperation might be fostered through legal arrangements for joint exploitation, or through joint protection under multilateral environmental agreements.

Introduction

South Asian states, traditionally agricultural nations, are at a high risk of water insecurity (Express Tribune News Network, 2014). Given their dependence on irrigation, this poses a serious threat to regional stability. The rivers that originate in the Himalayas act as the lifeline to Southern Asia. The Yarlung Zangbo/Brahmaputra River originates in the Himalayan Mountain Range (Tibet, China) and links China, India and Bangladesh as it flows into the Bay of Bengal. Three main tributaries originating in Bhutan join the Yarlung Zangbo/Brahmaputra River in India (Negi, 1991) and affect downstream states.[1] Of the 1350 billion cubic metres (BCM) annually discharged by the Ganges–Yarlung Zangbo/Brahmaputra–Meghna River Basin, the Yarlung Zangbo/Brahmaputra River contributes approximately 700 BCM (Salehin, Khan, Prakash, & Goodrich, 2011).

Since water resources are essential to agriculture, hydropower, sanitation, industrial development and the ecosystems of the region, the amount of water that is available can have a drastic impact on the social, political, economic and environmental conditions of a country. Although wars over water have not yet occurred, there is ample evidence showing that water scarcity has led to intense political instability (Wolf, Yoffe, & Giordano, 2003). This is also true for the Yarlung Zangbo/Brahmaputra River. China, referred by some as an 'upstream superpower' (Nickum, 2008), is building dams and hydropower stations on the upstream stretches of the Yarlung Zangbo River in order to meet its economic development goals.

India, situated at the mid-stream of the Yarlung Zangbo/Brahmaputra River, acts as an 'upstream superpower' to Bangladesh. Bhutan, the upstream state on several tributaries, is located at the southern side of the Himalayan Mountain Range and seems to be relatively free to develop its water resources. Although this may be the case, climate change as well as seasonal droughts and floods are still serious issues for Bhutan, and the construction of hydropower stations has caused many environmental problems.

Bangladesh, where the Yarlung Zangbo/Brahmaputra River discharges into the Bay of Bengal, faces multiple threats including annual floods, climate change and sea water encroachment. Currently, there is no comprehensive water agreement which involves all four riparian states. The majority of these states prefer to develop the water resources of the Yarlung Zangbo/Brahmaputra River on a unilateral basis, with a limited scope for bilateral cooperation primarily concerned with the exchange of hydrological information and the joint construction of hydropower stations.

Few studies exist that conduct a legal analysis of water agreements on the Yarlung Zangbo/Brahmaputra River. What is the overall legal framework of water cooperation on the Yarlung Zangbo/Brahmaputra River? This paper surveys historical and recent state practice related to this basin and analyses the relevant, primarily bilateral, transboundary water agreements. Following an overview of the current state of the Yarlung Zangbo/Brahmaputra River in Bangladesh, Bhutan, China and India respectively, and exploring the different practices of territorial sovereignty as a challenge for cooperation, the study will examine the transboundary water agreements relating to the Yarlung Zangbo/Brahmaputra River, focusing on Sino-Indian water agreements. The paper concludes by identifying the limitations and gaps in the state and treaty practice on the Yarlung Zangbo/Brahmaputra River and puts forward recommendations for an integrated hydropower development approach and environmental cooperation which aim at enhancing transboundary water cooperation (FAO, 2011b).

The Yarlung Zangbo/Brahmaputra River Basin: an overview

The Yarlung Zangbo/Brahmaputra River (2840 km) flows through China (1700 km) where it is called the Yarlung Zangbo River (雅鲁藏布江), India (760 km) and Bangladesh (337 km) and reaches into Bhutan through tributaries (Rahaman & Varis, 2009). The annual flow from China to India is 165.40 km^3, from Bhutan tributaries to India is 78 km^3, and from India to Bangladesh is 537.24 km^3 (FAO, 2011b). Though endowed with diverse ecosystems, flora and fauna, these resources are threatened (Hoq, 2009; Singh, Sharma, & Ojha, 2004). With massive hydropower potential and water flow, the Yarlung Zangbo/Brahmaputra River not only contributes to the economic development of its riparian countries, but also triggers tensions and disputes.[2] Climate change is also posing an increasing threat on the Yarlung Zangbo/Brahmaputra River.

Compared with other transboundary watercourses in China, the Yarlung Zangbo River is not highly developed, however with a push to develop generally across China, this is likely to change. For example, in 2009 the Yarlung Zangbo River diversion plan raised concerns in India and Bangladesh; but China denied this project (*Sohu News*, 2009). Nevertheless, the Zangmu Hydroelectric Power Station, proposed by the '11th Five-Year Plan of Tibet's Development of Society and Economy (2006–2010)' (Government of Tibet Autonomous Region, 2006), was put into operation on 23 November 2015 (*China Daily*, 2014) and became the first mega hydroelectric power station on the mainstream of the Yarlung Zangbo River (*Indiatoday*, 2014). Although China has pledged that water will continue to flow, the environmental impact assessment of the Zangmu Project has not been made available to the public (*China Daily*, 2010). Three more dams on the Yarlung Zangbo were approved by 'China's 12th Five Year Plan of Energy Development' on 1 January 2013 (Government of the People's Republic of China, 2013).

Agriculture is the dominant driver of the economy within the Yarlung Zangbo/ Brahmaputra River Basin in India, contributing more than 50% of the total gross domestic product (GDP) and employing about 70% of the regional workforce (Sharma, 2011). Taking this into consideration, the relationship between water resources and irrigation has a direct impact on India's security and socio-economic development. To increase food production and meet the needs of the population, India has proposed the National River Linking Project (NRLP) which would divert water from the Yarlung Zangbo/Brahmaputra into the Ganges. Of the total hydropower potential of India, 44.42% (66,065 MW) lies within the Yarlung Zangbo/Brahmaputra River Basin (Rahaman & Varis, 2009). However, According to paragraph 13 (Transboundary Rivers) of 2012 India's National Water Policy, both the exchange of hydrological data and the management of transboundary rivers should be conducted on a bilateral basis (Ministry of Water Resources, Government of India, 2012).

Bhutan (connected to the basin by tributaries) is located entirely within the Yarlung Zangbo/Brahmaputra River Basin (FAO, 2011a). Approximately 94% of water use in Bhutan in 2008 was used for agriculture (FAO, 2011a). Of the total electricity generated in Bhutan, 99% is from hydropower (Shrestha, Anal, Salam, & van der Valk, 2015). By cooperating with India on the construction of hydropower projects since 1974, Bhutan has been able to create large amounts of government revenue from this economic resource. Although this is the case, seasonal changes in water level and the lack of water storage capacity have the ability to impact its hydropower revenue (IMF, 2011) and could lead to seasonal water shortages for drinking and irrigation (FAO, 2011a).

The Yarlung Zangbo/Brahmaputra River, renamed three times in Bangladesh (*National Encyclopedia of Bangladesh* 2012; Rahaman, 2012), helps to make the Ganges–Brahmaputra Delta a highly productive agricultural area. Almost 67% of its annual water flow comes from upstream states and Bangladesh is heavily affected by tropical monsoons which result in drought and floods (FAO, 2011b). In addition, agriculture generates half of Bangladesh's employment and contributes 20% of its annual GDP (Asia Development Bank, 2011). Bangladesh is a strong proponent of regional and international cooperation on water issues and frequently enters into agreements with other countries as well as international and regional organizations. Through these organizations Bangladesh exchanges data and information, undertakes joint studies and joint management on transboundary waters (Government of Bangladesh, 2013).

Territorial sovereignty

At the core of the majority of international water conflicts is the question of 'equitable' allocation of the uses of transboundary waters (Wolf, 1999). Since state sovereignty is at the heart of international relations, the use and allocation of transboundary waters revolves around this fundamental notion. Among the challenges faced by China and India concerning their transboundary waters, their foreign policy approaches to territorial sovereignty plays an integral role in bilateral and multilateral water cooperation in Southern Asia.

There are four main theories of territorial sovereignty within water law: absolute territorial sovereignty, absolute territorial integrity, limited territorial sovereignty and the community of interests (McCaffrey, 1996; Rieu-Clarke, Moynihan, & Magsig, 2012).

China, India and Bangladesh adopt different approaches with respect to these theories, reflecting in large part their location on the transboundary watercourse.

The 'Harmon Doctrine' of absolute territorial sovereignty, generally favoured by upstream countries, issued approximately 100 years ago, holds that a country enjoys absolute territorial sovereignty over the waters of an international watercourse within its borders. This means that the riparian state is free to exploit the waters located within its territory as it pleases (McCaffrey, 1996). This doctrine has little support in state practice (Rahaman, 2009), and McCaffrey, (1996) deemed that it should 'be laid to a richly-deserved rest'. By way of contrast, the theory of absolute territorial integrity, favoured by downstream riparians, holds that each state is entitled to an uninterrupted flow of the water into its territory (Caponera & Nanni, 2007). These two opposing theories in international water law reflect the upstream–downstream dilemma, especially in the case of the Yarlung Zangbo/Brahmaputra River. Limited territorial sovereignty, which is recognized as the foundation of the UN Watercourses Convention (UNWC), stipulates that all watercourse states enjoy an equal right to the utilization of a shared water resource and that each watercourse state should respect the sovereignty and correlative rights of other watercourse states (Rieu-Clarke, 2005). This theory calls for a balancing of interests in the uses of the water resources of transboundary waters, an approach embedded in the notion of reciprocity, which is the bedrock of international law.

China, an 'upstream superpower' (Nickum, 2008) and 'a most enthusiastic champion' for the principle of sovereignty (Wang, 1990), voted against the UN Resolution adopting the UNWC, a move characterized as 'a significant modern reassertion of the absolute territorial sovereignty approach' (Seligman, 2008, p. 119). However, such generalizations need closer scrutiny. Two fundamental points need to be clarified in this regard: first, China's approach to international law, generally, and to the principle of sovereignty within this context, requires a more complete understanding (Xue, 2007); and secondly, China's approach to transboundary water cooperation must be considered within the broad context of China's overall approach to international law. The next part provides more details on these two matters.

China's foreign policy, directly influenced by historical events, has firmly embedded sovereignty as the governing principle. The 'Five Principles of Peaceful Coexistence' (1954), contained in China's constitution (Chen, 2013), accord with the fundamental principles of the UN Charter, including mutual respect of sovereignty and territorial integrity (Xue, 2007). China's position has found consistent expression in many instances, especially in its support for the United Nations. For example, in the Written Statement of China to the International Court of Justice on the Issue of Kosovo, China claimed, in accordance with the UN Charter, that respect for sovereignty and territorial integrity is a fundamental principle within international law, the practices of international organization and countries, and the judgments of international courts (People's Republic of China, 2009).

Saul (2013) argues that China's emphasis on state sovereignty is more in its use as a defensive, rather than as an aggressive concept, coherent with its overall approach to international law. China uses the notion in defence (in support) of its position on state sovereignty and not as a vehicle for intervention in the affairs of other sovereign nations. In line with the Five Principles of Peaceful Coexistence, China has devised a foreign policy based on good neighbourliness and non-intervention (Saul, 2013; Xue, 2011).

A survey of China's treaty practice on transboundary water cooperation reveals its adherence to the theory of limited territorial sovereignty (Vinogradov & Wouters, 2013; Wouters & Chen, 2013). Among the three transboundary water-specific agreements concluded by China with Mongolia, Kazakhstan and Russia, only the 2001 Agreement on Cooperation in the Use and Protection of Transboundary River between China and Kazakhstan incorporates the principle of mutual respect for sovereignty and territorial integrity into its text.

The above-mentioned agreements also accord with the fundamental principles of international water law (Wouters & Chen, 2013; Su, 2014). For example, article 4, paragraphs 6, 8 and 16, of the 2008 Sino-Russian Agreement on Reasonable Utilisation and Protection of Transboundary Waters regulates precaution, control and the reduction of transboundary impacts which may result from pollutant emissions and water-related projects (China & Russia, 2008); article 3 of the 2001 Agreement Between Kazakhstan and China on Cooperation in the Field of Use and Protection of Transboundary Rivers states that 'The Parties shall undertake appropriate measures and shall make efforts to prevent or mitigate serious harm caused to a State Party as a result of flooding disasters and man-made accidents' (China & Kazakhstan, 2001). China upholds the general principle of equitable and reasonable utilization within these three agreements.

Turning to India, as a middle riparian state, its approach to territorial sovereignty in terms of transboundary waters would appear to have two strands, consistent with upstream and downstream positions. In 1948, India temporarily interrupted the flow of water from India to Pakistan, reflecting an upstream absolute territorial sovereign approach (Upreti, 2006). In the early stages of negotiation on the Indus Waters Treaty between India and Pakistan, India relied on the principle of absolute territorial sovereignty in its claims (Wolf, 1999). Although absolute territorial sovereignty still seems to contribute to India's relations with Pakistan concerning the Indus Basin and the Ganges River with Bangladesh, in a statement in relation to the dispute over the construction by India of a barrage on the Ganges River at Farakka, India has expressly denied reliance on the Harmon Doctrine and confirmed that 'for its part, had always subscribed to the view that each riparian state was entitled to a reasonable and equitable share of the waters of an international river' (McCaffrey, 1986, p. 110). Nonetheless, India's practices and projects concerning dam construction and completing water diversion projects has led to claims that India is attempting to establish regional hegemony by controlling water resources (*Pakistan Observer*, 2011).

In terms of Indo-Nepal water relations, India has protested several Nepalese irrigation and hydropower projects by stating that these projects would violate the principle of territorial integrity, thus asserting a downstream riparian approach (Upreti, 2006). In a similar vein, India argues against China's exploitation of the upstream Yarlung Zangbo River. Some commentators go so far as to claim that China's idea of rerouting the Yarlung Zangbo River is 'the most dangerous idea' since this project implies the decrease of India's inflow with adverse impacts downstream (Chellaney, 2009). However, in 2009, Wang Shucheng, former Minister of China's Ministry of Water Resources, declared that China did not plan to divert water from the Yarlung Zangbo/Brahmaputra River (*Sohu News*, 2009). From this brief overview, India's position as a midstream state has resulted in two very different approaches to its riparian neighbours: absolute territorial sovereignty and absolute territorial integrity.

As far as the basin state Bhutan is concerned, although it voted for the adoption of the UNWC, it was not active in the process of drafting and commenting on the International Law Commission (ILC) Draft on the law of the non-navigational uses of international watercourses. Bhutan is active in its cooperation with India on the construction of various hydropower projects within its territory, but primarily in contractual type arrangements. Bhutan's National Water Policy of 2003 (Salehin et al., 2011) highlighted the importance of water cooperation at the basin level, thus reflecting Bhutan's open attitude towards transboundary water cooperation. Thus it appears to take a limited territorial sovereignty approach, which enhances the potential for transboundary water cooperation.

Bangladesh (formerly known as East Pakistan), downstream to all other countries on the Yarlung Zangbo/Brahmaputra River, favoured the theory of absolute territorial integrity in the 1950s (Caponera & Nanni, 2007). After independence, especially during the drafting of the law of the non-navigational uses of international watercourses, the country's attitudes toward transboundary waters appeared to consolidate around this notion, albeit with some flexibility. In the 35th session, 59th meeting of the Sixth Committee of the General Assembly, the representative of Bangladesh pointed out that it was:

> important to ensure that a watercourse passing from one State to another, or through many States, was given an international character, and that any diversion or any other use of water which was in any way detrimental to any State should be made absolutely illegal. (Schwebel, 1981, p. 70)

Bangladesh did not forbid absolutely the utilization of a river by upstream states, but instead argued that upstream uses should not cause adverse impacts to downstream states. This practice demonstrates how Bangladesh modified its stance and moved towards the theory of limited territorial sovereignty on its transboundary waters. Water negotiations between India and Bangladesh made notable progress as a result.[3]

Transboundary water cooperation can prevent regional conflicts over water, notably where a riparian nation adopts the theory of limited territorial sovereignty, recognizing the interdependencies across the basin. Riparian nations can modify their approaches over time. For example, in its response to the ILC's 1982 questionnaire, Bangladesh supported the use of the term 'international water-course basin' and 'international drainage basin', instead of solely using the terms 'international watercourse' or 'river system' (UN General Assembly, 1982), which reflect a holistic approach. Bangladesh incorporated regional cooperation into its 1999 National Water Policy by entering into agreements with co-riparian countries (Ministry of Water Resources of Bangladesh, 1999). In its water policies and state practice Bangladesh has attempted to maintain the integrity of the basin and to bring it under the control of international law, consistent with the theory of limited territorial sovereignty. Nevertheless, as the most downstream riparian, Bangladesh has limited influence on the water management policies of upstream states (Sood & Mathukumalli, 2011).

The above survey reveals how differing riparian nation approaches to theories of territorial sovereignty have influenced state practice with respect to transboundary water cooperation. The next part examines the legal aspects of the transboundary water cooperation on the Yarlung Zangbo/Brahmaputra.

Transboundary water cooperation on the Yarlung Zangbo/Brahmaputra River: a legal analysis

Both China and India prefer to engage riparian states at the bilateral level (Ho, 2014; Wouters, 2014), whereas Bhutan and Bangladesh appear to be open to multilateral cooperation. None of the four countries, however, is party to the 1997 UN Watercourse Convention nor the 1992 UNECE Water Convention. At the regional level, there are no multilateral water treaties among them. While some argue that, by rejecting a multilateral framework, China is pursuing an 'upstream strategy' (Ho, 2014), this fails to recognize regional state practice, which shows mostly bilateral agreements. In most cases China prefers to take a bilateral approach, however there are still some exceptions. For example, on the Tumen and Mekong River Basin China wishes to promote its regional economy or navigation. India has signed over 20 bilateral water agreements with Bhutan, Bangladesh, China, Nepal and Pakistan, the majority of which are concerned with the construction of water projects.

Sino-Indian water agreements

At the bilateral level, Sino-Indian water cooperation primarily concerns flood control. The two states have, since 2002, concluded five Memorandums of Understanding (MoUs) on the provision of hydrological information of shared rivers, three of which concern the Yarlung Zangbo/Brahmaputra River: the 2002 MoU on the Provision of Hydrological Information of Yarlung Zangbo/Brahmaputra River in Flood Season by China to India (expired in 2007) and its implementation plan, the 2008 MoU and its implementation plan on the same topic for the period of 2008–12, and the 2013 MoU on the Provision of Hydrological Information on the Yarlung Zangbo/Brahmaputra River in Flood Season by China to India (China & India, 2013a) and its Implementation Plan which is now in force. In 2006, the Expert-Level Mechanism on Trans-border Rivers was established. China and India have also concluded the 2013 MoU on Strengthening Cooperation on Trans-border Rivers (China & India, 2013b). In order to analyse the operation and effectiveness of these treaties and MoUs, this article will look at five core elements used to evaluate the legal aspects of transboundary water regimes: scope, substantive rules, procedural rules, institutional mechanisms and dispute settlement mechanisms (Wouters, 2013). The next parts examine each Sino-Indian transboundary water agreement by extracting these five elements taking China and India's significant role in cooperation on the Yarlung Zangbo/Brahmaputra River and the following water agreements into consideration.

The 2013 Memorandum of Understanding on the Provision of Hydrological Information on the Yarlung Zangbo/Brahmaputra River in Flood Season by China to India

A treaty is defined as 'an international agreement concluded between States in written form and governed by international law, whether embodied in a single instrument or in two or more related instruments and whatever its particular designation' (Vienna Convention on Law of Treaties, 1969). With this in mind, this article will treat the 2013 MoU and its Implementation Plan as a single treaty rather than two agreements.

Objective and scope. This agreement was concluded to meet the demands for flood control and disaster mitigation in the downstream regions of the Yarlung Zangbo/Brahmaputra River. Different from the Sino-Russian transboundary water agreements which cover a wide range of water-related issues from economic uses to protection (Vinogradov & Wouters, 2013), this agreement focuses on the provision of hydrological information of the during the flood season by China to India. Therefore, Sino-Indian water cooperation appears to be at an early phase, however long-term cooperation regarding the provision of hydrological information has proved invaluable to flood forecasting and mitigation and has enhanced mutual trust (Ministry of Foreign Affairs of China, 2014).

Substantive rules. Article 1 of the 2013 MoU lays down the foundation of this agreement, namely friendly cooperation, mutual benefit, good faith and equality. Actually, this foundation has been repeated in several of China's transboundary water agreements.[4] This accords with China's approach in this field, the focus on peaceful diplomacy (the duty to cooperate; duty of equitable and reasonable utilization) (Wouters & Chen, 2013). In addition, unlike the 2002 MoU which only regulates China's obligation to provide hydrological information concerning the Yarlung Zangbo River in flood season, the 2008 MoU also requires India to 'provide [the] Chinese side with information regarding data utilization in flood forecasting and mitigation'. This provision was also rewritten into the 2013 MoU. This not only reveals the subtle change of attitudes of each state towards the other, but also suggests an emerging closer Sino-Indian cooperation and more open interaction on water issues. According to articles 5 and 6, China agrees to providing documents concerning the state of the Yarlung Zangbo River's catchment area as well as historical information regarding floods, natural disasters and any abnormal rise and fall of water levels or discharge which might lead to sudden flood (China & India, 2013a). According to article 10 of the 2013 Implementation Plan upon Provision of Hydrological Information of the Yarlung Zangbo/Brahmaputra River in Flood Season by China to India, India should pay China RMB850,000 within the period of validity of the present Implementation Plan to compensate for China's data provision. By way of comparison, article 4 of the 2008 MoU on the Provision of Hydrological Information of the Yarlung Zangbo/Brahmaputra River in Flood Season by China to Bangladesh states that 'China accede to bear the fees of provision of hydrological information of Yarlung Zangbo/ Brahmaputra to Bangladesh' (China & Bangladesh, 2008). Therefore, it would appear that China's diplomatic strategies differ according to the capacity of the countries with which it interacts (Ho, 2014). Under the UNWC (article 9(2)) payment for data provision is allowed.

Procedural rules. To fulfil the abovementioned rights and obligations, articles 2 and 3 provide specific technical aspects of cooperation. Article 2 concerns three Chinese hydrological stations required to provide hydrological information including the Nugesha, Yangcun and Nuxia stations. Due to the dramatic rise in water levels on the Yarlung Zangbo/Brahmaputra River during the monsoon season, article 3 requires China to provide hydrological information concerning water level, discharge and rainfall twice a day from 1 June to 15 October (China & India, 2013a). These procedural rules detail the functional cooperation under the agreement.

Institutional mechanisms. Different from the 2008 Mekong River Agreement (which regulates the establishment of a Joint Working Group in article 4),[5] the implementing agencies of this Sino-Indian MoU are the official departments of both sides. For India this is the Central Water Commission, Ministry of Water Resources, whereas for China this includes the Bureau of Hydrology and Water Resources.

In the 2006 Joint Declaration between China and India, both sides agreed to establish an Expert-Level Mechanism to discuss interaction and cooperation regarding the provision of flood-season hydrological data, emergency management and other trans-border river issues (China & India, 2006). Article 11 of the Implementation Plan of the 2013 MoU refers to the Expert-Level Mechanism:

> At the meeting of the India–China Expert Level Mechanism on Trans-Border Rivers, the Indian side will also inform the Chinese side the information of the hydrological station which lies on the mainstream of the Yarlung Zangbo/Brahmaputra River and is close to China's Nuxia station.

Although eight meetings of the Expert-Level Mechanism have been held so far (Ministry of Water Resources of India, 2014), it is unclear how it operates or what progress has been made.

Compared with other Chinese practice, China's institutional engagement with India is less developed. For example, the 2008 Sino-Russian Agreement on Reasonable Utilisation and Protection of trans-border Rivers provides not only for the establishment of a Joint River Commission in article 4, but also rules for the 'Competent Authorities' of both parties for the coordination and organization of cooperation under its framework (article 3) (Vinogradov & Wouters, 2013). The 2001 Agreement on Cooperation in the Use and Protection of Transboundary River between China and Kazakhstan establishes joint bodies in article 8 (Vinogradov & Wouters, 2013); article 10 of the 1994 Agreement on Protection and Utilization of Boundary Waters between China and Mongolia mentions the creation of a joint committee by each party to appoint one representative. China's transboundary water cooperation with some of its co-riparian neighbours including Russia, Mongolia and Kazakhstan demonstrates possible models of institutional mechanisms that might be further developed with India and other Southeast Asian states. Border disputes between China and India continue to be one of the main areas of concern in diplomatic relations for both countries (Wiegand, 2011), but progress is being made there also (Wouters, 2014).

Dispute settlement. Dispute settlement appears to be neglected by the 2013 MoU, partly because of its narrow area of cooperation. The 2008 MoU on the Provision of Hydrological Information between China and Bangladesh and the 2002 MoU on the Provision of Hydrological Information of the upper stream of Zuo River during Flood Season by Vietnam to China also do not include any articles on dispute settlement. China generally prefers the principle of peaceful settlement of disputes which can be illustrated by China's water agreements with Kazakhstan, Mongolia and Russia, however China is inclined to address disputes through consultations or negotiation with other states (Wouters & Chen, 2013). This would appear to be the ongoing practice between China and India with respect to its transboundary waters.

2013 Memorandum of Understanding on Strengthening Cooperation on Trans-border Rivers between China and India

After the signature of the 2013 MoU on the Yarlung Zangbo/Brahmaputra River, China and India signed the 2013 MoU on Strengthening Cooperation on Trans-border Rivers. This was the first MoU governing all Sino-Indian transboundary rivers. The legal issues of scope, substantive rules, procedural rules, institutional mechanisms and dispute settlement are each covered to varying degrees in that agreement. On the matter of *scope*, the use of the term 'trans-border rivers' is quite specific, limiting the legal reach to only rivers, thus excluding other bodies of water; notably there is no definition of this term in the agreement. The scope here is narrower than China's water agreements with Russia and Mongolia,[6] and even narrower than the scope employed by the UNWC and the UNECE. Nonetheless, it is broader than earlier Sino-Indian water agreements which only relate to the provision of hydrological information. Significantly, the scope refers to the 'related natural resources and environment of all riparian countries', which adopts a more inclusive and perhaps more integrated approach (article 1). This provision opens the door for an integrated water resources management approach. China and India recognize transboundary rivers as well as their related natural resources and environment as a whole and connect them with socio-economic development. It implies that both sides wish to sustain the utilization of their shared rivers and resources.

On *substantive rules*, the agreement is quite limited (only five provisions), largely reaffirming past cooperation. Articles 2 and 5, *procedural provisions*, attach great importance to the Expert Level Mechanism (*institutional mechanism*), and the provision of flood-season hydrological data and emergency management, the exchange of views on issues of mutual interest, and agreement to enhance mutual cooperation through the Expert Level Mechanism. Article 3 shows India's appreciation for the provision of hydrological information by China and article 4 extends the period of data sharing concerning the Yarlung Zangbo/Brahmaputra River which was written in the 2013 MoU on the Provision of Hydrological Information on the Yarlung Zangbo/Brahmaputra River in Flood Season by China to India.

The 2013 MoU on Strengthening Cooperation is one more step towards more effective transboundary water cooperation between China and India. This incremental, step-by-step approach mitigates against the potential for conflict, which remains a possibility in the absence of more complete agreements (Vajpeyi, 2012).

Sino-Indian environmental agreements

Environmental sustainability was recognized as one of the four dimensions of water governance within the 2006 United Nations World Water Development report (UNESCO, 2006). Sino-India environmental agreements are examined here as part of the overall legal framework involving transboundary water cooperation. This part summarizes the key treaties in this area.

In 1993, India and China concluded the bilateral Agreement on Environmental Cooperation. Article 2 commits the parties to enhancing bilateral cooperation in all fields of environmental concern on the basis of equality and mutual benefit, echoing the Five Principles of Peaceful Co-existence (China & India, 1993). More details are provided in article 3, which includes 'water quality protection' (para. 3, article 3) as one of the high priority cooperation areas (China & India, 1993). (The scope of this provision, however, is narrower than the 1994 Sino-Russian Agreement on Environment Cooperation, which uses the 'comprehensive utilization of water resources and water body protection (including border rivers)' (article 2(2)) as one of the areas of environmental cooperation.)

While the 1993 Agreement makes provision for environmental impact assessment (article 3(4)), this is not linked with transboundary water issues. Instead it emphasizes 'procedures and experiences' rather than imposing an obligation to conduct an environmental impact assessment upon China and India.

The 2009 Agreement on Cooperation on Addressing Climate Change between China and India, aimed at enhancing bilateral cooperation on climate change (particularly in the areas of energy efficiency, renewable energy, clean technology, sustainable agriculture and afforestation) (China & India, 2009). Considering that 30% of the flow of the Yarlung Zangbo/Brahmaputra River comes from the melt water of Himalayan glaciers this agreement is very important for the utilization and protection of the river. On the one hand, the huge hydropower storage of the Yarlung Zangbo/Brahmaputra River creates a platform for promoting increased Sino-Indian cooperation; on the other hand, the sustainable utilization of the Yarlung Zangbo/Brahmaputra River calls for the sustainable management of water and land ecosystems. In article 3 of this agreement, the two parties agree to cooperate on the reduction of greenhouse gas emissions which further entails energy conservation and increasing efficiency, afforestation, and the sustainable management of forests and ecosystems – all potentially linked with effective transboundary water management. The agreement establishes an India–China Working Group on Climate Change to negotiate their respective domestic policies and relevant measures and to assist with implementation of cooperative projects (article 2(2)).

India and Bangladesh transboundary water cooperation

Sino-Bangladesh cooperation on the Yarlung Zangbo/Brahmaputra River deals primarily with flood control. The only agreement between China and Bangladesh on this river is the 2008 MoU on the Provision of Hydrological Information of the Yarlung Zangbo/ Brahmaputra River in Flood Season by China to Bangladesh.

Although Bangladesh and India have a long history of transboundary water tensions, with some progress, many issues remain. In 1972 the Joint Rivers Commission (JRC) was established under the Statute of Indo-Bangladesh Joint River Commission. The JRC is composed of a chairman and three members from each state (India & Bangladesh, 1972). According to article 4, the functions of the JRC are restricted to flood control. The JRC holds ordinary sessions four times a year and special meetings of Working Groups or Ad-hoc Expert Groups nominated by the parties (India & Bangladesh, 1972). All the meetings of the JRC 'shall be closed meetings unless the Commission desires otherwise' (India & Bangladesh, 1972). The JRC has no remit to settle disputes; differences should be referred to the two governments in order to reach a settlement (India & Bangladesh, 1972). Recommendations to restructure the JRC to improve its effectiveness might result in changes (Iyer, 2003; Strategic Foresight Group, 2013). It has been four years since the last JRC meeting (*Bdnews24*, 2014). Notwithstanding these drawbacks, the JRC has worked on reaching agreement on the water sharing on the Ganges and Teesta rivers, despite the sensitive political environment between India and Bangladesh.

Negotiations between India and Bangladesh on the Teesta River, a tributary of the Yarlung Zangbo/Brahmaputra River, can be traced back to the 1950s–60s before the independence of Bangladesh (The Asia Foundation, 2013). After the establishment of the JRC, it met on a variety of issues in relation to the Teesta River. Finally, in 1983, during the 25th Meeting of the JRC, India and Bangladesh signed the Agreement on Ad-hoc Sharing of the Teesta Waters, which expired in 1985. Under the ad-hoc agreement, Bangladesh receives 36% of the water, India 39% with the remaining 25% unallocated. This agreement was never signed as a formal instrument, but instead included in the minutes of the JRC (Salman & Uprety, 2002), without any implementation plan for realizing the water allocation. In 2004, the India–Bangladesh Joint Technical was formed, which drafted an agreement on the Teesta (The Asia Foundation, 2013). Although both sides were due to sign the draft agreement on the use of the Teesta River in 2011, this deal was eventually refused (GlobalVoices, 2012). To this day, India and Bangladesh have not reached consensus on the uses of Yarlung Zangbo/Brahmaputra River.

India and Bhutan transboundary water cooperation

India and Bhutan have a long history of cooperation on joint hydropower projects; electricity generation is the single most important source of revenue for Bhutan. This cooperation, beginning in 1961 with the signing of the Jaldhaka Agreement between Bhutan and India, has resulted in one of the most successful projects between India and Bhutan: the Chukha Hydro-Electric Project. According to the 1974 Agreement on the Chukha Hydro-Electric Project, which has acted as a foundation for transboundary cooperation and shared benefits between the two states (Shrestha et al., 2015), India provides financial resources for the project (article 3) and Bhutan sells the surplus electricity generated by this project to India (article 8). It is said that approximately 70% of the power generated by the Chukha project is exported to India (Banskota, 2012). Through this hydropower agreement Bhutan has achieved economic and social development gains (Biswas, 2004).

Several agreements have since been signed, including the 1996 Agreement on Tala Hydroelectric Project (revised in 2012), the 2006 Agreement on Punatsangchhu-I Hydroelectric Project, and the 2006 Agreement Concerning Cooperation in the Field of Hydroelectric Power. More recently, India has agreed to build four more hydropower projects in Bhutan (*The Times of India*, 2014). Largely funded by India, the hydropower cooperation between Bhutan and India is regarded as a model for how transboundary water bodies can be successfully jointly managed by basin-sharing countries to promote regional economic development (Biswas, 2011).

Take the 60-year 2006 Agreement Concerning Cooperation in the Field of Hydroelectric Power as an example. With 12 provisions, it builds on past joint exploitation experiences and establishes a legal framework for the furtherance of hydroelectric power cooperation. It provides for the construction of hydropower projects through both public and private sector participation (article 1). For the projects which are jointly implemented by the two governments, a joint group is to be established in order to facilitate the process and India agrees to a minimum import of 5000 MW electricity from Bhutan by 2020 (article 2). Both states wish to cooperate on renewable energy under the Clean Development Mechanism of the Kyoto Protocol (article 7), however none of the provisions takes environmental protection or environmental impact assessment into consideration even though hydropower stations drastically impact water and land ecosystems. Within the texts of the hydropower agreements, only article 10(3) of the 2006 Agreement on Punatsangchhu-I Hydroelectric Project provides that 'the Authority (the Punatsangchhu-I Hydroelectric Project Authority) shall ensure that the Project complies with all statutory environmental legislation of Bhutan'. Therefore, notwithstanding the gains that India and Bhutan will experience due to the economic benefits the joint development of hydropower resources, they have appeared to exclude environmental protection provisions.

Summary and observations

In sum, there are diverse, but no comprehensive, transboundary water agreements across Southern Asia (Chatterji & Jain, 2008). The above survey yields the following observations: First, in terms of the utilization of the water resources of the Yarlung Zangbo/Brahmaputra River, both China and India embrace theories of national sovereignty to justify building dams and hydropower stations within their own territory, largely through unilateral action. Legal frameworks requiring prior notification are not in place. Bangladesh and Bhutan have limited power or influence to influence more basin-wide cooperation. Secondly, none of the four basin riparians has signed any global or multilateral water sharing agreements, but instead they prefer bilateral arrangements. Third, most of the bilateral transboundary water cooperation is issue specific, for example focusing on flood control, data sharing or hydroelectric projects. Fourth the environmental of the transboundary watercourse has not attracted satisfactory consideration, despite the growing threats to the ecosystem of the Yarlung Zangbo/Brahmaputra River Basin. These four features of existing transboundary water cooperation on the Yarlung Zangbo/

Brahmaputra River illustrate a somewhat fragmented picture, an incomplete and fragile foundation for any future meaningful cooperation.

Several attempts at addressing these problems – such as expanding the mandate of the South Asia Association for Regional Cooperation (SAARC) (i.e. to act as a mediator in the settlement of water related disputes and facilitate the creation of water-sharing agreements in the region) (Chatterji & Jain, 2008; Puthucherril, 2014); or the idea of a South Asian Water Convention (University of Bangladesh, 2014) – have not yet yielded results. The following sections explore various possibilities for enhancing transboundary cooperation on the Yarlung Zangbo/Brahmaputra River.

The Yarlung Zangbo/Brahmaputra River: exploring a multilateral platform for enhanced transboundary water cooperation

The current legal approaches on the basin provide some insights into how future cooperation might be crafted. The next section examines possible methods of transboundary water cooperation that will assist the riparian states to move forward in accordance with the fundamental principles of international water law with respect to the uses of the Yarlung Zangbo/Brahmaputra River.

Sharing benefits through joint exploitation: integrated hydropower development approach as a platform for enhanced transboundary cooperation

As mentioned above, hydropower and flood control (data sharing) are currently the two fields of cooperation found within the Yarlung Zangbo/Brahmaputra River Basin. The Yarlung Zangbo/Brahmaputra River possesses a large potential for hydropower, which is unevenly distributed. China, India and Bhutan possess 53.4%, 32.0% and 14.6% of the river's hydropower potential respectively, while Bangladesh possesses 0% (Rahaman, 2012). Integrated hydropower development that takes into account downstream interests and ensures environmental protection could be a driving force for regional transboundary water cooperation (Rahaman, 2012). The 1961 Columbia River Treaty is an example of such a strategy. Canada, located upstream on the Columbia River, has constructed and operates three dams on the river. In return for the flood control and power benefits of the dams, the United States pays Canada agreed amounts (Columbia River Treaty, 1961). Building on existing practice (Bhutan/India) and supplementing this with legal requirements to address downstream and environmental concerns, a forward-looking strategy could include the three main features of an integrated hydropower development approach: transformation of interests from flowing water to electricity, the sharing of benefits of more electricity and peaceful regional relationships, and risk-sharing, meaning that states will share the burden of floods, droughts or even the adverse impacts of global warming and climate change. It is an approach based on the notion of limited territorial sovereignty, balancing upstream development with downstream interests, in accordance with the governing customary rule of equitable and reasonable utilization.

Given the current political conditions, the conclusion of a comprehensive water agreement involving all four riparian countries might not be feasible, but a phased bilateral step-by-step approach could be envisaged. India and Bhutan have developed a good example of how transboundary cooperation around joint hydropower projects might be implemented. China and Kazakhstan began the long-awaited China–Kazakhstan Friendship Joint Water Diversion Project on the Khorgos River, a 150-kilometre-long tributary of the Ili which forms the border between both countries (*China Daily*, 2011).

China also cooperates with North Korea on the construction of the Sup'ung Hydroelectric Plant, which included broader transboundary water cooperation as part of the arrangement (China & North Korea, 1955). Bilateral cooperation between China and Russia concern issues of navigation, fisheries, power generation, water resource utilization and ecosystem protection (Vinogradov & Wouters, 2013). Transboundary cooperation based on joint hydropower projects might be a way forward in the future.

Joint protection: environmental cooperation as a platform for enhanced transboundary water cooperation

Biodiversity is a cross-sectoral issue, and virtually all sectors have an interest in its conservation and the sustainable use of its components (Glowka, Burhenne-Guilmin, Synge, & McNeely, 1994). Transboundary waters represent high levels of biodiversity (Roy, Barr, & Venema, 2011), and the Yarlung Zangbo/Brahmaputra River is one of 17 international waters that has been deemed to be at risk and which lacks a river basin institutional arrangement (Wolf et al., 2003). Hydro projects and dams have significant influences on biodiversity in terrestrial, coastal and marine ecosystems (Secretariat of the Convention on Biological Diversity, 2010). There are no appropriate and applicable water agreements on the Yarlung Zangbo/Brahmaputra River which govern these issues. While China, India, Bhutan and Bangladesh are each parties to the following multilateral environmental agreements – Convention on Biological Diversity (CBD), Convention on Wetlands (RAMSAR), the Convention on Climate Change (UNFCC) and the Convention to Combat Desertification (UNCCD) – these multilateral platforms have not been considered as vehicles for enhancing transboundary water cooperation across the region. As only one example, this paper will examine the potential contribution of the CBD in this regard.

First, in light of article 2(1) CBD, 'aquatic ecosystems' are considered one of the sources of living organisms and thus linked directly with transboundary water resources (United Nations, 1992). Therefore, through the duty to protect biodiversity, the CBD directly applies to transboundary waters, an obligation agreed to by each of the four co-riparian states of the Yarlung Zangbo/Brahmaputra.

Second, the duty to cooperate and to conduct an environmental impact assessment (EIA) within the CBD covers transboundary waters. Article 5 expressly requires that each contracting party cooperate for the conservation and sustainable use of biodiversity in respect to areas beyond national jurisdiction and on other matters of mutual interest (Convention on Biological Diversity, 1992). Various duties to cooperate are also elaborated throughout the CBD – articles 8(m), 9(e), 12 (c), 13(b), 16(5) and 28. Article 14 requires parties to conduct an EIA for proposed projects and to reach bilateral, regional or multilateral arrangements in order to promote communication on activities under their jurisdiction or control which are likely to affect significantly and adversely the biodiversity of other states or areas (United Nations, 1992). Therefore, given that there are no water agreements that require China, India or Bhutan to assess the environmental impacts of proposed projects, article 14 of the CBD lays down duties and enhances the transparency of water projects.

Third, the CBD Programme of Work on Protected Areas (PoWPA), adopted by the Seventh Meeting of the Conference of the Parties (COP) to the CBD in 2004 (Conference of Parties to Convention on Biological Diversity, 2004), also takes the establishment of transboundary protected areas into account under the guidance of an ecosystem approach (Secretariat of the Convention on Biological Diversity, 2004). The creation of protected areas along the Yarlung Zangbo/Brahmaputra River will, to some extent, help to protect local ecosystems and this could be an operational platform for cooperation. China has already established two protected areas along the Yarlung Zangbo River, namely the Yarlung Zangbo River Great Canyon National Protected Area (916800 hectares) and the Yarlung Zangbo River Middle Reaches Black-necked Crane Nature Protected Area (614,350 hectares) (Government of China, 2012). In India, 12 protected areas lying along the alluvial plains of the Yarlung Zangbo/Brahmaputra River cover approximately 250,000 hectares of intact habitat (World Wild Life, 2005).

It is possible to establish a transboundary protected area along the Yarlung Zangbo/Brahmaputra River and there are examples of jointly protected areas between China and its neighbours. In 1994, China, Russia and Mongolia founded the Dauria International Protected Area. In 2010, China, India and Nepal created the first Transboundary Biodiversity Initiative for conservation of the Mount Kailash Sacred Himalayan Landscape, with the support of the International Center for Integrated Mountain Development. Again on June 2014, China, India and Myanmar agreed on a framework for regional cooperation for the Yarlung Zangbo/Brahmaputra–Sallween Transboundary Landscape in promoting biodiversity conservation. This cooperation is an application of the CBD at the regional level and will contribute to the protection of the Yarlung Zangbo/Brahmaputra River.

Fourth, through the CBD Programme on Inland Water Biodiversity (PoIWB) 'inland waters' has been identified as an important thematic work area (Revenga & Kura, 2003). Transboundary rivers, lakes and other water bodies are clearly part of inland waters. From a historical view, inland water biodiversity was first put on the agenda by the COP to the CBD in its second ordinary meeting (Secretary of the Convention on Biological Diversity, 1995). After continuous work on this issue, COP 4 adopted a work programme on inland waters biodiversity (Secretary of the Convention on Biological Diversity, 1998). These actions have attracted ever increasing attention at the national, regional and global levels.

Together, these multilateral duties linked directly to transboundary water resources issues and binding on all four riparians provide a legal platform for enhanced operational cooperation on the Yarlung Zangbo/Brahmaputra.

Observations and conclusions

The Yarlung Zangbo/Brahmaputra River is a rich resource that contributes directly to the social, economic and environmental well-being of its riparian countries. This paper has explored the basin state approaches to national territorial sovereignty with respect to their transboundary water practice. This revealed China's adherence to 'limited territorial sovereignty' has influenced its state practice. India as a mid-stream riparian nation state has vacillated between absolute territorial integrity and absolute territorial sovereignty in support of its stance, depending on its location on the respective watercourse. Bangladesh, downstream appears to have moderated its position and adopts a limited territorial sovereignty approach. Bhutan, a basin state, through its bilateral agreements with India (mostly hydropower projects), would appear to also embrace the limited territorial sovereignty theory.

The research also revealed a patchwork of bilateral agreements on the Yarlung Zangbo/Brahmaputra. There are no comprehensive basin-wide transboundary water cooperation agreements, but a series of ad hoc agreements, including MoUs that are either quite specific (i.e. contractual-type arrangements on hydropower projects, mostly between India and Bhutan) or more general agreements for data provision (i.e. between China and

India). This spectrum of typologies of transboundary water cooperation on this watercourse reveal the potential for local tailor-made solutions.

Are there existing models of cooperation on the Yarlung Zangbo/Brahmaputra that might enhance transboundary water cooperation? The author proposes two separate avenues that this might be achieved: through joint exploitation or joint protection legal regimes. In the first example, the legal agreements between India and Bhutan, classified as models of best practice (Biswas, 2011), might be replicated across the basin in a series of bilateral agreements, perhaps extending to a multilateral platform. Of course this depends upon a number of factors, but the legal framework is there for consideration. Another opportunity lies in the existing multilateral environmental cooperation already signed up by all the riparian countries. The author considers that the Convention on Biodiversity provides a solid vehicle for taking forward transboundary water cooperation based on the shared obligations under that instrument.

One major shortcoming from the state practice surveyed here is the lack of dispute settlement arrangements. The Sino-Indian Expert mechanism might be able to provide a forum for dispute prevention discussions, but the lack of transparency on the operation of that organ prevent more detailed observations.

Acknowledgements

The author like to thank Professor Patricia Wouters, supervisor, for her considerable support during the revision and finalizing of this paper. The author's gratitude to David Devlaeminck for assistance with the English editing of the paper. Many thanks to the China International Water Law team (CIWL, School of Law, Xiamen University) who supported the author in many ways, especially with constructive critical feedback on earlier drafts of this work. Thanks are due also to the anonymous peer reviewers.

Notes

1. For a comprehensive picture of the Brahmaputra river, see the FAO-AQUASTAT 2011 – Map of Ganges–Brahmaputra–Meghna Basin. Retrieved December 2, 2014, from http://www.fao.org/nr/water/aquastat/basins/gbm/gbm-map_detailed.pdf
2. China, India, Bhutan and Bangladesh are still agricultural countries which depend heavily on irrigation. Clean energy such as hydropower generated by the Yarlung Zangbo/Brahmaputra River provides momentum for their development.
3. The establishment of the Joint River Commission in 1972, the conclusion of the 1974 Agreement Concerning the Demarcation of the Land boundary, the 1977 Agreement on Sharing of Ganges Water at Farakka, and the 1982 MOU for Sharing the Flows of the Ganges During the Dry Seasons of 1983 and 1984.
4. Preamble of the 2008 Sino-Russian Agreement on Reasonable Utilisation and Protection of Transboundary Waters; article 1 of 2008 MoU on Provision of Hydrological Information of the Yarlung Zangbo/Brahmaputra River in Flood Season by China to Bangladesh; Preamble of 2001 Agreement Between Kazakhstan and China on Cooperation in the Field of Use and Protection of Transboundary Rivers.
5. 2008 Mekong Agreement on the Provision of Hydrological Information of Mekong River by China to Mekong River Commission.
6. The term 'transboundary waters' was used in the 2008 Sino-Russian Agreement on Reasonable Utilisation and Protection of Transboundary Waters and in the 1994 Agreement on the Protection and Utilization of Transboundary Waters.

References

Alam, B. M. (2008). Trans-boundary water sharing of the Ganges river: A regional planning approach. *World Environmental and Water Resources Congress, 2008*, 1–8. doi:10.1061/40976(316)261

Asia Development Bank. (2011). *Bangladesh: Country Partnership Strategy (2011–2015) – Sector Assessment (Summary): Agriculture and Natural Resources*. Retrieved December 2, 2014, from http://www.adb.org/sites/default/files/linked-documents/cps-ban-2011-2015-ssa-05.pdf

Banskota, N. P. (2012). *South Asia trade and energy security: The role of India*. Boca Raton: Universal-Publishers.

Bdnews24. (2014, September 20). India agrees to schedule JRC meeting suiting Bangladesh. *Bangladesh's First Internet Newspaper*. Retrieved December 7, 2014, from http://bdnews24.com/bangladesh/2014/09/20/india-agrees-to-schedule-jrc-meeting-suiting-bangladesh

Biswas, A. (2004). Water and regional development. In A. K. Biswas, O. Ünver, & C. Tortajada (Eds.), *Water as a focus for regional development* (pp. 1–13). New Delhi: Oxford University Press.

Biswas, A. K. (2011). Cooperation or conflict in transboundary water management: Case study of South Asia. *Hydrological Sciences Journal, 56*(4), 662–670. doi:10.1080/02626667.2011.572886

Brownlie, I. (2008). *Principles of public international law* (7th ed.). New York, NY: Oxford University Press.

Caponera, D. A., & Nanni, M. N. (2007). *Principles of water law and administration: National and international* (2nd ed.). London: Taylor & Francis.

Chatterji, M., & Jain, B. (2008). *Conflict and peace in South Asia*. Birmingham: Emerald Group Publishing Limited.

Chellaney, B. (2009). Coming water wars: Beware the future. *The International Economy*, 38–39. Retrieved December 2, 2014, from http://www.international-economy.com/TIE_F09_Chellaney.pdf

Chen, A. (2013) *The voice from China – An CHEN on international economic law*. Berlin and Heidelberg: Springer.

China & Bangladesh. (2008). Memorandum of understanding upon provision of hydrological information of the Brahmaputra/Yalunzangbu River in flood season by China to Bangladesh. Retrieved December 2, 2014, from http://shlx.chinalawinfo.com/newlaw2002/SLC/SLC.asp?Db=eag&Gid=100670807

China & India. (1993). Agreement on Environmental Cooperation between the Government of the Republic of India and the Government of the People's Republic of China. Retrieved December 2, 2014, from http://www.mea.gov.in/Portal/LegalTreatiesDoc/CHB1157.pdf

China & India (2006). Joint Declaration by the Republic of India and the People's Republic of China. Retrieved December 2, 2014, from http://pib.nic.in/newsite/erelease.aspx?relid=22168

China & India. (2009). Agreement on Cooperation on Addressing Climate Change between the Republic of China and the Government of the Republic of India. Retrieved December 2, 2014, from http://pib.nic.in/newsite/erelease.aspx?relid=53317

China & India. (2013a). 2013 Memorandum of Understanding on the Provision of Hydrological Information on the Yaluzangbu/Yaluzangbu/Brahmaputra River in Flood Season by China to India. Retrieved October 20, 2014, from http://mea.gov.in/Portal/LegalTreatiesDoc/CH13B0811.pdf

China & India. (2013b). Memorandum of Understanding between the Ministry of Water Resources, the Republic of India and the Ministry of Water Resources, the People's Republic of China on Strengthening Cooperation on Trans-border Rivers. Retrieved December 2, 2014, from http://www.indianembassy.org.cn/ArchiveNewsDetails.aspx?NewsId=438

China & Kazakhstan. (2001). Agreement between the Government of Kazakhstan and the Government of People's Republic of China on management and protection of transboundary rivers. Retrieved December 2, 2014, from http://faolex.fao.org/docs/texts/bi-65815e.doc

China & North Korea. (1955). 1955 Protocol on Yanlujiang River Sup'ung Hydroelectric Plant Between the People's Republic of China and North Korea. Retrieved December 2, 2014, from http://policy.mofcom.gov.cn/section/gjty!fetch.action?libcode=gjty&id=TYCX000104&industrycode=S09421;D04412&secondcode=563004;563006

China & Russia. (2008). Agreement on Reasonable Utilisation and Protection of Transboundary Waters between the People's Republic of China and the Government of Russian Federation. Retrieved December 2, 2014, from http://www.chinainternationalwaterlaw.org/pdf/treaties/russia/2008.pdf (in Chinese).

China Daily. (2010, November 19). China pledges water will still flow. (Q. Wang, Ed.) *China Daily*. Retrieved December 27, 2014, from http://www.chinadaily.com.cn/cndy/2010-11/19/content_11574260.htm

China Daily. (2011, April 15). *Joint Water Diversion Project Launched in Xinjiang*. Retrieved December 2, 2014, from *China Daily*: http://www.chinadaily.com.cn/china/2011-04/15/content_12336268.htm

China Daily. (2014, November 24). Major hydroplant begins operation in power-thirsty Tibet. *China Daily*. Retrieved December 2, 2014, from http://usa.chinadaily.com.cn/business/2014-11/24/content_18966002.htm

Express Tribune News Network. (2014, December 12). Vulnerable to climate change: South Asia at risk. *Express Tribune News Network*. Retrieved December 27, 2014, from http://tribune.com.pk/story/805585/vulnerable-to-climate-change-south-asia-at-risk-of-food-water-insecurity/

FAO. (2011a). *FAO Water Report 37-Irrigation in Southern and Eastern Asia in Figures: Bhutan*. Retrieved December 2, 2014, from http://www.fao.org/nr/water/aquastat/countries_regions/BTN/BTN-CP_eng.pdf

FAO. (2011b). *FAO Water Report 37-Irrigation in Southern and Eastern Asia in Figures: Ganges–Brahmaputra–Meghna River Basin*. Retrieved December 2, 2014, from http://www.fao.org/docrep/016/i2809e/i2809e.pdf

GlobalVoices. (2012, June 8). India, Bangladesh: Water disputes and Teesta river diplomacy. *GlobalVoices*. Retrieved December 11, 2014, from http://globalvoicesonline.org/2012/06/08/india-bangladesh-water-disputes-and-teesta-river-diplomacy/

Glowka, L., Burhenne-Guilmin, F., Synge, H., & McNeely, J. A. (1994). *A guide to the convention on biological diversity*. Bonn: IUCN. Retrieved December 2, 2014, from https://portals.iucn.org/library/efiles/documents/EPLP-no.030.pdf

Government of Bangladesh. (2013, November 6). Bangladesh Water Act 2013. Article 7. Retrieved December 2, 2014, from http://www.warpo.gov.bd/pdf/WaterActEnglish.pdf

Government of China. (2012). 2012 China Natural Reserve. Retrieved December 2, 2014, from http://sts.mep.gov.cn/zrbhq/zrbhq/201309/P020130927386454626867.pdf

Government of the People's Republic of China. (2013, January 1). China's 12th Five Year Plan of Energy Development in Chinese. Retrieved December 24, 2014, from http://www.gov.cn/zwgk/2013-01/23/content_2318554.htm

Government of Tibet Autonomous Region. (2006). 11th five-year plan of Tibet's development of society and economy (2006–2010). Retrieved December 24, 2014, from http://www.gov.cn/test/2006-02/08/content_181832.htm

Ho, S. (2014). River politics: China's policies in the Mekong and the Brahmaputra in comparative perspective. *Journal of Contemporary China*, *23*(85), 1–20. doi:10.1080/10670564.2013.809974

Hoq, M. E. (2009). *BOBLME (2009) Report of the Ganga Brahmaputra workshop*. Bangladesh Fisheries Research Institute, New Delhi. Retrieved December 2, 2014, from http://www.boblme.org/documentRepository/BOBLME-2009-Ecology-01.pdf

IMF. (2011). *IMF Country Report No. 11/123*. Retrieved December 29, 2014, from http://www.imf.org/external/pubs/ft/scr/2011/cr11123.pdf

India & Bangladesh. (1972). Statute of Indo-Bangladesh joint river commission. Retrieved December 2, 2014, from http://www.internationalwaterlaw.org/documents/regionaldocs/indo-bangladesh.html

Indiatoday. (2014, November 23). China puts first Yaluzangbu/Brahmaputra dam into operation. *Indiatoday*. Retrieved December 24, 2014, from http://indiatoday.intoday.in/story/brahmaputra-dam-india-vs-china-zangmu-yarlung-tsangpo-zangbo-hydropower-project/1/403379.html

Iyer, R. R. (2003). *Water: Perspectives, issues, concerns*. New Delhi: SAGE Publications.

McCaffrey, S. C. (1996). The Harmon Doctrine one hundred years later: Buried, not praised. *Natural Resources Journal*, *36*, 965–1007. Retrieved December 2, 2014, from http://lawschool.unm.edu/nrj/volumes/36/4/13_mccaffrey_harmon.pdf

McCaffrey, S. C. (1986). *A/CN.4/399 and Add.1 and 2, Second Report on the Law of the Non-navigational Uses of International Watercourses*. Retrieved December 2, 2014, from http://legal.un.org/ilc/documentation/english/a_cn4_399.pdf

McIntyre, O. (2006). The role of customary rules and principles of international environmental law in the protection of shared international freshwater resources. *Natural Resources Journal*, *46*, 157–210.

Ministry of Foreign Affairs of China. (2014, November 24). Foreign ministry spokesperson Hua Chunying's Regular Press Conference on November 24, 2014. China. Retrieved December 27, 2014, from http://www.fmprc.gov.cn/mfa_eng/xwfw_665399/s2510_665401/t1214040.shtml

Ministry of Water Resources of Bangladesh. (1999). 1999 National Water Policy. Retrieved December 27, 2014, from http://www.mowr.gov.bd/images/pdf/National%20Water%20Policy %20(English).pdf

Ministry of Water Resources of India. (2014, December 2). *India-China Co-Operation*. Retrieved from Ministry of Water Resources of India: http://wrmin.nic.in/forms/list.aspx?lid=349

Ministry of Water Resources, Government of India. (2012). National water policy. India. Retrieved December 27, 2014, from http://wrmin.nic.in/writereaddata/NationalWaterPolicy/ NWP2012Eng6495132651.pdf

National Encyclopedia of Bangladesh. (2012). *Brahmaputra–Jamuna River System*. Retrieved December 2, 2014, from http://www.banglapedia.org/HT/R_0265.htm and http://www.banglapedia.org/HT/B_0698.htm

Negi, S. S. (1991). *Himalayan rivers, lakes, and glaciers* (2004 ed.). New Delhi, India: Indus Publishing Company.

Nickum, E. J. (2008). The upstream superpower: China's international rivers. In O. B. Varis (Eds.), *Management of transboundary rivers and lakes* (pp. 227–244). Berlin and Heidelberg: Springer.

Paisley, R., & McDaniels, T. (1995). International water law, acceptable pollution risk and the Tatshenshini river. *Nature Resources Journal*, *35*, 111–132.

Pakistan Observer. (2011, March 7). India's water hegemony. *Pakistan Observer*. Retrieved December 3, 2014, from http://pakobserver.net/detailnews.asp?id=79655

People's Republic of China. (2009, April 16). *Written Statement of the People's Republic of China to the International Court of Justice on the Issue of Kosovo*. Retrieved December 27, 2014, from International Court of Justice: http://www.icj-cij.org/docket/files/141/15611.pdf

Puthucherril, T. G. (2014). *Towards sustainable coastal development: Institutionalizing integrated coastal zone management and coastal climate change adaption in South Asia*. Leiden, Netherlands: Brill Nijhoff.

Rahaman, M. M. (2009). Principles of transboundary water resources management and Ganges treaties: An analysis. *International Journal of Water Resources Development*, *25*(1), 159–173. Retrieved December 2, 2014, from http://www.internationalwaterlaw.org/bibliography/articles/ general/Rahaman-Ganges-Water_Res_Devel.pdf

Rahaman, M. M. (2012). Hydropower ambitions of South Asian nations and China: Ganges and Brahmaputra rivers basins. *International Journal of Sustainable Society*, January *4*(1–2), 131–157. doi:10.1504/IJSSOC.2012.044670

Rahaman, M. M., & Varis, O. (2009). Integrated water management of the Brahmaputra basin: Perspectives and hope for regional development. *Natural Resources Forum*, *33*, 60–75. Retrieved December 2, 2014, from https://www.academia.edu/2290064/Integrated_water_mana gement_of_the_Brahmaputra_basin_Perspectives_and_hope_for_regional_development

Revenga, C., & Kura, Y. (2003). *CBD technical series: Status and trends of biodiversity of inland water ecosystems*. Montreal: Secretariat of the Convention on Biological Diversity. Retrieved November 29, 2014, from https://www.cbd.int/doc/publications/cbd-ts-11.pdf

Rieu-Clarke, A. (2005). *International law and sustainable development: Lessons from the law of international watercourses* (1st ed.). London, UK: IWA Publishing.

Rieu-Clarke, A., Moynihan, R., & Magsig, B.-O. (2012). *UN Watercourses Convention: User's Guide*. Retrieved December 2, 2014, from http://www.gwp.org/Global/Our%20Approach/ Strategic%20Allies/User%27s%20Guide%20to%20the%20UN%20Watercourses%20Convent ion%20%282012%29.pdf

Roy, D., Barr, J., & Venema, H. D. (2011). *Ecosystem Approaches in Integrated Water Resources Management (IWRM)-A Review of Transboundary River Basins*. International institute for sustainable development (IISD). Retrieved December 2, 2014, from http://www.iisd.org/pdf/ 2011/iwrm_transboundary_river_basins.pdf

Salehin, M., Khan, M. S., Prakash, A., & Goodrich, C. G. (2011). Opportunities for trans-boundary water sharing in the Ganges, the Brahmaputra, and the Meghna basins. In Infrastructure Development Finance Company (Ed.), *India infrastructure report 2011: Water: Policy and performance for sustainable development* (pp. 29–43). New Delhi: Oxford University Press. Retrieved December 27, 2014, from http://www.idfc.com/pdf/report/iir-2011.pdf

Salman, S. M., & Uprety, K. (2002). *Conflict and cooperation on South Asia's international rivers – A legal perspective*. Hague: Kluwer Law International.

Saul, B. (2013). China, natural resources, sovereignty and international law, Asian studies review. *Asian Studies Review, 37*(2), 196–214. doi:10.1080/10357823.2013.769497

Schwebel, S. M. (1981). *A/CN.4/348, Third report on the law of the non-navigational uses of international watercourses*. Retrieved December 2, 2014, from http://legal.un.org/ilc/documentation/english/a_cn4_348.pdf

Secretariat of the Convention on Biological Diversity. (2004). *CBD Programme of Work on Protected Areas*. Retrieved December 2, 2014, from https://www.cbd.int/doc/publications/pa-text-en.pdf

Secretariat of the Convention on Biological Diversity. (2010). *Global Biodiversity Outlook 3*. Montréal. Retrieved December 2, 2014, from http://www.cbd.int/doc/publications/gbo/gbo3-final-en.pdf

Secretary of the Convention on Biological Diversity. (1995). Second Conference of Parties to Convention on Biodiversity (COP 2): Annex to Decision II/18. Retrieved December 2, 2014, from http://www.cbd.int/decision/cop/default.shtml?id=7091

Secretary of the Convention on Biological Diversity. (1998). *Fourth Conference of Parties to Convention on Biodiversity (COP 4): Decision IV/4*. Retrieved December 2, 2014, from http://www.cbd.int/decision/cop/default.shtml?id=7127

Seligman, D. (2008). *World's major rivers: An introduction to international water law with case studies*. Las Vegas: Colorado River Commission of Nevada. Retrieved December 2, 2014, from http://crc.nv.gov/docs/Worlds_Major_Rivers.pdf

Sharma, B. (2011). Changing pattern of agricultural productivity in the Brahmaputra Valley, Assam, India. *Indian Journal of Agricultural Economics, 62*(1), 139–151. Retrieved December 2, 2014, from http://dlc.dlib.indiana.edu/dlc/bitstream/handle/10535/7196/60.pdf?sequence=1

Shrestha, S., Anal, A. K., Salam, P. A., & Van der Valk, M. (2015). *Managing water resources under climate uncertainty*. Springer.

Singh, V., Sharma, N., & Ojha, C. S. P. (2004). *The Brahmaputra basin water resources*. Dordrecht: Kluwer.

Sohu News. (2009, March 25). Former Minister of the Ministry of Water Resources of China: China did not plan to divert water from Yaluzahngbu River. Retrieved November 29, 2014, from: http://news.sohu.com/20090525/n264162124.shtml (in Chinese).

Sood, A., & Mathukumalli, P. B. (2011). Managing international river basins: Reviewing India–Bangladesh transboundary water issues. *International Journal of River Basin Management, 9*(1), 43–52. doi:10.1080/15715124.2011.553832

Strategic Foresight Group. (2013). *Restructure India Bangladesh Relations*. Mumbai. Retrieved December 7, 2014, from http://www.strategicforesight.com/publication_pdf/22345riversof-peace-website.pdf

The Asia Foundation. (2013). *Political Economy Analysis of the Teesta River Basin*. Retrieved December 7, 2014, from http://asiafoundation.org/resources/pdfs/TheAsiaFoundation.PoliticalEconomyAnalysisoftheTeestaRiverBasin.March20131.pdf

The Times of India. (2014, May 28). *India to Build 4 New Hydropower Projects in Bhutan*. Retrieved December 2, 2014, from *The Times of India*: http://timesofindia.indiatimes.com/india/India-to-build-4-new-hydropower-projects-in-Bhutan/articleshow/35683039.cms

Treaty between Canada and the United States of America Relating to Cooperative Development of the Water Resources of the Columbia River. (1961). The Columbia River Treaty Protocol and Related Documents, (1964) p. 58. Retrieved December 2, 2014, from https://www.cbt.org/crt/assets/pdfs/1964_treaty_and_protocol.pdf

UN General Assembly. (1982). A/CN.4/352 and Add.l: Replies of Governments to the Commission's questionnaire. Retrieved December 2, 2014, from http://legal.un.org/ilc/documentation/english/a_cn4_352.pdf

UNESCO. (2006). *World Water Development Report 2: Water, A Shared Responsibility*. Retrieved November 29, 2014, from http://unesdoc.unesco.org/images/0014/001444/144409E.pdf

United Nations. (1969). Vienna Convention of the Law of Treaties.

United Nations. (1992). Convention on Biodiversity.

United Nations Economic Commission for Europe. (2009). *River basin commissions and other institutions for the transboundary water cooperation – Capacity for water cooperation in Eastern Europe, Caucasus and Central Asia*. New York, NY: United Nations Publications.

University of Bangladesh. (2014, November 22). *Call for SAARC Water Convention at Workshop*. Retrieved December 2, 2014, from University of Bangladesh: http://www.unb.com.bd/print/water-issue-ecological-approach

Upreti, T. (2006). *International watercourses law and its application in South Asia, Managing Director Padam Siwakoti, 2006*. Kathmandu: Managing Director Padam Siwakoti. Retrieved December 1, 2014, from http://www.internationalwaterlaw.org/bibliography/articles/general/Part-%203.pdf

Vajpeyi, D. K. (2012). *Water resource conflicts and international security: A global perspective*. Plymouth: Lexington Books.

Vinogradov, S., & Wouters, P. (2013, December). Sino-Russian transboundary waters: A legal perspective on cooperation. *Stockholm Paper*. Retrieved December 2, 2014, from http://www.chinainternationalwaterlaw.org/pdf/resources/20131216_001.pdf

Wang, T. (1990). International law in China: Historical and contemporary perspectives. *Recueil des Cours (Hague Academy)*, *221*, 195–369.

Wiegand, K. E. (2011). *Enduring territorial disputes: Strategies of bargaining, coercive diplomacy, and settlement*. Georgia: University of Georgia Press.

Wolf, A. T. (1999). Criteria for equitable allocations: The heart of international water conflict. *Natural Resources Forum*, February *23*(1), 3–30. 10.1111/j.1477-8947.1999.tb00235.x

Wolf, A. T., Yoffe, S. B., & Giordano, M. (2003). International water: Identifying basins at risk. *Water Policy*, *5*, 29–60.

World Wild Life. (2005). *South Asia: Eastern India*. Retrieved December 2, 2014, from World Wild Life: http://www.worldwildlife.org/ecoregions/im0105

Wouters, P. (2013). International law-facilitating transboundary water cooperation. *TEC Background Papers of Global Water Partnership*, *17*, 17.

Wouters, P. (2014). Comparing China's and India's approach to managing transboundary waters in light of contemporary developments in international water law, (manuscript on file with the author, not yet in press).

Wouters, P., & Chen, H. (2013). China's 'soft-path' to transboundary water cooperation examined in the light of two UN global water conventions: Exploring the 'Chinese way'. *Journal of Water Law*, *22*, 229–247.

Xue, H. (2007). Chinese observations on international law. *Chinese Journal of International Law*, *6*(1), 83–93. doi:10.1093/chinesejil/jml060

Xue, H. (2011). Chinese contemporary perspective on international law: History, culture and international law. *Recueil des Cours (Hague Academy)*, *355*, 41–234.

Yu, S. (2014). Contemporary legal analysis of China's transboundary water regimes: International law in practice. *Water International*, *39*(5), 705–724. doi:10.1080/02508060.2014.950856

Benefit sharing in the Mekong River basin

Seungho Lee

This study aims to evaluate the extent to which the cooperative relationship between China and the downstream countries in the Mekong River basin has evolved in relation to the various benefits shared between these riparian countries from the 1990s to the present. Benefit sharing is deployed as an analytical framework. Larger benefits obtained through cooperation have led China to become more cooperative with the downstream countries, as seen in the Greater Mekong Subregion programme. The establishment of a new form of cooperation in the river basin hinges upon the ability to share the socio-economic, political, and energy security benefits.

Introduction

This research analyzes the extent to which the cooperative relationship between China and the downstream countries in the Mekong River basin (Myanmar, Lao PDR, Thailand, Vietnam and Cambodia) has evolved thanks to a variety of benefits shared by the riparian countries from the 1990s to the present day. This study focuses on conflicts over water issues and deploys the benefit-sharing mechanism as an analytical framework, which is useful for evaluating the variety of options used to encourage cooperation between the countries rather than unilateral development. The theoretical approach helps anatomize the recent pattern of cooperation between China and the downstream countries during a time of growing criticism related to China's upstream dams.

Over the past few decades, the major discourse in the Mekong River basin has been the development of hydropower for socio-economic development. Unilateral action taken by China to build dams in the upper reaches of the river undermined the relationship between China and the downstream countries. The 1995 Mekong Agreement has served as a transboundary framework to facilitate cooperation through the Mekong River Commission (MRC). This framework has not necessarily been effective in transboundary water management, i.e. hydropower development, due to the lack of legally binding agreements and the absence of two riparian countries from the MRC: China and Myanmar. China's vote against the adoption of the UN Watercourses Convention in 1997 elucidates the country's willingness to remove potential interruption of its territorial sovereignty (Browder & Ortolano, 2000; Onishi, 2007). Therefore, it is very unlikely that China would alter its hegemonic attitude towards the lower riparian countries for transboundary water management.

Continued tension between these countries has been incrementally alleviated by China's growing involvement in regional integration frameworks, such as the Greater Mekong Subregion (GMS) programme of the Asian Development Bank (ADB), which began in 1992. Cooperation has increasingly been established in the course of implementation of the GMS programme. The downstream countries have gained a rare opportunity to harness China's unchallenged hegemony in the region through active engagement in the fields of trade and investment in relation to the GMS programme and the provision of energy import conduits for China.

The first part of the study reviews the relevant literature and explores benefit sharing as an analytical framework. Second, attention is paid to China's upstream dams and their possible impacts on hydrological regimes, ecosystems, and the livelihoods of local communities. The third part relates the positive impact of the GMS programme, which is highlighted as an exemplary case of socio-economic benefit sharing. Policy implications are discussed in the final part, and this evaluates how the application of benefit sharing in the river basin relates to socio-economic, political, and energy security–related benefits.

Literature review and theoretical framework

Literature review

The benefit-sharing mechanism is useful for assessing evolving trends in relationships between riparian countries, which focus on nonwater issues such as socio-economic, political and environmental benefits. Benefit sharing sheds light on the outputs derived from water use and has been regarded as a practical tool that can be utilized to promote cooperation between countries with transboundary rivers. This approach emphasizes hydro-interdependency and a trade-off between contemporary and future benefits (Alam, Dione, & Jeffrey, 2009). Benefit sharing is regarded as a conduit leading to multilateral cooperation between riparian states in transboundary rivers. However, even though diverse benefits are considered, most international practices under the benefit-sharing mechanism highlight socio-economic benefits (Alam et al., 2009; Haas, 2009; Hensengerth, Dombrowsky, & Scheumann, 2012; Sadoff & Grey, 2002; Qaddumi, 2008). Sadoff and Grey classify shared benefits into four types: benefits to the river; benefits from the river; benefits derived from the reduction of costs due to the river; and benefits beyond the river.

Successful cases of benefit sharing have been explored in a growing literature. The case of the Senegal River is closely explored from the perspective of a "sovereignty bargain" (Alam et al., 2009; Haas, 2009). Tarlock and Wouters (2007) discuss the cooperative relationship between Canada and the United States in relation to the Columbia River Treaty in 1964, and the Orange-Senqu River is shown to be a successful case for South Africa, Lesotho, Botswana and Namibia (Hensengerth et al., 2012; Yu, 2011). However, various challenges in the benefit-sharing mechanism have been identified, including its inability to address poverty alleviation and ecologically disastrous events (Tarlock & Wouters, 2007).

Since the 1990s, tensions have been mounting in relation to hydropower development in the Mekong River basin, particularly in the upper reaches. The enthusiasm for hydropower dams has been growing since the late 1950s (Hirsh, 2010). Hydropower-centred policies have continued to be bolstered by the willingness of the Chinese to exploit massive hydropower potential in the upper reaches and by the downstream countries to promote socio-economic development (Biba, 2012; Hirsh, 2010; Lennaerts, Southalack, & Phiromchai, 2012; Ward et al., 2012). Research outcomes indicate that Chinese upstream dams may have had negative impacts on the downstream countries (Chellaney, 2011; Gajaseni, Heal, & Edwards-Jones, 2006; Goh, 2004; International Center for Environmental Management [ICEM], 2010). China has largely resorted to unilateral action to meet the soaring domestic demand for electricity and to adapt to climate change (Urban, Nordensvärd, Khatri, & Wang, 2013).

ICEM (2010) and Roland-Holst and Heft-Neal (2012) provide comprehensive data on the potential for hydropower development. A constellation of negative impacts possibly related to the Chinese dams has been evaluated (Arias et al., 2014; Bird, 2012; Gajaseni et al., 2006; Lu & Siew, 2006). A growing number of research papers have taken a careful approach in assessing the possible impacts of the Chinese dams on downstream areas (Arias et al., 2014; Campbell, 2007; Chapman & He, 1996; Fu, He, & Lu, 2008; Goh, 2004; Li, He, & Feng, 2011). Evidence of a change in China's behaviour towards the countries downstream is mentioned by Hirsh (2010), Ogden (2012), and Onishi (2011). Hirsh (2010) and Onishi (2011) delineate how China has altered its policy to favour the countries downstream to achieve economic gains through the GMS programme.

The 1995 Mekong Agreement and the creation of the MRC were supposed to bring about cooperation and benefit sharing in terms of water resources management, but they appear to have been ineffective (Sneddon & Fox, 2006). Without participation in the MRC system, China has primarily opted for unilateral development to speed up hydropower generation. Despite a positive assessment of China's engagement in its transboundary water treaties, it is noted that China is one of the few countries to vote against the UN resolution to adopt the 1997 Convention on the Law of the Non-navigational Uses of International Watercourses (Chen, Rieu-Clarke, & Wouters, 2013; Choi & Lee, 2014). This clarifies China's reluctance to adopt the principle of equitable use of transboundary water against its own strategic interests, and epitomizes China's hegemonic approach towards water issues in the Mekong River basin (Browder & Ortolano, 2000; Onishi, 2007).

The GMS programme highlights the essence of benefit sharing in the Mekong River basin. The programme envisages the establishment of a power grid network to enlarge energy benefits for all the countries in the programme (Dosch & Hensengerth, 2005; Economic Consulting Associates [ECA], 2010; Goh, 2004; Lew & Choi, 2012; Schmeier, 2010). In the period from 1992 to 2011, considerable growth in GDP and poverty alleviation were achieved (ADB, 2012; Bolt, 2011; Choi & Lee, 2014; Dosch & Hensengerth, 2005; Eyler, 2013; Goh, 2004; Lew & Choi, 2012; Onishi, 2011; Park, 2013; Schmeier, 2010). Also, China's active engagement with the GMS programme has paved the way for cooperation (Schmeier, 2010). However, despite such achievements, Oehlers (2006) criticizes the fundamental challenges embedded in the programme. Ogden (2012) calls for a careful approach to costs related to externalities, particularly with regard to negative environmental impacts.

In addition to socio-economic benefits, the effect of Chinese development aid bestowed on the downstream countries, in particular Lao PDR and Cambodia, has helped bolster socio-economic development in the fields of transport, telecommunication, health, education, human resources development and construction, which can lead to regional cooperation (Ho, 2014; Reilly, 2012; Schmeier, 2010). Also, Chinese investment in downstream dams has softened complaints about the China's upstream dams (Ho, 2014). China's benign attitude towards the downstream countries has been driven partly by its efforts to secure more energy through a diversification of import routes, such as

Myanmar's gas pipeline and international seaports in Thailand, Vietnam and Cambodia (Eyler, 2013; Lee, 2013).

Theoretical framework

The prospect of larger benefits through cooperation rather than unilateral action encourages riparian countries to cooperate. The benefit-sharing approach revisits a way of possessing water by giving more credit to the benefits of joint rather than unilateral development of the shared resource. This type of joint management of water resources epitomizes 'hydro-interdependency' and deals with a trade-off between current and future benefits. The approach emphasizes water as an output, but not an input, by providing such commodities as electricity, food, and environmental services (Alam et al., 2009).

In this study, benefit sharing is defined as "any action designed to change the allocation of costs and benefits associated with cooperation" (Sadoff & Grey, 2002). This approach has been discussed not only in the field of transboundary water management but also in relation to other natural resources, e.g. petroleum and forestry. The term 'benefit' can be understood in different ways, from nationally administered revenue funds to revenue-sharing contracts between companies and local communities. The approach is regarded as a means of distributing a range of benefits derived from resource utilization across the economy and can facilitate the enlargement of growth with the support of socially inclusive policies (Haas, 2009). If this idea is applied to the realm of international water law, the principles of equitable and reasonable use of water without any major harm also apply. The idea of benefit sharing in transboundary water issues indicates a change of negotiation focus from the allocation of water volumes to the benefits gained by the riparian countries (Hensengerth et al., 2012).

Water is considered a productive resource for the realization of economic benefits. The maximization of economic benefits is legitimate only when environmental damage and disruption are avoided. The range of benefits often surpasses economic gains, including social, political and environmental benefits (Alam et al., 2009; Hensengerth et al., 2012; Qaddumi, 2008; Sadoff & Grey, 2002). Sadoff and Grey introduce four types of benefit sharing. Type 1 is related to increasing the benefits to the river, for example improving water quality or enhancing biodiversity. Type 2 is associated with increasing benefits *from* the river, such as enhanced management for hydropower or those related to agricultural use (irrigation). Type 3 encompasses a reduction of costs due to the river, including flood and drought management, and a decrease of international tensions; and type 4 indicates growing benefits beyond the river, such as integrated regional markets.

To lead to benefit sharing in transboundary rivers, the role of third-party involvement, namely, that of international donor agencies or bilateral agencies, is often imperative. Such involvement can help facilitate the establishment of cooperative mechanisms, as seen in the GMS programme, which was financed and instigated by the ADB. Third parties can also provide technical competence and best practices, and can provide riparian countries with assistance in negotiation and mediation skills, for example provision of legal and other water-related expertise. They also have the responsibility of facilitating investment within the transboundary setting (Quaddumi, 2008).

Successful cases of the benefit-sharing mechanism on a global level have been recorded. The Columbia River Treaty in 1964 delineates the way Canada and the United States interacted to reach a consensus based on a benefit-sharing mechanism (Tarlock & Wouters, 2007). Since 1972, Guinea, Mali, Mauritania and Senegal, in the Senegal River basin, have shared the benefits derived from water use of the river, such as irrigation, hydroelectricity generation, and inland navigation (Alam et al., 2009; Haas, 2009). In cases where a downstream country has built a dam upstream to increase aggregate net benefits, both the downstream and upstream countries have shared the project's benefits, rather than pursuing a unilateral alternative. An example can be found in the Orange-Senqu River basin, which is shared by South Africa, Lesotho, Botswana and Namibia (Hensengerth et al., 2012; Yu, 2011).

However, there are challenging issues related to the benefit-sharing mechanism. First, this mechanism does not necessarily guarantee the alleviation of poverty, nor rectify problems caused by transboundary water projects. In many countries, particularly developing countries, the benefits of economic growth are often reaped by the rich and powerful rather than the poor and marginalized. Second, issues related to sustainable development are often de-emphasized. For instance, ecological aspects of the Columbia River were not considered, thereby triggering the environmental degradation of the river (Tarlock & Wouters, 2007). These points are also associated with the challenges faced by the GMS programme (Oehlers, 2006; Ogden, 2012).

This mechanism is applicable to the Mekong River basin. China's unilateral action of building hydropower dams in the upper reaches spawned tensions between China and the downstream countries, but these have gradually been alleviated thanks to the GMS programme and Chinese development aid. Although Chinese dam building has continued since the 1990s, socio-economic benefits derived from the GMS programme have been increasingly attributed to China's greater engagement in talks with the downstream countries to exploit additional advantages. This case illustrates that hydro-interdependency between the countries in the river basin has gradually been established thanks to the success of economic benefit sharing.

China's benefit monopoly: dams in the upstream of the Mekong River basin

Overview

The overall hydropower capacity in the Mekong River basin can be estimated using figures on three geographical areas: China, the lower Mekong Basin tributaries, and the lower Mekong Basin mainstream. China has continued to plan and construct dams since the 1980s. China possesses eight dams, with a combined hydropower potential of 15,450 MW. If the hydropower-generating capacity of future dams is considered, China's hydropower output could be even larger (between 7550 MW and 13,480 MW). In the lower Mekong River basin tributaries, 70 hydropower projects have been explored, accounting for 9,364 MW; a hydropower potential of 14,697 MW exists in the lower Mekong River basin (ICEM, 2010).

A myriad of Chinese hydropower projects were launched in the upper reaches in the late 1980s, and six dams have been completed as of 2014. There has been sharp criticism from the downstream countries, as well as from the international community, related to the potentially devastating impacts of these dams. The primary motive for China's growing investment in hydropower projects has been an increasing domestic demand for hydroelectricity. The country seeks to achieve energy security (electricity) through exploitation of its hydropower potential and increasing imports from Lao PDR and Myanmar. In addition, China appears to have the intention to reduce environmental impacts and to eschew resettlement issues (Urban et al., 2013).

Chinese dams

Although the dams in the lower basin have a large hydropower potential, the Chinese dams are more controversial because they are upstream and can result in various profound impacts to the downstream countries, including reduction of fisheries and crop production, and degradation of ecosystems. So far, the country has shown reluctance to fully cooperate with its neighbouring riparian states, which have concerns related to the long-term impacts on ecosystems, flow rate, flooding and silting downstream and have urged China to share information (Chellaney, 2011; Gajaseni et al., 2006; Goh, 2004; ICEM, 2010). Since 2001, China has shown its commitment to cooperation in the fields of information sharing, environmental impacts and inland navigation. In addition, China became dialogue partner of the MRC. However, the overall response from downstream countries has not been entirely positive.

The total installed capacity of the eight planned dams in China will be over 15,400 MW when complete, with a generation of around 74 kWh of electricity per annum (Figure 1). The Xiaowan Dam is an example of the possible devastating impacts of upstream dams. The second-largest of the eight dams in terms of total storage capacity

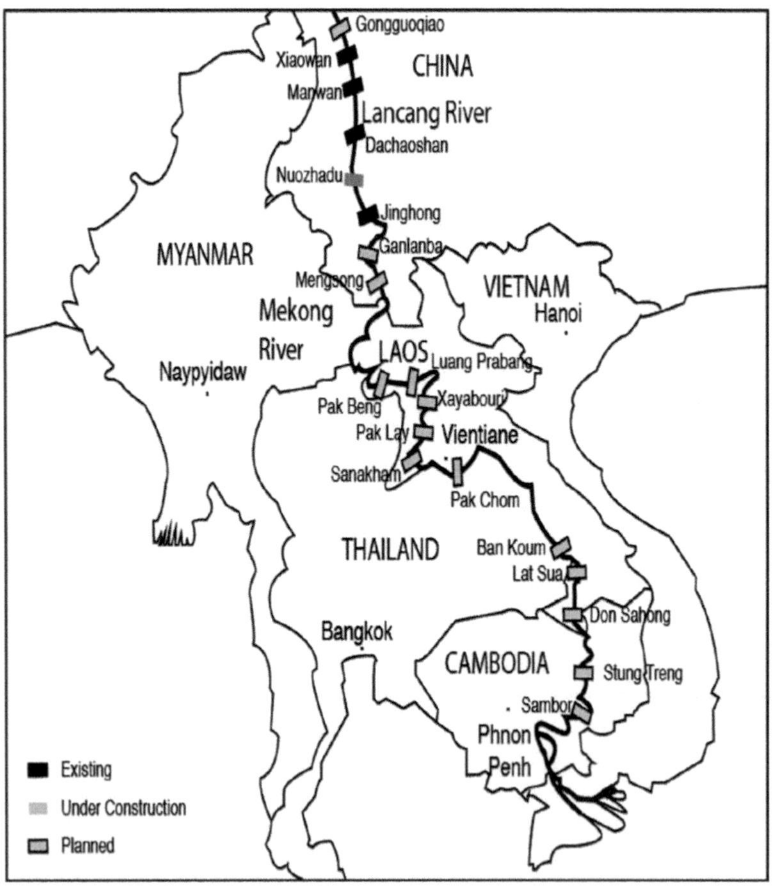

Figure 1. Mainstream dams in the Mekong River basin.

Source: Biba (2012).

(15,130 million m^3), it was completed in 2010 (Kuenzer et al., 2013). The two primary concerns in relation to the construction of this dam are: (1) it may remove up to 35% of the silt needed to fertilize the floodplains in the lower basin; and (2) it has triggered the relocation of 33,000–35,000 people. The blockage of silt by the dam is causing the downstream countries to be worried about the reduction of their fisheries and crop production (Lennaerts et al., 2012).

Lu and Siew (2006) explore the possible impacts of the Manwan Dam on the lower reaches before (1962–1992) and after (1993–2000) completion of the dam. Their research reveals that the water discharge regime of the river has been affected by the dam, and that the sediment flux had decreased as a whole. The authors warn that such a change could cause detrimental impacts to the soil, riparian vegetation, and aquatic ecosystems. However, the study displays a careful approach to the degree of impact, pointing out that the negative impacts would reach only as far as Vientiane, Lao PDR.

The Chinese dams are likely to have negative impacts on the downstream countries in many respects. However, it may be too early to judge the range of these devastating impacts. A growing number of experts agree that it is imperative to conduct more in-depth and long-term research to understand the complexity of ecosystems in the river basin and the causes of ecosystem disruption (Campbell, 2007; Chapman & He, 1996; Fu et al., 2008; Goh, 2004; Lee, 2013; Li et al., 2011).

The Greater Mekong Subregion programme

Overview

Various regional initiatives exist for economic integration in Southeast Asia, among them the ADB's GMS programme, the ASEAN Mekong Basin Development Cooperation and the Initiative for ASEAN Integration. Such initiatives deal with the water resources sector; however, their attention has been paid more to other sectors such as transport, energy, trade, telecommunications, tourism and the environment. One of the most sensitive issues in transboundary water management, water allocation, is not discussed in these regional initiatives because the 1995 Mekong Agreement was exclusively focused on transboundary water management issues (Sneddon & Fox, 2006). Amongst the regional initiatives, the GMS programme has been hailed as a successful case in terms of the degree of implementation. It is useful to investigate the programme as a representative case in relation to socio-economic benefit sharing.

The primary purpose of the GMS programme, launched in 1992, is to enhance economic relations in the subregion: that is, the two provinces of China (Yunnan and Guangxi), Myanmar, Lao PDR, Thailand, Vietnam and Cambodia. There are nine priority sectors: agriculture, energy, environment, human resources development, telecommunications, tourism, transport infrastructure, transport and trade facilitation, and investment. The highest priority has been given to transport and energy. The funding for the programme has been channelled from the ADB as well as from other development partners alongside the GMS countries, and is anticipated to be as much as USD 40 billion (ADB, 2012a; Park, 2013). The GMS programme consists of five phases; the second phase is presently underway. Around 180 projects in the programme have already been completed or are in progress (ADB, 2012b; Park, 2013). The programme has been well implemented, and this is particularly due to China's active involvement in the sectors of transport and energy.

Transport and energy

As mentioned above, the GMS programme embraces two priority areas: transport and energy. With regard to the field of transport, the programme has taken the initiative of creating economic corridors, for example the North-South Economic Corridor, the East-West Economic Corridor, and the Southern Economic Corridor (Park, 2013; Schmeier, 2010). The 1800 km Kunming-Bangkok Highway (North-South Corridor) was built in 2008. China, Myanmar, Lao PDR and Thailand agreed to blast rapids for navigation improvement in the basin, which stretches along the Thai–Lao border near the Thai province of Chiang Rai (Lee, 2013). The transport network within the programme has paved the way for the hinterlands of Yunnan Province to be connected with the other riparian countries, providing more opportunities for socio-economic development in the province.

The other priority area is the energy sector, especially hydropower, which relates to the establishment of the Mekong power grid within the GMS programme. It is anticipated that the regional grid will help establish a system for regional power trade and private investment in the power sector, with the participation of the World Bank, the Japan Bank for International Cooperation, and ASEAN. In the grid, transmission lines will be installed to link Thailand with the Jinghong and Nuozhadu Dams on the upper basin, dams on the Salween River in Myanmar, and the Nam Theun 2 Dam in Lao PDR. In addition, Cambodia will benefit from hydroelectricity imported from Thailand and Vietnam. China has promised to support Myanmar with funding and technologies to construct the dam (Dosch & Hensengerth 2005; ECA, 2010; Goh, 2004; Lew & Choi, 2012; Schmeier, 2010).

An important issue within the energy sector is the trade of electricity within the GMS. In the near future, Lao PDR wishes to become an energy hub in the river basin, 'the battery of Asia', and the country is expected to export at least 10 GW to Thailand and Vietnam by 2025. Myanmar competes with Lao PDR in exporting an estimated 28 GW of hydroelectricity to China and Thailand. This situation shows that the GMS programme will bring collective benefits for the GMS countries through hydropower development projects, not only through the building of hydropower dams and electricity generation but also through electricity trade. Consequently, some of the most difficult challenges of this region, such as the eradication of poverty, could be addressed (Bird, 2012).

This analysis of the GMS programme delineates the extent to which the role of a third party (the ADB) is instrumental in terms of facilitating the relationship between China and the other riparian countries. Also, this shows the significance of how nonwater issues can resolve conflicting issues on water (Onishi, 2011; Park, 2013).

Implications

Socio-economic benefit sharing

The Chinese dams have led to distrust and uneasiness in the downstream countries since the 1990s. Although still debatable, the possible negative impacts from the upstream dams might become more apparent over time. Meanwhile, the discontent continues in the downstream countries. Chen et al. (2013) stress that the principle of cooperation has been well reflected in China's bilateral treaties on international rivers. However, the country does not seem to have shown its commitment to taking an active part in the MRC system. This means no legally binding agreement or treaty in the river basin to accommodate the significant principles for transboundary water management, such as equitable use and no harm, as embodied in the UN Watercourses Convention (Browder & Ortolano, 2000; Choi & Lee, 2014; Onishi, 2007). The current MRC system looks obsolete with respect to generating cooperation.

However, the relationship between China and the downstream countries has not deteriorated further, thanks to the development of a cooperative relationship between the riparian countries, driven particularly by socio-economic benefit sharing. The four types of benefit sharing outlined by Sadoff and Grey can be applied to the Mekong case. The continuous dam building in the river basin (not only by China but also by the other riparian countries) is a good example of Type 2 benefits (increasing benefits obtained from the river, such as improving management for hydropower or agricultural use). Dam building in the 1990s and the 2000s was dominated by China. However, the coming decades might witness rapid catching-up by the downstream countries, in particular Myanmar and Lao PDR. The GMS programme is representative of a case delivering Type 4 benefits (increasing benefits beyond the river, derived from integrated regional markets). In particular, the ADB has initiated, financed, and facilitated the implementation of the GMS programme, together with the GMS countries and development partners.

The programme has played a pivotal role in establishing favourable conditions for the riparian countries to cooperate in socio-economic development. The rapid development of economic ties among the GMS countries implies the creation of a regional market in the basin – a Type 4 benefit, representing increasing benefit sharing beyond river management issues. Activities of Type 1 (increasing benefits to the river, such as improving water quality and enhancing biodiversity) have, to some extent, been conducted by the MRC. Activities relevant to Type 3 (reducing costs to the river, flood and drought management, and reducing international tensions) have not occurred in the Mekong River. The MRC has not been able to establish a joint management system for floods and droughts or to succeed in compiling international treaties for reducing tension, but has conducted scientific research and studies on water resources management, floods, and drought issues.

The significance of the GMS programme lies in the provision of large benefits, which includes the fields of water, energy, transport and natural resources development. This implies that all Mekong countries can share more benefits than before, and can begin to recognize the salience of negotiations in nonwater issues, particularly with regard to China (ADB, 2012b; Park, 2013). It is necessary to examine the extent to which the GMS has brought about positive outcomes from socio-economic benefit sharing, for instance by accelerating growth in terms of the annual average GDP per capita among the GMS countries. Between 1992 and 2011 there was a dramatic change in the GDP per capita in these countries. For instance, in 1992, the GDP per capita was USD 63 in Myanmar and USD 1894 in Thailand; by 2011 these had jumped to USD 832 and USD 5394, respectively. The per capita incomes of many countries grew 10-fold in the same period (ADB, 2012b).

The GMS programme's emphasis on poverty alleviation, such as a rise of per capita income, has also helped reduce the overall poverty levels in the GMS, as measured by the number of people living on less than USD 1.25 per day. In all the GMS countries except Thailand, the poverty level was near or over 50% in 1991. By 2011, poverty incidence had been reduced by a large margin: from 45% to 23% in Cambodia; 8.6% to 0.4% in Thailand; 64% to 13% in China; 56% to 34% in Lao PDR; and 64% to 17% in Vietnam (ADB, 2012b).

Consequently, most of the GMS countries appear to be on track to meet the Millennium Development Goal (MDG) for poverty, and other MDG and social development goals can be met by many GMS countries, including infant mortality rates, life expectancy, primary and secondary school enrolment rates, and literacy levels (ADB, 2012b).

China's volume of trade and its outward investment in the downstream countries was estimated to be larger than that in ASEAN countries in 2012. In addition, the Yunnan and Guangxi Provinces of China developed rapidly, with a GDP growth rate of 12–15% between 2010 and 2012. This growth has brought considerable benefits to the region. The soaring economic growth of the two south-western provinces of China has resulted from their close economic ties with the Mekong countries (Eyler, 2013). And in return, the economic growth of these provinces has enabled economic development in the Mekong countries (Bolt, 2011).

However, Oehlers (2006) questions the viability of the programme, in relation to its de-emphasis of the promotion of institutional and political reforms which could culminate in a fundamental transformation of industrial and production structures. He suggests that the programme should facilitate the development of financial markets and markets for various products using networking between diverse sectors, and pay more attention to enhancing governance, by concentrating on the rule of law, property, and political rights.

Addressing the costs derived from the development of the GMS is another issue. These costs are associated with sustainability, i.e., threats to livelihoods and degradation of the environment. It is essential to establish a framework to undertake rigorous environmental impact assessments for various projects throughout the GMS countries. However, it is impossible to conduct transboundary environmental impact assessments because there are no transboundary institutional frameworks to impose such rules on the associated countries. Therefore, the ADB might need to encourage the GMS countries to conduct environmental impact assessments beyond the boundary of sovereignty rights (Oehlers, 2006; Ogden, 2012).

Hydro-interdependency beyond socio-economic benefit sharing

One example of China's willingness to cooperate with the downstream countries is its agreement to provide hydrological information from the flood seasons of 2002. In the agreement, China promised to provide hydrological information in relation to the upper reaches. Such information and data from China have been useful in improving the accuracy of flood forecasts for stations in Thailand and Lao PDR. China's new approach has been interpreted as a cooperative gesture towards the downstream countries (Biba, 2012; Onishi, 2011). Another example of information sharing occurred in 2009, when China decided to suspend the construction of the Mensong Dam in order to create an alternative passage for fish. Although the dam project was not running on schedule and was set to be discontinued, the timely decision did much to alleviate tension with the downstream countries. China's increasing commitment to cooperation culminated in the MRC Summit in Hua Hin in April 2010, where the Chinese delegates pledged to unleash dry-season flows to help the downstream countries prepare for low flows (Hirsh, 2010).

It is also of note that in 2009, Vietnam and Cambodia signed a treaty on freedom of navigation on the Mekong between the countries. An agreement on navigation was then signed in 2011 between China and Myanmar, Lao PDR and Thailand, and was aimed at opening several ports on the Chinese side (Guanlei, Menghan, Jinghong and Simao). These agreements have improved inland navigation in the Mekong River (Onishi, 2011).

The flow of development aid from China to the downstream countries is a by-product of China's Going-Out Strategy (2001–2005) and its distinctive foreign policy towards developing countries, which are often politically unstable and undemocratic (McDonald,

Bosshard, & Brewer, 2009). Developing countries in South-East Asia, particularly those in the Mekong River basin, hailed China's non-interference in politics and its no-strings-attached policy in the provision of aid. Chinese aid has become an alternative conduit for boosting their socio-economic development without resorting to development aid from the West (Biba, 2012; Ho, 2014; Urban et al., 2013). The provision of development aid to the downstream countries can be regarded as an example of benefit sharing beyond utilization of water resources.

It is evident that China has been strategically motivated in its cooperation with the downstream countries. China's rapid economic development has prompted an upsurge in its society's demand for energy. The central government in China succeeded in constructing strategic oil and natural gas pipelines throughout Asia during its diversification scheme. PetroChina, one of the major state-owned enterprises in the oil and gas sector, completed the pipeline projects, which run from Myanmar's Indian Ocean coastline to Kunming in Yunnan. In addition, inland navigation through the river gives China access to international seaports such as Bangkok in Thailand, Hanoi, Haiphong and Ho Chi Minh in Vietnam, and Phnom Penh in Cambodia, for its oil and gas import (Lee, 2013).

China has striven to import sufficient electricity through a hydropower energy grid from the Mekong River basin, by its connection with Myanmar and Lao PDR. It is imperative that China retain a peaceful relationship with its neighbours to secure a stable supply of energy. For instance, Myanmar is able to open or close the significant flow of energy to China. This means that the position of China as the upstream country in the Mekong River basin is no longer hegemonic. In contrast, it now appears that the nonhegemonic countries, Myanmar and Lao PDR, could be more influential than China in terms of energy-security issues (Eyler, 2012).

Cooperation has been accelerated between China and the downstream countries in terms of a gradual socio-economic tie, development aid flow, and increasing energy security aspects. This has been achieved through a wider cooperation framework, related to the GMS programme. China's active involvement in the regional framework can offset the benefits that could be gained through unilateral development. The downstream countries may embrace the bargaining power to influence Chinese manoeuvres in the river basin, because China needs the shared benefits through cooperation for its economic expansion, political influence and energy security.

Conclusions

This research has evaluated the extent to which thc rclationships among China, Myanmar, Lao PDR, Thailand, Vietnam and Cambodia have evolved cooperatively within the Mekong River basin thanks to the introduction of benefit sharing in the 1990s. An array of benefits (socio-economic, political, and energy security) have served as a catalyst to invite China to the negotiation table for discussions with its counterparts in relation to development paths. The usefulness of the benefit-sharing approach lies in unlocking a list of common agendas in the form of nonwater issues, which have had the effect of providing fresh approaches to solving conflicting issues related to volumes of water and hydropower development within the water sector.

Since the 1990s, China's extensive dam building in the upper reaches has triggered numerous complaints from the downstream countries, and the power asymmetry between China and its downstream neighbours has intensified such complaints. Although it is generally assumed that China has shown a reluctance to cooperate, this study has shown that China has been making efforts to be actively engaged in dialogue on information sharing and navigation improvements. Though the degree of cooperation from China is not highly regarded by its counterparts, the country seems to be showing a willingness to cooperate. The GMS programme has brought about a fundamental shift in China's policy towards the downstream countries. The enlarged scope of benefits has continued to encourage China to take an active part in numerous projects within the sectors of transport and energy (hydropower). Yunnan Province in China has received vast socio-economic benefits from the GMS programme, as can be observed from the various ADB indicators, including the growth of GDP per capita, and poverty alleviation.

In addition to these socio-economic benefits, China has become closer to the downstream countries in order to expand its political influence with the provision of development aid, particularly towards Lao PDR and Cambodia. A series of ports in Thailand, Vietnam and Cambodia, and the pipeline for oil and gas in Myanmar, have become increasingly imperative for securing China's energy, and this is another reason for China to retain a favourable relationship with the downstream countries.

Although this is still not established, the relationships between the riparian countries seem to be cooperative, in particular with reference to socio-economic, political, and energy security benefits. The trend appears to be continuing. The next step is to establish a new regime for sustainable water management in the river basin. To do that, the riparian countries should seek consensus on the introduction of transboundary cumulative impact assessment, including long-term environmental impact assessment and social impact assessment of structures built in the tributaries and mainstream of the river. In addition, the UN Watercourses Convention, which came into effect in May 2014, can be a catalyst to encourage all member countries to negotiate and start discussing the future of the Mekong River basin. These efforts might have a positive influence on the architecture of sustainable development in the Mekong River basin.

References

Alam, U., Dione, O., & Jeffrey, P. (2009). The benefit-sharing principle: Implementing sovereignty bargains on water. *Political Geography, 28*, 90–100. doi:10.1016/j.polgeo.2008.12.006

Arias, M. E., Cochrane, T. A., Kummu, M., Lauri, H., Holtgrieve, G. W., Koponen, J., & Piman, T. (2014). Impacts of hydropower and climate change on drivers of ecological productivity of Southeast Asia's most important wetland. *Ecological Modelling, 272*, 252–263. doi:10.1016/j.ecolmodel.2013.10.015

Asian Development Bank. (2012a). *Overview: Greater Mekong Subregion, economic cooperation program*. Manila: ADB.

Asian Development Bank. (2012b). *The Greater Mekong Subregion at 20: Progress and prospects*. Manila: ADB.

Biba, S. (2012). China's continuous dam-building on the Mekong River. *Journal of Contemporary Asia, 42*(4), 603–628. doi:10.1080/00472336.2012.712257

Bird, J. (2012). Water resources management in the Greater Mekong Subregion: linkages to hydropower planning for a sustainable future. In H. Moinuddin & J. Maclean (Eds.), *Conference Proceedings: International Conference on GMS 2020: Balancing economic growth and environmental sustainability: focusing on food-water-energy nexus* (pp. 386–398). 20–21 February 2012. Royal Orchid Sheraton Hotel and Towers, Bangkok. Thailand.

Bolt, P. (2011). Contemporary Sino-Southeast Asian relations. *China: An International Journal, 9*(2), 276–295.

Browder, G., & Ortolano, L. (2000). The evaluation of an international water resources management regime in the Mekong River Basin. *Natural Resources Journal, 40*, 499–531.

Campbell, I. C. (2007). Perceptions, data, and river management: Lessons from the Mekong River. *Water Resources Research, 43*(2). doi:10.1029/2006WR005130

Chapman, C., & He, D. (1996). Downstream implications of China's dams on the Lancang Jiang (Upper Mekong) and their potential significance for Greater Regional Cooperation Basin-wide.

In B. Stensholt (Ed.), *Development dilemmas in the Mekong Subregion* (pp. 16–24). Workshop Proceedings, Monash Asia Institute.

Chellaney, B. (2011, August 30). Water is the new weapon in Beijing's armoury. *Financial Times*. Retrieved from http://www.ft.com/cms/s/0/4f19a01e-d2f1-11e0-9aae-00144feab49a.html#axzz2hHZhMJ60

Chen, H., Rieu-Clarke, A., & Wouters, P. (2013). Exploring China's transboundary water treaty practice through the prism of the UN Watercourses Convention. *Water International, 38*(2), 217–230. doi:10.1080/02508060.2013.782134

Choi, S., & Lee, S. (2014). Patterns of international relations on hydropower development in the Mekong River Basin. *Peace Studies, 22*(1), 389–434.

Dorsch, J., & Hensengerth, O. (2005). Sub-Regional cooperation in Southeast Asia: The Mekong Basin. *European Journal of East Asian Studies, 4*(2), 263–286. doi:10.1163/157006105774711422

Economic Consulting Associates. (2010). The potential of regional power sector integration: greater mekong subregion (GMS) transmission and trading case study. January 2010.

Eyler, B. (2013, July 16). China needs to change its energy strategy in the Mekong Region. *Chinadialogue*. Retrieved from https://www.chinadialogue.net/article/show/single/en/6208-China-needs-to-change-its-energy-strategy-in-the-Mekong-region

Fu, K. D., He, D. M., & Lu, X. X. (2008). Sedimentation in the Manwan reservoir in the Upper Mekong and its downstream impacts. *Quaternary International, 186*, 91–99. doi:10.1016/j.quaint.2007.09.041

Gajaseni, N., Heal, O., & Edwards-Jones, G. (2006). Chapter 2: The Mekong River Basin. In M. Finger,. L. Tamiotti, & J. Allouche (Eds.), *The multi-governance of water*. Albany, NY: SUNY Press.

Goh, E. (2004). *China in the Mekong River Basin: the regional security implications of resource development on the Lancang Jiang* (Working Paper, No. 69). Institute of Singapore: Defence and Strategic Studies.

Haas, L. (2009). *Introducing local benefit sharing around large dams in West Africa*. London: International Institute for Environment and Development and the Global Water Initiative.

Hensengerth, O., Dombrowsky, I., & Scheumann, W. (2012). Benefit-sharing in dam projects on shared rivers. In a Discussion Paper at the German Development Institute, June 2012.

Hirsh, P. (2010). The changing political dynamics of dam building on the Mekong. *Water Alternatives, 3*(2), 312–323.

Ho, S. (2014). River Politics: China's policies in the Mekong and the Brahmaputra in comparative perspective. *Journal of Contemporary China, 23*(85), 1–20. doi:10.1080/10670564.2013.809974

International Center for Environmental Management. (2010). *Strategic environment assessment (SEA) of hydropower in Mekong mainstream*. October 2010, Phnom Penh: Mekong River Commission.

Kuenzer, C., Campbell, I., Roch, M., Leinenkugel, P., Tuan, V. Q., & Dech, S. (2013). Understanding the impact of hydropower developments in the context of upstream-downstream relations in the Mekong river basin. *Sustainability Science, 8*(4), 565–584. doi:10.1007/s11625-012-0195-z

Lee, S. (2013). Hydropower development in the Mekong River Basin – analysis through the hydro-hegemony approach. *Southeast Asia Research, 23*(1), 227–264.

Lennaerts, T., Southalack, P., & Phiromchai, S. (2012). Water and development in the Lower Mekong Basin. In H. Moinuddin & J. Maclean (Eds.), *Conference Proceedings at the International Conference on GMS 2020: Balancing economic growth and environmental sustainability: focusing on food-water-energy nexus* (pp. 302–311). 20–21 February 2012. Royal Orchid Sheraton Hotel and Towers, Bangkok. Thailand.

Lew, S., & Choi, J. (2012). A study on China's regional cooperation efforts with Southeast Asia: The Greater Mekong Subregion economic cooperation program. *Journal of International Studies, 21*(2), 137–173.

Li, Z., He, D., & Feng, Y. (2011). Regional hydropolitics of the transboundary impacts of the Lancang cascade dams. *Water International, 36*(3), 328–339. doi:10.1080/02508060.2011.585447

Lu, X. X., & Siew, R. Y. (2006). Water discharge and sediment flux changes over the past decades in the Lower Mekong River: Possible impacts of the Chinese dams. *Hydrology and Earth System Sciences, 10*, 181–195. doi:10.5194/hess-10-181-2006

McDonald, K., Bosshard, P., & Brewer, N. (2009). Exporting dams: China's hydropower industry goes global. *Journal of Environmental Management, 90*, S294–302. doi:10.1016/j.jenvman.2008.07.023

Oehlers, A. (2006). A critique of ADB policies towards the Greater Mekong Sub-region. *Journal of Contemporary Asia, 36*(4), 464–478. doi:10.1080/00472330680000291

Ogden, S. (2012). Challenges facing cooperation and sustainability on water security and hydropower development in the Mekong River Basin: the GMS response. In H. Moinuddin & J. Maclean (Eds.), *Conference Proceedings at the International Conference on GMS 2020: Balancing economic growth and environmental sustainability: focusing on food-water-energy nexus* (pp. 276–286). 20–21 February 2012. Royal Orchid Sheraton Hotel and Towers, Bangkok. Thailand .

Onishi, K. (2007). Interstate negotiation mechanisms for cooperation in the Mekong River Basin. *Water International, 32*(4), 524–537. doi:10.1080/02508060.2007.9709686

Onishi, K. (2011). Reassessing water security in the Mekong: The Chinese rapprochement with Southeast Asia. *Journal of Natural Resources Policy Research, 3*(4), 393–412. doi:10.1080/19390459.2011.607962

Park, J. (2013). *Strategy of strengthening foreign policy of the Republic of Korea with the Mekong River Basin countries*. Seoul: Institute of Foreign Affairs and National Security, Korea.

Qaddumi, H. (2008). *Practical approaches to transboundary water benefit sharing* (Working Paper No. 292). London: Overseas Development Institute.

Reilly, J. (2012). A norm-taker or a norm-maker? Chinese aid in Southeast Asia. *Journal of Contemporary China, 21*(73), 71–91. doi:10.1080/10670564.2012.627667

Roland-Holst, D., & Heft-Neal, S. (2012). Food-water-energy nexus: GMS challenges of growth for 2020 and beyond. In H. Moinuddin & J. Maclean (Eds.), *Conference Proceedings at the International Conference on GMS 2020: Balancing economic growth and environmental sustainability: focusing on food-water-energy nexus* (pp. 484–500). 20–21 February 2012. Royal Orchid Sheraton Hotel and Towers, Bangkok. Thailand.

Sadoff, C., & Grey, D. (2002). Beyond the river: the benefit of cooperation on international rivers. *Water Policy, 4*, 389–493.

Schmeier, S. (2010). Regional cooperation efforts in the Mekong River Basin: Mitigating river-related security threats and promoting regional development. *Australian Journal of Southeast Asian Studies, 2*(2), 28–52.

Sneddon, C., & Fox, C. (2006). Rethinking transboundary waters: A critical hydropolitics of the Mekong Basin. *Political Geography, 25*, 181–202. doi:10.1016/j.polgeo.2005.11.002

Tarlock, A., & Wouters, P. (2007). Are shared benefits of international waters an equitable apportionment? *Colorado Journal of International Environmental Law and Policy, 18*(3), 523–536.

Urban, F., Nordensvärd, J., Khatri, D., & Wang, Y. (2013). An analysis of China's investment in the hydropower sector in the Greater Mekong Sub-Region. *Environment, Development, and Sustainability, 15*(2), 301–324. doi:10.1007/s10668-012-9415-z

Ward, J., Foran, T., Samjgl, A., Bouapao, L., Pech, S., & Xing, L. (2012). Mekong energy metabolism: connecting energy demand into the nexus of food-water-energy security. In H. Moinuddin & J. Maclean (Eds.), *Conference Proceedings at the International Conference on GMS 2020: Balancing economic growth and environmental sustainability: focusing on food-water-energy nexus* (pp. 364–385). 20–21 February 2012. Royal Orchid Sheraton Hotel and Towers, Bangkok. Thailand.

Yu, W. (2011). *Benefit sharing in international rivers: Findings from the Senegal River Basin, the Columbia River Basin, and the Lesotho Highlands Water Project*. Washington, DC: World Bank.

A new perspective on water governance in China: Captain of the River

Liping Dai

Although formal law plays an increasing role in water governance in China, the political arena has a large influence upon it. This article seeks to provide a new perspective to understand water governance and what role formal laws play during China's transition phase through the lens of the 'Captain of the River', a newly developed water governance instrument in China.

The Captain of the River

Wuxi City (in Jiangsu province) is the founder of the 'Captain of the River' (*Hezhang*) instrument. As one of the most developed cities in eastern China, water quality is a pressing issue. In 2007, a devastating blue algae bloom in Lake Tai seriously affected Wuxi City. Two million residents could not access clean tap water for five days, causing frustration among them (Tai & Ellis, 2008). Afterwards, the city government adopted a series of measures to tackle water pollution, and thus the Captain of the River instrument came into being.

The Wuxi (Communist) Party Committee and Wuxi City government require party heads and government officials at all administrative levels (city, county and district) to be 'Captains' of 64 rivers by signing responsibility contracts. The results from water quality monitoring in these 64 rivers are taken as a reference to assess the captains' administrative achievements. The Wuxi government created a special account to implement the Captain of the River instrument – a Guarantee Deposit. For example, in Huishan District (of Wuxi City), the captain of each river is required to deposit RMB3000 (US$491) into the account at the beginning of each year. The fund is exclusively used for rewarding and penalizing the captain in question. At the end of the year, the captain whose river quality is assessed as 'improved' is eligible to have their deposit doubled as a reward. Those who have maintained the quality status quo can have their money returned, but for those whose water quality has worsened, their deposit will be confiscated.

This incentive policy of the Grantee Deposit may play a role in motivating captains to protect their rivers, but this may not be the main factor. The amount of RMB3000, which is comparable with the annual per capita disposable income (RMB38,999/US$6378 during 2013) (Chen & Tan, 2014), is not a great incentive to persuade captains to go to extraordinary, yet necessary, lengths to control pollution, especially given the pressure of gross domestic product (GDP) growth targets (discussed below in the section entitled 'Relationship between economic development and environmental protection'). Moreover, the deposits typically come from the government to which the captains belong, rather than being the captains' own funds, which can also undermine their motivation.

In spite of this, the Captain of the River instrument has achieved great success. The water quality in monitored sectors has improved significantly: 74.7% had reached the established standards in 2008, 50% more when compared with the previous year when the system had not yet been adopted (Ministry of Environmental Protection, n.d.). What contributed to this success was actually the Veto System (*yipiao foujue*), which is a core operating tool in China's (Communist) Party and governmental responsibility system. Under the system, a failure to attain any of the assigned targets will unilaterally lead to the disqualification of the official.

The Communist Party of China (CPC) Wuxi Committee incorporated environmental protection into its veto system. It states that:

> For those who poorly govern environmental protection, those who cannot achieve the established objectives for energy-saving and emission reduction [...], the Organizational Department (of the CPC), after an investigation, will 'veto the relevant leadership when they participate in a city-level competition for an effective leading group, or veto the relevant responsible officials when they participate in the competition for advanced or excellent individuals, or veto the promotion of those directly responsible officials [...]. (*Xinhua News*, 2007)

Significant career pressure persuades captains to devote themselves to controlling water pollution. The reported improvement in the short-term, as stated above, is tremendous. The Captain of the River instrument, therefore, is rapidly expanding. As of September 2008, Wuxi City had promoted the instrument in its 815 rivers (Zhang, n. d.). Up to 2012, the city had 1284 captains (Environmental Protection Department, 2012). The system has since gone through a series of updates and iterations. For example, the 'Joint Captains of the River' instrument, where officials of the party and governments at both the city and county levels are asked to be 'Joint Captains' of selected rivers. In some areas, such joint captains extend four administrative levels (city, county, town and village). To date, the rivers in Wuxi City have been covered by captains without difficulty and each part of the river has a specific person in charge.

After seeing the success of the Captain of the River instrument in Wuxi City, the provincial Jiangsu government has begun to promote and improve this system since 2008. The province selected 15 rivers to implement the Joint Captains of the River instrument, connecting provincial and city levels. The names of the responsible officials are published by government. Up to 2014, 727 main rivers in Jiangsu province have their own captains (Wang, 2014).

The Captain of the River plays a significant role in the improvement of river water quality of Jiangsu today, and many other provinces are following suit. The Ministry of Water Resources encourages the development of this new instrument and plans to promote it nationwide. This is because that the Captain of the River happens to become an efficient tool to implement the 'Most Stringent Standards' water management mechanism promoted by central government, which assigns tasks to local governments through a Government Objective Responsibility System (GORS) (Dai, 2015). In other words, the Captain of the River renders the GORS more operational.

Why Captain of the River?

One of the key strengths of Chinese socialism is its capacity for long-term, national-level planning (Hu, 2011). China establishes its national development priority in its Five-Year

Plans, and normally implements them through a GORS. Through the GORS, central government assigns its overall established targets to provincial governments, which then reassign their quotas to lower levels (municipal and county levels). Party leaders and government officials at all levels are required to sign responsibility contracts and their performance will be assessed by the upper levels, sometimes through the veto system. The Captain of the River instrument is an implementation tool of the GORS.

Government Objective Responsibility Systems (GORS) in China

China's civil servant system includes two types of positions, leading cadres and non-leading cadres (Civil Servant Law, 2006), each managed differently. Non-leading cadres fall under the personnel departments and are subject to civil service regulations; they are admitted by examination, which is open to the public. The leading cadres are appointed and elected by the Organization Department of the CPC Committee, but are also regulated by the civil service regulations. The CPC Committee exercises authority over the leading cadres' promotion, dismissal and transfer one step down the administrative hierarchy, and the lower level is accountable to the next level above it (Edin, 2003).

Higher-level governments set targets and assign them to lower levels by signing performance contracts with the leading cadres, which form the GORS. In the contracts leading cadres pledge to achieve, and are personally held responsible for achieving, certain targets laid down by the higher levels (Edin, 2003). Non-leading cadres may also sign contracts, but these contracts are with their own working units. Incentives play a central role in motivating the cadres to fulfil their targets. Outstanding performances in the annual cadre evaluation are rewarded through promotion (in rank or position), additional wages or bonus payments or other material benefits, including administrative benefits (e.g., free transport, entertainment, training and travel), and other allowances for cadres (e.g., subsidized housing, healthcare and opportunities for further education) (Genia, 2014). Conversely, failing to meet some targets will result in censure and the denial of promotion.

The relative importance of different targets is made explicit in the GORS: soft ('guidance') targets, hard targets and 'targets with a veto power' (Minzner, 2009; Wang Y., 2013). Targets with a veto power are exclusively used for key policies of the central levels and sometimes also used for key policies at local levels (Edin, 2003). These targets are the most important and a failure to meet them automatically results in censure. Poor performance concerning veto targets cannot be remedied by good performance on other targets (Wang, 2013, 12 18). Hard targets are also important mandatory targets. Soft targets are lower priority objectives and are usually applied to those areas that are difficult to measure and quantify as well as to policies that are not deemed important by higher levels, such as cultural development (Edin, 2003).

Relationship between economic development and environmental protection

In 2014 China became the second largest economy in the world (Yueh, 2014). This success actually largely benefits from the GORS system. Economic goals have long been placed as nationwide hard targets in China. The country has had clear targets to boost the economy since reform and opening-up policies were introduced. The overall economic construction objectives were clearly laid down in the 'Three-Step Development Strategy' set out in 1987: Step One – to double the 1980 gross national product (GNP) by the end of the 1980s; Step Two – to quadruple the 1980 GNP by the end of the 20th century; and Step Three – to increase per-capita GNP to the level of medium-developed countries by the mid-21st century (China.org.cn, n.d.). These targets have prioritized economic development at all administrative levels throughout the past three decades. GDP achievement have long been the focus of local governments and the performance of a government cadre has been solely based on its economic success (Easterlin, 2014). Cadres are extremely sensitive to their GDP indicator as it largely determines their career path. Research has found that provinces with higher revenue contributions and more rapid economic growth have gained more representation in the central committees compared with other provinces (Edin, 2003), however things have recently begun to change.

Speeches delivered by President Xi Jinping earlier in 2014 have stressed that 'we should no longer evaluate the performance of leaders simply by GDP growth. Instead, we should look at welfare improvement, social development and environmental indicators to evaluate leaders' and 'the CPC should adopt more comprehensive criteria for assessing the performance of its officials' (Rapoza, 2013). The Organization Department of the CPC Central Committee also stated that 'GDP will no longer be the most important factor when evaluating an official's performance' (An & Zheng, 2013). These are healthy signs that China is switching its focus from 'GDP worship' to a more sustainable development mode. To date, more than 70 counties and cities (but only around 3% of the total) have bid farewell to the 'only the GDP' era, have cancelled GDP assessment, and have adopted an environmental and livelihood assessment guide instead (China has canceled over 70 counties GDP assessment, 2014).

Government Objective Responsibility Systems (GORS) in environmental pollution management

Environmental degradation has always been a concern for the central level. For example, as early as 1983 during the Second National Conference on Environmental Protection, the need for a coordinated advancement of both economic development and environmental protection was explicitly emphasized (CCICED, 2013). Environmental protection, however, has certainly not been a priority for both central and local governments. The idea of a 'Green GDP' was once explored by central government, but its 2004 Chinese Green National Economic Evaluation Research Report was not released until 2006, and subsequently ran aground (An end to GDP worship, 2011). Environmental targets have been determined to be soft targets and have long been devalued on the agendas of local governments. While prioritization was primarily given to economic growth, an insufficient effort and expenditure was devoted to environmental protection and social development (CCICED, 2013).

A historical change occurred in the 11th Five-Year Plan (2006–10), which elevated environmental goals to the same level as economic goals, i.e. from soft to hard goals (CCICED, 2013). The Plan lists eight 'binding targets' (hard targets) in which the 'Major Pollutants Control Rate' was included for the first time (Outline of the 11th Five-Year Plan, 2006). Its sub-plan, the 11th National Environment Protection Plan, was also, for the first time, issued as a State Council Document. It emphasized that 'the CPC Central Committee and the Central Government take environmental protection to be a national development strategy' (11th Five-Year Plan on Environmental Protection, 2007). The most prominent environmental targets of this sub-plan were the goals to reduce the major pollutant sulphur dioxide (SO_2) and chemical oxygen demand (COD) pollution by 10%, respectively (Wang, 2013, 12 18). It explicitly states that 'tasks are assigned according to levels through the GORS, thereby incorporating the environmental indicator into the performance achievement assessment of the Party leaders and the governments' officials and establishing a reward and sanctioning system'. Later that year the National Environmental Protection Bureau (which subsequently became the Ministry of Environmental Protection in 2008) signed responsibility contracts with seven major SO_2 and nine major COD emission provinces, and gradually with other provinces, to specify the provincial emission targets. Since then, environmental protection has been given unprecedented attention in China.

Government Objective Responsibility Systems (GORS) in water pollution control

The 12th Five-Year Plan (2011–15) highlights the need to 'construct a resource-conserving and environmentally-friendly society' (Hu, 2011). It incorporates 'energy saving and emission reduction' as one of its social development goals, and goes a step further towards water quality protection by initiating 'the most stringent water management' mechanism. It adds two more major pollutants to its binding goals, and again emphasizes target responsibility assessment.

To apply the 'the most stringent water management' mechanism, the CPC Central Committee outlined a 10-year water management plan, specified implementing policies, and developed national and provincial water quality objectives. The State Council specified objectives and timeframes and assigned them to each province. The implementation of each of province will be evaluated using a scoring system (Dai, 2015). At the beginning of 2014, the Ministry of Water Resources, jointly with 10 other ministries, issued an Implementation Plan Concerning the Most Stringent Water Management Assessment. The process, contents and scoring methods of the assessment are further specified by the plan. The GORS for water quality control have so far formed.

The veto system and the Captain of the River instrument

Veto targets, as stated above, are the most important targets for governments, meaning that if these targets are not met, all other achievements by a local leader will be rendered null and void (Genia, 2014). The veto power was originally applied to birth control in Hunan province in 1984 and was then updated as a national tool to ensure 'social stability' after 1991. Today, maintaining social stability and meeting birth control targets are almost universally covered by the veto, which may also cover items such as fiscal income, workplace safety and attracting investment (Wang, 2013). Energy saving and emission reduction fell under the veto power in 2007 (Notice of Energy Saving and Emission Reduction, 2007). That is why in the following year the CPC Wuxi Committee clearly declared that it would apply the veto system to assess leadership according to their achievements in energy saving and emission reductions.

Thus the 11th Five-Year Plan incorporated the 'Major Pollutants Control Rate' as a binding government target, the 12th Five-Year Plan added 'the Most Stringent Standards' to water resource management, and central government declared that it would apply the 'veto power' in emission reductions. It is against this background that the Captain of the River instrument rapidly became an adaptive instrument that can concentrate governmental manpower and personnel resources in a short period of time to deal with urgent pressures. It has been successful to an extent and has been approved by Wuxi City. However, it is not without problems in practice, which are discussed below in the section on the pros and cons of the Captain of the River instrument.

Where does the law stand in this respect?

Since 1979, the China National People's Congress (NPC) and its Standing Committee have passed as many as 280 pieces of legislation, of which 29 (almost 10%) relate to environmental issues. However, environmental pollution in China has reached a critical level, thus the implementation of these laws appears to have failed (Wang J., 2010).

The legislative body of the NPC, China's parliament, has grown in importance since 1978 when the domestic gates were opened. The NPC is often criticized as being a 'rubber stamp' for party decisions (How China is ruled, n.d.). Since approximately 75% of NPC deputies are CCP members, by the time a bill is placed before the NPC the major content of the bill has already been decided and members are expected simply to vote it through; bills are never rejected (Truex, 2013).

However, through the last decade there have been signs that the NPC is striving to be more independent. Today, it is playing an increasingly important role in China's governance and tremendous progress has been made by the NPC. For example, it approved the Administrative Procedure Law in 1989, which marked the first time that legal standards were applied requiring every level of government and all government employees to operate in accordance with the law (How China is ruled, n.d.). It added 'ruling the country in accordance with the law and building a socialist country of law' to the third amendment of the Constitution in 1999. To date, the NPC has passed hundreds of pieces of legislation, covering all aspects of society. As a country with 200 years of bureaucratic history, the construction of a legal system cannot be easy.

Not only is the NPC striving to create a country governed by the rule of law, but so is the party. The party created the ideology of 'ruling the country in accordance with the law' for the first time at the 15th Party Congress in 1998 and formally added the phrase 'ruling the country according to law' to the Party Constitution at the 16th Party Congress in 2002. It also included the requirement of 'exercising power scientifically, democratically and in accordance with the law' in 2003 (Cai & Wang, 2010).

The fact of the matter is that at China's present stage, neither the political nor the legal developments are a linear impetus to develop a law-ruled country. They are deeply intertwined and develop following their own dynamics; they cooperate with, yet constrain each other.

This cooperation shows that law has sometimes served as a lagging indicator, memorizing and 'legalizing' priorities established by the party state (Sunstein, 1996; Wang, 2013). For example, after the party incorporated pollution control as a governmental 'binding target' in the 11th Five-Year Plan, the Water Pollution and Prevention Law added GORS as a new provision in 2008, and this was also emphasized in the revisions of the Environmental Protection Law in 2014. The laws symbolically confirmed the GORS' legitimacy.

Law and party policy occasionally constrain each other. In practice, law has played a secondary role in China, being readily breached when it is in conflict with key bureaucratic targets (Minzner, 2009; Wang, 2013). In a random check of 126 industrial parks in 11 provinces (autonomous regions) in 2007, 87.3% of these parks were shown to have been approved by governments illegally or *ultra vires*, or by lowering the standards of environmental impact assessment. In more than half of the governmental urban sewage treatment projects corners had been cut or work was simply stopped (Geng, 2013). The enthusiastic pursuit of economic growth led governments to ignore the legal norms of environmental protection on a wide scale, and as one scholar has stated, 'a lack of good legislation is not the cause of China's ecological crisis, but rather a failure by government to prioritize the environment and enforce laws effectively' (Liu, 2014).

Law enforcement by the judicial system is weak, as local judiciaries are dependent on provincial and local governments for their funding. By controlling local judicial salaries and court finances, local governments often interfere with judicial decisions in order to protect local industries or litigants, or, in the case of administrative lawsuits, to shield themselves from liability (Brown, n.d.). For example, in Beijing City, the success rate of plaintiffs in administrative cases was 12.1% in 2013, and the national average success rate was less than 10% (*Xinhua News*, 2014). Some 87% of the public, when they encounter environmental disputes, choose to report the issue to governmental departments rather than bring the case before the courts (Huang, 2014). For example, the Wuxi Intermediate People's Court created a special tribunal for environmental protection in 2008; however, up to the end of that year, not a single conflict had been brought before the tribunal (Huang, 2014).

The fate of the 'failure of environmental law' is to a large extent exacerbated by the GORS, as the environmental protection therein has been included as one of the soft guidance targets and it has been given secondary importance to the GDP targets determined by the party (Wang, 2013). Moreover, environmental achievements were not an indicator for assessing officials' performance until 2006. This largely constrains the motivation of lower governments to conduct environmental protection.

Although the GORS, as well as the Captain of the River instrument, has been criticized as a 'rule of man' (Minzner, 2009), as being 'against the law' (Minzner, 2011), or as 'fixing the symptom instead of the root causes', it does not seem to be a better way of controlling rapid environmental degradation and achieving immediate results against the backdrop of a country undergoing a political and legal transition from an 'only GDP' era to a sustainable development era. Since the environmental problem is unique and must be quickly addressed, 'fixing' the symptoms can save time for 'fixing' the root causes.

Pros and cons of the Captain of the River instrument

A useful way of thinking about the Captain of the River instrument is that it is an adaptive tool to deal with the urgent pressures of water pollution. It has its strength, but it also has its shortcomings.

Benefits of the Captain of the River instrument

The Captain of the River instrument largely improves administrative efficiency. Fragmentation has been determined to be the primary weak point in China's institutional arrangement and legal framework for water resources management. It is often referred to as a multi-headed dragon since so many departments are involved therein (Dai, 2012). The Captain of the River instrument, to some extent, solves this problem. It contractually compels the official leaders of the party and governments as the first responsible actors to take charge of water quality, and directly correlates the consequences with their career possibilities. As the captains are the immediate leaders of different governmental sectors, and have direct influence on these fragmented departments, under their guidance it is easier to combine sectorial strength concerning pollution control and resource management in a short period of time.

The Captain of the River instrument promotes more site-specific policies. Water pollution sources vary on a regional basis, and especially in a country like China, industrial and agricultural pollution, the two main sources of water pollution, are unevenly distributed. National or provincial regulations are not always suitable for local conditions. The captains, instead, can compensate for this defect. By diagnosing the specific causes of water pollution in their rivers, the captains 'relate the remedy to the specific situation', so that pollution problems are dealt with in a river-based perspective; the means, therefore, tends to be more practical.

Although the GORS occasionally constrains the implementation of formal law, it can also promote specific laws when setting hard targets in specific fields. For example, following the assignment of water quality targets by State Council to local governments, the NPC further strengthened the legal framework of environmental protection. Not only can the implementation of existing laws and regulations facilitate the pursuance of governmental goals, but also the designation of specific targets can also promote the implementation of formal laws and regulations.

Shortcomings of the Captain of the River instrument

The veto system is not only applied in the Captain of the River instrument, but also in other fields. For example, GDP achievements, birth control, safe production control and food safety, etc. According to a news report, in Jiangxi province alone there were 62 different items under the veto system applied to its local governments up to 2011 (Pan, 2013). Local cadres have a wide range of responsibilities, water pollution control (if they happen to be 'captains') being only one of them. Officials can be overburdened with pressure, especially when each of the responsibilities plays a decisive role in their promotion, reward or sanction. This pressure tends to lead to favouritism and manipulation, which are unavoidable under certain circumstances especially when competition for promotion is fierce. Many strategies have been adopted by lower governments in order to cope with the overwhelming pressure, for example, selective reports, selective implementation or the manipulation of data, etc. (Dai, 2015).

The Captain of the River instrument has achieved a level of success in the short-term; however, it might be the consequence of campaign-style enforcement. This enforcement style forces governments to resolve outstanding issues in the short-term by concentrating their manpower and material resources. An example of this is the 'Hard Strike' anti-crime campaign in 1983, which targeted a particular crime, and ramped up the law enforcement and state propaganda machines to obtain mass arrests, rapid trials and mass executions (Nestor, 2007). Another example is an anti-prostitution campaign in Dongguan City at the beginning of 2014. In one afternoon alone, the city dispatched 6525 police officers in raids to arrest 67 people and to close 12 entertainment venues (Kaiman, 2014). Both the 'Hard Strike' and the 'anti-prostitution' campaigns achieved immediate success, but neither had a long-term effect (Nestor, 2007).

Although the Captain of the River instrument tends to be a long-term strategy, it is not immune to the allure of campaign-style enforcement. Every captain has a limited tenure and everyone has his own understanding of career achievement. It cannot ensure that every cadre has the same enthusiasm concerning the pursuit of his career, or more directly in this case concerning water quality protection, especially when the captain may be close to the age limit for promotion. As one local official once said, he was 'too old for promotion, too young to stop working, but just right for mahjong and drinking' (Smith, 2013). It is, therefore, possible to have tight control on water quality in one period or place and to have a loose policy in others due to the individual character and situation of captains.

Another shortcoming of the Captain of the River instrument is that environmental effects tend to appear after a certain period of time, meaning that the effects may only emerge after a number of years and may not become apparent during the five-year tenure of the leadership at the township level. This time lapse can result in either officials ignoring the high-polluting source with slow consequences, like agricultural pollution, or with the captains being unfairly assessed, i.e. unjustly rewarded or unjustly sanctioned due to the previous tenure's efforts or ignorance. In the case of trans-regional rivers, the captains may be misjudged since the river is shared by different captains and the monitoring result in fact is a joint effort.

Transparency and public participation are poor in the veto system of the Captain of the River. In the process of assessing the captains in question, the public at large are generally excluded, since the veto system is the Communist Party's internal personnel system. Decisions are made by the Organizational Department and approved by the CPC Committee, independent monitoring is not necessary and the public does not need to be involved therein. Public participation, promoted by central government, is more applicable to non-leading cadres at the level of local governments. However, the opinions of the public are typically regarded as 'references' and have little practical effect on a cadre's career prospects (O'Brien & Li, 1999). Moreover, the Captain of the River instrument is result-oriented: it can easily lead to the lack of transparency and public participation and which may further lead to the illegitimate implementation and the ignorance of public rights.

Suggestions and conclusions

Although the Captain of the River instrument is largely ruled by man and seems to conflict with the country's objective of building a law-ruled society, it is still an efficient tool to promote environmental protection during this transitional period. While strengthening the legal framework, there is certainly room for and a need to improve this instrument.

One of the legislative tasks declared by the Communist Party in 1978 is that 'there must be laws for people to follow'. To date, although there has been major progress in water-related legislation by the NPC, its legal framework appears 'fragmented' and 'insufficient'. Approaches to tackle water pollution caused by point source (e.g. industry) and non-point source (e.g. agriculture) pollution has not yet been differentiated (Dai, 2014a). The integration of different water-related legislation needs to be further developed. In other words, there are still insufficient legal norms or obligations for either the captains or the public to follow. In order to correct this, developing an integrated and concrete legal framework through legal reform, as the author has observed in previous studies (Dai, 2014a, 2015), should be a priority.

Another task of the legislation initiated by the party is to ensure that 'the existing laws are observed'. In the 'GDP' era, the violation of environmental law by government was not uncommon. Since the party has now laid down water quality objectives by means of the GORS and has demoted the GDP goals, local governments may therefore be more inclined to respect environmental law as they are subjected to fewer pressures regarding economic development. However, if there is no sound monitoring system, the GORS, as well as the Captain of the River instrument, may lend weight to the failure of environmental compliance. As both the GORS and the Captain of the River instrument are results-oriented, countermeasures from local governments tend to vary, as discussed above, and, without monitoring, effective and legitimate implementation cannot always be ensured.

'Accepting the public monitor' is a rule that has been adopted by the party over the years; however, it was not well implemented. It is a promising signal that, since the government of Xi Jinping, more attention has been paid to public participation, as Xi has said that 'winning or losing public support is an issue that concerns the CPC's survival or extinction'. Therefore the Political Bureau of the CPC Central Committee has decided to launch a campaign to boost ties between CPC members and the public in the near future (Xinhuanet, 2013). However, Xi's new policies will be challenged by poor and ineffective levels of coordination between central government and local leadership and the inability of the country's top supervisory body to regulate local cadres (Wang Y., 2013). Recently, the party has taken the unprecedented step of establishing an Online Public Opinion Monitoring Centre to monitor local-level government members on a daily basis (and also to keep a watchful eye on citizens in general) by collecting information from citizens (FlorCruz, 2013). This is a good start for the party in accepting public monitoring; however, given that only one-twelfth of China's total population use the internet, other channels and more instruments may need to be further developed.

During the past few years, central government has expanded public participation in all processes, from access to information, participation in decision-making, to access to justice (Dai, 2014). Examples include, from a legal perspective, the publication of the Environmental Impact Assessment Law in 2002, the Implementation Measures of public participation in the Environmental Impact Assessment Process in 2004, the State Council Open Government Information Regulations in 2008, and similar local regulations and administrative measures (Wang, 2013). A skeleton structure for public participation has already been basically built at the national level; however, there is still a considerable gap between what is proposed by central government and what is actually implemented and enforced at the local level. This is because, historically, China's governance is subject to a lack of public participation, 'a wealth of laws with shallow roots', as one scholar once described it (Wu, 2008). A suggestion would be to apply the existing norms more seriously by closed supervision with both a top-down and bottom-up approach, and parallel to this, strengthening general legal awareness and education.

Specific to the Captain of the River instrument, concerns arise with regard to the process of how targets are established, how the captains fulfil their assigned tasks and how they are assessed.

The establishment of water quality targets has been discussed in the author's previous research (Dai, 2015). When central government designs water quality objectives, ecosystem indicators are not taken into account, and no attention is given to issues of groundwater quality. At the local level there is no mechanism to ensure that the captains' tasks are assigned scientifically. With regard to this, more interdisciplinary research is needed, as neither legal nor political reform alone can fully correct this.

To fulfil the assigned tasks, captains are required to act in accordance with the law. Legitimacy should be fundamental for the system, which, however, has been largely ignored by the evaluators. Public participation can further improve this. For the time being, Zhejiang province is trying to establish an online monitoring system to implement its Captain of the River instrument, in which the public can upload photographs of polluted water and address complaints about misconduct by the captains. This bottom-up channel, once initiated, will largely improve governmental transparency and will also assist central authorities to tighten control over lower levels of government.

The veto system of the Captain of the River, as the party's personnel management tool, calls for the internal reform of the party. Although increasing transparency and public participation has been suggested, its implementation requires a change in party mentality. From a legal perspective, tremendous efforts have been made by central government, as discussed above, and are still made. For instance, at the beginning of 2014, the Ministry of Water Resources, jointly with 10 other ministries, issued an Implementation Plan on the Most Stringent Water Management Assessment, in which the process, contents and scoring methods of how to assess local governments are specified. According to the plan, the Ministry of Water Resources is to undertake the overall assessment. It is not yet clear how this plan can be harmonized with the veto system. Public participation needs to be further developed both within the party system and the formal law system, as neither is able to achieve the state's final goal – building a country ruled by law – on its own.

In this transition period, the Captain of the River instrument is an adaptive tool to curb emerging environmental pressures. It has been developed against the background that the party runs the state through a political system with a tight top-down responsibility system – GORS – in which the formal law is less important. Along with both the political and legal reform, the Captain of the River instrument faces the challenges, from a legal perspective, that the enforcement of available formal laws has not been effectively ensured and that there is a lack of legal instruments to enable governments to address water pollution, particularly non-point source pollution. With regard to this, the author proposes that while strengthening the enforcement of the existing laws, the legal framework that suits local conditions also be further improved.

Acknowledgements

The author would like to thank Professor Marleen van Rijswick, Dr Andrea Keessen, Professor Patricia Wouters, Professor James E. Nickum, members of the CIWL team (Xiamen University), and two anonymous reviewers who together supported the author in many ways to allow her continued study of water law.

Funding

This project was undertaken with funding support from the China Scholarship Council (CSC).

References

11th Five-Year Plan on Environmental Protection. (2007). Retrieved from http://www.reformdata.org/content/20071122/7936.html

An, B., & Zheng, J. (2013, December 11). *GDP no longer focus of officials' evaluations*. Retrieved 11 5, 2014, from Chinadaily: usa.chinadaily.com.cn/epaper/2013-12/11/content_17167398.htm

An end to GDP worship. (2011, August 17). Retrieved 11 5, 2014, from Chinadialogue: https://www.chinadialogue.net/blog/4467-An-end-to-GDP-worship/en

Brown, S. (n.d.). *Judicial Independence in the PRC*. Retrieved 11 5, 2014, from http://www.cecc.gov/judicial-independence-in-the-prc

Cai, D., & Wang, C. (2010). *China's Journey Toward the Rule of Law: Legal Reform, 1978-2008*. Leiden, Netherlands: Brill.

CCICED. (2013). *China's environmental protection and social development*. CCICED 2013 Annual General Meeting. Retrieved from: http://www.cciced.net/encciced/event/AGM_1/2013agm/speeches2011/201311/P020131106443218117388.pdf

Chen, Y., & Tan, D. (2014, January 27). *Wuxi City achieved 8,000 billion GDP in 2013*. Retrieved 8 17, 2014, from Jinagsu News Window: http://js.people.com.cn/html/2014/01/27/285116.html

China has canceled over 70 counties GDP assessment. (2014, August 25). Retrieved 11 5, 2014, from EN News163: http://www.enews163.com/2014/08/25/china-has-canceled-over-70-counties-gdp-assessment-50298.html

China.org.cn. (n.d.). *Three-Step Development Strategy.* Retrieved 11 5, 2014, from About China: http://english.people.com.cn/92824/92846/93014/6448494.html

Civil Servant Law. (2006, January 1). adopted by the Fifteenth Meeting of the Standing Committee of the Tenth National People's Congress. Retrieved from http://law.lawtime.cn/d353351358445_1_p3.html

Dai, L. (2012). Recovering the costs of water services in the People's Republic of China: Lessons from Article 9 of the European Union Water Framework Directive. *Utrecht Law Review, 8*(3), 102–118.

Dai, L. (2014a). Something old, something new, something borrowed and something blue - Tackling diffuse water pollution from agriculture in China: Drawing inspiration from the European Union. *Utrecht Law Review, 10*(2), 136–154. Retrieved from http://www.utrechtlawreview.org/index.php/ulr/article/view/URN%3ANBN%3ANL%3AUI%3A10-1-115816

Dai, L. (2014b). Water Resources Management in China – a legal approach and overview. Forthcoming.

Dai, L. (2015). Regulating water pollution in China and the European Union in terms of agricultural pollution. *Journal of Water Law,* forthcoming.

Easterlin, R. (2014, October 27). *What matters more: GDP or happiness?* Retrieved 11 5, 2014, from Goodbye GDP: http://goodbyegdp.org/2014/10/27/what-matters-more-gdp-or-happiness/

Edin, M. (2003). Remaking the Communist party-state: the cadre responsibility system at the local level in China. *China: An International Journal, 1*(1), 1–15. Retrieved from http://muse.jhu.edu/journals/china/v001/1.1edin.html

Environmental Protection Department. (2012, August 6). *The Captain of the River of Wuxi ensures Lake Tai's safety for five consecutive years.* Retrieved 11 5, 2014, from Environmental Protection Department of Jiangsu Province: http://www.jshb.gov.cn/jshbw/xwdt/sxxx/201208/t20120806_210991.html

FlorCruz, M. (2013, October 3). *China's Government Employs 2 Million 'Public Opinion Analysts'.* Retrieved 11 5, 2014, from International Business Times: http://www.ibtimes.com/chinas-government-employs-2-million-public-opinion-analysts-1414576

Geng, H. (2013, June 13). *Can the Captain of the River save China's environmental crisis?* Retrieved 11 5, 2014, from Huanshan News: http://www.newshs.com/a/20130613/00203.htm

Genia, K. (2014, August 1). Barriers to the implementation of environmental policies at the local level in China. *World Bank Policy Research Working Paper.*

How China is ruled. (n.d.). Retrieved from BBC News: http://news.bbc.co.uk/2/shared/spl/hi/in_depth/china_politics/government/html/7.stm

Hu, A. (2011, March 28). *Green light for hard targets.* Retrieved 11 5, 2014, from Chinadaily: http://www.chinadaily.com.cn/opinion/2011-03/28/content_12234557.htm

Huang, S. (2014, July 11). *The dilemma of environmental tribunals in China.* Retrieved 11 5, 2014, from Legaldaily: http://www.legaldaily.com.cn/Frontier_of_law/content/2014-07/11/content_5667242.htm

Kaiman, J. (2014, February 11). *Prostitution crackdown in China province after TV investigation.* Retrieved 11 5, 2014, from The Guardian: http://www.theguardian.com/world/2014/feb/11/china-province-prostitution-crack-down-guangdong-sex-workers

Liu, J. (2014, March 24). *China's new environmental law looks good on paper.* Retrieved 11 5, 2014, from Chinadialogue: https://www.chinadialogue.net/blog/6937-China-s-new-environmental-law-looks-good-on-paper/en

Ministry of Environmental Protection. (n.d.). *Wuxi: developing 'the Captain of the River' and strengthening river protection.* Retrieved 8 18, 2014, from Ministry of Environmental Protection of People's Republic of China: http://www.zhb.gov.cn/ztbd/rdzl/hzhzh/gdsj/200905/t20090526_152010.htm

Minzner, C. (2009). Riots and cover-ups: counterproductive control of local agents in China. *Journal of International Law, 31*(1), 53–71.

Minzner, C. (2011). China's Turn Against Law. *American Journal of Comparative Law, 2011; Washington University in St. Louis Legal Studies Research Paper No. 11-03-01*, 937–984.

Nestor, P. (2007). When the price is too high: rethinking China's deterrence strategy for robbery. *Pac. Rim L. & Pol'y J, 16*(27), 525.

Notice of Energy Saving and Emission Reduction. (2007). Retrieved from http://www.gov.cn/jrzg/2007-06/03/content_634545.htm

O'Brien, K., & Li, L. (1999). Selective policy implementation in rural China. *Comparative Politics, 32*(2), 167–186. doi:10.2307/422143

Outline of the 11th Five- Year Plan. (2006). Retrieved from http://www.sszzxx.com/html/llyzc/zc/2010/0927/766.html

Pan, J. (2013, November 7). *Veto should not become a norm of cadre evaluation.* Retrieved 11 5, 2014, from Guanmingdaily: http://epaper.gmw.cn/gmrb/html/2013-11/07/nw.D110000gmrb_20131107_2-02.htm

Rapoza, K. (2013, July 1). *China's Pres Xi: GDP No Longer The Measure Of Success.* Retrieved 11 5, 2014, from Forbes: http://www.forbes.com/sites/kenrapoza/2013/07/01/chinas-pres-xi-gdp-no-longer-the-measure-of-success/

Smith, G. (2013). Measurement, promotions and patterns of behavior in Chinese local government. *The Journal of Peasant Studies, 40*(6), 1027–1050. doi:10.1080/03066150.2013.866095

Sunstein, C. R. (1996). On the expressive function of law. *University of Pennsylvania Law Review, 144*(2021), 2021–2053. doi:10.2307/3312647

Tai, C., & Ellis, L. (2008, October). Taihu: Green wash or green clean? *China Environmental Health Project Research Brief.*

Truex, R. (2013). The Returns to Office in a "Rubber Stamp" Parliament. *American Political Science Review, 108*(2), 235–251. doi: 10.1017/S0003055414000112.

Wang, A. (2013). The Search for Sustainable Legitimacy: Environmental Law and Bureaucracy in China. *Harvard Environmental Law Review, 37*, 366–440.

Wang, J. (2010, September 23). *China's green laws are useless.* Retrieved 11 5, 2014, from Chinadialogue: https://www.chinadialogue.net/article/show/single/en/3831–China-s-green-laws-are-useless-

Wang, J. (2014, March 21). *Jiangsu appointed Captains of its 727 main rivers.* Retrieved 11 5, 2014, from Sinanews: http://news.sina.com.cn/c/2014-03-21/180629763406.shtml

Wang, Y. (2013, December 18). *Public participation and legal reform key to creating ecological civilisation.* Retrieved 11 5, 2014, from Chinadialogue: https://www.chinadialogue.net/article/show/single/en/6583-Public-participation-and-legal-reform-key-to-creating-ecological-civilisation

Wu, J. (2008). Public participation in the enforcement of China's anti-pollution laws. *LEAD Journal (Law, Environment and Development Journal)*, 35–48.

Xinhua News. (2007, December 7). *Organizational Department of Jiangsu Provincial Party Committee.* Retrieved 11 5, 2014, from Wuxi: Veto on deficient implementation of major decisions and arrangements: http://www.jszzb.gov.cn/dt2111111132.asp?DocID=2111118486

Xinhua News. (2014, September 21). *Success rate of the plaintiff in administrative cases is less than 10%.* Retrieved 11 5, 2014, from Sina News: http://news.sina.com.cn/c/2014-09-21/145630891138.shtml

Xinhuanet. (2013, June 18). *Xi: Upcoming CPC campaign a "thorough cleanup" of undesirable practices.* Retrieved 11 5, 2014, from Xinhuanet: http://news.xinhuanet.com/english/china/2013-06/18/c_132465115.htm

Yueh, L. (2014, April 30). *Is China about to overtake US as the world's largest economy?* Retrieved from BBC: http://www.bbc.com/news/business-27216705

Zhang, Y. (n.d.). *Whether contract system can save China's rivers.* Retrieved 11 5, 2014, from Ministry of Environmental Protection of the People's Republic of China: http://www.mep.gov.cn/ztbd/rdzl/hzhzh/gfpl/200905/t20090526_152020.htm

Exploring China's approach to implementing 'eco-compensation' schemes: the Lake Tai watershed as case study considered through a legal lens

Liping Dai

For close to a decade China has been implementing 'eco-compensation' mechanisms to address water-related ecosystem issues. This paper examines China's approach to eco-compensation through experience in the Lake Tai watershed. Four typologies of eco-compensation schemes are identified and analysed, primarily through a legal lens. It is concluded that while progress has been made, there is need for improved legal approaches to this complex topic.

Ecosystem services

Well-functioning ecosystems provide a broad range of services to myriad communities, anthropogenic and others. With growing populations the demands for natural resources are continually increasing, often resulting in dwindling healthy ecosystems. Identifying the significance of ecosystem services and assessing their monetary value are not new or novel ideas; these can be traced back to an influential article published in *Nature* (Costanza et al., 1997). In that seminal study the authors classified ecosystem services into 17 major categories: gas regulation, climate regulation, disturbance regulation, water regulation, water supply, erosion control and sediment retention, soil formation, nutrient cycling, waste treatment, pollination, biological control, refugia, food production, raw materials, genetic resources, recreation and cultural services (p. 254).

The notion of 'ecosystem services' has been mainstreamed over the past decade, referred to as 'environmental services,' 'ecological services' or simply 'investing in nature'. In the first wave of research in this field (1990–96), a survey of the legal literature reveals under 20 important studies that refer to the term 'ecosystem services'. However, during the following seven years, 1997–2003, over 10 times that number of law review articles referred to ecosystem services reflecting the growing legal discourse in this field (Ruhl & Salzman, 2007).

The most influential categorization of ecosystem services comes from the Millennium Ecosystem Assessment (MA). It follows Costanza et al. (1997) in taking both natural and man-made ecosystems as sources of ecosystem services, and follows Daily (1997) in using the term 'services' to encompass both the tangible and intangible benefits that humans obtain from ecosystems, which are sometimes separated into 'goods' and 'services' respectively (Alcamo & Bennett, 2003; Daily, 1997). MA categorizes ecosystem services into four groups based on their functional lines: provisioning services, regulating services, cultural services and supporting services (Alcamo & Bennett, 2003).

Perhaps the most important basis for supporting a policy that would protect otherwise threatened ecosystem services is the growing evidence that society gains more value from such protection than it gives up (Goulder & Kennedy, 2011). However, valuing ecosystem services is an extraordinarily difficult undertaking, and the economic valuation of ecosystem services at the global level has been a hot topic of debate (Simpson, 2011). For example, Costanza et al.'s (1997) suggestion that a 'minimum estimate' of such values was US$33 trillion has given rise to a great deal of criticism, particularly from economists (Ayres, 1997; Simpson, 2011; Toman, 1998). A more fundamental criticism of the Costanza et al.'s work is that it confused marginal and total values (Simpson, 2011). Nonetheless, the universal consensus is that ecosystems are tremendously valuable, regardless of whether or not there are challenges in accurately assessing such value. Thus, while we continue to attempt to calculate the actual value, ecosystems should be managed wisely before they are lost. It is within this context that the legal underpinnings of such schemes are so important, as robust implementing vehicles for policies in this field.

What needs to be kept in mind is that ecosystem processes do not yield ecosystem services until they are used by human beings (Ruhl, Kraft, & Lant, 2007). In other words, without human beings, 'ecosystem services' would not be enjoyed: the food supplied or the water purification processes provided by the ecosystems are simply ecosystem processes in and of themselves. Thus, any law or policy formulated to address ecosystem services should not deal with relations between humans and nature directly, but instead must focus on adjusting relations between humans. For example, compensation for ecosystems cannot be paid to ecosystems directly, but to the person who protects or repairs them. Thus, in this article, although the term 'eco-compensation' – compensating the ecosystem – is adopted, it refers in fact to compensation schemes among people. Only by adjusting the different interests among different stakeholders of an ecosystem can ecosystem services compensation mechanisms be formulated.

Eco-compensation in China

At the international level, incentive-based approaches for conserving ecological services have been devised as vehicles to contribute to achieving environmental sustainability. In this context the notion of 'payment for ecosystem services' (PES) has become an important strategy in dealing with various challenges in environmental management (United Nations Economic Commission for Europe (UNECE), 2007, p.5). According to the UNECE (2007), PES refers to a contractual transaction between a buyer and a seller for an ecosystem service or for a land use/management practice which is likely to secure that service.

The Chinese term 'eco-compensation' (*sheng tai bu chang*) has often been used interchangeably with the international term 'PES' – especially in comparative studies in this field examining China and global approaches (Zhang, Bennett, Kannan, & Jin, 2010). However, although the two schemes share some similarities, they are quite different. The Chinese notion of eco-compensation is broader, encompassing both PES-like policies and also a wide range of other policies and programme types (Bennett, 2009). The Chinese approach is elaborated further below.

PES schemes generally refer to voluntary transactions between service providers and service buyers. When successful, PES creates economic incentives for landholders to conserve or even improve the function of their lands for services as varied as watershed protection, carbon sequestration and biodiversity conservation (Zhang, Lin, Bennett, & Jin, 2010).

Eco-compensation schemes generally create not only incentives but also disincentives. Incentives refer to a reward or compensation for a right that is foregone in order to maintain a certain ecosystem service. Disincentives refer to charges for the loss of or damage to ecosystems and natural resources (China Council for International Cooperation on Environment and Development (CCICED), 2010; Zhang et al., 2010).

Therefore, eco-compensation in China is defined in both narrow and broad terms. The narrow definition, which is comparable with PES, refers to rewards for protecting the environment and natural resources; the broad definition covers not only rewards but also environmental pollution charges (Li & Liu, 2010), e.g. a pollution discharge fee. As there are already a series of laws and regulations that deal with pollution charges in China, this article focuses more specifically on the narrow definition of eco-compensation, as it is a relatively new and innovative development being implemented in China.

In order to develop internal ecosystem services markets, China's central and local governments have rapidly expanded their environmental protection policies, especially during the past few years, largely under the heading of 'eco-compensation' (see Appendix 1). The first official document to stimulate an eco-compensation mechanism (ECM) was a 'Decision regarding Strengthening Environmental Protection' issued by the State Council in 2005, which states that the government 'should improve eco-compensation policy, and develop an eco-compensation mechanism as soon as possible [...] pilot projects can be launched at both local and national level'. Following this, many provinces enacted their own regulations and eco-compensation projects. The 'win–win development' principle was later laid down as one of the cornerstones of the ECM by the Ministry of Environmental Protection, which recommended carrying out pilot projects in four fields:

- Eco-compensation for nature reserves.
- Eco-compensation for eco-function areas.
- Eco-compensation for the development of mineral resources.
- Eco-compensation for watersheds.

The number of eco-compensation projects in watersheds alone has increased significantly in only a decade – from eight in 1999 to more than 47 in 2008, with an estimated transacted value of roughly US$7.8 billion, covering some 290 million ha (Stanton, Echavarria, Hamilton, & Ott, 2010; Zhang & Radstake, 2010).

For the purposes of this study, the Lake Tai watershed in Jiangsu province has been selected as a case study. Through a legal analysis of the ECM in this watershed, this article aims to provide an overview of how this scheme has been applied in China. The Lake Tai case study is particularly well suited to this examination – it is one of the eco-compensation pilot schemes in China, one of the most developed and polluted regions, and one of the watersheds where a number and variety of eco-compensation schemes are being applied.

Although eco-compensation schemes are diverse in different watersheds, there are still some common rules that can be distilled from state practice. The Lake Tai example provides ample opportunity for harvesting valuable lessons for national water management regimes in the provision of water-related ecosystem services.

Eco-compensation in the Lake Tai watershed

Lake Tai is the third largest freshwater lake in China. The watershed occupies an area of some 36 500 km^2 and extends across multiple jurisdictions: Jiangsu province (52.6%), Zhejiang province (32.8%), Shanghai municipality (14%) and Anhui province (0.6%) (Monitor Center, 2013). As one of the most developed regions in China, with only 0.4% of the land territory but 4.4% of the population, the Lake Tai watershed produced 10.3% of gross domestic product (GDP); per capita GDP in this region was 2.4 times more than the national average in 2012 (Taihu Basin and Southeast Rivers Water Resources Bulletin 2011, 2012). The lake connects seven large cities across East China, including Shanghai and Hangzhou, which have populations of 23.8 million and 8.8 million, respectively.

Serious water pollution has been caused by unprecedented economic growth and rapid urbanization in the Lake Tai watershed region (Figure 1). The entire lake has suffered from eutrophication since 1993, the most serious crisis coming in 2007 when dozens of centimetres-thick algal blooms covered the entire lake and tap water turned yellow and was foul-smelling (Liang & He, 2012). Transboundary water pollution problems ('transboundary' in this article refers to as water bodies that cross two or more administrative regions) are persistent issues in this watershed as it extends across three provinces and one municipality.

In 2008 alone, China's Central Government allocated more than RMB111 billion (US $17.9 billion) to improve national lake water quality from Class V to Class IV, with an overall goal to achieve Class III status by 2020 (Liang & He, 2012). The government of Jiangsu province – which is covered by more than half of Tai's total watershed (52.6%) – has worked to improve its legal framework in order to improve water quality in the lake. Over the past decade eco-compensation schemes have been significantly developed across this region.

Four types of eco-compensation have been developed and applied in the Lake Tai watershed: eco-compensation between governments, eco-compensation between governments and farmers, eco-compensation between governments and industry (Luo, Qu, Feng, Shi, & Jiang, 2011), and eco-compensation among industries. These are explored in more detail below.

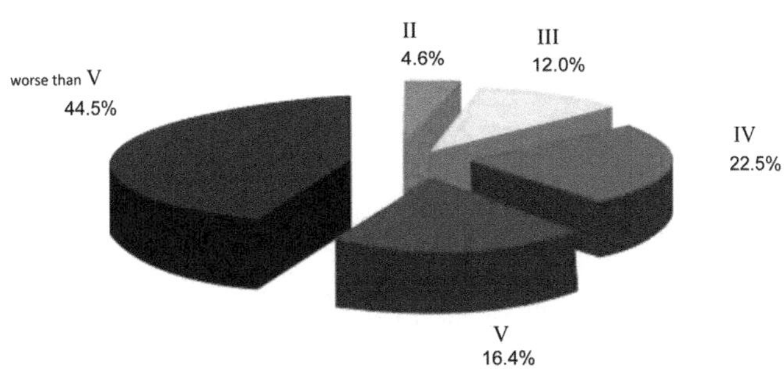

Figure 1. Water quality of Lake Tai in 2011. Note: There are five classifications of water quality in China: Class I, water source, national protection areas; Class II, centralized drinking water supply spawn grounds for rare fishes and shrimps, nursery areas for larvae, juvenile and young fishes; Class III, grounds and migration paths for common fishes and shrimps, aquaculture areas and swimming areas; Class IV, general industrial water areas, entertainment areas; and Class V, farmland areas, general landscape. Source: Taihu Basin and Southeast Rivers Water Resources Bulletin 2011 (2012).

Eco-compensation between governments

Bidirectional intergovernmental eco-compensation between upstream and downstream areas within one watershed is a newly developed mode of eco-compensation, aimed primarily at addressing transboundary water pollution problems. It can motivate both the upstream and downstream jurisdictions to act jointly in protecting their shared water resources.

Jiangsu province selected four cities – Nanjing, Changzhou, Wuxi and Zhenjiang – as pilot schemes for applying governmental eco-compensation instruments, beginning in 2007. Seven monitoring areas were selected in the four cities, where water quality standards were set by the provincial government. Using these standards as baselines, the provincial government combines the environmental protection responsibility of city governments with financial incentives. For example, in the Xu River in Changzhou City (one of the sub-watersheds of Lake Tai), a monitoring site was established by the provincial Administrative Department of Environmental Protection. The department records the water quality on a weekly basis and calculates the monthly average. If the result exceeds the baseline, meaning the water quality is below the standard set, the upstream city (Nanjing) has to compensate the loss suffered by the downstream city (Changzhou) in accordance with Jiangsu provincial regulation. The rationale for this approach is that the extra pollution caused by Nanjing City results in extra expenditure on pollution control for Changzhou City. Up to 2008, Nanjing City had compensated Changzhou City by RMB18,000 (US$29,032), and Changzhou City had compensated its downstream city Wuxi City by RMB180,000 (US$29,032) due to the recorded water quality results in the monitoring areas below the standard set (Internation Consumer Rights Forum, 2012).

In order to enhance the motivation for water quality protection, the compensation level is set at twice the pollution control cost. The compensation is incorporated into special environmental protection funds or pollution prevention and control funds for water pollution control and ecosystem restoration (The Compensation Method of Regional Environmental Resources of Jiangsu Province (Trial Implementation), 2007). In another case, if the recorded results in the monitoring areas between the upstream and downstream cities are above the designated baselines, the downstream city has compensate the upstream city, which stated by the State Council: 'if the upstream cities achieve the water quality targets in the monitoring areas of administrative boundaries, the downstream regions should compensate the upstream regions' (Regulation on the Administration of the Tai Lake Basin, 2011). However, the legal nature of such compensation gives rise to further discussion (see the fourth section).

Eco-compensation between governments and farmers

Diffuse water pollution is a main contributor to water pollution in the Lake Tai watershed (see Figure 2). If total nitrogen (TN) and total phosphorus (TP) – two main pollutants in diffuse agricultural pollution – had not been included in the evaluation of water quality, most of the surface water in Lake Tai would have reached Class III (Figure 2–2). However, when TN and TP are taken into account, the water quality drops to a level worse than Class V (Figure 2–1), especially in Jiangsu province.

In 2011, the State Council introduced a special regulation aimed at tackling diffuse water pollution in Lake Tai (Regulation on the Administration of the Tai Lake Basin, 2011), which required local governments to take measures such as constructing an ecological protection forest within:

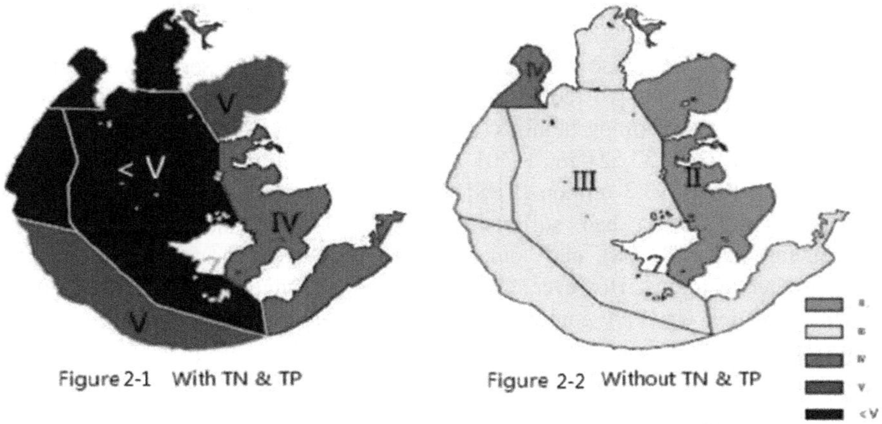

Figure 2. Diffuse water pollution of Lake Tai in 2011. Source: Taihu Basin and Southeast Rivers Water Resources Bulletin 2011 (2012).

- a 500-m area around the shoreline of the lake;
- a 1500-m area around the drinking water source protection zones; and
- a 200-m area along both of the river banks of the shore of the lake.

Local county governments

- Should provide subsidies and support to farmers who have to change their jobs due to the ban on aquaculture and livestock breeding, and the projects of returning the cultivated land or fishery to the lake.
- Should guarantee a basic life for those farmers by skill training or incorporating them into the social security system.
- Should provide subsidies for farmers whose income has decreased or whose expenditure has increased due to the projects to reduce pesticide and fertilizer use.

In fact, Jiangsu province had already formulated its own regulation in 2007 to address pollution problems caused by algal blooms (Opinions on Energy Conservation and Emission Reduction in Jiangsu Province, 2007). It requires the cities within its jurisdiction 'to return the cultivated land to the lake, to plant forests and to remove livestock breeding and traditional planting within 5-kilometres around the first-grade protection zones of Lake Tai'. This proved to be a difficult exercise in practice. For example, the East Lake Tai in Suzhou City, an 180,000 mu (12,000 ha) bay on Lake Tai, was occupied by enclosed fish farms with 165,700 mu (11,048 ha) (Han, 2010), which accounted for more than 90% of the surface water of the East Lake Tai, and more than 80% of the total enclosed fish farm area in the Lake Tai. The intensive enclosed fish farms were one of the main causes of the algal blooms due mainly to the excessive use of fish feed. In order to achieve its water quality target for 2012 (from Class V to Class IV), the government of Suzhou City reorganized its intensive enclosed aquaculture. The city's governmental policy requires the decrease of enclosed aquaculture from 300,000 mu (20,000 ha) of water areas to 45,000 mu (3000 ha) (Suggestions on Implementation of

Reorganizing Intensive Enclosed Aquaculture, 2008), which resulted in significantly improved water quality.

However, problems arose since the rural fish farmers were seriously affected as a consequence of this massive reorganization. For example, in the Wuzhong District of Suzhou City, 426 fish farming families (252 professional and 174 non-professional) were directly affected when 22,521 mu (1501 ha) water areas were reclassified. The government provided RMB793.3 million (US$128 million) in total as compensation subsidies for those farmers who had suffered financial losses (Fisheries Supervision and Management Station, 2008), and some of them were compensated by resettling fish farms in other locations. However, the compensation system did not run smoothly, as the actual situation was very complicated, with some unsatisfactory outcomes, discussed in more detail in the fourth section below.

Eco-compensation between governments and industry

According to the Water Environment Comprehensive Management Plan for the Lake Tai Basin, there are some 2.10 million industries in the Comprehensive Treatment Region of the lake. Of this total, around 1.04 million are in Jiangsu province and 1.06 million in Zhejiang province; 0.56 million of these industries belong to the six major pollution industries (textile industry; manufacture of paper and paper products; petroleum processing industry; coking and nuclear fuel processing; manufacture of raw chemical materials and chemical products; manufacture of medicines and the manufacture of chemical fibres), which also contribute significantly to the economic development in the Lake Tai region.

To control the water pollution caused by its intensive polluting industries, Jiangsu provincial government has implemented an approach that evaluates the receiving capacity of the surface water in water environmental function zones, applies a scheme of pollutants loading cap control, and a scheme of the discharge credits paid-use (limited to chemical oxygen demand (COD) discharge so far). The Price Bureau of Jiangsu province set different charging standards for emission credits for different industries. Under the pollutants loading cap control system, the amount of the pollution discharge credits is limited, which means that once the government has allocated all the credits, new applicants cannot purchase any from the government, but can only either buy surplus credits from other dischargers via an emissions trading platform (see the third section) or improve their own pollution prevention facilities to save credits themselves. It is a 'bubble policy', where polluters are free, within an imaginary bubble, to offset excess emissions from one source by a reduction made in another source, as long as the overall quantity is not exceeded (Kraemer, Kampa, & Interwies, 2004).

In the Lake Tai watershed, 1357 dischargers (annual emission > COD 100 tons) have been selected in the programme of discharge credits paid-use until 2010. The purchase amounts of COD achieved 49,700 tons per year during 2009–10, and the collected payments from discharge permits reached RMB175 million (US$28.2 million) (Li, Fan, Yan, & Gao, 2010). The revenue, which is managed as governmental non-tax revenue, allocates 10% to a provincial special fund for environmental protection, and 90% as local (Price Bureau, 2008). This special fund is used exclusively for environmental governance, the establishment of environmental monitoring, and the construction and maintenance of the emission credits trading platform in the Lake Tai watershed within Jiangsu jurisdiction.

Eco-compensation among industries

On the basis of the scheme of the discharge credits paid-use, the emissions trading system has been initiated in a few pilot cities in the Lake Tai watershed since 2008, but limited to COD emissions too. The governments of local cities set maximum limits on the total allowable emissions of COD, and then allocate these to the governments at county levels, which allocate their credits to selected industrial dischargers for a specified period of time. After receiving a written notice from the local environmental protection bureau, the selected industrial dischargers can buy discharge credits from governments which are embodied in discharge permits. With these permits comes the right to use the environmental capacity resources and to buy or sell their discharge credits.

Emissions trading occurs only in one 'bubble' – in which the total maximum amount of pollutants is determined, which means that purchasing from out of the region is not allowed for the city or county whose total discharge pollutants have already exceeded the control targets, or where the receiving water body has failed to reach the required water quality standards. Trading can be initiated between the dischargers and the Regulatory Authority of Emissions Trading or among the dischargers themselves on a specified trading platform monitored by the Provincial Regulatory Authority (Figure 3).

Jiangyin City is one of the pilot cities for emissions trading in the Lake Tai watershed. In 2010, 158 dischargers (annual COD emission > 100 tons) discharged 6930.7 tons COD, and paid RMB18.7 million (US$3 million) for discharge permits. Among these 158 industrial dischargers, 68 received extra discharge credits by emission trading with a total turnover achieved of RMB6.7 million (US$1.2 million) (Li et al., 2010). As well as the collection from discharge credits paid-use, the revenue from the trading is used exclusively for environmental protection measures, the establishment of environmental monitoring facilities and the maintenance of the emission credit trading platform.

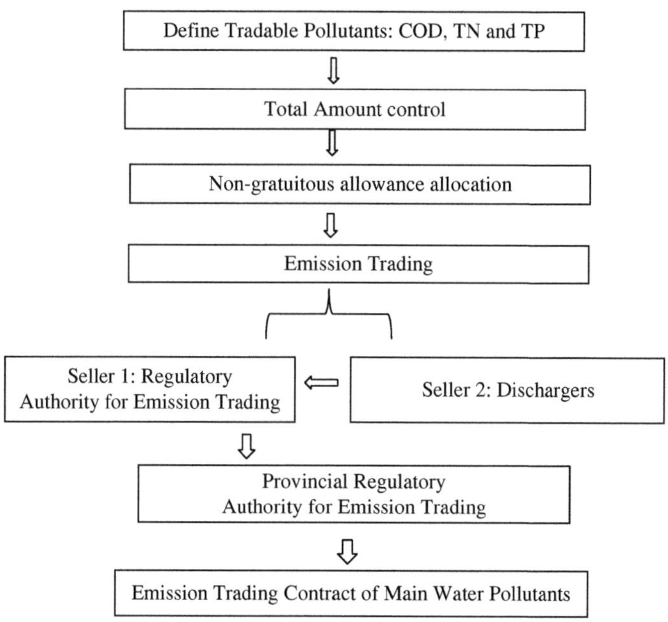

Figure 3. Process of emission trading in the Lake Tai watershed. Source: Interim measures of emission trading of main pollutants in Tai Lake Basin of Jiangsu province.

Legal issues arising from eco-compensation schemes in the Lake Tai case study

This section considers each of the four typologies of eco-compensation implemented in the Lake Tai case study.

In the first type of scheme (government to government), the upstream jurisdiction is required by law to compensate the losses of the downstream jurisdiction when the monitoring data show that the water quality is below the legally defined standards in the monitoring areas. But is this a true 'eco-compensation' scheme – in fact, from a legal perspective, this compensation is more akin to payments for illegal water pollution, a legal liability approach, not as compensation per se for ecosystem services.

The national Environmental Protection Law states that:

> Enterprises and institutions discharging pollutants in excess of the predetermined national or local discharge standards shall pay a fee for excessive discharge according to the state provisions and shall assume responsibility for eliminating and controlling the pollution.

Therefore, the 'designated discharge standards' are actually compulsory standards regulated by the law; polluters that discharge pollutants exceeding the standards should bear legal liability (Du & Chen, 2013). Given this reading, the compensation paid by the upstream city to the downstream city is not really eco-compensation. It neither provides rewards for protecting the environment and natural resources nor does it introduce pollution charges, i.e. water use fee or waste water discharge fee and thus cannot be considered to qualify as a true 'eco-compensation' mechanism.

Under this same line of reasoning, asking the downstream city to compensate the upstream city when the water quality does not exceed the standards also lacks legal support as a true ECM, because again the standards imposed are compulsory regulations – nobody should be compensated for merely abiding by the law.

One approach to transforming this approach into a true eco-compensation scheme would be to establish a 'negotiable water quality' (Du & Chen, 2013) instead of referring simply to the compulsory regulatory standards. By agreeing on a certain water quality (must be better than the compulsory quality) in monitored areas, the upstream and downstream cities may voluntarily agree to an eco-compensation contract: if the recorded results in the monitoring areas are above the contractual water quality, the party who puts efforts into making this should be compensated by the other one. Through such means, supplementing and building upon the existing regulatory requirements, a more holistic and functional eco-compensatory scheme can be formulated and implemented.

For this type of voluntary eco-compensation to work in practice, however, more scientific and legal research is needed in order to address a broad range of complex issues, such as monetizing the target ecosystem services, governance mechanisms for stakeholder involvement, adequate legal frameworks, as just some of the most pertinent examples.

In the case study examining the farmer compensation schemes, the governments compensate the farmers for changing their water-use practices, which is aimed at improved water quality. The eco-compensation relationship seems to be comparatively clear – the ecosystem service buyers are the Jiangsu provincial government, the Suzhou City government and the related district/county governments, and the ecosystem service providers are the fish farmers. The compensation payments include compensation through direct cash payments and fish farm resettlement.

However, these schemes have proved to be problematic in practice, with apparent divergent approaches for professional and non-professional farmers. While the former category is permitted to select their type of compensation – either cash compensation or resettlement, non-local fish farmers and non-professional fish farmers have only one choice – direct cash payments. Thus, this category of farmers is required to give up their primary livelihoods. This unequal treatment led to protests by some non-local farmers, who challenged this discriminatory approach by the governments. Another shortcoming of this scheme is that the city government dominated the entire compensation process, with a marked absence of market party participation, with no third-party evaluations and assessments. This resulted in some poor decision-making, i.e. many of the newly resettled areas were not suitable for aquaculture (Han, 2010). This situation meant that farmers had limited options because signing the contract was a precondition for the new farm resettlement, with the new aquaculture zones already planned by government, making the cost of reorganization too high in many respects. Given this reality, the compensation for resettlement made no sense at all for those farmers whose newly allocated farms produced substantially lower yields; it was made even worse in light of the fact that they had given up the option of cash compensation. This has given rise to new social conflicts, although water pollution has been improved to a certain extent.

Another issue relates to the compensation criteria that is used. For example, in the forest rehabilitation project in Lake Tai region, 68.18% of farmers interviewed were not satisfied with the government compensation because the farmland was productive as the irrigation was sufficient and the soil was fertile. Before rehabilitation farmers could get RMB13,890/hm^2 (US\$2240/hm^2) income per year by growing ordinary vegetables, but after rehabilitation they could only get RMB6000–9000/hm^2 (US \$968–1452/hm^2) from the government as compensation (Luo et al., 2011). This is not a minor loss for a farmer whose per capita disposable income is RMB38,459 (US \$6203) in 2012 (National Bureau of Statistics, 2014). If rational decision-makers are assumed to be participants, they would be unlikely to accept a payment unless it exceeds the sum of the opportunity costs they face (Wunder, Engel, & Pagiola, 2008). In light of all this, it seems that the 'win–win' objective set forth in the regulations has not been achieved.

In the third and fourth typologies, the eco-compensation between the government and industry and among the industries, these have succeeded in making considerable contributions to various environmental protection funds. The scheme of discharge paid-use works appears to work quite efficiently. Nonetheless, it must be noted that this system is actually different from the scheme of national pollution discharge fees. Under the scheme of discharge paid-use, governments set pollutants loading cap for a 'bubble' and allocate discharge credits. Dischargers buy credits guided by the principle of the 'user pays'; it reflects the dischargers' right to use natural resources. Under the latter scheme of national pollution discharge fees, dischargers pay fees whether they discharge pollutants into the water in excess of discharge standards or not. The difference from the former scheme is that instead of governments setting pollutants loading cap and allocating discharge credits, dischargers in the latter scheme report to and register with the local governments about the variety, quantity and density of discharged pollutants and wait for the governments' approval. Dischargers pay fees based on the principle of the 'polluter pays'; it reflects the dischargers' liability for using the natural resources.

Under the former scheme of discharge paid-use, dischargers are more motivated than under the scheme of national pollution discharge fees, as once they save discharge credits, they can keep them for the following year or sell them on the market. Dischargers themselves are the main pollution control bodies; governments only design and control the 'bubble'. Under the scheme of national pollution discharge fees, dischargers normally do not have enough motivation to reduce emissions if their discharges do not exceed the discharge standards approved by the governments. Governments are the main pollution control bodies. It is less cost-efficient than under the former scheme.

Although dischargers who have legally purchased the emission credits still have to undertake the legal responsibility of pollution control, but the two different charges should not be repetitively collected, i.e. who buys the discharge credits should not pay pollution discharge fees. However, in practice there are no published legal guidelines to address this problem, leaving it unclear how the governments have managed this in practice.

In the scheme of discharge paid-use, the governments play the role of ecosystem service providers for the purpose of maintaining a healthy water ecosystem and ensuring that the ecosystem can provide continuous eco-services, they set the pollutants loading cap for a 'bubble', monetize the pollutants and allocate the discharge credits. The selected dischargers are service buyers. In the COD emission trading system, those selected dischargers become service providers, who save discharge credits and provide certain ecosystem services by improving their pollution prevention facilities or inputting some other efforts, those who buy credits from other dischargers are service buyers.

As new and experimental instruments, both the scheme of discharge paid-use and emissions trading have some shortcomings. For example, it is uncertain how the provincial governments adjust their pollutant discharge targets and how they allocate or set prices for the emissions in the next five years, while central government adjusts national pollutant targets every 5 years. This lack of transparency leads to considerable uncertainties for the key actors in these schemes; as a result industrial dischargers face considerable risks in making decisions such as whether or not to buy the discharge credits, or how many to buy. In addition, the current emissions trading in the Lake Tai watershed within Jiangsu province is limited to COD emissions only; while the prices for TN and TP emissions trading were announced in 2011, there is not yet a specific legal regulation covering these. Furthermore, it is also very difficult to evaluate the environmental benefits from the emissions trading alone as it is generally applied together with many other policy instruments. According to research, tradable discharge permits are actually among the most challenging regulatory policies in terms of both their design and implementation (Kraemer et al., 2004).

In summary, the case study undertaken here reveals that four types of ECMs have been deployed across the Lake Tai region (Table 1). The common feature in each case is the dominant role played by governments (especially in the first three types). The main financial source for compensation is governmental payment. For example, at the time of the algal bloom in 2007, Jiangsu was spending RMB2 billion (US$322 million) per year to address Lake Tai's pollution problems (Liang & He, 2012). Since 2008, Jiangsu provincial government has contributed RMB0.2 billion (US$32 million) per year to a special fund to control water pollution in Lake Tai, with local governments asked to contribute 10–20% (He, 2014). Governments are the main actors in formulating and implementing eco-compensation schemes. Although commercial actors also contribute to the fund (e.g. the revenue of COD trading), this amount is insignificant when compared with the level of governmental payments. The single financial source from government might weaken the expectations of the eco-compensation projects. An example is the 'Three-North' Shelterbelt Project (see Appendix 2).

Table 1. Types of Eco-compensation mechanisms (ECMs) in the Lake Tai region within Jiangsu province.

Types	Water problems	Water-related eco-services	Providers	Buyers	Payment	Laws/regulations
ECM between governments	Pollution, disputes	Provision of higher water quality than compulsory standards	Upstream city/ downstream city	Downstream city/ upstream city	Cash, others	None
ECM between governments and farmers	Pollution	Improved water quality	Farmers	Governments	Cash, farm resettlement	Opinions of Energy Conservation and Emission Reduction in Jiangsu Province, No. 63 (2007) Regulation on the Administration of the Lake Tai Basin, State Council, No. 604 (2011)
ECM between governments and industry	pollution	Pollution control in certain cap	Governments	Industries	Cash	Implementing measures for Main Pollutants Discharge and Emission Trading in Pilots of Lake Tai Watershed in Jiangsu Province, No. 8 (2008) Administration for Charges for Credits of Main Pollutants Discharge in the Lake Tai Watershed of Jiangsu Province (Trial Implementation), (2008) Regulation of Jiangsu Province on Prevention and Control of Water Pollution in the Lake Tai region, No. 113 (2012)
ECM among industries	Pollution	Improved water quality	Industries	Industries	Cash	Implementing Measures for Main Pollutants Discharge and Emission Trading in Pilots of the Lake Tai Watershed region in Jiangsu Province, No. 8 (2008) Interim Measures for Main Pollutants Emission Trading in the Lake Tai Watershed of Jiangsu Province, No. 4 (2010)

Conclusions

Well-functioning ecosystems provide human beings with a broad range of important services, many fundamental to sustainable development. Effective eco-compensation schemes can contribute to the preservation of ecosystem services and lead to more sustainable development both within and outside China, the subject of this study.

Through examining the four types of eco-compensation schemes applied across the Lake Tai watershed, a number of observations can be made.

Eco-compensation schemes in Lake Tai watersheds are dominated primarily by governments through primarily governmental-sourced financial transfers. Although market-based eco-compensation, e.g. the emissions trading of COD, has been experimented with, it is still at a very early stage and needs to be further developed. The single source of governmental financial transfers might lead to a risk of a fund shortage in the future. A shortfall could, in turn, weaken the sustainability of the mechanism itself revealing a critical overall risk.

Constructing effective ECMs in watersheds is a long-term project requiring multi-disciplinary expertise. As has been discussed here, designing a robust legal framework capable of anchoring true eco-compensation schemes (as opposed to pollution liability regimes) requires careful consideration of a range of issues, focusing only on the mechanism itself is far from sufficient. Attention must also be paid to the preconditions in each case, such as: water management system details; the public's willingness to participate; and the collaboration between or among provinces and regions and such other conditionalities that might support or impede the mechanism. Even across the legal domain, eco-compensation schemes cross a complex matrix of legal regimes – a multidimensional construct of rules, laws and regulations, including (but not limited to!) administrative, corporate, contractual, public, private, regulatory and trade matters (Wouters, 2007).

Despite these challenges, the eco-compensation schemes being implemented in China provide a meaningful platform for addressing the complex issues related to eco-system services. More legal research is required to address the gaps identified in current domestic practice.

Acknowledgements

The author would like to thank Professor Marleen van Rijswick, Dr Andrea Keessen, Professor Patricia Wouters, Dr Owen McIntyre and members of the CIWL team at Xiamen University who together have supported the author in many ways in his continued study of comparative water law.

References

Alcamo, J., & Bennett, E. 2003. *Ecosystems and human well-being: A framework for assessment.* Washington, DC: Island Press. Retrieved from http://www.millenniumassessment.org/en/Framework.html

Ayres, R. U. (1997). *On the economic valuation of ecosystem services.* Fontainebleau: INSEAD.

Bennett, M. T. (2009). *Markets for ecoystem services in China: An exploration of China's' eco-compensation' and other market-BASED environmental policies: A report from phase I work on an inventory of initiatives for payments and markets for ecosystem services in China.* Forest Trends. Retrieved from http://www.forest-trends.org/documents/files/doc_2317.pdf

Costanza, R., d'Arge, R., Groot, R., Farber, S., Grasso, M., Hannon, B., & Belt, M. (1997). The value of the world's ecosystem services and natural capital. *Nature, 387*, 253–260. Retrieved from http://www.esd.ornl.gov/benefits_conference/nature_paper.pdf

Daily, G. C. (1997). *Nature's services. Societal dependence on natural ecosystems*. Washington, DC: Island Press.

Du, Q., & Chen, Z. (2013). On common but differentiated responsibilities of eco-compensation in river basin-based on legal analysis of water quality objectives (Chinese). *Journal of China University of Geo-sciences* (Social Sciences Edition), 01: 9–16. Retrieved from http://wx.xast.cn/KCMS/detail/detail.aspx?filename=DDXS201401003&dbcode=CJFD&dbname=CJFD2014

Fisheries Supervision and Management Station. (2008). *Ocean and fisheries in Jiangsu province*. Retrieved from re-organizaion of fishery farms ended successfully in Wuzhong district (in Chinese): http://www.jsof.gov.cn/art/2008/3/28/art_58_19385.html

Goulder, L., & Kennedy, D. (2011). Interpreting and estimating the value of ecosystem services. *Natural Capital–Theory and Practice of Mapping Ecosystem Services,* 15–33. doi:10.1093/acprof:oso/9780199588992.003.0002. Retrieved from: http://web.stanford.edu/~goulder/Papers/Published%20Papers/Interpreting%20and%20Estimating%20Value%20of%20Ecosystem%20Services%20(Goulder-Kennedy,%202011).pdf

Han, L. (2010). *Study of ecological compensation of the fence removal in Taihu Lake*. Retrieved from http://www.globethesis.com/?t=2121360305484392

He, G. (2014). *A Dilemma of water pollution control in Lake Tai*. Retrieved from Time-weekly: http://time-weekly.com/story/2014-01-16/132082.html

Internation Consumer Rights Forum. (2012). *Eco-compensation enters a substantive phase*. Retrieved from Internation Consumer Rights Forum: http://www.fj42.com/54067.html

Kraemer, R., Kampa, E., & Interwies, E. (2004). The role of tradable permits in water pollution control. *Inter-American Development Bank*, 11. Retrieved from http://idbdocs.iadb.org/wsdocs/getdocument.aspx?docnum=1481959

Li, N., Ding, S., & Zhao, W. (2010). Study on limitations of financial policies of regional ecological compensation in China. *China Population, Resources and Environment, 20*(6), 74–19.

Li, L., Fan, S., Yan, Y., & Gao, J. (2010). *The official opening of emission trading Market in Tai Like Basin (Chinese)*. Retrieved from Environment-China: http://www.cenews.com.cn/xwzx/cysc/cyqs/201010/t20101024_677617.html

Li, W., & Liu, M. (2010). Reflections on the development of eco-compensation mechanisms in the People's Republic of China. *Resources Science*, *32*(5), 791–796. http://wenku.baidu.com/view/95671 6c108a1284ac85043b2

Liang, G., & He, H. (2012). *Long struggle for a cleaner Lake Tai*. Retrieved from Chinadialogue: http://www.chinadialogue.net/article/show/single/en/4767-Long-struggle-for-a-cleaner-Lake-Tai-

Luo, X., Qu, F., Feng, S., Shi, X., & Jiang, H. (2011). Ecological compensation mechanism in Taihu Lake Basin: Based on theory and experience of watershed ecological compensation in and out of China (in Chinese). *Journal of Nanjing Agricultural University (Social Sciences Edition),* 82–89. http://118.145.16.217/magsci/article/article?id=901926

Monitor Center. (2013). *Introduction of Tai Lake (in Chinese)*. Retrieved from Monitor Center of Hydrological and Water Resources: http://www.thwb.gov.cn/aboutth/intro.asp

National Bureau of Statistics. (2014). *National data*. Retrieved from http://data.stats.gov.cn/search/keywordlist2?keyword=gdp

Opinions on Energy Conservation and Emission Reduction in Jiangsu Province. (2007). No. 63. Jiangsu Provincial Government.

Price Bureau, F. D. (2008). *Administration of charges for main pollutants discharge credits in the Tai Lake watershed of Jiangsu province (Trial Implementation)*. No. 18. http://www.jscz.gov.cn/pub/jscz/zcfg/czgfxwj/czzh/200910/t20091010_10112.html

Regulation on the Administration of the Tai Lake Basin. (2011). No. 604. State Council. http://www.tba.gov.cn/thlygltl/page1.asp?s=13094

Ruhl, J., Kraft, S., & Lant, C. (2007). *The law and policy of ecosystem services*. Washington, DC: Island Press.

Ruhl, J., & Salzman, J. (2007). The law and policy beginnings of ecosystem services. *Journal of Land Use and Environmental Law, 22*(2). http://papers.ssrn.com/sol3/papers.cfm?abstract_id=1028759

Simpson, D. (2011). *The 'Ecosystem Service Framework': A critical assessment. Ecosystem Services Economics Working Paper Series.*

Stanton, T., Echavarria, M., Hamilton, K., & Ott, C. (2010). *State of watershed payments: An emerging marketplace*. Ecosystem Marketplace. Retrieved from: http://www.foresttrends.org/documents/files/doc_2438.pdf

Suggestions on Implementation of Reorganizing Intensive Enclosed Aquaculture in East Tai Lake. (2008). Government of Suzhou City, No. 3. http://www.law-lib.com/law/law_view.asp?id=271518

Taihu Basin and Southeast Rivers Water Resources Bulletin 2011. (2012). Shanghai: Bureau of Taihu Lake Basin Ministry of Water Resources. Retrieved from http://218.1.102.107:9001//tba/content/TBA/lygb/szygb/0000000000003585.html

The Compensation Method of Regional Environmental Resources of Jiangsu Province (Trial Implementation). (2007). Governmental Office of Jiangsu Province.

Toman, M. (1998). Why not to caculate the valure of the world's ecosystem services and natural capital. *Ecological Economics, 78*

United Nations Economic Commission for Europe (UNECE). (2007). *Recommendations on payments for ecosystem services in integrated water resources management.* New York, NY: United Nations Publications.

Wouters, P. (2007). A new generation of local water leaders. *Colo. J. Int'l Envtl. L. & Pol'y 513, 18 (3)*, 513–522.

Wunder, S., Engel, S., & Pagiola, S. (2008). Taking stock: A comparative analysis of payments for environmental services programs in developed and developing countries. *Ecological Economics, 65*, 834–852, at 846. doi:10.1016/j.ecolecon.2008.03.010

Zhang, Q., Bennett, M., Kannan, K., & Jin, L. (2010). *Payments for ecological services and eco-compensation: Practices and innovations in the People's Republic of China.* Philippines: Asian Development Bank. Retrieved from http://www.adb.org/publications/payments-ecological-services-and-eco-compensation-practices-and-innovations-peoples-rep

Zhang, Q., Lin, T., Bennett, M., & Jin, L. (2010). *An eco-compensation policy framework for the People's Republic of China: Challenges and opportunities.* Mandaluyong: Asian Development Bank.

Zhang, Q., & Radstake, F. (2010). Payments for ecological services: Future prospects for the Asian Development Bank operations in the People's Republic of China. *Payments for ecological services and eco-compensation: Practices and innovations in the People's Republic of China* (pp. 118–129). Mandaluyong: Asian Development Bank. http://www.perpustakaan.depkeu.go.id/FOLDEREBOOK/payments-ecological-services-prc.pdf

Appendix 1: Development of the eco-compensation mechanism (ECM) in China.

Political speeches			
Year	Date	Congresses	Recommendations
2005	8 December	5th Session of the Sixteenth Central Committee of the CPC	Recommends that ECMs should be established soon, based on the principle of developer protects, and beneficiary compensates
2006	5 March	Central Government 2006 Work Report	Comprehensively mobilizes all means available, and in particular economic instruments, such as pricing and taxation, to promote the rational and sparing use of natural resources and to develop an ECM as soon as possible
2007	5 March	Central Government 2007 Work Report	Calls for a reform of important natural resources prices and pollutant charges and an acceleration in the development of the ECM
	15 December	17th National Congress of the CPC	Establishes and improves the resource use paid system and ECMs
2008	5 March	Central Government 2008 Work Report	Reforms the natural resources tax and fees system, and improves the system of paid natural resource use and of ECMs
2009	5 March	Central Government 2009 Work Report	Accelerates the establishment of sound ECMs, and reforms and improves the natural resource tax system
2010	5 March	Central Government 2010 Work Report	Accelerates important watersheds management
2011	5 March	Central Government 2011 Work Report	Improves water conservation facilities; makes progress in better controlling important tributaries of large rivers as well as lakes and small and medium-sized rivers
2012	5 March	Central Government 2012 Work Report	Establishes a sound system of compensation for ecological damage, strengthens ecological protection and restoration and strictly protects river sources, wetlands, lakes, and other priority functional ecological zones

(Continued)

Appendix 1: (Continued).

Policies			
Year	Date	Documents	
2005	3 December	State Council Decision Regarding Implementing the Scientific Development View to Strengthen Environmental Protection. National Issue [2005] No. 39	Calls for the improvement of eco-compensation policies and the establishment of an ECM; recommends the consideration of eco-compensation factors in central and provincial fiscal transfer payment systems
2006	14 March	Outline of the 11th Five-Year Plan	Calls for policy-makers to innovate in environmental policy, establishment of ECMs (especially intraregional and watershed-related ECMs) based on the principle of 'who develops protects, who benefits subsidizes'
	19 March	State Council 2006 Work Outline. National Issue [2006] No. 12	Develops ECMs, and gradually resolves and improves the evolving mechanisms for pricing natural resources and primary commodities
2007	24 August	Guiding Opinions on the Development of Eco-compensation Pilot Work. Ministry of Environmental Protection (MEP) Issue [2007] No. 130	Basic principles of eco-compensation: those who develop and exploit resources should also protect the environment; those who benefit from it should subsidize it; and who pollutes should pay; recommends carrying out eco-compensation pilot projects preferentially within four areas: natural protection areas, key ecological areas, mining resource areas and watersheds.
	22 November	The National 11th Five-Year Plan for Environmental Protection (2006–2010). National Issue [2007] No. 37	Proposes a target responsibility system for watershed pollution control and a water quality examination system for trans-province boundary waters, and speeds up the establishment of ECMs
2008	29 March	State Council 2008 Work Outline. National Issue [2008] No. 15	Improves the paid natural resource use system and ECM
	22 July	Views Regarding Work on Deepening Economic Structural Reforms for 2008. General Office of the State Council Issue [2008] No. 103	Reforms the system for distributing the benefits of natural resources; establishes a system for natural resource compensation for underdeveloped areas; and promotes the establishment of pilot projects for inter-provincial watershed eco-compensation

(Continued)

Appendix 1: (Continued).

Policies

Year	Date	Documents	
2011	16 March	Outline of the 12th Five-Year Plan	Improves the system of equalization transfer for key ecological function areas; sets up a special fund for national eco-compensation; promotes a reverse system for the sustainable development of resource-oriented enterprises; encourages, guides and explores an ECM whereby downstream areas compensate upstream areas, development areas compensate protection areas and ecological beneficiary areas compensate ecological protection areas; explores a market-based ECM; and speeds up the formulation and implementation of an Eco-compensation Regulation
2011/12	15 December	The National 12th Five-Year Plan for Environmental Protection (2011–2015). National Issue [2011] No. 42	Accelerates the research on, the formulation and implementation of the Eco-compensation Regulation; establishes an ECM for watersheds and important ecological function zones
	12 January	Opinions of the State Council on Applying the Strictest Water Resources Control System. National Issue [2012] No. 3	Conserves and restores the ecological system; conducts health assessment for national key rivers and lakes; establishes and improves eco-compensation mechanisms for water ecology

Laws and regulations

Year	Authorities	Name	
Adopted in 1984, revised in 2008	Standing Committee of the National People's Congress	Water Pollution Prevention and Control Law	Article 7: the state sets up an ECM, with a public financial transfer, to protect the drinking water source areas and headwaters of rivers, lakes, and reservoirs
Adopted in 1991, revised in 2010	Standing Committee of the National People's Congress	Water and Soil Conservation Law	Article 31: the state strengthens the prevention and treatment of soil erosion in river source areas, drinking water source protection zones and headwaters; raises funds through different channels; eco-benefits compensation for water and soil conservation into state ECM
Drafting		Eco-compensation Regulation	

Appendix 2: 'Three-North' Shelterbelt Project

The 'Three-North' Shelterbelt Project is the biggest eco-compensation/PES project in the world, with a total plan area of 6103 billion mu (406.9 million ha), some 42% of China's total land area (Bennett, 2009). The project runs from 1978 to 2050 and aims to control desertification in northern China. The funds for compensating farmers who return farmland to forest rely mainly on state financial transfers. During the past 30 years the 'Three-North' project has demonstrated certain ecosystem benefits, like farmland protection, soil and water conservation, wind reduction and sand dune fixation. However, the project is now in a difficult stage as the compensation fund is not sufficient. Since the implementation of the project in 1978, the state has not increased the compensation rate while overall living expenses are continually increasing. The cost of afforestation is RMB250–310 (US$30–38)/mu (0.07 ha), but the subsidy from the state is only 2–4% of the actual cost. For example, the Inner-Mongolia Autonomous Region during the past few decades has invested at least RMB2250/ha (US$363/ha) to control the soil desertification of around 2298 ha; by comparison, the state has only invested less than RMB150/ha (US$24/ha), 15 times less than the actual cost (Li, Ding, & Zhao, 2010). In this case, it is very difficult or unfeasible to ask for more local government investment as those provinces/autonomous regions are comparatively underdeveloped.

In the field of eco-compensation for watersheds, the same problem exists, as many water function zones and water sources are located in the west of China. Many of them are relatively underdeveloped, therefore relying on large-scale local government investments is almost impractical. It is essential to promote further research on other types of eco-compensation, e.g. market-based compensation, as governmental compensation alone is never be enough in the long run.

Lessons and observations

Patricia Wouters, Huiping Chen, James E. Nickum and David Devlaeminck

Increasing global insecurity – economic, political and social – sees national governments redefining foreign policy in a number of ways. Nonetheless, in an interconnected world, cooperation over shared resources is no less imperative. China, the primary country of interest in this book, continues to expand its international outreach, albeit in line with been the 'Five Principles of Peaceful Co-existence', the foundation of its foreign policy. From increased financial engagement around the globe – more bilateral investment, strong role in the establishment of the Asian Infrastructure Investment Bank (AIIB) and ratification of the Paris Agreement (Xinhuanet, 2016) – to its continued and enhanced support for the United Nations (UN) (i.e., increased support for UN peacekeeping as only one example) and investing US$1 billion peace and development fund (Xinhuanet, 2015), China's role and leadership is evolving. China's management of the uses of its transboundary waters appear to follow this trend, with cooperation growing in substantive and procedural/institutional areas, discussed in this collection. As this cooperation continues to develop, it can be informed from the emerging lessons from this book, considered under three broad themes:

(1) The incremental and evolutionary nature of transboundary water cooperation.
(2) The difficult challenges of the so-called 'upstream dilemma' in the light of the international legal duty to cooperate.
(3) The untapped potential contribution of the rules of international law that apply in this field, including the norms and practice embodied under the UN's global water-related treaties – the Convention on the Law of the Non-Navigational Uses of International Watercourses (commonly referred to as United Nations Watercourse Convention [UNWC]) and the United Nations Economic Commission for Europe Convention on the Protection and Use of Transboundary Watercourses and International Lakes (UNECE TWC).[1]

The incremental and evolutionary nature of transboundary water cooperation

Especially in the case of international relations, the bonds of cooperation generally take years to consolidate and implement. In the field of transboundary waters, where the resource is irreplaceable and directly linked to economic, social and environmental security issues, the journey can be long and episodic. International law is founded upon two fundamental principles: state sovereignty and the duty to cooperate. These elements are elaborated in the UN Charter and in the extensive body of treaty and customary law, including state practice that applies to the use of international freshwater resources. Consistent with these bedrock notions, riparian countries are encouraged to cooperate by taking into account each other's needs, albeit within the context of domestic (national sovereign) needs. The general duty to cooperate as formulated in the UNWC reflects this approach, urging watercourse states to cooperate 'on the basis of sovereign equality, territorial integrity, mutual benefit and good faith in order to attain optimal utilization and adequate protection of an international watercourse'. How this is to be operationalized in practice finds further elaboration in the two UN multilateral water conventions (UNWC and UNECE TWC), which are each now available to be adopted and applied as relevant in the China/Asian region (discussed below). Further guidance can also be found in the diverse state practice surveyed, in part, in this collection. For example, Leb, Libert and Magsig elaborate various incremental approaches to transboundary water cooperation in selected case studies, revealing, inter alia, how local circumstances can influence regional win–win solutions. Kibaroglu explores how Turkey, also facing an 'upstream dilemma', evolved its state practice in the light of EU and international legal frameworks, despite ongoing national development challenges. The Canada–United States example, examined in Chapters 8 and 9, reveals how even long-standing treaty and institutional regimes between cooperative riparian neighbours require adjustments over time, changes that can be addressed on a case-by-case basis under overarching treaty arrangements. Thus, legal norms and agreements provide malleable frameworks for cooperation capable of evolving to meet changing circumstances.

The difficult challenge of the 'upstream dilemma' in light of the international legal duty to cooperate

Upstream riparian nations are both advantaged and disadvantaged by their hydro-geographic situation. On the one hand, controlling the headwaters offers considerable leverage in developing the water resources to meet national needs. On the other hand, given the interdependence of the transboundary water riparian communities (broadly defined), and the overarching international legal obligation to cooperate peacefully, upstream riparian states face the challenge of taking into account not only their own domestic needs but also those of riparian neighbours (Leb, 2015).

China's upstream geographical location on so many transboundary waters with such a diverse collection of riparian neighbours faces the 'upstream dilemma' on a number of fronts. Both domestic and international issues arise (Nickum, 2008). At home, China must cope with challenging domestic agendas, complicated by a myriad of factors. Internationally, in line with its express foreign policy (firmly embedded in the Five Principles of Peaceful Co-existence), China, as an emerging global actor, seeks to be 'the good neighbour', especially at the regional level, where it engages more fully in multilateral forums, such as the Shanghai Cooperation Organisation and Association of Southeast Asian Nations (ASEAN).[2] Nonetheless, attempting to meet regional and international economic, social, cultural and environmental imperatives through win–win solutions is a daunting task for any nation.

In the transboundary water context, upstream riparians, such as China, need to find the operational means to implement this duty to cooperate. As such, 'states establish rules of cooperation and agree to limit the scope of their sovereignty because cooperation creates new opportunities. The existence of commonly accepted rules of conduct and the establishment of joint mechanisms can be mutually beneficial' (Leb, 2015). Domestic and international demands require an ongoing balancing of interests, sought through win–win solutions, based on meaningful dialogue and consultation. Given China's express foreign policy objective of being 'the good neighbour', keen to share its benefits at home and abroad, how might this self-declared role play out in the transboundary water context?

Transboundary waters have long been characterized as a 'community of interests'.[3] In the international context, this notion was first introduced by the Permanent Court of International Justice (PCIJ) in the *River Oder* case,[4] decided almost a century ago (Territorial Jurisdiction of the International Commission of the River Oder, 1929). The PCIJ decision explained:

> When consideration is given to the manner in which States have regarded the concrete situations arising out of the fact that a single waterway traverses or separates the territory of more than one State, and the possibility of fulfilling the requirements of justice and the considerations of utility which this fact places in relief, it is at once seen that a solution of the problem has been sought not in the idea of a right of passage in favour of upstream States, but in that of a community of interest of riparian States. This *community of interest in a navigable river becomes the basis of a common legal right, the essential features of which are the perfect equality of all riparian States in the use of the whole course of the river and the exclusion of any preferential privilege of any riparian State in relation to others.*
>
> (emphasis added)

The PCIJ noted that:

> [i]n the present case, it is enough to go back to the general principles of international river law to find that, if the right of upstream States to free access to the sea has, as Poland maintains, played a considerable part in the formation of river law, *the basic concept which dominates this area of law is that of a community of interests of riparian States which in itself leads to a common legal right.* [. . .] This concept is found already in the Act of the Congress of Vienna of June 9th, 1815 and it has inspired subsequent instruments. Such a *community of interests*, however, extends to the whole navigable course of the river and *does not stop at the last political frontier.* As for the Treaty of Versailles, it has *only enlarged the community of riparian States to cover all users of a river, riparians or not, thus making it completely international.*
>
> (emphasis added)

Although this case dealt with navigational uses, its approach has been endorsed in the broader context of transboundary waters, including in subsequent International Court of Justice (ICJ, 2010) cases, notably the Pulp Mills case, 2010, which also looked at community of interests in an international watercourse dispute (Leb, 2015; Wouters & Tarlock, 2013).

The notion of 'taking others needs into account' is reflected in numerous global instruments, including Principle 3 of the 1992 Rio Declaration, which adopted the Brundtland Commission's definition of sustainable development and applied it to the exploitation of all natural resources including water: 'The right to development must be fulfilled so as to equitably meet developmental and environmental needs of present and future generations'. Chapter 18 set out specific targets for equitable and sustainable water management. Since Rio, the promotion of sustainable water development and use has been a major objective of the UN. A UN Secretary General report elaborates

> In a world of interconnected threats and challenges, *it is in each country's self-interest* that all of them are addressed effectively. Hence, the cause of larger freedom can only be advanced by broad, deep and sustained global cooperation among states. *Such cooperation is possible if every country's policies take into account not only the needs of its own citizens but also the needs of others.* This kind of cooperation not only advances everyone's interests but also recognizes our common humanity.
>
> (Report of the UN Secretary General, 2005, emphasis added)[5]

The 'global-community approach' finds further expression in the significant body of global environmental treaties that comprise what can be considered to be a global governance framework (Lee, J., 2015). As such, 'the Ramsar framework of governance could be a window of opportunity for the integration of competing demands and interests of stakeholders in the alleviation of conflict where the convention represents the consensus of the international community' (Lee, J., 2015). China is party to a number of multilateral environmental agreements – Convention on Wetlands of International Importance (Ramsar Convention, 1976); the UN Convention on Biological Diversity (CBD) (Convention on Biological Diversity, 1992); UN Framework Convention on Climate Change; UN Convention to Combat Desertification in those Countries Experiencing Serious Drought and/or Desertification, particularly in Africa – to name some of the most relevant. These multilateral treaties provide a significant cross-cutting platform for transboundary cooperation in a number of areas that relate directly to the management of China's transboundary water resources. For example, the Ramsar Convention has the potential to offer 'diplomatic opportunit[ies] for cooperation on the joint management of water resources and wetlands between China and its neighbouring countries with which the former shares transboundary Ramsar sites' (Lee, J., 2015).

These strands of 'global community cooperation' help to address China's 'upstream dilemma' in several important ways. As identified by McIntyre (2015) ('it appears that benefit-sharing can rely for support upon all three of the key rules of general international water resources law, i.e. the principle of equitable and reasonable utilization, the duty to prevent significant transboundary harm, and the general duty to cooperate in the utilization and protection of an international watercourse'.), Leb ('The existence of commonly accepted rules of conduct and the establishment of joint mechanisms can be mutually beneficial. Experience in transboundary basins shows that, more often than not, cooperation between riparian states leads to greater benefits for all basin states'.) and Seungho Lee (Benefit sharing is 'regarded as a means of distributing a range of benefits derived from resource utilization across the economy and can facilitate the enlargement of growth with the support of socially inclusive policies. If this idea is applied to the realm of international water law, the principles of equitable and reasonable use of water without any major harm also apply'.), international law provides for the development of a shared-benefits approach that can take into account upstream and downstream needs and interests. The Canada–United States examples demonstrate how this can be developed and transformed in a bilateral context, taking on new dimensions as the century-old treaty regime evolved to address changing circumstances related to water quality concerns and new levels and types of governance (Grover & Krantzberg, 2015). Over the past 30 years, water governance between Canada and the United States has undergone 'rescaling' through province–state agreements, non-federal mechanisms, increasing local participation and indigenous-led mechanisms (Norman & Bakker, 2015). This evolutionary nature of the Canada–United

States boundary waters treaty over the past 100 years demonstrates the inherent flexibility possible in international water agreements.

Seungho Lee elaborates on how China's upstream dilemma is being transformed through a shared-benefits approach to its regional international relations on the Mekong and across the Greater Mekong Subregion. Noting China's increasing energy imports via the Mekong's hydropower energy grid, Lee notes that in order to receive mutual benefits 'it is imperative that China retain a peaceful relationship with its neighbours to secure a stable supply of energy'. Furthermore, 'this means that the position of China as the upstream country in the Mekong River basin is no longer hegemonic. In contrast, it now appears that the non-hegemonic countries, Myanmar and Lao PDR, could be more influential than China in terms of energy-security issues' (Lee, S., 2015).

Thus, there are concrete examples of how the 'upstream dilemma' might be addressed in light of the duty to cooperate. The studies presented here demonstrate how multilateral environmental governance arrangements offer opportunities for enhanced transboundary water cooperation, and also suggest that win–win solutions might arise from a benefits-sharing approach. These approaches provide inroads for mitigating the 'upstream dilemma' through redefining the nature of benefits available to both upstream and downstream riparian nations.

The untapped potential of international water law, including the UN Water Conventions

The third theme emerging from this book is the untapped potential of the rules of international law that govern this field, including the norms and state practice under the UN Water Conventions for promoting improved transboundary water cooperation. In the case of China, releasing this untapped potential could support its overarching global foreign policy objectives in three possible areas: (1) implementing China's 'good neighbour' foreign policy through more complete transboundary water cooperation, (2) enhancing transboundary governance regimes in ways that facilitate meaningful transboundary water cooperation across the region and (3) adding to China's growing role as emerging global leader, including its consistent endorsement of the UN.[6] This is not a revolutionary proposal, but instead, part of the 'step-by-step' approach described by many of the authors in this collection (Feng, He, & Wang, 2015; Leb, 2015; Libert, 2015; Magsig, 2015; Su, 2014; Van Rijswick et al., 2014). This might help to catalyse new approaches to the current concerns related to China's southern transboundary waters, currently characterized as 'A Case of Asymmetry' (Kattelus et al., 2015).

The UN Water Conventions provide a suite of norms of international law and processes that contribute to facilitating transboundary cooperation, demonstrated in extensive state practice under these and other regional legal regimes. Leb, Libert, Kibaroglu, McIntyre, Su, and Wouters and Chen each offer insights into how the two UN Water Conventions might be used in the Asian context. This collection of works also demonstrates how regional arrangements can be agreed, based upon the principles of the UN Water Conventions but tailor-made to address particular regional circumstances. Cross-fertilization amongst these rules of international law that governs the uses of transboundary waters can provide 'important normative connectivity and legal clarity for fundamental rules of IWL' (Su, 2014). As identified in multiple chapters, such cross-fertilization across various legal regimes can assist to 'support the more complete realization of the human right to water'

(Chen, 2015), 'reveal the potential for local tailor-made solutions' (Liu, 2015), 'promote cooperation, confidence building, and ultimately peace-making between states sharing transboundary basins through the collaborative management of transboundary wetland ecosystems' (Lee, J., 2015).

Considering how regional transboundary cooperation across Asia might be embedded in treaty regimes, it is helpful to revisit and consider more closely the five key legal elements of a transboundary water-sharing regime – scope, substantive rules, procedural rules, institutional mechanisms, dispute settlement; aspects addressed in varying detail in each of the UN Water Conventions. Covering these issues in an international water agreement goes a long way to providing a comprehensive foundation for meaningful cooperation. Table 1 summarizes these elements.

Wouters and Chen (2013) and Su (2014) have explored China's treaty practice (Appendix A). China also has good transboundary water cooperation with Mongolia, and North Korea, although regional differences (and issues) mark China's transboundary water regimes (Nickum, 2008; Su, 2014; Wouters & Chen, 2013).

China's southern transboundary water resources, yet to be covered by treaty regimes (apart from any water-related obligations China has agreed to under its multilateral environmental agreement engagements) continue to be the subject of bilateral discussions in most cases. China and India recently signed a series of agreements, including hydrological data exchange arrangements (Wouters, 2014). China has also signed a data-sharing agreement with the Mekong River Commission (Mekong River

Table 1. Five key legal elements of a transboundary water-sharing regime

Key elements	Details	UN water conventions
1. Scope	- Legal reach (what waters?) - Definitions (watercourse; uses) - Parties (States REIOs)	UNWC: Art. 2 UNECE: Art. 1(1)
2. Substantive Rules	- Legal duties and entitlements (equitable and reasonable utilization; due diligence; protection) - Rules of substance (generator precise)	UNWC: Art. 5, 6, 7, 8, 10, 21 UNECE: Art. 2, 17
3. Procedural Rules	- Rules of procedure (duty to cooperate as bridge) - Notification/exchange of information	UNWC: Part III UNECE: Part II
4. Institutional Mechanisms	- Joint bodies (RBOs) - Conference of the Parties (MoP) - Organizations/organs (ministerial level; other)	UNWC: recommended (Art. 24) UNECE: MoP
5. Dispute Settlement	- Dispute avoidance (consultation and negotiation) - Dispute settlement (Art. 33 UNWC; other) - Compliance verification (reporting; facilitation)	UNWC: Art. 33 UNECE: Art. 22

MoP, meeting of the parties; RBO, river basin organisation; REIO, Regional Economic Integration Organization; UNECE, United Nations Economic Commission for Europe; UNWC, United Nations Watercourse Convention.

Commission, 2013; Wouters & Chen, 2013), and has established the Lancang-Mekong Cooperation Framework covering five key cooperation areas, one of which is water resources (Ministry of Foreign Affairs, 2016). These incremental steps can be fortified through more detailed agreements and joint institutional arrangements (step-by-step approach), which could begin by focusing on technical cooperation, an approach featured in some of the case studies discussed in this collection (Feng, He, & Wang, 2015; Leb, 2015; Magsig, 2015; Van Rijswick et al., 2014).

Where bilateral consultations, negotiations and dialogue are unable to resolve transboundary water challenges, China might want to explore new avenues for dispute avoidance and settlement, possibly under other existing treaty engagements, such as environmental (Lee, J., 2015; Liu, 2015), human rights (Chen, 2015) and bilateral investment regimes (Han, 2015). There are a number of options that might be considered. On the dispute avoidance side, there are several approaches that can be developed, including but not limited to joint river basin commissions, more detailed joint cooperative undertakings and increased involvement of civil society. Examples exist of how these might be developed in accordance with the UNECE TWC, such as the list of areas of cooperation (Part II). There is also Sino-Russian and Sino-Kazak transboundary water commission practice, which could offer possible models for bilateral arrangements in the south. Where transboundary water disputes cannot be prevented, peaceful settlement might be dealt with through fact-finding commissions, conciliation or arbitration; each of these procedures provide ample legal space for sovereign inputs that would be consistent with China's approach to international law (Saul, 2013). China's express support for the UN Charter, the work of the UN and the ICJ, underlines this position (Xue, 2005). Additionally, China's growing body of bilateral investment treaties (BIT) and its participation in the World Trade Organization (WTO) provide dispute settlement measures that include arbitration. Analysing China's BIT practice could provide new insights in how to move forward with its transboundary water issues (Chen, 2015; Han, 2015).

Some final thoughts

In 2015, following extensive global debate and study, the international community came together to adopt the '2030 Transforming our World' document, and with it the Sustainable Development Goals. The Sustainable Development Agenda,[7] which will guide the global development agenda until 2030, provides 17 goals and 169 accompanying targets, aimed at setting the development agenda for the next 15 years. Goal 6 directly relates to water, set to 'ensure availability and sustainable management of water and sanitation for all' (UN, 2016). Such a goal will require cooperative management of shared water resources across the globe. While Asia has a long tradition of managing its shared water resources (Benvenisti, 2008; Wouters, 2011), future Sustainable Development Goals will have an impact on future activities across the region. How Asia, and especially China, addresses growing water demands, especially across the water/energy/food/environment (WEFE) nexus will have a direct impact on regional (and possibly global) economic, social and environmental health and security.

Transboundary water resources have the potential to connect nations in a number of ways through a variety of mechanisms. Legal frameworks and governance approaches that meet regional needs go a long way to facilitating transboundary water cooperation, including in addressing the difficult challenges of the 'upstream dilemma'.

Development of benefit-sharing arrangements crafted in upstream–downstream scenarios that protect ecosystems downstream and still meet needs upstream, provide only one example (Tarlock, 2014). Devising such 'win–win' transboundary water solutions across Asia could advance China's foreign policy objectives and contribute to realizing the Chinese dream at home and abroad.

Achieving effective transboundary water cooperation is a long-term game, one that is costly and fraught with complications arising from diverse, often uncontrollable, sources. International law can help to mitigate the uncertainties involved in this task through its substantive and procedural norms and practices.

Notes

1. See UNWC (1997), United Nations Convention on the Law of Non-Navigational Uses of International Watercourses (UNGA 51/229). Retrieved from http://www.un.org/ga/search/view_doc.asp?symbol=A/res/51/229, accessed May 2, 2017. UNECE-TWC (1992), United Nations Economic Commission for Europe, Convention on the Protection and Use of Transboundary Watercourses and International Lakes. Retrieved from https://treaties.un.org/doc/Treaties/1992/03/19920317%2005-46%20AM/Ch_XXVII_05p.pdf, accessed May 2, 2017.
2. China's policy of good neighbourliness is 'anchored in targeted international high-level outreach and deeper engagement with regional organization and express support for the UN'. In her 2013 article, Wouters identifies visits to India, Pakistan, Russia, ASEAN and the Shanghai Cooperation Organization as examples of such outreach and engagement. See Wouters (2014). The Yin and Yang of international water law: China's transboundary water practice and the changing Contours of Sovereignty *RECIEL, 23*(1), 67–75. This trend of high-level, targeted visits has continued, and to this list we can add recent regional visits by President Xi Jinping to: the Republic of Korea (July, 2014), Mongolia (August, 2014), Tajikistan, Maldives and Sri Lanka (September, 2014), Australia, New Zealand and Fiji (November, 2014), Pakistan (April, 2015), Vietnam and Singapore (November, 2015), as well as Cambodia and Bangladesh (October, 2016). President Xi Jinping has also made high-level visits to various organizations and conferences, some of which have been hosted in China, including: the 6th BRICS Summit (July 2014), 14th Shanghai Cooperation Organization Meeting (September, 2014), 9th G20 Leaders' Summit (November, 2014), Boao Forum for Asia (March, 2015), Asia-Africa Summit and 60th Anniversary of the Bandung Conference (April, 2015), 15th Meeting of the Council of the Heads of State of the Shanghai Cooperation Organization (June, 2015), 7th BRICS Summit (July, 2015), 70th Session of the UN General Assembly (September, 2015), the Paris Climate Change Conference (December, 2015), G20 Summit (Hangzhou, China – September, 2016), the World Economic Forum (January, 2017) and will host the 9th BRICS Summit (Xiamen, China – September, 2017). This is a non-exhaustive list and China's leadership has been especially active and engaged over the past few years. For press releases and speeches from these visits, see Ministry of Foreign Affairs of the Peoples Republic of China, *Topics*, available at: http://www.fmprc.gov.cn/mfa_eng/topics_665678/
3. Rocha Loures (2015) identifies five characteristics, one or more of which may identify cooperation within a community of interests framework. They include: (1) the promotion of a drainage basin approach, (2) a view of watercourses as common property, (3) mutual consent, (4) joint management of the shared water resources and (5) sub-basin agreements and institutions. See Roche Loures, F. (2105). The dragon as part of transboundary basin 'villages': community of interests and international law in China. *Journal of Water Law, 24*, 24.
4. *Territorial Jurisdiction of the International Commission of the River Oder, Judgment No. 16, 1929, P.C.I.J., Series A, No. 23*, p. 27.
5. UN Secretary-General Report, *In Larger Freedom: Towards Development, Security and Human Rights for All*, UN Doc A/59/2005, 6 (para. 18).
6. Wouters, P. (2014). The Yin and Yang of international water law: China's transboundary water practice and the changing Contours of Sovereignty *RECIEL, 23*(1), 72.
7. See UN. (2016). Transforming Our World: The 2030 Agenda for Sustainable Development. Retrieved from https://sustainabledevelopment.un.org/content/documents/21252030%20Agenda%20for%20Sustainable%20Development%20web.pdf, accessed April 12, 2017.

References

Benvenisti, E. (2008). Asian traditions and contemporary international law on the management of natural resources. *Chinese Journal of International Law*, *7*, 273–283. doi:10.1093/chinesejil/jmn008

Chen, H. (2015). The human right to water and foreign investment: friends or foes? *Water International*, *40*(2), 297–311. doi: 10.1080/02508060.2015.1012784

Convention on Biological Diversity. (1992). United Nations Treaty Series, Vol. 1760 (2001) p. 79. Retrieved from www.cbd.int/convention/text/. Rio de Janeiro, 5 June 1992.

Feng, Y., He, D., & Wang, W. (2015). Identifying China's transboundary water risks and vulnerabilities – a multidisciplinary analysis using hydrological data and legal/institutional settings. *Water International*, *40*(2), 328–341. doi: 10.1080/02508060.2015.1012251

Grover, V., & Krantzberg, G. (2015). Transboundary water management: lessons learnt from North America. *Water International*, *40*(1). doi: 10.1080/02508060.2014.984962

Han, X. (2015). Approaches to investment in Chinese transboundary waters. *Water International*, *40*(1), 71–86. doi: 10.1080/02508060.2014.978970

International Court of Justice (ICJ). (2010). Case concerning Pulp Mills on the River Uruguay (Argentina v. Uruguay). ICJ Reports, Judgment of 20 April 2010. Retrieved from www.icj-cij.org/docket/files/135/15877.pdf

Kattelus, M., Kummu, M., Keskinen, M., Salmivaara, A., & Varis, O. (2015). China's southbound transboundary river basins: a case of asymmetry. *Water International*, *40*(1), 113–138. doi: 10.1080/02508060.2014.980029

Leb, C. (2015). One step at a time: international law and the duty to cooperate in the management of shared water resources. *Water International*, *40*(1), 21–32. doi: 10.1080/02508060.2014.978972

Lee, J. (2015). The governance of wetland ecosystems and the promotion of transboundary water cooperation: opportunities presented by the Ramsar Convention. *Water International*, *40*(1), 33–47. doi: 10.1080/02508060.2014.989681

Lee, S. (2015). Benefit-sharing in the Mekong River Basin. *Water International*, *40*(1), 139–152. doi: 10.1080/02508060.2014.978974

Libert, B. (2015). The UNECE Water Convention and the development of transboundary cooperation in the Chu-Talas, Kura, Drin and Dniester River Basins. *Water International*, *40*(1), 168–182. doi: 10.1080/02508060.2014.990202

Liu, Y. (2015). Transboundary water cooperation on the Yarlung Zangbo/Brahmaputra – a legal analysis of riparian state practice. *Water International*, *30*(2), 354–374. doi: 10.1080/02508060.2015.1015319

Magsig, B.O. (2015). Water security in Himalayan Asia: first stirrings of regional cooperation?. *Water International*, *40*(2), 342–353. doi: 10.1080/02508060.2015.1011463

McIntyre, O. (2015). Benefit-sharing and upstream/downstream cooperation for ecological protection of transboundary waters: opportunities for China as an upstream state. *Water International*, *40*(1), 48–70. doi: 10.1080/02508060.2014.990350

Mekong River Commission. (2013). Mekong River Commission and China Boost Water Data Exchange. 14 Nov. 14. Retrieved from www.mrcmekong.org/news-and-events/news/mekong-river-commission-and-china-boost-water-data-exchange/

Ministry of Foreign Affairs, the People's Republic of China. (2016). Sanya Declaration of the First Lancang-Mekong Cooperation (LMC) Leaders' Meeting--For a Community of Shared Future of Peace and Proseprity among Lancang-Mekong Countries.

Nickum, J. E. (2008). The Upstream Superpower: China's International Rivers. In O. Varis, A. K. Biswas, & C. Tortajada, (Eds.), *Management of Transboundary Rivers and Lakes* (pp. 227–244). Berlin: Springer.

Norman, E. S., & Bakker, K. (2015). Do good fences make good neighbours? Canada–United States transboundary water governance, the Boundary Waters Treaty, and twenty-first-century challenges. *Water International*, *40*(1). doi: 10.1080/02508060.2014.978973

Ramsar Convention. (1976). Convention on Wetlands of International Importance Especially as Waterfowl Habitat (Ramsar, 2 February 1971) United Nations Treaty Series, Vol. 996, 1–14583 (1976) p. 247. On China's commitments under RAMSAR. Retrieved from www.ramsar.org/cda/en/ramsar-documents-list-anno-china/main/ramsar/1-31-218%5E16477_4000_0

Report of the UN Secretary General. (2005). K. Annan, In Larger Freedom: Towards Development, Security and Human Rights for All, UN Doc A/59/2005.

Saul, B. (2013). China, natural resources, sovereignty and international law. *Asian Studies Review, 37*(2), 196–214. doi: 10.1080/10357823.2013.769497

Su, Y. (2014). Contemporary legal analysis of China's transboundary water regimes: international law in practice. *Water International, 39*(5), 705–724. doi: 10.1080/02508060.2014.950856

Tarlock, A. D. (2014). Mexico and the United States assume a legal duty to provide Colorado River delta restoration flows: an important international environmental and water law precedent. *Review of European Comparative & International Environmental Law (RECIEL), 23*, 76. doi: 10.1111/reel.12066

Territorial Jurisdiction of the International Commission of the River Oder. (1929). Judgment No. 16, 1929, P.C.I.J., Series A, No. 23, p. 27.

UN. (2016). Transforming our world: the 2030 Agenda for Sustainable Development. Retrieved from https://sustainabledevelopment.un.org/content/documents/21252030%20Agenda%20for%20 Sustainable%20Development%20web.pdf, accessed April 12, 2017.

Van Rijswick, M., Edelenbos, J., Hellegers, P., Kok, M., & Kuks, S. (2014). Ten building blocks for sustainable water governance: an integrated method to assess the governance of water. *Water International, 39*(5). doi: 10.1080/02508060.2014.951828

Wouters, P., & Chen, H. (2013). China's 'soft-path' to transboundary water cooperation examined in the light of two UN global water conventions: exploring the 'Chinese way'. *Journal of Water Law, 22*, 232.

Wouters, P., & Tarlock, A. D. (2013). The third wave of normativity in global water law: the duty to cooperate in the peaceful management of the world's water resources: an emerging obligation erga omnes?. *Journal of Water Law, 23*, 51.

Wouters, P. (2011). The International Law of Watercourses: New Dimensions. In *Collected Courses of the Xiamen Academy of International Law* (Vol. 3). Leiden: Martinus Nijhoff.

Wouters, P. (2014). The Yin and Yang of international water law: China's transboundary water practice and the changing Contours of State Sovereignty. *RECIEL, 23*(1), ISSN2050–0386. doi: 10.1111/reel.12069

Xinhuanet. (2015, September 29). Xi Announces Major Measures to Support UN, *Xinhuanet*. Retrieved from http://news.xinhuanet.com/english/2015-09/29/c_134669393.htm, accessed April 12, 2017.

Xinhuanet. (2016, September 3). China's Legislature Ratifies Paris Agreement on Climate Change, *Xinhuanet*. Retrieved from http://news.xinhuanet.com/english/2016-09/03/c_135656703.htm, accessed April 12, 2017.

Xue, H. (2005). China's Open Policy and International Law. *Chinese Journal of International Law, 4*(1), 133–139. doi: 10.1093/chinesejil/jmi004

Appendix A: China's major international watercourses (listed by country) and relevant treaties

Country	*Shared Watercourse*		*Agreements*
Afghanistan	Aksu (Agsu) River Indus (Pakistan, India, Kashmir) Karachukur Su River Wakhjir River	None	
Bangladesh	Yarlung Zangbo/Brahmaputra River	2008	China-Bangladesh Memorandum of Understanding upon Provision of Hydrological Information of the Brahmaputra/Yaluzangbu River in Flood Season by China to Bangladesh
		2007	Joint Communiqué Between the Government of the People's Republic of China and the Government of the People's Republic of Bangladesh[i]
Bhutan	Yarlung Zangbo/Brahmaputra River (India)	None	
India	Ayeyarwady (Myanmar) Ganges (Nepal, Bangladesh) Indus (Pakistan, Afghanistan, Kashmir) Shiquan River/Indus (Pakistan) Yarlung Zangbo/Brahmaputra River (Bangladesh)	2013	Memorandum of Understanding (MOU) between the Ministry of Water Resources, the Republic of India, and the Ministry of Water Resources, the People's Republic of China on Strengthening Cooperation on Trans-border Rivers[ii]
		2010	Joint Communiqué (Para. 9)
		2008	Memorandum of Understanding (MoU) upon the Provision of Hydrological Information of the Brahmaputra/Yaluzangbu River in the Flood Season[iii]
		2005	Memorandum the Provision of Hydrological Data on Langqen Zangbo/Sutlej during Flood Season[iv]
		2005	Joint Statement (Para. 9)
		2002	Memorandum of Understanding (MoU) on the Sharing of Hydrological Information on the Yaluzangbu/Brahmaputra River during the Flood Season
Kazakhstan	Emin River Irtysh-Ob River (Mongolia; Russia) Yili River	2015	Joint Declaration on New Stage of Comprehensive Strategic Partnership Between the People's Republic of China and the Republic of Kazakhstan[v]
		2013	Sino-Kazakhstan Joint Declaration on Further Deepening Comprehensive Strategic Partnership[vi]
		2013	Agreement on Management and Operation of Dostyk (YouYi) Joint Hydro United on the Khorgos River
		2011	Cooperation Agreement on Environmental Protection
		2011	Agreement on Protection of Water Quality of Transboundary Rivers
		2010	Cooperation Agreement on the Joint Construction of Horgos River Friendship Joint Diversion Control Project
		2007	Joint Communiqué (Para. 6)
		2006	Agreement on Management System of China–Kazakhstan Border (Art. 1(14) and (15), Chap. 4 (Arts 8–11), Art. 15(2))[vii]
		2006	Agreement on Exchange of Hydrological and Hydro-chemical Information (Data) of Border Gauging Stations on Major Transboundary Rivers
		2006	Agreement on Development of Research Cooperation on Transboundary Rivers
		2005	Agreement on Emergency Notification of the Parties of Natural Disasters on Transboundary Rivers
		2001	Cooperation Agreement on Cooperation in the Field of the Use and Protection of Transboundary Rivers[viii]
Kyrgyzstan	Aksu River Tarim (Tajikistan, Kashmir)	None	

Appendix A: China's major international watercourses (listed by country) and relevant treaties

Laos	Lancang/Mekong River (Myanmar, Thailand, Cambodia, Vietnam)	2011	Joint Statement on Law Enforcement Cooperation along the Mekong River among China, Laos, Myanmar, Thailand
	Yuan/Red River (Vietnam)	2000	Agreement on Commercial Navigation on Lancang/Mekong River among the governments of China, Laos, Burma (Myanmar) and Thailand[ix]
		1994	Agreement on Freight and Passenger Transport along the Lancang/Mekong River
		1993	Treaty on the Border System (Article 6)
Mongolia	Heilong/Amur River (North Korea; Russia)	2010	Treaty on the Management of the Boundary (Art. 1(11), Chapter 4)
		1994	Agreement on Protection and Utilisation of Border Waters[x]
	Kherlen River (Russia)	1988	Treaty on Common Boundaries and Settlement of Boundary Matters (Part 3
	Ulungur River (Russia)	1962	Boundary Treaty (Art. 2)[xi]
Myanmar	Ayeyarwady River (India)	2011	Joint Statement on Law Enforcement Cooperation along the Mekong River among China, Laos, Myanmar, Thailand
	Lancang/Mekong River(Cambodia, Laos, Thailand, Vietnam)	2000	Agreement on Commercial Navigation on Lancang/Mekong River among the Governments of China, Laos, Burma (Myanmar) and Thailand[xii]
	Nu/Salween River (Thailand)	1997	Agreement on Border Management and Cooperation (Art. 1(5), Arts. 9 and 10)
		1994	Agreement on Freight and Passenger Transport along the Lancang/Mekong River[xiii]
Nepal	Ganges (India, Bangladesh)	None	
North Korea	Heilong/Amur River (Mongolia; Russia)	2012	Agreement on Joint Construction, Management and Maintenance of Ji'an-Manpu Border - River Road Bridge on Yalu River
	Tianchi/Heavenly Lake	2011	Cooperation Agreement on Maritime Administration of Yalu River Water Area
	Tumen River (Russia)	2010	Agreement on Joint Construction, Management and Maintenance of Border-River Road Bridge on the Yalu River
	Yalu River	2002	The Protocol on Cooperation on Navigation and Shipping in Boundary Rivers[xiv]
		1995	Memorandum of Understanding on Environmental Principles Governing the Tumen River Economic Development Area and Northeast Asia
		1995	Agreement on the Establishment of the Consultative Commission for the Development of the Tumen River Economic Development Area and Northeast Asia
		1982	Rules for Vessel's Navigation in the Border Rivers
		1978	Cooperation Agreement on Hydrologic Work of Yalu River and Tumen River[xv]
		1964	The Protocol on Borderline (Art. 17)[xvi]
		1962	The Treaty of Borderline (Art. 3)
		1960	The Agreement on Cooperation on Navigation and Shipping in Boundary Rivers[xvii]
Pakistan	Shiquan River/Indus (Afghanistan, India, Kashmir)	None	
Russia	Argun River	2008	Agreement on Reasonable Utilisation and Protection of Transboundary Waters XVII
	Heilong/Amur River (Mongolia; North Korea)	2006	Agreement on Management System of China–Russia Border
	Kherlen River (also Mongolia)	2001	Treaty of Good-Neighborliness and Friendly Cooperation[xviii]
	Irtysh-Ob River (Kazakhstan; Mongolia)	1999	Agreement on the Joint Economic Use of Separate Islands and Surrounding Water Areas in Frontier Rivers[xix]
	Shuifeng River	1998	The Protocol to the Cargo Transportation between Coastal Ports and Inner Water Ports by Chinese Ships via the Russian Part of Heilong River[xx]
	Tumen River (also North Korea)		
	Wusuli/Ussuri River	1996	Agreement on Khanka/Xingkai Lake International Nature Reserve[xxi]
		1995	Memorandum of Understanding on Environmental Principles Governing the

Appendix A: China's major international watercourses (listed by country) and relevant treaties

	Xingkai/Khanka Lake		Tumen River Economic Development Area and Northeast Asia[xxii]
		1995	Agreement on the Establishment of the Consultative Commission for the Development of the Tumen River Economic Development Area and Northeast Asia[xxiii]
		1994	Agreement Concerning Protection, Regulation and Reproduction of Living Water Resources in Frontier Waters of Rivers Amur and Ussuri[xxiv]
		1992	Agreement on the Organizing of Shipping of Foreign Trade Goods by Ships of China and Russia in Amur and Songhua River[xxv]
		1988	Agreement between China and Soviet Union on Cooperation in Fisheries
		1956	Agreement on Joint Research Operations to Determine the Natural Resources of the Amur River Basin and the Prospects for Development of its Productive Potentialities and on Planning and Survey Operations to Prepare a Scheme for the Multi-Purpose Exploitation of the Argun River and the Upper Amur River[xxvi]
		1951	Agreement on the Navigation and Construction of Border Waters including Amur, Ussuri, Argun, Sungacha River and Khanka Lake[xxvii]
Tajikistan	Tarim (Kyrgyzstan, Kashmir)	None	
Vietnam	Beilun/Ka Long River	2009	China–Vietnam Land Border Management System Agreement (Art. 1(11), Chap. 4)
	Lancang/Mekong River (Cambodia, Laos, Myanmar, Thailand)	1991	Provisional Agreement on the Handling of Border Affairs (Art. 2)[xxviii]
	Pearl River		
	Yuan/Red River		
Mekong River Commission		2011	Joint Statement on Law Enforcement Cooperation along the Mekong River among the People's Republic of China, The Lao People's Democratic Republic, the Republic of the Union of Myanmar and the Kingdom of Thailand
		2002	Agreement on Provision of Hydrological Information on the Lancang/Mekong River during Flood Season
		2000	Agreement on Commercial Navigation on Lancang/Mekong River among the Goverments of China, Laos, Burma (Myanmar) and Thailand

---

i The Chinese title of the communiqué is 中华人民共和国政府和孟加拉人民共和国政府联合公报, available at: http://news.china.com/zh_cn/domestic/945/20050409/12229933.html, accessed April 10, 2017. The English version is available at: http://bd.mofcom.gov.cn/aarticle/zxhz/hzjj/200705/20070504698156.html, accessed April 10, 2017.

ii The English version is available at: http://mea.gov.in/bilateral-documents.htm?dtl/22368, accessed April 11, 2017.

iii The Chinese title of the MoU is 关于中方向印方提供雅鲁藏布江 - 布拉马普特拉河汛期水文资料的谅解备忘录. This MoU is a second one based on the first MoU (2002) and was valid for five years. The first MoU is available at: http://www.fsou.com/html/text/eag/1006692/100669213.html, accessed April 10, 2017.

iv The Chinese title of the MoU is 关于中方向孟方提供雅鲁藏布江 - 布拉马普特拉河汛期水文资料的谅解备忘录. Neither the Chinese nor the English version of the MoU is available online thus far.

v Available in English at: http://www.fmprc.gov.cn/mfa_eng/wjdt_665385/2649_665393/t1293114.shtml, accessed April 11, 2017.

vi Available in English at: http://www.internationalwaterlaw.org/documents/regionaldocs/ussr-china.html, accessed April 11, 2017.

vii The 2006 Sino-Kazakh Agreement on Management Systems of China-Kazakhstan Border, Art. 1, 8, 11, 155 translated from the Chinese by the Authors.

viii The Chinese title of the agreement is 中哈 - 关于利用和保护跨界河流的合作协定, available at: http://www.fsou.com/html/text/eag/1006692/100669255.html, accessed April 10, 2017. The English version of the agreement is available at: Ronald B. Mitchell and the IEA Database Project, 2002–2012: http://iea.uoregon.edu/treaty-text/2001-useprotectiontransboundaryriversentxt , accessed April 10, 2017.

ix The Chinese title of the agreement is 澜沧江-湄公河商船通航协定, available at: http://baike.baidu.com/view/6632573.htm, accessed April 10, 2017.

x The Chinese title of the treaty is 关于中蒙边界制度和处理边境问题的条约, available at: http://www.law-lib.com/law/law_view.asp?id=77528 , accessed April 10, 2017.

xi The Chinese title of the treaty is 中华人民共和国和蒙古人民共和国边界条约, available at: http://www.npc.gov.cn/wxzl/gongbao/1963-03/04/content_1480962.htm, accessed April 10, 2017.

xii The Chinese title of the agreement is 中老缅泰 - 澜沧江-湄公河商船通航协定, available at: http://baike.baidu.com/view/6632573.htm, accessed April 10, 2017.

xiii The Chinese title of the agreement is 中缅 - 关于澜沧江—湄公河客货运输协定, available at: http://www.86148.com/onews.asp?id=17871, accessed April 10, 2017.

xiv The Chinese title of the protocol is 中朝 - 国境河流航运合作协议书, available at: http://code.fabao365.com/law_34983.html, accessed April 10, 2017.

xv The Chinese title is 中华人民共和国水利电力部和朝鲜民主主义人民共和国气象水文局关于鸭绿江和图门江水文工作合作协定, available at: http://www.law-lib.com/lawhtm/1949-1979/75714.htm, accessed April 11, 2017.

xvi The Chinese title of the protocol is 中朝边界议定书, available at: http://www.cfdd.org.cn/bbs/thread-53260-1-1.html, accessed April 10, 2017.

xvii The Chinese title of the agreement is 中朝 - 关于国境河流航运合作的协定, available at: http://www.86148.com/onews.asp?id=18091, accessed April 10, 2017.

xviii The Chinese title of the treaty is 中俄睦邻友好合作条约. The English version is available at: http://www.fmprc.gov.cn/mfa_eng/wjdt_665385/2649_665393/t15771.shtml , accessed April 10, 2017.

xix The Chinese title of the agreement is 中俄 - 关于对界河中个别岛屿及其附近水域进行共同经济利用的协定, available at: http://sifaku.com/falvfagui/38/zcedppa15p0b.html, accessed April 10, 2017.

xx The Chinese title of the protocol is 中俄 - 关于中国船舶经黑龙江俄罗斯河段从事中国沿海港口和内河港口之间货物运输的议定书, available at: http://www.chinaruslaw.com/CN/CnRuTreaty/53o92p5c/20091130154658_217281.htm, accessed April 10, 2017.

xxi The Chinese title of the agreement is中俄 - 关于兴凯湖自然保护区协定, available at: http://www.hflib.gov.cn/law/law/falvfagui/GJTY/HJ/HJ1017.htm, accessed April 10, 2017.

xxii Memorandum of Understanding on Environmental Principles Governing the Tumen River Economic Development Area and Northeast Asia (New York, 6 December 1995), available at: http://nautilus.org/aprenet/memorandum-of-understanding-on-environmental-principles-governing-the-tumen-river-economic-development-area-and-northeast-asia-1994/, April 10, 2017.

xxiii Agreement on the Establishment of the Consultative Commission for the Development of the Tumen River Economic Development Area and Northeast Asia, New York, 6 December 1995, available at: http://en.pkulaw.cn/display.aspx?cgid=100668800&lib=tax, accessed April 11, 2017.

xxiv The Chinese title of the agreement is 中俄关于黑龙江、乌苏里江边境水域开展渔业资源保护、调整和增殖议定书, available at: http://www.xunyu.com/html/xxbk/zcfg/wm/19940527.htm, accessed April 10, 2017.

xxv The Chinese title of the agreement is中俄 - 关于在黑龙江和松花江利用中俄船舶组织外贸货物运输协议, available at: http://www.chinaruslaw.com/CN/CnRuTreaty/53o92p5c/20091130211214_211145.htm, accessed April 10, 2017.

xxvi Available in English at: http://www.internationalwaterlaw.org/documents/regionaldocs/ussr-china.html, accessed April 11, 2017.

xxvii The Chinese title of the agreement is中苏 - 关于黑龙江、乌苏里江、额尔古纳河、松阿察河及兴凯湖之国境河流航行及建设协定, available at: http://www.lawxp.com/statute/s1012721.html, accessed April 10, 2017.

xxviii The Chinese title of the agreement is 中越处理两国边境事务的临时协定, available at: http://www.chinabaike.com/law/gjt/1427939.html, accessed April 10, 2017.

This list of treaties was first published in the *Journal of Water Law, 22*(6), 229–247.

Appendix B

China International Water Law (CIWL) School of Law, Xiamen University

China International Water Law (CIWL) Programme is located at the School of Law, Xiamen University in Xiamen, China. Lead by Professor Chen Huiping and Professor Patricia Wouters under the Chinese Ministry of Education Thousand Talents Programme, CIWL and the School of Law are building expertise in international water law. CIWL researchers and PhD students work on a wide range of topics in international water law, theory and practice, and CIWL cooperates and collaborates with a series of other institutions, universities and professionals to develop water leaders through research excellence.

CIWL Publications

Chen, H. et al. (2013). Exploring China's Transboundary Water Treaty Practice Through the Prism of the UN Watercourses Convention. *Water International,* 38 (2013).

Chen, H. (2015) The Human Right to Water and Foreign Investment: Friends or Foes?. *Water International,* 40.

Devlaeminck, D., et al.,(eds.) The Human Face of Water Security (Springer Publishing 2017).

Devlaeminck, D. (2016). Reciprocity and the Development of International Water Law. *Journal of Water Law,* 25.

Devlaeminck, D. (2015). Transboundary Water Cooperation and the Responsibility to Protect, *Journal of Water Law,* 24.

Kang, K. (2017). Making Use of Paradoxes: Law, Transboundary Hydropower Dams and Beyond the Technical, *Law and Critique.*

Liu, Y. (2015). Transboundary Water Cooperation on the Yarlung Zangbo/Brahmaputra - a Legal Analysis of Riparian State Practice, *Water International,* 40(2).

Mahbub, N. T. (2016). River Law and Governance. In A. Farazmand (Ed.), *Global Encyclopedia of Public Administration, Public Policy and Governance.* Springer.

Mahbub, N. T. (2016). Sustainable Development and its Evolution in the Realm of International Environmental Law, *Nnamdi Azikiwe University Journal of International Law and Jurisprudence,* 7.

Mahbub, N. T. (2016). Environmental Protection in China: An Analysis and Assessment of Policy and Law, *Quarterly Journal of Chinese Studies,* 4(4).

Rocha Loures, F. (2015). The Dragon as Part of Transboundary Basin 'Villages': Community of Interests and International Law in China, *Journal of Water Law,* 24, 220.

Rocha Loures, F. (2015). History and Development of the Community-of-Interests Doctrine. In Tvedt, Tjere, et al. (Eds.) *A History of Water: Sovereignty and International Water Law (2)*. I.B. Taurus.

Shirani, F. (2017). Cooperation in International Water Law and State's Human Rights Obligation to Fulfill the Human Right to Water, *Global Journal of Politics and Law Research, 5(2).*

Su, Y. (2014). Contemporary Legal Analysis of China's Transboundary Water Regimes: International Law in Practice, *Water International, 39(5).*

Wouters, P. (2011). The International Law of Watercourses: New Dimensions, *Collected Courses of the Xiamen Academy of International Law* (Vol. 3). Leiden: Martinus Nijhoff.

Wouters, P. (2013). 'Dynamic Cooperation' in International Law and the Shadow of State Sovereignty in the Context of Transboundary Waters - Part 1, *Environmental Liability,* 3.

Wouters, P. (2013). 'Dynamic Cooperation' in International Law and the Shadow of State Sovereignty in the Context of Transboundary Waters - Part 2, *Environmental Liability,* 4.

Wouters, P. (2014). The Yin and Yang of International Water Law: China's Transboundary Water Practice and the Changing Contours of Sovereignty, *RECIEL* 23.

Wouters, P. & Chen, H. (2013). China's 'Soft-Path' to Transboundary Water Cooperation Examined in the Light of Two UN Global Water Conventions–Exploring the 'Chinese Way', *Journal of Water Law,* 22.

Wouters, P. & Tarlock, A. D. (2013). The Third Wave of Normativity in Global Water Law: The Duty to Cooperate in the Peaceful Management of the World's Water Resources: An Emerging Obligation Erga Omnes?, *Journal of Water Law,* 23.

CIWL Dissertations

**Working titles and dates of completion where applicable.*

Devlaeminck, D. *Reciprocity in Transboundary Waters: The Function and Role of Reciprocity in the International Law of Transboundary Waters.* (2018 Expected).

Kang, K. *Embracing the Study of Paradoxes: A Social Systems Perspective of International Law's Regulation of Transboundary Waters.* (2017 Expected).

Mahbub, N. T. *Unpacking Equity: Its Role and Significance in Transboundary Water Cooperation.* (2018 Expected).

Rocha Loures, F. *International Water Law and the Community-of-Interests Doctrine: Content, Status and Role in the 21st Century.* (2018 Expected).

Shirani, F. *International Human Rights Law and International Water Law: Coherence and Mutual Support.* (2018 Expected).

Su, Y. *The Communicative Dimension of International Law: With a Special Focus on the Normativity in International Law.* (2015).

Vangh, A. *Amicus Curiae in International Transboundary Freshwater Arbitration.* (2019 Expected).

Index

For Product Safety Concerns and Information please contact our EU representative GPSR@taylorandfrancis.com Taylor & Francis Verlag GmbH, Kaufingerstraße 24, 80331 München, Germany

Batch number: 10726526

Printed by Printforce, the Netherlands